Innovative Technologies in Engineering Production (ITEP'18)

MATEC Web of Conferences Volume 244 (2018)

Bojnice, Slovakia
11 – 13 September 2018

Editors:

Dana Stancekova
Milan Vasko
Anna Rudawska
Jakub Szabelski
Nadezda Cubonova
Alzbeta Sapietova
Jozef Mrazik
Vladimir Tlach

ISBN: 978-1-5108-7598-2

Printed from e-media with permission by:

Curran Associates, Inc.
57 Morehouse Lane
Red Hook, NY 12571

Some format issues inherent in the e-media version may also appear in this print version.

No changes have been made to the content of these proceedings. There may be changes to pagination and minor adjustments for aesthetics.

Printed by Curran Associates, Inc. (2019)

For additional information, please contact EDP Sciences – Web of Conferences at the address below.

EDP Sciences – Web of Conferences
17, Avenue du Hoggar
Parc d'Activité de Courtabœuf
BP 112
F-91944 Les Ulis Cedex A
France

Phone: +33 (0) 1 69 18 75 75
Fax: +33 (0) 1 69 28 84 91

contact-edps@webofconferences.org

Additional copies of this publication are available from:

Curran Associates, Inc.
57 Morehouse Lane
Red Hook, NY 12571 USA
Phone: 845-758-0400
Fax: 845-758-2633
Email: curran@proceedings.com
Web: www.proceedings.com

TABLE OF CONTENTS

Organizing committee

assoc. prof. Ing. Dana Stančeková, PhD. - *(chairmen)*
assoc. prof. Ing. Anna Rudawska, PhD. - *(chairmen)*
assoc. prof. Ing. Dalibor Barta, PhD.
Ing. Vladimír Bulej, PhD.
prof. Ing. Nadežda Čuboňová, PhD.
Ing. Tomáš Dodok, PhD.
Ing. Jozef Holubják, PhD.
prof. Dr. Ing. Ivan Kuric
Ing. Izabela Miturska
Ing. Jozef Mrazik, PhD.
Ing. Iwona Rybicka
Ing. Jakub Szabelski
Ing. Vladimír Tlach
assoc. prof. Ing. Milan Vaško, PhD.

Editors

assoc. prof. Ing. Dana Stančeková, PhD. - *chairman*
assoc. prof. Ing. Milan Vaško, PhD.
assoc. prof. Ing. Anna Rudawska, PhD.
prof. Ing. Nadežda Čuboňová, PhD.
assoc. prof. Ing. Alžbeta Sapietová, PhD.
Ing. Jozef Mrázik
Ing. Jakub Szabelski
Ing. Vladimír Tlach

Conference Secretariat

assoc. prof. Ing. Dana Stančeková, PhD.
Department of Machining and Production Technologies
Faculty of Mechanical Engineering
and Scientific - technical Society at ŽU
University of Zilina
Univerzitná 1
010 26 Zilina
e-mail: dana.stancekova@fstroj.uniza.sk
https://www.itpe.eu

Scientific Committee and Reviewers

Additional Scientific Reviewers

prof. Ing. Milan Edl, PhD.
prof. Ing. Peter Demeč, PhD.
assoc. prof. Ing. Robert Grega, PhD.
Dr.h.c. prof. Ing. Karol Vasilko, DrSc.
prof. Ing. Augustín Sládek, PhD.
prof. Ing. Ivo Hlavatý, PhD.
assoc. prof. Dr. Ing. Josef Soukup
prof. Ing. Jozef Bocko, PhD.
assoc. prof. Ing. Milan Naď, PhD.
prof. Dr. Ing. Radu Cotetiu
prof. Ing. Jozef Meško, PhD.
assoc. prof. Ing. Mária Čilliková, PhD.
prof. Ing. Jozef Jurko, PhD.
assoc. prof. Ing. Sylvia Kusmierczak, PhD.
Ing. Ingrid Görögová, PhD.
assoc. prof. Ing. Kateryna Kravchenko, PhD.
Ing. Tomasz Kmita, PhD.
assoc. prof. Ing. Radim Halama, PhD.
assoc. prof. Ing. Michal Hatala, PhD.
prof. Ing. Anna Mičietová, PhD.
prof. Ing. Sergej Hloch, PhD.
prof. Ing. Ildiko Maňková, PhD.
prof. Dr. Ing. Miroslav Neslušan
assoc. prof. Ing. Peter Pokorný, PhD.
prof. Ing. Stanislav Beroun, PhD.
assoc. prof. Ing. Plamen Punov, PhD.
prof. D.Sc. Olexandr Kravchenko, PhD.
prof. Ing. Pavol Kukuča, PhD.

Presentation of the ITEP 2018

Gala opening of the conference with organizational guarantors assoc. prof. Ing. Dana Stančeková, Ph.D. - *University of Zilina* and assoc. prof. Ing. Anna Rudawska, Ph.D. - *Lublin University of Technology* (*Photo by Jozef Mrázik*)

Conference opening ceremony with vice-rector prof. Ing. Pawel Dzoździel, Ph.D. - *Lublin University of Technology* (*Photo by Jozef Mrázik*)

Conference participants (*Photo by Jozef Mrázik*)

FOREWORD

The International Scientific Conference *Innovative Technologies in Engineering Production – ITEP* is organized as a joint scientific event under auspices of Faculty of Mechanical Engineering at the University of Žilina (UNIZA), Scientific-technical Society by UNIZA and Faculty of Mechanical Engineering at the Lublin University of Technology.

ITEP conference was established on the basis of long-standing cooperation between these two upper mentioned universities and from the year 2016 it is organized alternately in Slovakia or in Poland.

The main organizer of the 3rd conference ITEP 2018 is Faculty of Mechanical Engineering - *Department of Machining and Production Technologies* and *Department of Automation and Production Systems.*

The main aim of the ITEP conference is to provide the communication platform for researchers, specialists, colleagues and partners with similar professional focus and to provide space for mutual presentations and sharing their scientific results and achievements. The conference should also give the possibility to discuss innovation matters as well as to establish new contacts and coordination channels between cooperating universities and companies.

All accepted papers deal with subjects related to current development trends in machining technology, progressive technologies, automated assembly, trends in CAx systems, intelligent manufacturing systems, CAD/CAPP/CAM/CIM technologies, automotive production, as well as current and future trends in industrial machine and equipment design.

This scientific publication presents the results of authors from chosen international research centres and universities, mainly from Slovakia and Poland, but also from other countries such as Czech Republic, Ukraine, Lithuania, Austria and Romania. The publication contains 45 articles in total divided to 3 chapters, which summarize the research activities of more than 70 authors in upper mentioned conference topics.

The publishing and presentations of contributions have some important objectives as for example to stimulate the specialists participation at the international conferences, to establish contacts especially between the doctoral students and young assistant professors in order to develop their personal skills and knowledge or to help during doctoral thesis working, and to promote new techniques of manufacturing approaches in Mechanical Engineering area.

organizing committee of the conference

Statement of Peer review

In submitting conference proceedings to *Web of Conferences*, the editors of the proceedings certify to the Publisher that

1. They adhere to its **Policy on Publishing Integrity** in order to safeguard good scientific practice in publishing.
2. All articles have been subjected to peer review administered by the proceedings editors.
3. Reviews have been conducted by expert referees, who have been requested to provide unbiased and constructive comments aimed, whenever possible, at improving the work.
4. Proceedings editors have taken all reasonable steps to ensure the quality of the materials they publish and their decision to accept or reject a paper for publication has been based only on the merits of the work and the relevance to the journal.

Title, date and place of the conference

ITEP´18 - Innovative Technologies in Engineering Production
September 11 – 13, 2018
Bojnice, Slovak Republic

Proceedings editor(s):

assoc. prof. Ing. Dana Stančeková, PhD. - *chairman*
assoc. prof. Ing. Milan Vaško, PhD.
assoc. prof. Ing. Anna Rudawska, PhD.
prof. Ing. Nadežda Čuboňová, PhD.
assoc. prof. Ing. Alžbeta Sapietová, PhD.
Ing. Jozef Mrázik
Ing. Jakub Szabelski
Ing. Vladimír Tlach

Date and editor's signature

8.11.2018

Hydrodynamics features in cavitation devices under tangent medium input

Tetiana Vitenko[1,*], *Paweł* Droździel[2], *Nazar* Horodyskyi[1]

[1]Ivan Puluj TNTU, Faculty of Industrial Machinery Engineering, str. Ruska 56, 2-80 Teronpil, vitenko@tntu.edu.ua, nazar_horodysky@ukr.net, Ukraine
[2]Lublin University of Technology, Faculty of Mechanical Engineering, Nadbystrzycka 36 Street, 20-618 Lublin, p.drozdziel@pollub.pl, a.rudawska@pollub.pl, Poland

Abstract. Theoretically analysed design features of the working section of a hydrodynamic cavitation device with swirling input. Using the numerical methods of SolidWorks software and the Flow simulation module, fluid flow modeling was performed to describe similar processes. The comparison of the results of visual and experimental studies has been carried out. Visual observations indicate that the life cycle of vapor-gas bubbles and cavities in the design under study can be divided into several periods, in particular the formation of vapor-gas bubbles, their growth, generation of cavities, their volume increase, transformation and destruction. The influence of input and output hydrodynamic and technological parameters of the flow and design features on the character of the cavitation area and the form of cavitation were studied. According to the results of theoretical and experimental research, rational mode of operation of the cavitation device is proposed.

Keywords: cavitation, hydrodynamics, bubbles, collapse, SolidWorks.

1 Introduction

The question of finding effective intensification methods for the technological processes of the chemical, food or pharmaceutical industries today does not lose its relevance. Particular attention is attracted to the dynamic effects arising from the development of bubble systems during cavitation. The basis of one of the concepts on which the explanation of the mechanism of cavitation influence on technological systems is based, is the theory of discrete-impulse energy input (DIEI) introduced into the apparatus. The references describe in detail general principles of DIEI, describe the energy and thermodynamic aspects and the main intensification mechanisms that can be initiated on the basis of this principle [1, 2]. DIEI mechanisms conditionally can be divided into hard ones and soft ones. The former should be used to stimulate hydromechanical processes, and the latter to accelerate the processes of phase heat and mass transfer, or for the purpose of intensive mixing of multicomponent media, when the level of intensification of the process can be smoothly changed within certain limits. In practice, such conditions are achieved by ultrasonic influence on the liquid [3] or by creating special hydrodynamic conditions in the flow in propellant or rotary cavitation devices. The cavitation mechanism is also implemented in

* Corresponding author: vitenko@tntu.edu.ua

Reviewers: *Vladimír* Vrba, *Dušan* Šekláč

rotary-pulsating homogenizers and in industrial apparatuses. The phenomenon of steam cavitation was studied under injection of superheated steam into a cold liquid [4].
Propellant cavitation devices using flow energy for the formation of a tear-off flow are effectively used in industry for the intensification of heat-mass transfer processes. The idea of their use is based on the cavitation effects, which are accompanied by the cold boiling of the liquid. The difference is that when increasing the relative flow velocity relative to the body up to the critical value, the pressure of the flow decreases up to the saturation vapor pressure. In this case, the liquid boils, forming cavitation gas-vapor bubbles of various sizes and caverns, which get into the area of high pressure, where they collapse (close up, condense). It was investigated that at these points, local pressures and temperatures are significantly increased [1-3]. If the bubbles collapse directly at the surface or the disperse particle, then the kinetic energy of the radial motion of the fluid is converted into the mechanical energy of a liquid cumulative jet that has a very high velocity and moves to this surface. This mechanism is described in the authors' studies [5].

Significant interest is caused by dynamic effects of varying intensity, which are determined by the stage of cavitation bubbles collapse and identify the effectiveness of the cavitation activity on the process media. The life cycle of cavitation bubbles was studied by the authors [5]. The method of high-speed filming shows the development of a spherical bubble from its formation to destruction. The theoretical analysis of these experiments was carried out using Rayleigh equation in the Plesset's work. Subsequent studies of the dynamics of gas-vapor bubbles took into account the kinetics of heat and mass transfer processes. The complex of studies and their generalization was carried out in papers [6, 7], which set out universal mathematical models that, from single thermodynamic positions, adequately describe the dynamics of individual bubbles and bubble ensembles under cavitation conditions.

Numerical simulation methods are widely used for planning and analyzing the results of experiments, research and study of such processes. Numerical models of cavitation are divided into three groups: Conditional, Mass, Euler. In [8], the author analyzed the "Conditional" models, which are divided into two types, depending on the hydrodynamic conditions that determine the cavitation number, the saturation vapor pressure, and the local flow velocity. Mass transfer in this class of models is not considered. The advantage of such models is the speed of obtaining a solution, but there is a certain difficulty in indicating the cavitation number. Such models are usually used for initial engineering calculations, under the condition that the cavitation occurs under the indicated cavitation number. At the same time, this does not abolish the need for an experiment to determine the exact value of the cavitation number.

Euler model allows to calculate steam and liquid characteristics. The possibility of simulation hydrodynamic cavitation under conditions of slow and high-speed fluid flows, developed cavitation, non-stationary currents are the advantages of this model. The disadvantages of Euler model should include the duration of the calculation. These models are used for simulation cavitation in hydropower plants and high-speed centrifugal pumps. It is also possible to model cavitation in vortex tubes with the connection of thermodynamic ratios. Models of this type are present in ANSYS Fluent and ANSYS CFX [9].

It should be noted that the cavitation processes described in accordance with the laws of mathematical simulation are practically impossible to describe by the behaviour of individual bubbles or even their groups that define certain parameters. On the other hand, it is possible to describe some of the processes in a particular situation using the dimension theory and general parameters such as geometry, external pressure, temperature, saturation vapor pressure at a certain temperature, properties of the working fluid (viscosity, density, compressibility). In this case, the simulation of the cavitation flow is based on the fundamental laws of fluid/gas mechanics, in particular on the equations of continuity, momentum equations, and the law of variation of turbulent viscosity from the velocity (turbulence model). In this context, it should be noted that the above mentioned theoretical and experimental studies of cavitation flows hydrodynamics features, depending on the design of the equipment, have allowed to solve an important task of forecasting its efficiency and prospects for further application.

2 Description of the main research material

In this paper, the simulation was conducted to study cavitational areas characteristics and the choice of rational regime parameters of the system's influence on the proposed design. The effect of input and output hydrodynamic flow parameters and geometric design features on the character of the cavitation area was studied. The work of the system was considered under the condition of the previous flow swirling, which provides acceleration of the flow at the inlet, additional fluid pulsations and increase the intensity of the impact on the technological environment. The photo of the working area "cone-working body" of the experimental installation is shown in Fig. 1.

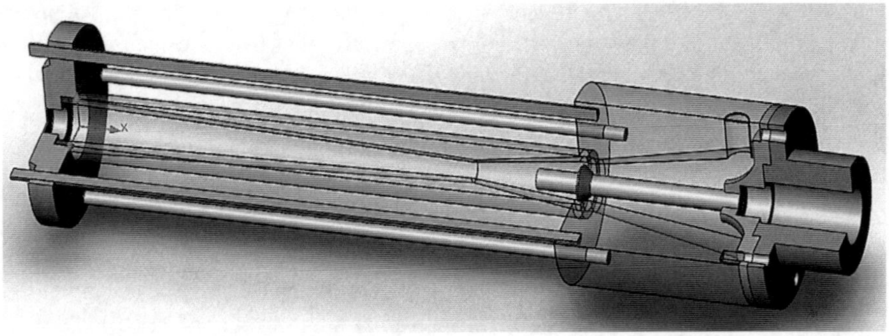

Fig. 1. Installation section in the cone area

Due to the rotation in the inlet pipe, the fluid flow becomes much more mobile (moving) along the central axis and acquires increased kinetic energy. Excessive pressure created in the inlet pipework provides fluid transfer through the narrowing and expansion part of the system. An increase in the fluid velocity in the narrow section of the Venturi tube causes a drop in local static pressure. The gradual increase in the flow of liquid leads to a reduction in pressure to a critical, equal to or close to the pressure of saturated vapor of the liquid, and the emergence of cavitation phenomena, which can be explained by the processes occurring around and within the vapor-gas phase.

Visual observations have shown that life cycle of vapor-gas bubbles and cavities in such a design can be conditionally divided into several periods. Consistent formation of vapor-gas bubbles, their growth, generation of caverns, increase of their volume, transformation and destruction. It was established that for the structure under study, the appearance of cavitation and the appearance of separate cavitation bubbles occurs under $\delta = \delta_{crit} = 2$mm.

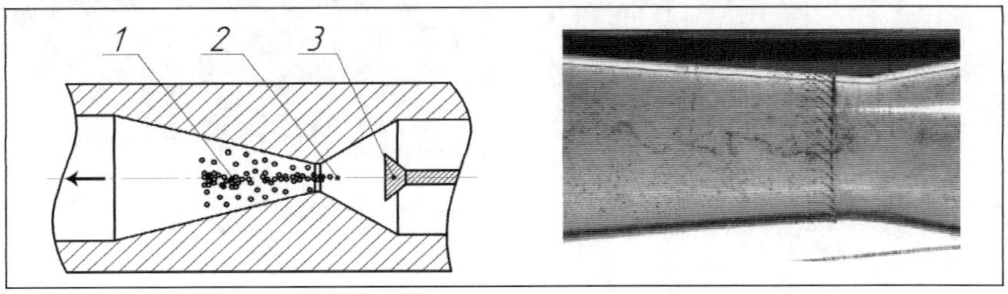

Fig. 2. Emergence of separate cavitation bubbles

At the corresponding value of pressure P_2 at the output of the tube, occurs the formation and growth of the cavity over the obstacle 3, the length of which increases over time, it extends and takes channel 2 and goes out to the diffuser 1 (Fig. 3).

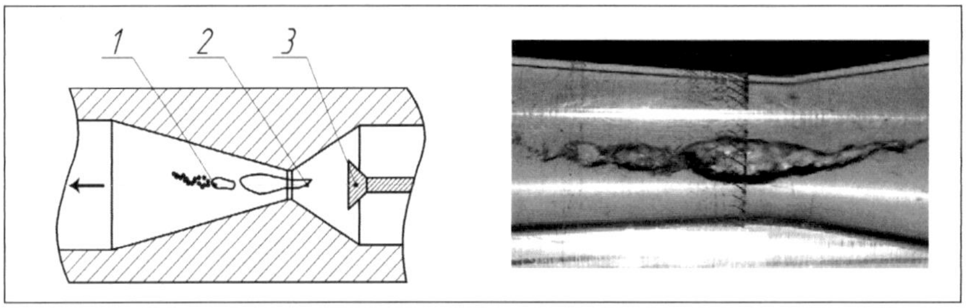

Fig. 3. Formation of the cavity

After one cavity separates into several ones, the new one starts to emerge on its place; the previous one goes down the stream and collapses. As the studies have shown, the process of cavity`s part separating takes place strictly periodically, that is, with a stable frequency *f*. There is a formation of periodically tearing-off flow, the distinctive feature of which is the stability of the separation frequency and collapse (in the stream, and not on the wall) of caverns (Fig. 4).

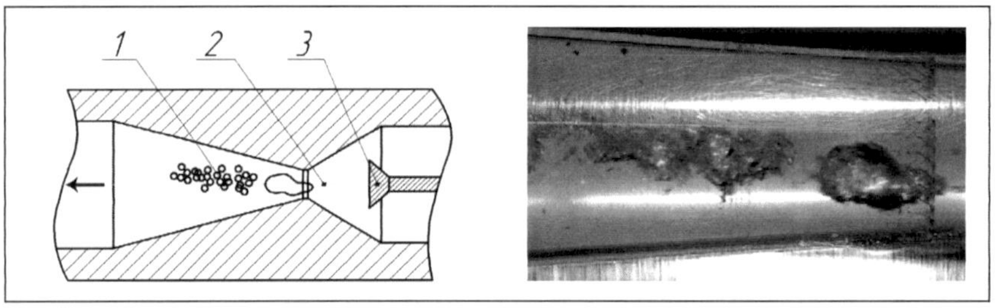

Fig. 4. Partition and separation of the cavity

Visual observations also showed that the cavitational cavity begins to form at the critical value of the cavitation number $\sigma = 1.6$. At the beginning, cavitation is of the bubble type, with a decrease of σ up to 1 is observed film flow, the size of the cavity is 0.005 - 0.007m. With a decrease of σ up to 0.9, the type of cavitation changes, and the average length of the cavity increases up to 0.015-0.02m. Cavities of small size are destroyed and form bubbles (Fig. 5). The destruction of the cavity is also facilitated by the local energy release, the formation of local pressure gradients that occur as a result of collapse of cavitation bubbles.

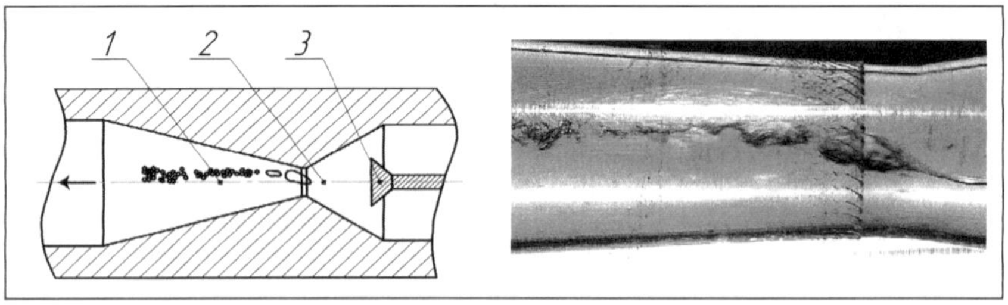

Fig. 5. Evolution of small size cavities

To study the possibility of using numerical methods to describe such processes, simulation was performed in the SolidWorks software complex using the Flow simulation module, and the results were compared with visual and experimental studies. Changes in pressure P within the area behind the diffuser are shown in Fig. 6. According to the obtained results under the input pressure P_1 = 0.2 MPa, the pressure at the outlet from the pipeline P_2 is 0.5 from P_1 (0.1 MPa). The pressure discontinuity within the area of the diffuser and behind it are Pmax = 1.2 – 1.3P_1, i.e. 0.24 – 0.26 MPa, Pmin = 0.1P_2, therefore, it is about 0.03 MPa. According to the obtained calculations, the pressure in the volume of liquid P falls in the diffuser from 0.2 MPa up to 0.17 MPa in front of the cone, then decreases up to 0.12 MPa at the entrance to the cylindrical area. Next comes the second wave of pressure drop from 0.12 MPa to 0.026 MPa, which is confirmed by the result of the experiment. In general, the static pressure along the flow line in the working chamber varies: the pressure on the periphery is higher than along the axis (Fig. 6), as the vortex flow has centrifugal forces directed from the center to the outer wall of the tube. Then the pressure increases from 0.02 up to 0.1 MPa.

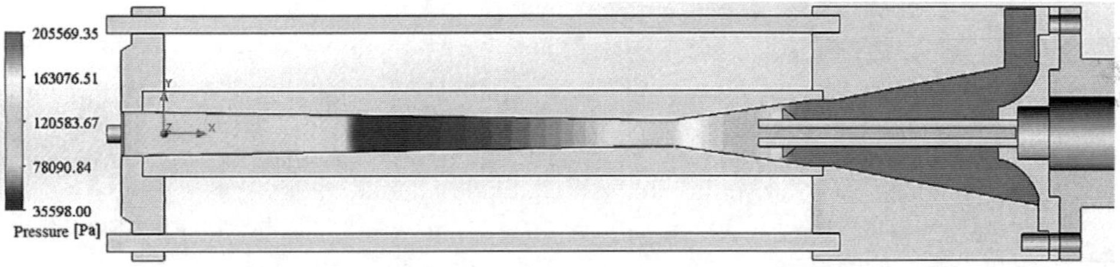

Fig. 6. Pressure distribution (δ_{crit} = 2mm) pressure at the input 0.2 MPa

Similar calculations are performed under other input pressures within the limits P_1=0.3-0.6 MPa at different values of $\delta_{кр}$. Adequate results are obtained according to the experiment measured better than 10%.

In Fig. 7 is shown the distribution of vapor volume concentration in the working channel for different modes (P_1 = 0.2 and P_1 = 0.4 MPa, δ_{crit} = 2mm). The smallest cavitation area (Fig. 7 (b)) corresponds to the input pressure P_1 = 0.4 MPa.

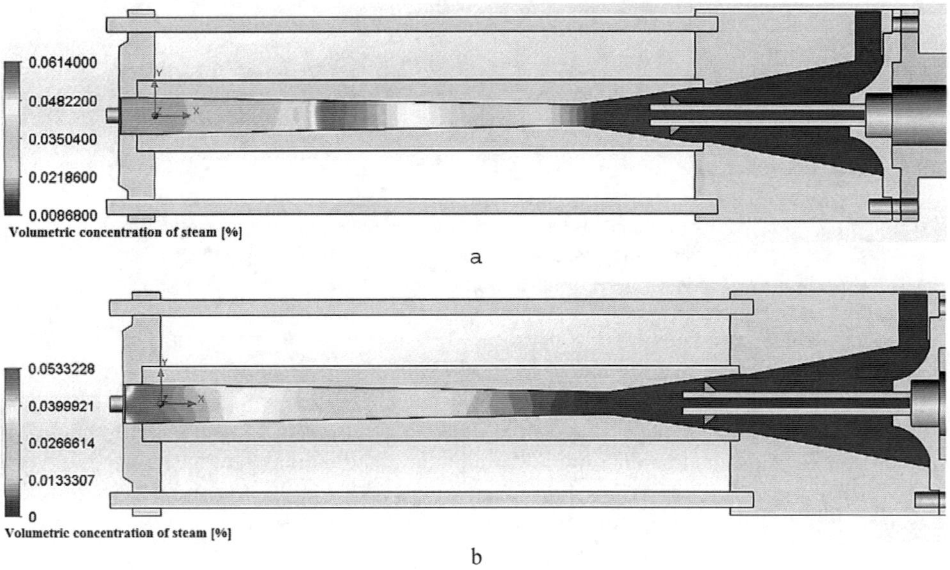

Fig. 7. Distribution of vapor volume concentration ($\delta_{кр}$ = 2mm):
a - input pressure 0.2MPa, b - input pressure 0.4MPa

There is also a shift of the cavitation area towards the outlet with input pressure increase (Fig. 8), which confirms the results of experimental and visual studies. Among the studied gaps, the largest volume of the vapor-gas phase is observed for the construction with δ_{crit} = 2mm (Fig. 8), and the smallest for δ_{crit} = 1mm.

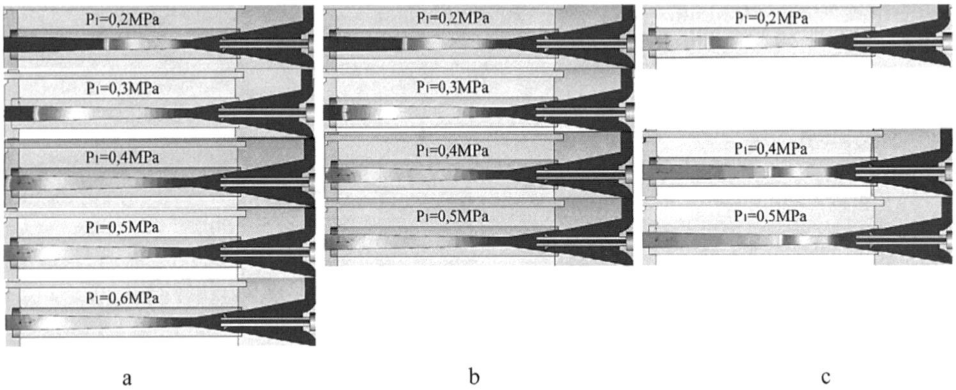

Fig. 8. Distribution of volume vapor concentration: a – δ_{crit} = 1mm; b - δ_{crit} = 2mm; c- δ_{crit} = 3mm

The analysis of areas with vapor-gas phase (in Fig. 8) shows that the construction with δ_{crit} = 2mm, with a maximum volume concentration of vapor (Fig. 9), is rational.

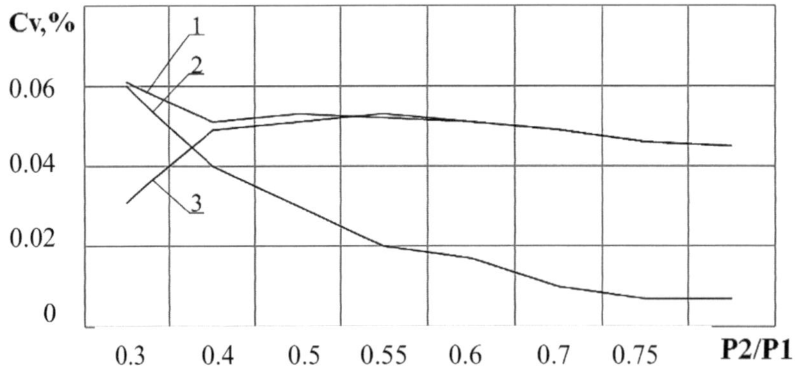

Fig. 9. The dependence of the maximum volume density of the vapor-gas phase on the ratio of pressures P_2/P_1: 1 - gap between the obstacle and the wall 2 mm; 2 - gap between the obstacle and the wall 3 mm; 3 - the gap between the obstacle and the wall 4 mm

An important parameter that determines the intensity of cavitation impact on the processing medium is the frequency of oscillations. It corresponds to the frequency of separation of cavitation cavities (*f*) and is determined by the form, geometric and regime parameters of local narrowing. The nature of these frequencies is similar to the frequency of the vortex collapse in the tear-off flow of the cylinder in the normal direction to its axis, which are called Strouhal frequencies of fluid oscillations, and for the hydraulic system with a Venturi tube are defined as the following [10]:

$$f = Sh_m \cdot \sqrt{\frac{2\cdot(P_1-P_5)}{\rho_{liq}}} \cdot \frac{tg\frac{\beta}{2}}{R_{kr}\left(\sqrt{\frac{\mu}{1-\sqrt{1-\tau}}}-1\right)} \quad (1)$$

where f - pulsation of cavitation bubble;

Sh_m - the modified Strouhal number;

P_1 and Ps - input pressure within the cavitation area;

$R_{кр}$ - the radii of the critical intersection;

μ - the cost coefficient;

τ - the ratio of the outlet pressure to the inlet pressure;

β - the angle of the diffuser opening;

ρ_{liq} - liquid density.

According to the results of studies [11-14] for the implementation of a current characterized by high amplitudes of pressure fluctuations in the pipeline along the Venturi tube, it is necessary that the angle of diffusor opening was β > 15°. The most rational angle of diffusior opening is considered to be β = 20 ÷ 30°. In this case, under the condition that the outlet pressure P_2 is in the range of 0.1 ÷ 0.2 MPa, the input pressure - 0.12 ... 10 MPa. Such conditions are fully realized in the device under study.

Experimental dependences of the oscillations frequency f on the parameter P_2/P_1 under the pressure P_{in} = 0.2 – 0.5 MPa, are shown in Fig. 10. From the figure below, it can be observed that under the change in the ratio P_2/P_1 changes the frequency f. Thus, with the value of the parameter P_2/P_1 = 0.4, with the increase in the input pressure from 0.2 up to 0.4 MPa, the oscillation frequency increases from about 400 up to 750 Hz. The nature of the dependencies is close to the linear one and can be described by the equations shown in Figure 10. The nature of oscillations frequency change significantly dependent on the Reynolds number and input/output pressure change, which is determined by the peculiarities of the design, the characteristics of the pump and determines a certain dependence of the Strouhal criterion on Re and P_2/P_1.

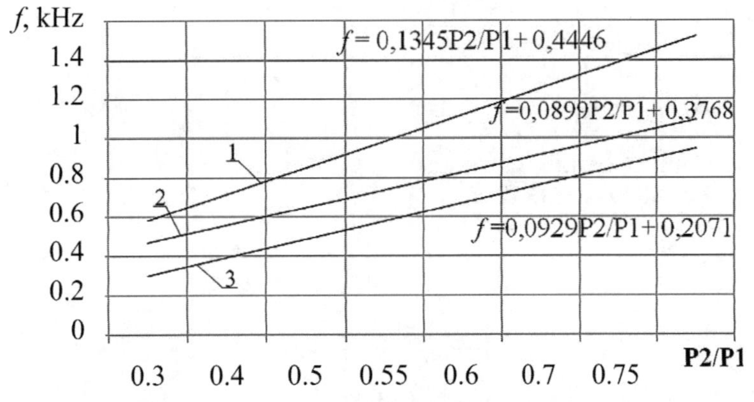

Fig. 10. The dependence of cavity tear-off frequency on pressure difference P_2/P_1: 1 - gap between the obstacle and the wall 2 mm; 2 - gap between the obstacle and the wall 3 mm; 3 - the gap between the obstacle and the wall 4 mm

On the other hand, the frequency of oscillations affects the fluid flow, the outlet pressure P_2, the hydraulic resistance of the area, which is important for determining the energy consumption and productivity.

The experimental dependences of the Sh_m change on the dimensionless Reynolds criterion are shown in Fig. 11.

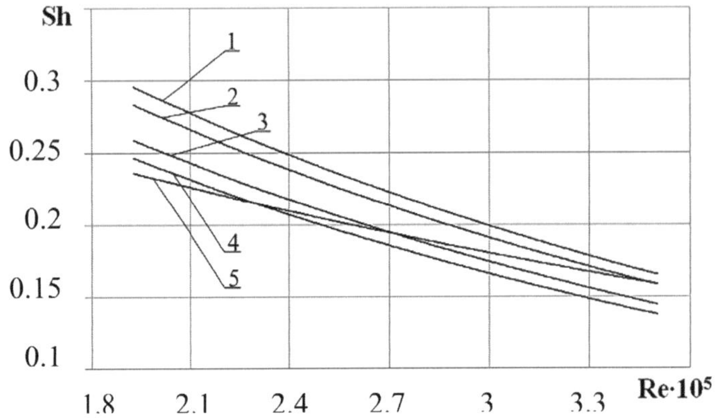

Fig. 11. The dependence of Strouhal number on Reynolds number: 1 - gap between the obstacle and the wall 1 mm; 2 - gap between the obstacle and the wall 2 mm; 3 - gap between the obstacle and the wall 3 mm; 4 - gap between the obstacle and the wall 4 mm; 5 - gap between the obstacle and the wall 5 mm

From the shown results, Strouhal number in the range of Reynolds number $2.2 \cdot 10^5$ - $4 \cdot 10^5$ is sufficiently stable and decreases with increasing flow velocity in the studied range of velocities. Analyzing the obtained data, it can be assumed that the smaller the diameter of the channel is, the more influence on the flow have hydro-effects near the walls of the working area and the formation of vortices, which contributes to the outlet pressure increase.

In this regard, it is expedient to establish the dependence Sh $= f$(Re, P_2/P_1). The obtained results provide a practical possibility to determine the nature of this dependence in a certain range of pressure ratios and flow velocity, which ensures the greatest efficiency. Let`s consider modes in the range of input pressure from 0.2 MPa up to 0.5 MPa. To solve the problem of the complex influence of Re; P_2/P_1 it is necessary to obtain an interpolation model (regression equation), which determines the influence of these factors on the Strouhal number. Frequency, Reynolds number and pressure difference are selected as independent variables for obtaining this regression equation.

X_1 - Reynolds number (Re);

X_2 - pressure difference (P_2/P_1);

The collected statistical information is partially shown in the table.

Table. 1. Input data

	X_0	X_1	X_2	Y
1	1	731099.4184	0.82	0.079442
2	1	350318.4713	0.78	0.165792
3	1	224203.8217	0.70	0.25905
4	1	207596.1312	0.55	0.2798
5	1	193279.1566	0.53	0.3005
Sum	5	170649`6.999	3.37	1.0846

In order to calculate the coefficients, in this case, a system of normal equations was used which was created on the basis of the least squares method, which provides a sufficiently high accuracy of approximation.

The regression equation for this case has the form:

$$Y=b_0+b_1X_1+b_2X_2 \quad (2)$$

According to the rule, the system of normal equations gets the form:

$$\begin{aligned} b_0\sum X_0^2+b_1\sum X_1X_0+b_2\sum X_2X_0=\sum X_0Y \\ b_0\sum X_0X_1+b_1\sum X_1^2+b_2\sum X_2X_1=\sum X_1Y \\ b_0\sum X_0X_2+b_1\sum X_1X_2+b_2\sum X_2^2=\sum X_2Y \end{aligned} \quad (3)$$

In the system of equations we calculate all values under sums based on experimental data. Errors of experiments were determined by the standard equation:

$$S_0^2 = \frac{1}{N_0-1}\sum_{k=1}^{N_0}(y_{0k} - \bar{y}_0)^2 \quad (4)$$

where S_0 - mean-root square error;
N_0 - is the number of parallel experiments;
y_{0k} - k-th term sequence;
y_0 - mean forecast error.

The described system of equations was solved according to the standard program. As a result, the numerical values of the coefficients b_0, b_1, b_2 were obtained for different conditions.

Table. 2. The coefficients value

b_0	b_1	b_2
1.89E-01	1.84E-10	3.68E-03
-5.94E-02	7.99E-11	1.53E-01
4.61E-02	5.33E-10	2.13E-01
1.54E-01	1.48E-10	5.85E-02
9.45E-02	2.52E-10	1.62E-01

These coefficients vary depending on the three parameters: the input pressure, the gap between the obstacle and the wall, and the cavity pulsation frequency.

The results of the solution of the inverse problem are given in Table 3.

Table. 3. Results of calculations

Y_1 experiment	Y_2 experiment	Y_3 experiment	Y_4 experiment	Y_5 experiment
0.079	0.076	0.148	0069	0.066
0.166	0.159	0.164	0.145	0.138
0.259	0.248	0.187	0.226	0.216
0.280	0.268	0.220	0.245	0.233
0.301	0.288	0.275	0.263	0.250
1.085	1.039	0.994	0.949	0.904
Y_1 calculation	Y_2 calculation	Y_3 calculation	Y_4 calculation	Y_5 calculation
0.208	0.118	0.148	0.178	0.152
0.211	0.147	0.164	0.182	0.161
0.215	0.187	0.187	0.187	0.174
0.221	0.246	0.220	0.195	0.193
0.231	0.342	0.275	0.208	0.224
1.085	1.039	0.994	0.949	0.904

Error of the theoretical calculated results does not exceed 10%.

3 Conclusions

The obtained results of experimental studies of the proposed construction of a device of a hydrodynamic type with a previous input swirling confirmed the formation of a cavitation regime at lower values of flow velocity at the inlet compared to the basic structure without tangential flow input. The dimensionless Reynolds criterion (Re) and gap (δ) between the wall and the cone were selected as the parameters that determine the volume density of the vapor-gas phase and the size of the cavitation area. It is established that there is a rational regime characterized by a certain ratio of Re and outlet pressure, under which short cavities with a significant volume of the vapor-gas phase are formed. It is investigated that for the proposed design, the emergence of separate cavitation bubbles occurs when the gap at the entrance to the diffuser δ_{crit}= 2mm. In this case, the critical value of the cavitation number is $\sigma = 1.6$. As the flow velocity increases, over time, the film flow that transfers into mixed cavitation mode is observed.

Based on simulation results in the SolidWorks software complex using the Flow simulation module, the results for pressure variations along the working chamber and the vapor-gas phase volume were obtained. Together with the change in pressure, displacement of the cavitation area towards the exit with increasing input pressure was observed, which was confirmed by the results of experimental studies. The analysis of areas with the content of the vapor-gas phase confirmed that the design with the gap δ_{crit} = 2mm, for which there is a maximum volume density of steam, is rational. The obtained dependences of the frequency of oscillations f on the parameter P_2/ P_1 at the input pressures P_{in} = 0.2-0.5 MPa, have shown that with the value of the parameter $P_2/ P_1 = 0.4$ with increasing input pressure from 0.2 up to 0,4 MPa the frequency of oscillations increases from 400 to 750Hz. The character of the obtained mathematical dependences is close to the linear one. It has also been established that the Strouhal number in the Reynolds number range $2.2\cdot10^5$ - $4\cdot10^5$ is sufficiently stable and decreases with increasing flow velocity in the studied range of velocities. The comparison of the obtained experimental and theoretical results obtained with respect to the volume of the vapor-gas phase and the dependence of $Sh = f(Re, P_2 / P_1)$ confirms their sufficient adequacy (within 10%).

References

1. A. A. Dolinsky, G.K. *Ivanitsky, The principle of discrete-pulse energy input-new approach to the development of efficient power-saving technologies*. Ann. Review of Heat Transfer,Vol. **13**, 151-194 (2003).
2. A. A. Dolinsky, G.K. Ivanitsky, *Theoretical substantiation of the principle of discrete impulse energy input*. Prom. Heat engineering, Vol. **17**, 3-28 (1995).
3. A. F Lugovskoy, N.V Chukhraev, *Ultrasonic cavitation in modern technologies*. Publishing and printing center "Kyiv University", Vol. **13**, p. 244 (2007).
4. L. D. Chen, G.M. Faeth, *Condensation of Submerged Vapor Jets in Subcooled Liquid*. Ibid, Vol. **7**, 179-187 (1982).
5. T. M. Vitenko, Ya.M. Humnitsky, *Effect of the vapor-gas phase on the kinetics of dissolution under cavitation conditions*. East-European magazine for cutting-edge technology, Vol. **4**, 4-8 (2009).
6. P. Cerone, J.R. Blake, *An ante is the instantaneous streamlines, pathlines and pressure constants for a cavity bubble near the boundary*. J. Austral. Math Soc.Ser B, Vol. **8**, 41-44 (1984).

7. Y. Tomita, A. Shima, *The Machanism of impulsive pressure generation and damage pit formation by bubble collaps*. J. Fluid Mach., Vol. **169**, 535-564 (1986).
8. Yu. M. Akhmetov, R. R. Kalimullin, V. A. *Tselischev, Numerical and physical simulation of fluid flow in a vortex heat generator.* Vol. 8, 42-49 (2010).
9. O. V. Baturin., N. V. Baturin., V.N. Matveev, Calculation of flows of liquids and gases with the help of the universal program complex Fluent. Samara: Publishung agency, Vol. **146**, p. 151 (2009).
10. V.V. Pilipenko, *Cavitation self-oscillations*. Kyiv: Scientific Opinion, Vol. **316**, 1-316 (1989).
11. V.V. Pilipenko, I.K. Man'ko, V.A. Zadontsev, Cavitation self-oscillations intensify technological processes. Proceedings of a Fluid Dynamics Panel Workshop, Vol. **4**, 32-1-32-4 (2/1998).
12. V.V. Pilipenko, V.A. Zadontsev, I.K. Man'ko, *About high-frequency oscillations in the hydraulic system behind Venturi cavitating tube*. Cavitation self-oscillations in pumping systems. Part 2, Naukova Dumka,Vol. **10**, 104-113 (1976).
13. V.V. Pilipenko, V.A. Zadontsev, I.K. Man'ko, Yu. Zhulai, N. A. Dzoz, *Oscillation generator of fluid pressure*. A.c. 1232299 USSR, MKI V06V 1/18, No. 3772504, Vol. **2**, p.24-28 (23/05/1986).
14. I. K. Manko, *Visual studies of cavitation self-oscillations in a hydraulic system with a transparent Venturi tube*. Cavitation self-oscillations in pumping systems. Part 2, Naukova Dumka,Vol. **6**, 113-118 (1976).

Welding materials used to increase service life of agricultural machinery processing soil

Petr Hrabě[1*], *Viktor* Kolář[1], *Abraham* Kabutey[2], *Aleš* Sedláček[2]

[1]Czech University of Life Sciences Prague, Faculty of Engineering, Department of Material Science and Manufacturing Technology, Kamýcká 129, Praha 6 - Suchdol, Czech Republic
[2]Czech University of Life Sciences Prague, Faculty of Engineering, Department of Mechanical Engineering, Kamýcká 129, Praha 6 - Suchdol, Czech Republic

Abstract. Intensive abrasive wear occurs in soil-treatment machines. This article is focused on increasing the service life of ploughs by the welding material. The welding material is applied at a 45° angle to the tool. This material is abrasive wear resistant. The welding material was applied parallel to the head of ploughshares with spacing of 60 mm. Carbide materials were used (SK 258 TiC-O, SK 900-O, SK A43-O, SK 299-O, SK A45-O, OK TUBRODUR 15.82, SK 258 TiC-O). The tested ploughshare variants were wearing the same when were used the welding material and the standard.

Keywords: abrasive wear, ploughshare, statistics

1 Introduction

Intensive abrasive wear occurs in soil-treatment machines. Abrasive wear is characterized by a reduction in tool volume and a change his shape that gradually reduces the work quality of the tool. The tool must be replaced after a certain limit has been exceeded. Renovation is one of the possible ways to keep the tool in a workable condition [1, 2].

The properties of the functional surface of tools and components can be effectively changed. Classical methods include welding, which makes it possible to successfully reduce materials and economic costs [1]. The success of the renovation was proven by ploughs [2].

The ploughshare is one of the most stressed parts of the plough and is subject to considerable demands. The ploughshare must meet relatively high strength requirements (especially toughness) and high wear resistance due to soil abrasive. The main cause of excessive wear of ploughshares are the hard particles contained in the soil (siliceous sand, stones). The size, shape and hardness of soil particles, climatic conditions, soil compaction and soil moisture are the main factors of abrasion intensity [3].

Abrasive wear can be reduced to an acceptable degree by appropriate technology and by selecting material to produce the entire tool or its part in greatest wear. The latest research attempts to find and apply such techniques in the production of ploughshares to reduce the friction between the tool and the soil. It would increase the service life of the most stressed parts and reduce soil resistance, which would save fuel [1-5].

* Corresponding author: hrabe@tf.czu.cz

Reviewers: *Daniela* Kalincová, *Janusz* Zarajczyk

Increasing of the service life of ploughshares has become increasingly topical in recent years. Creating better soil technology would ensure a significant improvement in the economy of the operation of these facilities. Potential savings from friction and wear reduction through improved tribology of agricultural tools would amount to almost $ 337 million per year [6].

A significant problem is the change in shape geometry, which also changes the vertical force response, which effects the deepening of the plough from the furrow. Decrease of the quality of the plough, reduction of the ability of the processing of plant residues etc. are negative consequences [7].

In the field of soil technology, it is possible to reduce the energy intensity as well as to increase the life of the tools by optimizing the material design solution. The wear of the tool can be reduced not only by used material, which is characterized by increased resistance to wear but also by the design, i.e. by shaping using shapes or design solutions combining different materials that effectively solve the problem. Welding is one way to create a new functional surface. Changing geometry, increasing the cross-section including energy intensity are a potential negative [8, 12].

The aim of this research is to evaluate the experiment of selected welding materials in the field test of wear of ploughshares.

2 Materials and methods

A field test in 136 hectares of field was performed for the experiment with a deep plough (25-30 cm) of a seven-shares Kverneland PB 100 plough. A welding machine with continuously served wire by ESAB with the LAF 635 DC source and the A2-A6 Process Controller was used for the welding of selected welding materials to ploughshares. The welding parameters for all materials are in Tab. 2. The welding parameters were chosen so that the width and height of the weld were the same for all types of welding materials. Two sets (14 shares) were welded due to shorter downtime during demounting, statistical evaluation and re-assembly of ploughshares. The mounds were placed on a share with spacing 60 mm and at an angle of 45° to the share edge. Layout of shares with the welding material was the same for both sets. The welding materials for the experiment were selected in the following order from the tractor (Table 1). Table 1 shows a guiding chemical composition of the used welding electrodes too.

Table 1. Welding materials used for the experiment and their guiding chemical composition

Order	Welding material	Guiding chemical composition [wt%]
1	SK 258 TiC-O	C-1.8; Mn-0.9; Si-0.2; Cr-6.1; Mo-1.4; Ti-5.5; Fe-base
2	SK 900-O	C-2.9; Mn-0.5; Si-0.4; W-42; Cr-5.8; Fe-base
3	SK A43-O	C-5.6; Mn-0.2; Si-1.3; Nb-6.7; Cr-20.2; Fe-base
4	SK 299-O	C-4.9; Mn-0.3; Si-1.0; Cr-11.30; Nb-6.8; V-6.0; B-0.7; Fe-base
5	SK A45-O	C-5.3; Mn-0.2; Si-0.7; Nb-6.0; Cr-21; V-0.8; Mo-6.3; W-1.9; Fe-base
6	OK TUBRODUR 15.82	C-5.0; Mn-0.9; Si-0.9; Nb-5.4; Cr-19; V-1.2; Mo-1.1; W-1.2; Fe-base
7	SK 258 TiC-O	C-1.8; Mn-0.9; Si-0.2; Cr-6.1; Mo-1.4; Ti-5.5; Fe-base

Table 2. Welding parameters for all materials

Current (A)	250
Voltage (V)	25
Speed (cm/min)	25
Energy (kJ/cm^3)	13

Dimensional and mass analysis were used to evaluate the service life of the ploughshares. The individual dimensions and weights of welded and standard ploughshares were measured after every seventeen hectares during the field test. Dimension A was measured at the tip of the share, dimensions B, C, D through screw holes and E in the back of the share (Fig. 1). A millimeter tape measure was used to measure individual dimensions. The weight of the shares was measured on a Portable Electronic Scale weight with accuracy to one thousandth of a kilogram.

The measured data were evaluated in STATISTICA program by using ANOVA F-test statistical method. Hypothesis H0 shows that there is no statistically significant difference ($p > 0.05$) between the test data. Hypothesis H1 denies the H0 hypothesis and states that there is a statistically significant difference between the test data ($p < 0.05$).

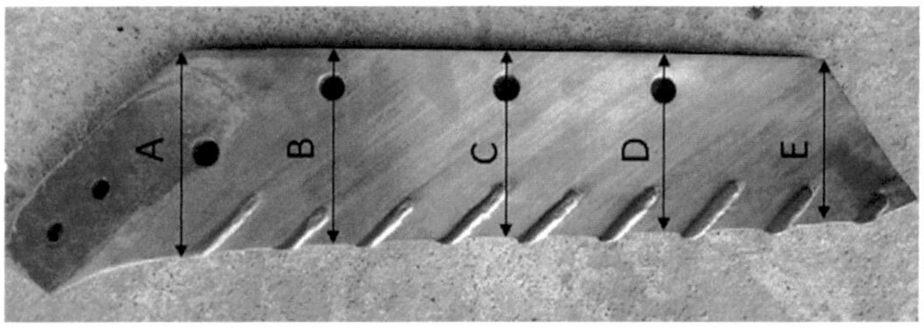

Fig. 1. Methodology of measuring individual dimensions of ploughshares

3 Results and discussion

From the point of view of statistical testing of the influence of different welding materials, it can be stated that they are statistically homogeneous groups, i.e. there is no difference between variants of the experiment, i.e. different welding materials and standards (Fig. 2). The H_0 hypothesis was confirmed, i.e. there was no difference in significance level of 0.05 between the individual tested variants.

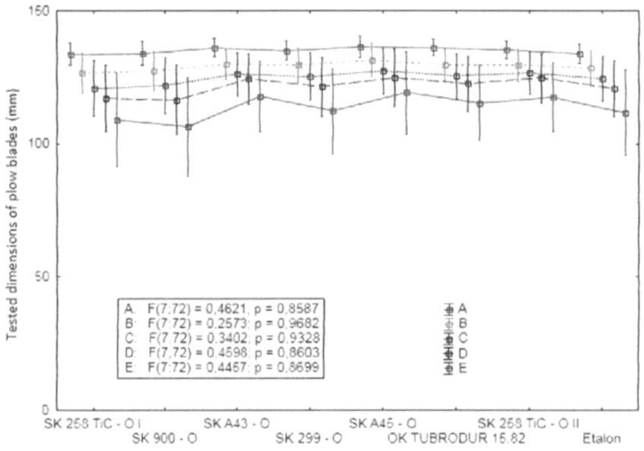

Fig. 2. Statistical testing of the influence of the different welding materials

From the point of view of statistical testing of uniformity of wear of the ploughshare over whole length of the share, i.e. changes in the geometric shape, it can be stated that it is statistically inhomogeneous groups, i.e. there is the difference between the tested parameters (measured points A, B, C, D and E).

The H_0 hypothesis was not confirmed, i.e. there was a difference in the significance level of 0.05 between the individual tested variants of the experiment (SK 258 TiC-O I, SK 900-O, SK A43-0, SK 299-O, OK TUBRODUR 15.82, SK 258 TiC-O II and standard). It follows from the above that the ploughshare is unevenly worn. This statistical conclusion of the measured data is supported by the results shown in Fig. 3.

This conclusion was not confirmed for SK A45-O, where p=0.0848. The H_0 hypothesis was not confirmed for this weld, i.e. there was no difference in significance level of 0.05 between the individual measured points A to E. Statistically difference was not found between the standard and the various welding materials.

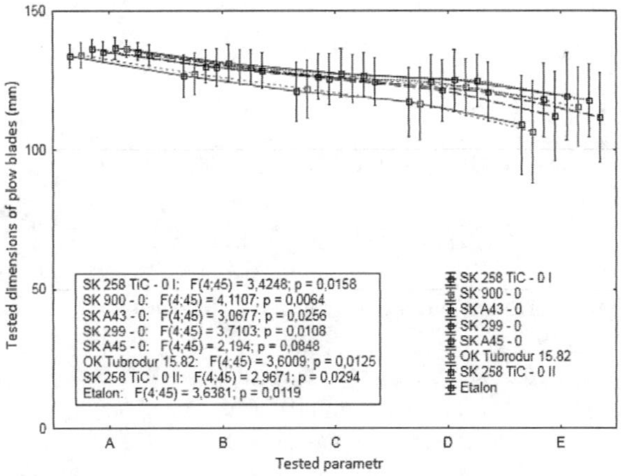

Fig. 3. Statistical testing of uniformity of wear of the ploughshares

From the point of view of statistical testing of weight loss in individual variants it can be stated that these are statistically homogeneous groups, i.e. there is no difference between tested parameters of experimental variants.

The H_0 hypothesis was confirmed, i.e. there was no difference in the significance level of 0.05 between the individual tested variants of experiment (SK 258 TiC-O I, SK 900-O, SK A43-O, SK 299-O, OK TUBRODUR 15.82, SK 258 TiC -OII and standard) p = 0.1628 (Fig. 4). It follows from the above that the ploughshare is worn identically with the use of welding materials and standards.

The dimensional loss of welded shares was only 0.41 % less than in the standards at the end of the experiment. Determination coefficient is higher for standards (R^2 = 0.9765) than for welded shares (R^2 = 0.9666).

The decisive factor in increasing the service life of the welded share relative to the standard is the dimension measured on the heel of share (in our case dimension E). The dimensional loss was 17.6% lower for the SK A43-O welding material than for the standard. It is possible to say that the service life of the welded share has increased relative to the standard. The dimensional loss was 1.1 % lower for the SK A45-O than for the standard and for OK TUBRODUR 15.82 3.7% lower than for standard. The dimensional loss was 7.8% higher for SK 900-O than for standard and 7.9 % higher for SK 258 TiC-O than for standard. Welded shares have 5.97 % higher weight loss than standards. Determination coefficient is higher for standards (R^2 = 0.9912) than for welded shares (R^2 = 0.9886).

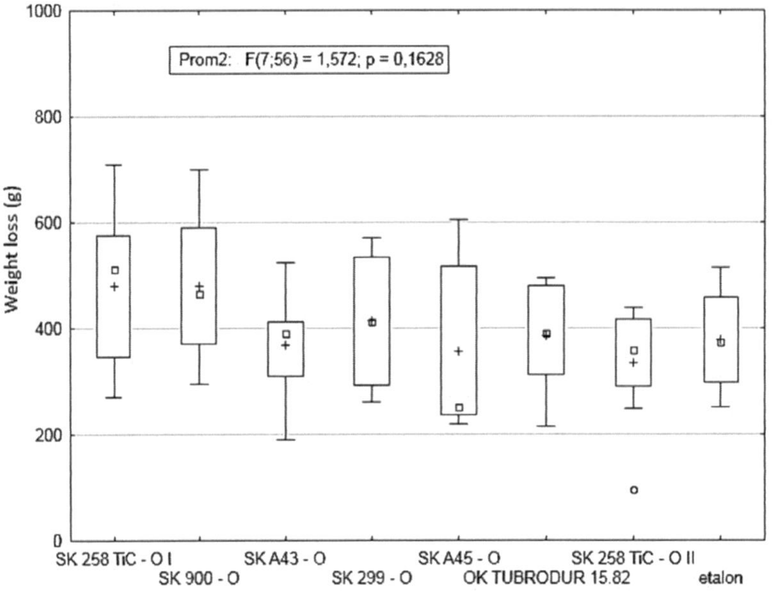

Fig. 4. Statistical testing of weight loss

The results of the experiment have confirmed the need for material and design research in a soil processing area [9, 11].

Technique of soil processing as a plough can be mentioned as a sample in which wear causes fuel surge, a fall in work efficiency etc. [9].

The experiment of ploughshares wear has shown lower fuel consumption and higher ploughing efficiency in shares with hard welding material compared to standards. The results of the research proved an increase in the service life of the ploughshare with the welding material OK TUBRODUR 15.82. The wear was 16.9 % lower than for the standard after 87 ha of ploughing [10].

The weight loss was 130.45 ± 22.28 g/ha in dry soil and 98.06 ± 12.85 g/ha in wet soil. Ploughshares are worn at an average of 90 to 210 g / ha [6].

4 Conclusions

The research was focused at increasing the service life of ploughshares. An obliquely applied welding technique was used due to the high cost of wearable parts. Dimensional and mass analysis were used to evaluate the service life of the ploughshares. These conclusions can be deduced from the results of the research:

1. Changes in share height, bottom edge length at share tip, share front length and share weight clearly did not show added value by welding. The course of partial dependencies was similar with minimal differences. By statistically comparing the mean values of weight, a statistically significant difference was not found between the tested variants.
2. Efficiency of ploughing is from the point of view of a practical user the most important (i.e. speed, fuel consumption, quality, etc.). Efficiency of ploughing has

confirmed the efficiency of the creation of a "tooth share" due to gradual wear. These conclusions can be determined on the basis of the cutting conditions of the "tooth share".

3. It is possible to create an abrasion-resistant surface, but also to use different geometrical arrangement of the welding layer. The aim of this arrangement is to follow the course of processed soil. The ploughshare with this functional surface is wearing unevenly during ploughing.

4. The experiment has confirmed the effectiveness of ploughing. The "tooth share" has always kept a distinctive self-sharpening effect. This effect greatly improves the sinking of the plough into the soil especially in the dry conditions.

References

1. S. Petrásek, M. Müller, *Setting of angle of soil flow on ploughshare at traditional processing of soil.* Manufacturing Technology, vol. **3**, 407-412 (2014)
2. P. Čičo, M. Kotus, M. Vysočanská, *Renovation of sugar beet harvesters - prolongation of service life.* Listy cukrovarnické a řepařské, vol. **9-10**, 280-283 (2012)
3. M. Müller, P. Hrabě, *Overlay materials used for increasing lifetime of machine parts working under conditions of intensive abrasion.* Research in Agricultural Engineering, vol. **59**, 16-22 (2013)
4. M. Müller, P. Novák, P. Hrabě, *Inovace materiálově konstrukčních řešení plužní čepele v oblasti konvenčního zpracování půdy při pěstování cukrové řepy.* Listy cukrovarnické a řepařské, **vol. 3**, 94-99 (2014)
5. M. Müller, R. Chotěborský, P. Hrabě, *Aplikace návarového materiálu na nástroje zpracovávající půdu za účelem zvýšení jejich odolnosti vůči abrazivnímu opotřebení,* Listy cukrovarnické a řepařské, **vol. 9-10**, 279-283 (2015)
6. Y. Bayhan, *Reduction of wear via hardfacing of chisel ploughshare.* Tribology International, **vol. 39**, 570-574 (2006)
7. V. Legát, V. Jurča, Z. Aleš, Contribution to plough shares and chisels useful life optimization. Scientia Agriculturae Bohemica, vol. 42, 73-78 (2011)
8. M. Müller, P. Novák, R. Chotěborský, P. Hrabě, *Efektivní zvýšení životnosti plužní čepele.* Mechanizace zemědělství, **vol. 62**, 68-70 (2012)
9. A. Natsis, G. Petropoulos, C. Pandazaras, *Influence of local soil conditions on mouldboard ploughshare abrasive wear.* Tribology International, **vol. 41**, 151–157 (2008)
10. Z. Horvat, D. Filipovi, S. Kostic, R. Emert, *Reduction of mouldboard plough share wear by a combination tech¬nique of hardfacing.* Tribology International, **vol. 41**, 778–782 (2008)
11. P. Doubek, J. Filípek, *Abrasive and erosive wear of technical materials.* Acta Universitatis Agriculturae et Silviculturae Mendelianae Brunensis, **vol. 59**, 13–21 (2011)
12. D. Kalincová, M. Ťavodová, V. Ľuptáčiková, Application of the weld deposits on function surfaces of the forest machines components. Manufacturing Technology, **vol. 18, No. 3,** 400-405 (2018)

Determination of the force required for the hydro-forming of al 99,5

Vasile Ceclan[1,*], *Sorin* Grozav[1]

[1]Technical University of Cluj Napoca, Faculty of Machine Building, Blv Muncii 103-105, Cluj Napoca, Romania

Abstract. In this paper the authors present the hydroforming of the tubular parts. It also presents the technology of the tubular parts' hydroforming. In this paper we determine the forces necessary for the hydroforming process. They are also presented the mechanical parameters of the material that are used in the deformation process. Also, the variation in wall thickness of the aluminium specimen is determined before the deformation process

Keywords: Hydroforming, Aluminium, Force, Wall thickness

1 Introduction

Hydroforming is a manufacturing process that uses the pressure exerted by a hydraulic medium to produce deformations of a sheet or tubular blank.

Tube hydroforming is a method commonly used in the manufacture of complex components in the automobile resistance structure and a manufacturing technology that is being assimilated in many other industrial sectors (aerospace, household goods).

Coordinating the internal pressure with the plunger axial feed is one of the basic requirements of the tube hydroforming process, which must ensure that there are no defects (breaking or breaking the walls).

The process of hydroforming the tubes with high internal pressure is one of the newest manufacturing processes for tubular components. Prior to hydroforming, the tubes may be performed. During hydroforming, the material is inflated by increasing the internal pressure by means of two plungers acting on the ends of the tube. The plungers perform the pressure of the tube heads by introducing the pressurized fluid through the plunger holes.

The advantage of this method of hydroforming consists in:

- the hydrostatic pressure of the liquid uniformly presses the preform using plunging, preventing the formation of creases, stretching and thinning of the processed material;
- the stress and deformation state is more favourable in the case of the hydroforming and it is possible to obtain parts in a single operation;
- the economic efficiency of the hydroforming process by introducing the pressure of liquid is more obvious and more productive than other classical processes deformation, since the deformation is accomplished in a single operation.

* Corresponding author: vasile.ceclan@tcm.utcluj.ro

Reviewers: *Róbert* Čep, *Jozef* Pilc

2 Calculations of pressure and forces required for hydroforming

2.1 Estimation of the hydroforming pressure

The maximum pressure of the hydroforming fluid can be determined by the relationship Approximate [Singh, Neugebauer]:

$$p_{max} = Rm.s/R \ [N/mm^2], \quad (1)$$

Where: Rm = 77.1 N/mm^2 is the maximum breaking strength of the material hydroformed (Al 99.5 ENAW 1050 H0 SR EN 10088-1: 2005);
s = 1.5 mm - nominal wall thickness;
R = 8 mm - the radius of connection of the vertical branch of the tube.

From the relation (1) results:

$$p_{max} = 14.45 \ N/mm^2 = 144.5 \ bar. \quad (2)$$

Consider the approximate character of the formula (1). Numerical simulation of the hydroforming process will allow verification that this maximum value of pressure hydroforming allows complete connection.

2.2 Determination of mould closing force

The force of closing the mould (exerted by the press ram) must balance the effect mechanical pressure acting within the tubular blank. Considering the fact that, in the case of the hydroforming process, the internal pressure increases extremely rapidly at the level maximum, the ram force can be calculated using the relationship:

$$F_{ram} = p_{max} \cdot l_{sf} \cdot d_{sf} = 144.5 \cdot 99.5 \cdot 35 = 503221.21 \ N = 503{,}22 \ KN \ . \quad (3)$$

The maximum pressure was multiplied by the area of the horizontal projection of the inner surface of the still unreformed blank.

2.3 Determination of sealing force

The sealing force (exerted by the plungers that axially press the ends of the emissive) performs a double function: plasticizing the material by bringing it closer to the tear strength, respectively balancing the opening effect of the pressure acting inside the tube. By summing the forces corresponding to the two axial plunger functions, the sealing force is obtained:

$$F_s = [\pi(d\text{-}s) \cdot s \cdot R_m] + [p_{max} \cdot \pi(d\text{-}2s)^2/4]$$
$$F_s = [3{,}14 \cdot (35\text{-}1.5) \cdot 2 \cdot 77.1] + [144.5 \cdot 3{,}14 \cdot (35\text{-}2 \cdot 1.5)^2/4] = 117371{,}1N = 117{,}37 \ KN \quad (4)$$

In order to determine all these forces it was necessary to determine the parameters of the materials used in the actual hydroforming process.

3 Experimental determinations of material parameters

Figure 1 shows the state of stresses in the tube wall during the process and the following notations are inserted: R - outer tube radius, respectively s - thickness of the wall, σ1 - represents a circumferential stretching stress, and σ2 - represents an axial compression stress.

The two non - zero voltage components in the tube wall are: negative and acts in the axial direction (corresponds to the compressive force F) and the other is positive and acts in the circumferential direction (represents the effect of the pressure stretching p). The hydroforming process is based, for example, on the swelling of a tubular blank, accompanied by axial or radial compression [1- 7]. This process typically consists of the following steps: expansion with compression, calibration [5]. Hydroforming is part of the cold processing industry.

The process is used for the construction of highly complex geometry cavities from high deformability tubes or boards.

Table 1. Material Parameter of Al 99.5 ENAW 1050 H0

Parameter of material	Al 99,5
Tensile strength Rm	77,1 MPa
Conventional Yield stress Rp0,2	58,9 MPa
Proportional elongation under maximum force Ag	13,5 %
Total elongation below maximum force Agt	18,7 %

Figure 2 shows the curve of tension-percentage elongation for trial test of the material according to the standard SR EN 10002-1.

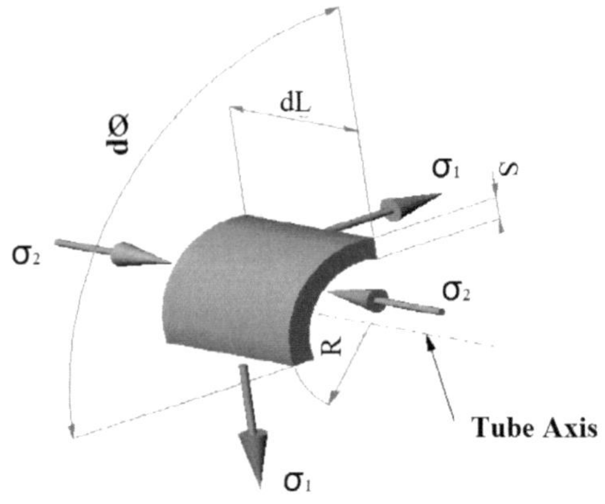

Fig. 1. The state of tension in the tube

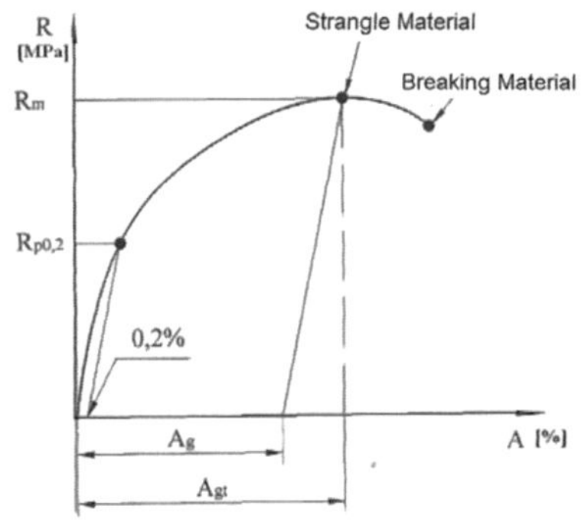

Fig. 2. Curve of tension-percentage elongation for traction test

One important fact on production of the tubular parts, it is the change of wall thickness. Because of the deformation and the tension inside of the material of the tubes, the parameters are change. On different specimens of aluminium tubes was scanned with very precis machine. Figure 3 it is present the variation of the wall thickness of aluminium AL 99.5 which was tested in this paper. This was measurement on 3D scanner with optical devises. The measurements of the specimens were performed in the same way they are presented by other authors [8,9]. In this figure it can be observed that it is a big variation of the thickness, between 1.46 mm and 1.58 mm. The theoretical thickness must be 1.5 mm.

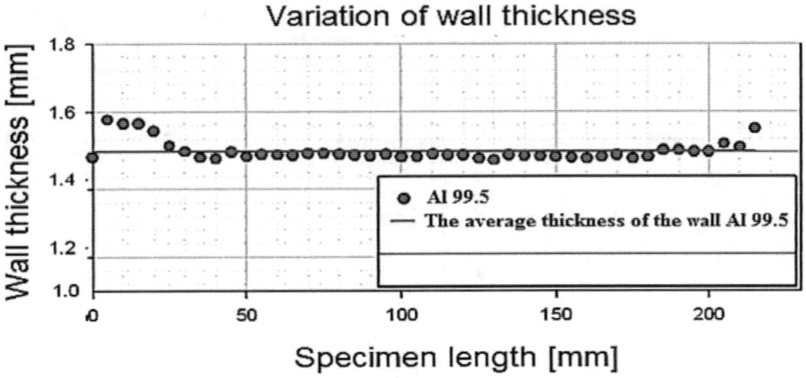

Fig. 3. Variation of wall thickness

These specimens were cut on the water jet machine OMAX, to don`t be influenced by material parameters, because with this techniques the properties of mechanical parameter are not influenced [10-12].

The standard EN 10002-1: 1995 and EN 10002-2: 1995 show the shape and dimensions of the test specimens for uniaxial tensile strength.

After extracting these specimens, these were tested for uniaxial traction on machine Zwick - Roel Z150 which is in Technical University of Cluj-Napoca, as show in figure 4.

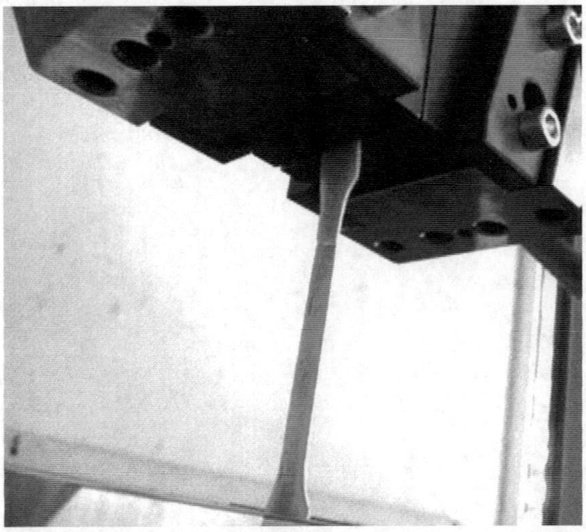

Fig. 4. Traction tests on the machine Zwick - Roel Z150.

To determine the mechanical properties characteristics it is good to make uniaxial tensile strength testing. This test involves the application of a tensile force on a specimen until fracture.

Figure 5 presents the stress-strain curves in the case of Al 99.5 ENAW 1050 H0

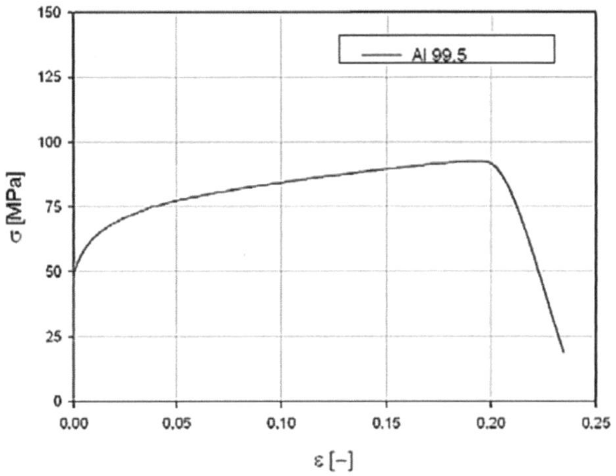

Fig. 5. Material curve for Al 99.5

4 The own achievement of hydroforming

In order to carry out the hydroforming process it was necessary to design and realize a deformation stand. This stand was made from: (1) - the hydroforming mould; (2) - hydraulic high pressure pump; (3) – Tinius - Olsen traction-compression test machine with rated force of 100 kN. In this stand it was necessary to control parameters of hydroforming.

Fig. 6. Stand of hydroforming

In this stand it was necessary to control certain parameters. The most important parameters were the internal pressure of the liquid this was done by means of the pump (2 of figure 6). Another important parameter is the ram force; this can be controlled with the help Tinius - Olsen traction-compression test machine (3 of figure 6). The sealing force is given by the hydroforming mould (1 of figure 6).

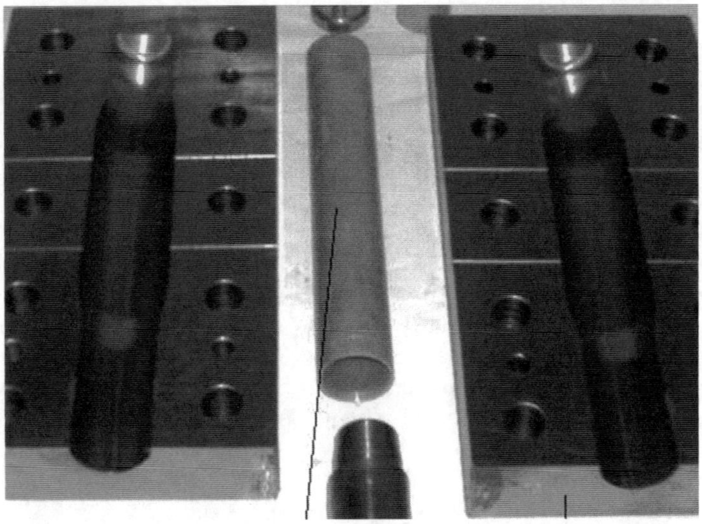

Fig. 7. Mould of hydroforming

The shape of the hydroforming mould is identical to the shape of the deformed tube. In figure 7 it is presented mould of hydroforming. This mould was made by CNC machine. [13-15]

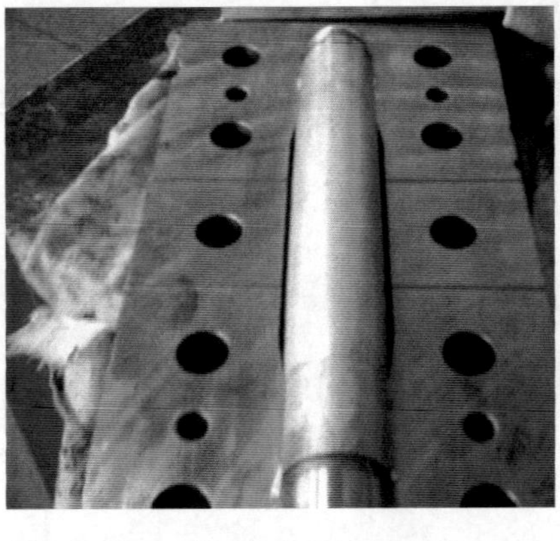

Fig. 8. Steps of hydroforming tubular part

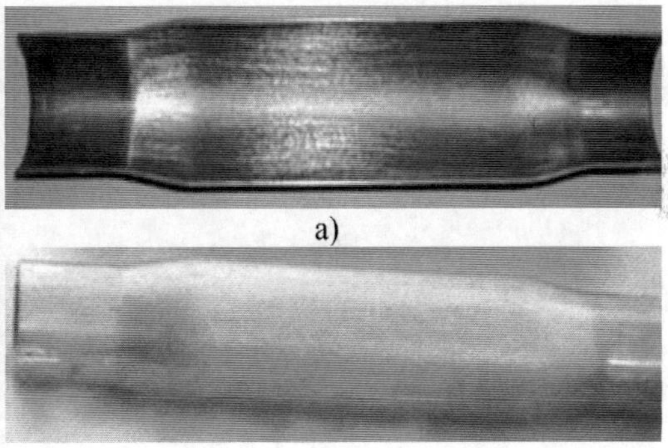

a)

b)

Fig. 9. Tubular parts hydroforming

Figure 8 shows the steps of the hydroforming process inside the mould. Hydraulic oil is used to deform the tubes. The specimen length was 220 mm. It can be seen that depending on the shape of the mould, you get the shape of the hydro former tube, a procedure very similar to vacuum casting [16-18]. Figure 9 shows the tubes made by hydroforming. It is possible to observe the inner and outer surface of the tubes [19-21]. Figure 9a shows the difference in roughness caused by the hydroforming process. For the deformation process, tubes manufactured by the drawing process are suitable [22,23]. In figure 9.b it can be seen that the surface of the tube in the deformed area is better, also having a better roughness.

5 Conclusions

This paper focused on the determination of the main factors influencing the hydroforming process of the tubular parts. The experiments in this paper focused on determining the main factors that influence the hydroforming process. Also took into account the quality of the tubular parts obtained.

As a result of the hydroforming process it can be observed that the obtained parts have much better strength and rigidity. Another effect of this process is to reduce the number of operations in order to obtain the finished piece, as well as to reduce its weight. Another important feature is the realization of complex shapes, and the optimal use of space. This process significantly reduces subsequent assembly operations. It is recommended to use hydraulic oil to reduce the friction forces between the piece and the die. After the laboratory tests, a theoretical model of calibration of the hydroforming process was made for different materials used. A great advantage of this simple device is that it is possible to install on universal cold pressing equipment.

This work was supported by a grant of the Romanian National Authority for Scientific Research and Innovation, CNCS/CCCDI–UEFISCDI, project number PN-III-P2-2.1-BG-2016-0216, within PNCDI III.

References

1. T. Jansson, *Optimization of Sheet Metal Forming Processes*. PhD thesis, Universitet Linkoping, Linkoping, Sweden (2002).
2. M. Jansson, *Hydro-mechanical forming of aluminium tubes – On constitutive modelling and process design*. PhD Thesis. Linkoping University (2006)
3. S, Jirathearanat, *Advanced Methods for Finite Element Simulation for Part and Process Design in Tube Hydroforming*. PhD Thesis. Ohio State University (2004)
4. J. Lunqvist, *Numerical Simulation of Tube Hydroforming*. Adaptive Loading Paths. Licentiate Thesis. Lulea University of Technology (2004)
5. Schuler Gmbh (ed.) *Metal Forming Handbook*. Berlin: Springer (1998)
6. K. Siegert, M. Haessermann, B. Losch, R. Rieger, *Recent Developments in Hydroforming Tehnology*. Journal of Materials Processing Technology 98, 251-258 (2000)
7. A. Szabo Ponce, *Experimental Modeling and Control of Tube Hydroforming Process*. MSc Thesis. University of Sao Paulo (2006)
8. L. Morovič and P. Pokorný, *Optical 3D Scanning of Small Parts*. Advanced Materials Research, Vols. 468-471, pp. 2269-2273 (2012)
9. J. Peterka et al., *Optical 3D Scanning of Cutting Tools*. Applied Mechanics and Materials, Vol. 421, pp. 663-667 (2013)
10. A. Popan et al., *Research on Abrasive Water Jet Milling of the Planar Surfaces, Slots and Profiles*. Applied Mechanics and Materials, Vol. 760, pp. 409-414 (2015)
11. A. Popan, N. Bâlc, I. Popan, *Preliminary study on occurrence of composite material delamination processed by abrasive water jet cutting*. MATEC Web of Conferences. EDP Sciences, 121, 02010 (2017)

12. A. Popan, N. Balc, A. Carean, A. Luca, A. Miron, *Research to improve the surface roughness of the parts made by water jet milling.* AJME-2013, Vol. 11, Issue 2, 105 – 109 (2013)
13. M.Sadílek, F. Fojtík, Z. Sadílková, K. Kolařík, J. Petrů, J. *A Study of Effects of Changing the Position of the Tool Axis to the Machined Surface*. Transaction of FAMENA, Vol. 39, No. 2, 33-46 (2015)
14. L. Kyncl, M. Sadílek, R. Čep, J. Petrů, D. Stanvekova, J. Procházka, P. Nováček, *Tests of drills during drilling holes into alloy wheels.* Manufacturing Technology, Vol. 14, No. 4, 554-561 (2014)
15. M. Sadílek, R. Čep, Z. Sadílková, J. Valíček, L. Petřkovská, *Increasing tool life during turning with variable depth of cut.* Materiali in tehnologije/Materials and technology. vol.47, no.2, 199-203 (2013)
16. A. Luca, N. Balc, I. Drstvensek, A. Popan, *Analysis of aluminum parts for accuracy improvement in vacuumcasting process.* Academic Journal of Manufacturing Engineering, Vol. IX, Issue 3, 75-80 (2011)
17. A. Luca, N. Balc, A. Popan, V. Ceclan, N. Panc, *Improving the quality of the parts made by rapid metalcasting process.* Academic Journal of Manufacturing Engineering, Vol. 12, Issue 1, 82-86 (2014)
18. A. Luca, N. Balc, A. Popan, N. Panc, *The influence of investment casting process parameters of complex metal parts.* Academic Journal of Manufacturing Engineering, Vol. 10, Issue 3, 57-62 (2012)
19. V.A. Ceclan et al. *Quality of the hydroformed tubular parts*. Advanced Engineering Forum 8-9: 215–224 (2013)
20. V. Ceclan, S.Grozav, E.Sabau, I. A. Popan, C. Borzan, *Structural analysis of tubes hydroforming*. Academic Journal of Manufacturing Engineering, Vol. 11, Issue 4, 56 - 59, (2013)
21. V. Ceclan, N. Balc, A. Miron, C. Borzan, I.A. Popan, *Numerical simulation of the tube bending process and validation of the results.* Academic Journal of Manufacturing Engineering, Vol. IX, Issue 3, 32-37 (2011)
22. Ridzoň M., *The Effect of Technological Parameters Influencing the Properties of Seamless Cold-Drawn Tubes*. - 1. ed. - (Hochschule Anhalt, Köthen). 89 (2012)
23. M. Ridzoň, Ľ. Eleková, J. Lišková, J. Bílik: *Effect of reduction in three-drawn and two-drawn single-run technology on roughness of inner surface of tubes*, Volume 18, Number 28 (2010)

Analysis of the properties of bone cement with respect to its manufacturing and typical service lifetime conditions

Robert Karpinski[1,*], *Jakub* Szabelski[1], *Jacek* Maksymiuk[2]

[1] Lublin University of Technology, Faculty of Mechanical Engineering, Nadbystrzycka 36, 20-618 Lublin, Poland
[2] Orthopaedic Department, Łęczna Hospital, Krasnystawska 52, 21-010 Łęczna, Poland

Abstract. This paper analyses the impact of exposure to Ringer's solution, an isotonic solution relative to the body fluids, on the strength of bone cement. Research was conducted in order to determine moisture uptake and strength degradation. Prior to mechanical tests, the samples were seasoned in the solution simulating the human body conditions and subsequently measured to determine their increase in weight. Samples were then tested, using MTS Bionix testing system, for compressive strength according to ISO Standard. The data obtained from the measurements allowed to develop a statistical model for predicting cement absorption rate over a longer period of time. Compression strength of both seasoned and unseasoned samples was measured over time. The results were analysed statistically to show an effect of seasoning on cement strength. Cement samples show moderate moisture uptake over time, together with some decrease of compressive strength. Correlation between two above mentioned, measured factors was proven and linear model prepared. Using obtained models, one can estimate the cement strength and moisture uptake over time.

Keywords: bone cement, absorption, compression strength, statistical analysis, modelling, regression analysis

1 Introduction

The prevalence of degenerative bone and joint diseases is increasing at an alarming rate, in fact, in developed countries they constitute a half of all chronic diseases in patients over 50. Analysis of trends shows that the percentage of patients with degenerative bone and joint diseases is likely to double by 2020. These conditions frequently require surgical procedures, which may include complete reconstruction of joints when their function has been lost. However, bone implants may also prove indispensable in other procedures, connected with: bone fracture, spinal problems, osteoporosis, or other conditions of human motor organs which require permanent, non-permanent or biodegradable implants. Complete or partial replacement of joint or bone tissue with orthopaedic implants may only be carried out with

* Corresponding author: r.karpinski@pollub.pl

Reviewers: *František* Holešovský, *Augustín* Görög

materials capable of performing biological functions of various tissues, e.g. bones, cartilages, ligaments or tendons, and if necessary aid bone regeneration [1, 2, 3].

1.1 Biomaterials

In modern dentistry and bone surgery, biomaterials, and typically cements, are commonly implemented. Surgical bone cements have found application in motor organ surgeries as a material in alloplastic joint replacement. Cements are predominantly used in alloplastic procedures of joints as a fixture for endoprosthesis, or to repair bone defects in considerable bone loss. Among other applications, the most significant ones are vertebroplasty and kyphoplasty, consisting in injecting bone cement into fractured vertebra and in fracture stabilisation.

Bone cements constitute a highly desirable material for surgeons. In surgical applications bone cements are commonly used as fixtures for hip joint endoprosthesis and to repair bone defects in considerable bone loss, in which case they are injected into the original bone. They also provide a common solution in stabilisation of knee or shoulder joint endoprosthesis, whereas low-viscosity bone cements are successfully implemented in spine surgeries following vertebral compression fracture, joining shattered bone fragments, and bone loss treatment. Their good plastic properties enable purpose-specific shaping the material in situ, to meet the demands of a particular application, providing exact filling and stable fixture of prosthesis [3, 4, 5, 6, 7, 8].

A healthy joint, whether hip or shoulder one, is characterised by high vibration damping capacity. In surgery, the natural structures, which have partly or fully lost their anatomical functions, are replaced by artificial components of high rigidity and low damping properties. In order to ensure that the mechanical properties of cement are not inhibited, the material must be implemented following a correct procedure and with observance of such conditions as: proper preparation (utmost accuracy in joining components), appropriate time of application (if implemented before the highly-exothermic process of polymerisation has finished, it may cause damage to tissues in the close area of application), or disruption of polymerisation by appearance of undesirable bone or fluid debris. Inhibited mechanical properties of cement may lead to complications during the procedure or loosening of the prosthesis in use [9, 10, 11, 12, 13, 14, 15].

2 Research Methods

Bone cement, similarly to the majority of other polymer materials, comes into reaction with the environment, which may take the form of absorption of water or bodily fluids. In contact with cement, water will act as a plasticiser and may affect its characteristic properties [16], and therefore, bone cement must be tested in the environment resembling the human body conditions to a greatest extent. It is of utmost importance regarding long-term viscoelastic properties of cement material [17, 18]. Our tests employed Ringer's solution, i.e. an isotonic solution, of identical osmotic potential as plasma, 1000 ml of which contains 8.6 g of sodium chloride, 0.3 g of potassium chloride, 0.33 g of calcium chloride dihydrate which corresponds to the following electrolyte levels: sodium - 147 mmol/l, potassium – 4 mmol/l, calcium - 2.2 mmol/l, chlorides - 156 mmol/l [19]. The effect of time of immersion in Ringer's solution on compressive strength of bone cement samples was determined in physical tests, consisting of linear compressive strength tests. The samples exposed and not exposed to Ringer's solution were tested in compression upon 1, 10, 20 and 30 days of seasoning in ambient temperature. The samples exposed to the solution were tested to determine the relative increase in weight, in the function of exposure time.

The study was carried out on DePuy CMW 3 GENTAMICIN Bone Cement, containing bactericidal antibiotic, gentamicin, (Fig. 1), whose composition is shown in Table 1. The test samples were formed to desired dimensions in special moulds, which enabled producing a large number of repeatable samples. The counter-mould was designed by means of Solid Edge ST9 software, and subsequently printed with the 3D uPrint SE Plus printer with the use of durable and high-strength ABS. The proper mould was produced with POLATOSIL® M-33 two-component silicone rubber, by filling the counter-mould. The selection of silicone rubber as the material of the mould was dictated by the mechanical properties of rubber, which facilitate the removal of samples from the mould and enable repeated use of the mould. The dimensions and tolerance of samples were prepared according to the International Standard ISO 5388:2002, as confirmed by the measurements of finished samples. Prior to mixing the components of the bone cement, the powder and the liquid were cooled down to 16°C to prolong the setting time, hence enabling proper filling of moulds before cement polymerisation would start. It ought to be mentioned that in surgical practice temperature of cement is similarly reduced before its components are mixed and also during polymerisation, which is an exothermic reaction generating relatively high heat.

Table 1. Chemical composition of CMW3 GENTAMYCIN bone cement.

	Percentage by weight
Bone cement powder components:	
Gentamicin Sulphate	4.22
Polymethyl Methacrylate	83.88
Benzoyl Peroxide	1.90
Barium Sulphate	10.00
Bone cement liquid components:	
Methyl Methacrylate	97.5
N,N-Dimethyl-p-toluidine	< 2.50
Hydroquinone	75 ppm

The compression strength tests were carried out on a custom-built test stand, consisting of: MTS Bionix - Servohydraulic Test System for biomedical applications, and special instrumentation in the form of customised grips. The MTS TestWorks software was used to programme and perform the experiment. Compression strength test parameters, including compression time and rate, were specified in accordance with the ISO standard.

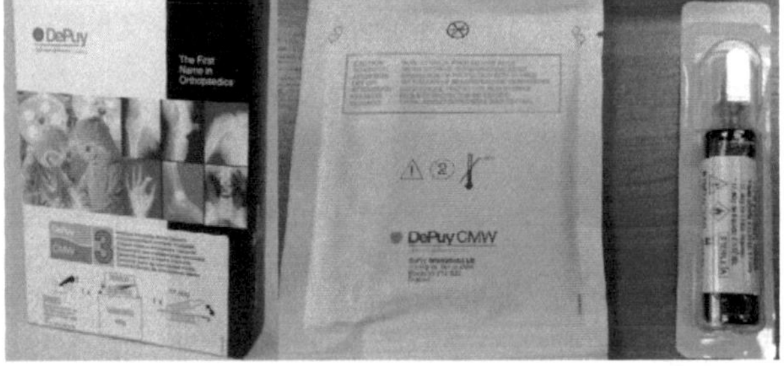

Fig. 1. CMW3 GENTAMYCIN bone cement by DePuy CMW

3 Moisture Uptake

Mean weight increase of cement samples as a result of moisture uptake from Ringer's solution was estimated by weighing the entire sample lots prior to and after exposure to the solution. The samples prior to weighing were gently wiped to remove potential micro drops off the surface of samples. As anticipated, progressing absorption processes led to weight increase of samples. Table 2 shows relative mean weight increase of cement material in the function of time of exposure to Ringer's solution.

Table 2. Relative mean weight increase of cement after exposure to Ringer's solution.

Seasoning time [days]	**Mean weight of unseasoned samples** [g]	**Mean weight increase of samples** [g]	**Relative weight increase of samples** [%]
1	0.358	0.0020	0.6%
10	0.354	0.0075	2.1%
20	0.348	0.0089	2.6%
30	0.351	0.0110	3.1%

To provide the best fit of the mathematical model to the results from the tests, the data was analysed statistically in order to establish qualitative coefficients of determination. As it was only the logarithmic model that was suitable for the purpose of describing diffusion in porous media, it was resolved that this would be the model analysed. Fig. 2 and Table 3 show the results of the analyses.

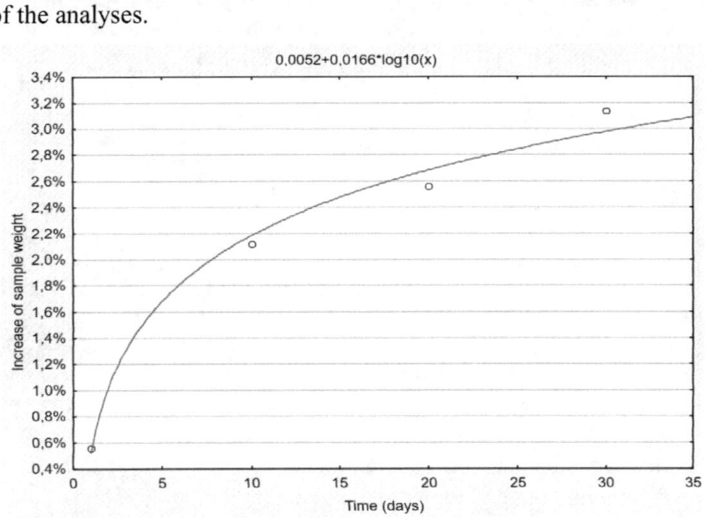

Fig. 2. Increase of sample weight over time

The conducted statistical analysis of the data obtained from the study and the verification of the adjusted model in the form of the equation: weight=0.52%+1.66%·log10 (time), indicate a good fit of the model (Fig. 2). The p-value for test F < α=0.05, therefore the regression equation is significant. The correlation coefficient R = 0.993 indicates a strong correlation between the variables. Coefficient of determination R2 explains the weight increase of seasoned samples in 98%.

The developed model was employed to project the weight increase of samples in the subsequent days. The estimated results are shown in Table 3 below.

Table 3. Projected weight increase of bone cement samples.

Seasoning time [days]	Estimated weight increase [%]
40	3.2%
60	3.5%
100	3.8%
200	4.3%

The logarithmic character of weight increase in samples suggests that the liquid absorption rate is the highest at the beginning of seasoning and decreases over time.

4 Compression Strength

Bone cement samples seasoned in Ringer's solution were subjected to static compression tests, mean results are shown in Table 4.

Table 4. Test results

Seasoning time [days]	Compression strength [MPa]	Standard deviation [MPa]	Coefficient of variation [%]
0	73.88	3.91	5.3%
1	71.10	3.96	5.6%
10	69.06	3.66	5.3%
20	63.68	2.93	4.6%
30	64.28	3.17	4.9%

Based on statistical analysis of compression strength test results for cement samples, it was observed that all values were of normal distribution (for the Kolmogorov-Smirnov, Lilliefors and Shapiro-Wilk tests $p > \alpha = 0.05$) (Tab.5.). There is no reason to reject the hypothesis of the normal distribution of the obtained results, which makes it possible to employ t-Student tests to evaluate the differences between mean results (Table 6)

Table 5. Normal distribution test result: max D – max. D value for Lillefors test, K-S p - Kolmogorov-Smirnov test, Lillef p. - Lillefors test, W - Shapiro-Wilk test, p - p value for Shapiro-Wilk test).

Seasoning time	max D	K-S p	Lillief. P	W	p
0 days	0.214150	p > .20	p > .20	0.926018	0.549718
1 day	0.164024	p > .20	p > .20	0.930915	0.456951
10 days	0.159465	p > .20	p > .20	0.929232	0.440382
20 days	0.165803	p > .20	p > .20	0.949603	0.707229
30 days	0.134917	p > .20	p > .20	0.969350	0.884752

Table 6. Testing for differences between mean values

	Tests for equality of variances			**Test for equality of means**		
Samples acc. to seasoning time [days]	**P F test**	**P Levene**	**P Brn-Fors**	**P Cochran-Cox**		
0 - 1	0.737	0.974	0.971	0.122	> α (0.05)	non-significant difference
0 - 10	0.433	0.571	0.569	0.048	< α (0.05)	significant difference
0 - 20	0.488	0.720	0.714	0.001	< α (0.05)	significant difference
0 - 30	0.215	0.494	0.512	0.001	< α (0.05)	significant difference
1 - 10	0.105	0.286	0.310	0.781	> α (0.05)	non-significant difference
1 - 20	0.102	0.320	0.352	0.015	< α (0.05)	significant difference
1 - 30	0.593	0.433	0.501	0.025	< α (0.05)	significant difference
10 - 20	0.677	0.607	0.682	0.005	< α (0.05)	significant difference
10 - 30	0.879	0.737	0.732	0.008	< α (0.05)	significant difference
20 - 30	0.737	0.974	0.971	0.700	> α (0.05)	non-significant difference

According to results obtained from statistical analyses regarding the mean values equality the following chart, presenting overall test results, was prepared (Fig.3).

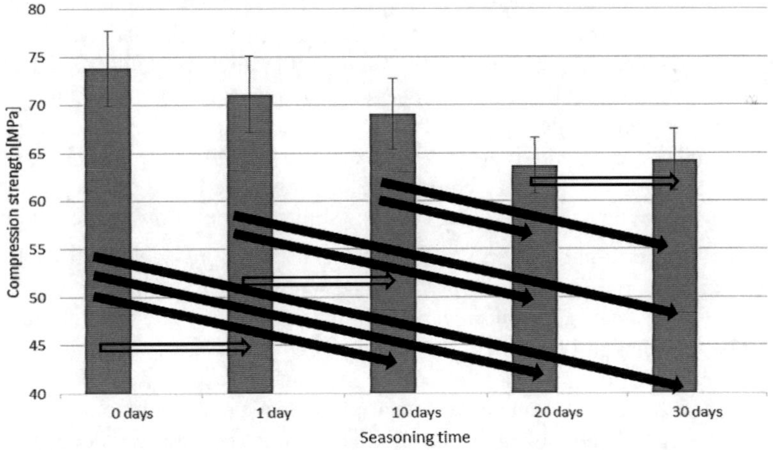

Fig. 3. Equality of mean values of compression strength after seasoning

The results of tests conducted on seasoned bone cement samples lead to the following observations:

- often comparison of strength test results for cements between two consecutive series show no statistically significant differences between the strength of either of cement sample, see: 0 – 1 day, 1 – 10 days, 20 – 30 days.
- however statistical analysis of strength test results shows substantial differences between series of samples when comparing wider combinations of seasoning time, for example between 0 days and 10 days, etc.
- following the analysis of obtained values, at a standard significance level (α = 0.05), for consecutive sample series, arrows were placed in Fig. 3 to illustrate the change. In

the case of no substantial differences, the arrows between the bars representing consecutive series were horizontal, decreasing strength was represented by solid arrow.

5 Conclusions

The results of tests determining weight increase of bone cement samples exposed to Ringer's solution (Fig 2, Tab. 2-3) show the rise in the weight of samples over the seasoning time. The statistical part of tests was conducted by means of Statistica 13.1 software. The statistical results allowed us to obtain a mathematical model that showed good fit with the results of the data from the compression strength tests, and the data found in relevant literature as well as from logarithmic models for porous media moisture uptake. The model was subsequently verified to show the certainty level of $R^2 = 98\%$.

The statistical software was used to analyse and correlate individual sample lots, accounting for the coefficient of variation of the lots. The coefficient of variation for all analysed sample lots was in the range of 4.5% and 5.6%, hence indicating low variance of results according to accepted criteria. The statistical analysis of compression strength of bone cement samples relative to the seasoning time in Ringer's solution, carried out at the specified level of significance, showed that the highest disparity between the sample lots was observed for samples not exposed to Ringer's solution and samples seasoned for at least 1 day. The difference between the sample lots analysed after subsequent periods of time showed a general tendency for samples seasoned for a shorter time to develop higher mean strength. Maximum mean compression strength was observed in samples that were not exposed to Ringer's solution, amounting to 73.88 MPa. The results of statistical analysis show decline in strength of the samples in almost all analysed sample lots, *i.e.* after 10, 20 and 30-day exposure. What is more, significant differences were shown in compression strength between samples after 1-day and 20-day and 30-day exposure to the solution. The lowest observed strength of samples was recorded after 20-day exposure (63.68 MPa). The analysis of results indicates that the time of the study could have been extended as the samples analysed in subsequent periods of time (*e.g.* 1-10 days or 20-30 days) did not show statistical differences.

The lack of correlation between our results and the data provided by the manufacturer may be explained by the fact that in our study the samples were prepared, stored, seasoned in Ringer's solution, and tested in normal conditions, *i.e.* at 22°C. The real conditions of their operation (37°C and slightly different environment) were not represented. However, it was shown that the created conditions do affect the mechanical properties of bone cements: the presence of bodily fluids has a negative effect on the mechanical characteristics of cement. One relevant finding of the study is that in certain conditions cements did not exhibit the limit of the compression strength of materials for clinical applications, 66.3 MPa, as specified by the standard (ISO 5833:2002). Nevertheless, in the majority of tests the bone cement samples exhibited sufficient strength.

The analysis of correlation between the results obtained from the study, *i.e.* the weight increase of cement samples seasoned in Ringer's solution and the mean compression strength of these samples, provides an 88% explanation for the aforementioned relationship by means of the following linear model: *strength=73.561-308.55*weight_increase* (Fig.4).

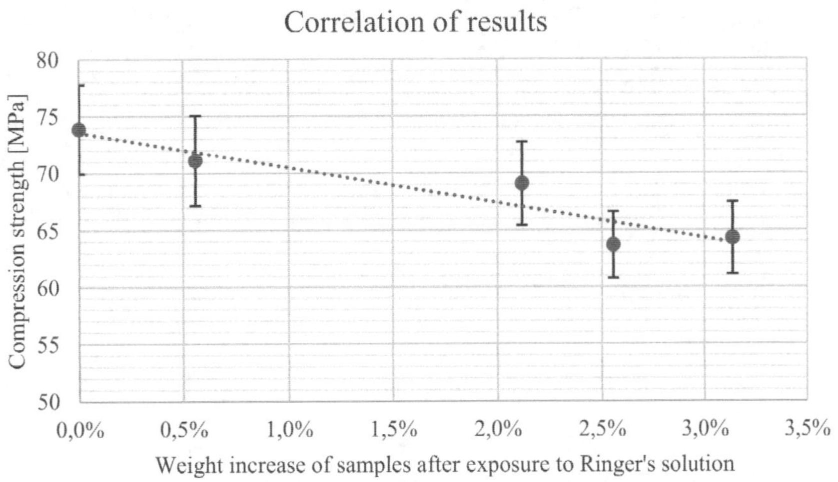

Fig. 4. Correlation between the weight increase of cement samples seasoned in Ringer's solution and the mean compression strength of samples

Thus developed model is prepared for further analysis into the strength behaviour of cement absorbing water; furthermore, if supplemented by the model of absorption in time, it may be employed to estimate strength of cement over time, which suggests direction for future research.

Bone cements are crucial in modern orthopaedics. The mechanical characteristics of these biomaterials require further exploration, and therefore future studies are recommended in order to determine the strength of cement in the presence of physiological fluid debris injected into the material, as well as the impact of elevated human body temperature conditions (approx. 37°C).

References

1. R. Karpiński, Ł. Jaworski, J. Zubrzycki, *Structural analysis of articular cartilage of the hip joint using finite element method.* Advances in Science and Technology Research Journal (2016)
2. J. Zubrzycki, R. Karpinski, L. Jaworski, A.M. Ausiyevich, N. Smidova, *Structural analysis of the pelvic girdle before and after hip replacement procedure.* Science & Technique. 17(2), 165-172 (2018) [on line - https://doi.org/10.21122/2227-1031-2018-17-2-165-172]
3. M. Navarro, A. Michiardi, O. Castano, J. A. Planell, Biomaterials in orthopaedics. Journal of the Royal Society Interface, 5(27), 1137-1158 (2008)
4. S. Błażewicz, L. Stoch, *Biomateriały.* Warszawa, Akademicka Oficyna Wydawnicza Exit (2003)
5. M. Nałęcz, H. Kuś, *Wydawnictwa Komunikacji i Łączności, eds. Problemy biocybernetyki i inżynierii biomedycznej*. Biomateriały, Wydawnictwa Komunikacji i Łączności (1990)
6. S. R. Knight, R. Aujla, S. P. Biswas, *Total hip arthroplasty–over 100 years of operative history.* Orthopedic reviews, 3(2), 16. (2011)

7. J. Caban, R. Karpiński, D. Barta, *Road traffic accident injuries—Causes and biomaterial related treatment.* Automotive Safety, 2018 XI International Science-Technical Conference, 1-7 (2018)
8. A., Massazza, G., Verné, E., Massè, A., Deledda, D., Ferraris, S. ... & Crova, M. *Antibiotic-loaded cement in orthopedic surgery: a review.* ISRN orthopaedics (2011)
9. H. Lelovics, T. Liptakova, *Time and mixing techniquedependent changes in bone cement SmartSet (R) HV.* Acta of bioengineering and biomechanics. 1;12(4):63-7 (2010)
10. T. Liptáková, H. Lelovics, L. Necas, *Variations of temperature of acrylic bone cements prepared by hand and vacuum mixing during their polymerization.* Acta Bioeng Biomech. 1;11:47-51 (2009)
11. H. Lelovics, T. Liptáková, *Comparison of some mechanical and rheological properties of bone cements.* 25th Danubia – Adria Symposium on Advances in Experimental Mechanics, Czech Republic, 157–158 (2008)
12. R. Vaishya, M. Chauhan, A. Vaish, *Bone cement. Journal of Clinical Orthopaedics and Trauma*, 4(4), 157–163 (2013), http://doi.org/10.1016/j.jcot.2013.11.005
13. M. Sarlinova, E. Halasova, P. Palcek, R. Seewald, *Fracture properties of bone cements.* ELEKTRO, 616-620 (2014)
14. Þ. Pétursson, K. J. Edmunds, M. K. Gíslason, B. Magnússon, G. Magnúsdóttir, G. Halldórsson,... & P. Gargiulo, *Bone mineral density and fracture risk assessment to optimize prosthesis selection in total hip replacement.* Computational and mathematical methods in medicine (2015)
15. P. Gargiulo, T. Pétursson, B. Magnússon, P. Bifulco, M. Cesarelli, G. M. Izzo,... & H. Jónsson, *Assessment of total hip arthroplasty by means of computed tomography 3D models and fracture risk evaluation.* Artificial organs, 37(6), 567-573. (2013)
16. A.J.C Lee, R.S.M Ling, S.S Vangala, *Some clinically relevant variables affecting the mechanical behavior of bone cement.* Arch Orthop Traumat Surg, 92: 1-18 (1978)
17. A.J.C Lee, R.S.M Ling, S. Gheduzzi, J-P Simon, R.J. Renfro. *Factors affecting the mechanical and viscoelastic properties of acrylic bone cement.* J Mater Sci - Mater in med., 13: 723-733 (2002)
18. A.J.C. Lee, *The mechanical properties of recovered PMMA bone cement.* Hip International 2004; 14 2 p.79 (2004)
19. R. Karpiński, J. Szabelski, J. Maksymiuk, *Effect of Ringer's Solution on Tensile Strength of Non-Absorbable, Medium-and Long-Term Absorbable Sutures.* Advances in Science and Technology. Research Journal, 11/4, (2017)

Optimization and design of the vulcanization press sandwich pressure plate parameters

Milan Vaško[1,*], *Marián* Handrik[1], *Peter* Kopas[1], *Bohuš* Leitner[2]

[1]University of Žilina, Faculty of Mechanical Engineering, Department of Applied Mechanics, Univerzitná 8215/1, Žilina, Slovak Republic

[2]University of Žilina, Faculty of Security Engineering, Department of Technical Sciences and Informatics, 1. mája 32, Žilina, Slovak Republic

Abstract. The goal of this paper is to present the optimization of the vulcanization compactor pressure plate construction design. The basic factor for the mechanism functionality is the required stiffness compliance. The total construction stiffness has an influence on the vulcanization form tightness. Several solutions of the pressure plate have been proposed in the initial phase. Each proposal has been optimized in order to minimize the pressure plate weight meeting the conditions of the maximum displacement and the maximum allowed stress. The best solution has been chosen on the base of the evaluation afterwards. The multisoftware approach using ADINA and MATLAB software was used to perform FE analyzes.

Keywords: ADINA, MATLAB, optimization algorithms, Nelder-Mead optimizing algorithm, sandwich pressure plate, FE analysis

1 Introduction

Structures exposed by extreme load are often characterized by a large weight of the carrier parts. This is currently removed by various design modifications. These adjustments require consistent control in terms of strength and fatigue life. By using FEM, it is possible to obtain deformations and stresses distribution in the structure and identify the critical locations. The results of the analysis are closer to reality, assuming a suitable computing finite element mesh [1, 2].

Contradictory requirements are placed on the pressure plates in the production of compactor parts. The plates must satisfy strength and deformation conditions in relation to the operational requirements. These often lead to the design of massive high weight plates. This deficiency can be eliminated by designing a sandwich plate that retains the same stiffness and deformation properties and meets the strength requirements. This approach will be applied to a specific construction in the article [3, 4, 5].

* Corresponding author: milan.vasko@fstroj.uniza.sk

Reviewers: *Tomasz* Domański, *Michal* Šofer

2 Formulation of the optimization problem

The aim of the contribution will be based on the optimization process to design a suitable thickness of the upper and lower plates of sandwich construction X_1, the thickness of the ribs X_2 and the distance between plates (rib height) X_3 (Fig. 1) [4, 6].

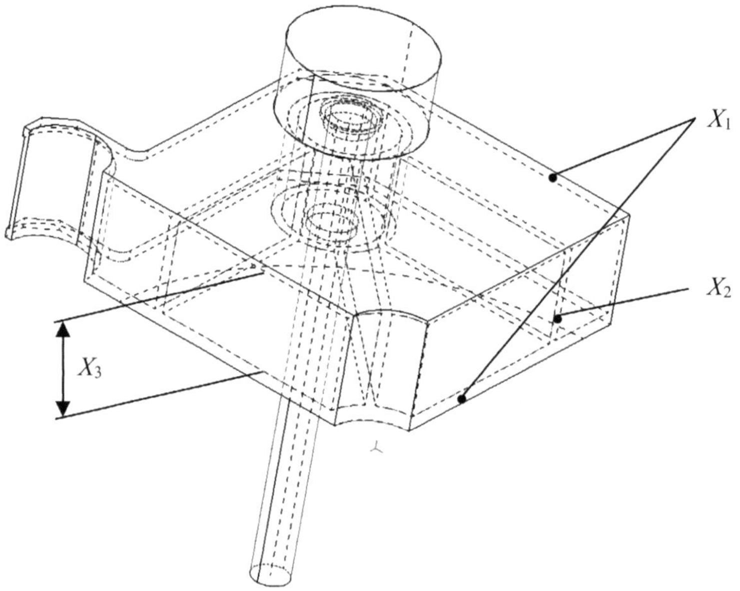

Fig. 1. Model of sandwich pressure plate with optimization parameters

Applying the above-mentioned requirements for the weight, stiffness (deformation) and strength (stress), the optimization problem was formulated as follows

$$F(X_1, X_2, X_3) = m = \rho \cdot \sum_{i=1}^{n} V_i (X_1, X_2, X_3) \rightarrow \min, \tag{1}$$

where m is the weight of the structure determined by FEM analysis from the volume of the model and the density of the material under the following conditions:

- the deformation and stiffness condition:

$$U_{max}(X_1, X_2, X_3) - U_{DOV} \leq 0, \tag{2}$$

- the strength condition:

$$\sigma_{max}(X_1, X_2, X_3) - \sigma_{DOV} \leq 0, \tag{3}$$

- the geometrical condition:

$$\begin{aligned} 0.01 \leq X_1 \leq 0.2; \\ 0.01 \leq X_2 \leq 0.2; \\ 0.10 \leq X_3 \leq 0.4. \end{aligned} \tag{4}$$

The U_{max}, U_{DOV} parameters represent the maximum calculated and allowed deflection. The σ_{max}, σ_{DOV} parameters represent the maximum calculated equivalent stress, respectively its permissible value.

The new objective function has been built with respect to restrictive conditions on the principle of the method of penalty functions in the form

$$F(X_1, X_2, X_3) = \begin{cases} \rho \cdot \sum_{i=1}^{n} V_i \rightarrow \min \text{ (if conditions 2, 3 and 4 are valid);} \\ \lambda \text{ (in other case),} \end{cases} \quad (5)$$

where the penalizing coefficient is $\lambda = 2000$ if one of the constraints is not met, otherwise $\lambda = 0$. The objective function (5) was optimized by an improved simplex Nelder-Mead algorithm programmed in MATLAB [7, 8].

3 Modified simplex optimization method

The original sequential simplex method was designed by Spendley, Hext and Himsworth [9] and then enhanced by Nelder and Mead [10]. The original proposal of the method consisted of applying a regular geometrical shape called simplex with $n + 1$ vertices in n–dimensional space R^n, as shown Fig. 2.

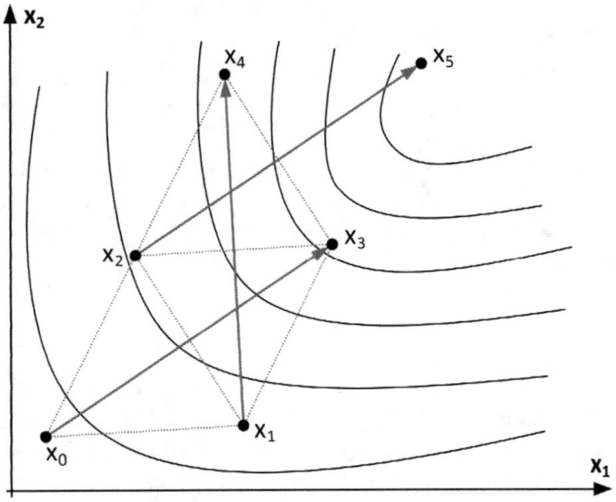

Fig. 2. Graphical representation of the method's algorithm

The improvements suggested by Nelder and Mead also begin with a regular simplex that, however, changes during the computation according to the convenience of the direction in which the search process is executed.

The simplex vertices can be defined by the beginning and the points along each of the n coordinate directions. Such simplex cannot be geometrically symmetrical. The following equations are suggested in reference [9]. To calculate the position of vertices of a uniform simplex of the size and in n–dimensional space, we use the following relations

$$\mathbf{x}_j = \mathbf{x}_0 + p \cdot \mathbf{e}_j + \sum_{\substack{k=1 \\ k \neq j}}^{n} q \cdot \mathbf{e}_k, \quad j = 1, \ldots, n, \quad (6)$$

with

$$p = \frac{a}{n\sqrt{2}}\left(\sqrt{n+1} + n - 1\right); \qquad q = \frac{a}{n\sqrt{2}}\left(\sqrt{n+1} - 1\right), \quad (7)$$

where $\mathbf{e}_k$ is the unit vector along the kth coordinate direction, and $\mathbf{x}_0$ is the initial reference point. For example, for a problem in the assumed two-dimensional space, equations (6) and (7) determine an equilateral triangle with sides a. When the simplex is defined, the function F is evaluated in each of the $n + 1$ vertices $\mathbf{x}_0$, $\mathbf{x}_1$, K, $\mathbf{x}_n$. Let $\mathbf{x}_h$ and $\mathbf{x}_l$ denote the vertices where the function F acquires its maximum and minimum in this sequence, and $\mathbf{x}_s$ denote the place where the vertex acquires the second highest value. The simplex method deletes the vertex $\mathbf{x}_h$ and replaces it with the point where the function F has the lowest value. This is achieved in three operations called *reflection*, *contraction* and *expansion*. Operation reflection generates a new point $\mathbf{x}_r$ along the line of the passing from $\mathbf{x}_h$ to the centre of gravity $\bar{\mathbf{x}}$ of the remaining points, which is defined as follows:

$$\bar{\mathbf{x}} = \frac{1}{n}\sum_{i=0}^{n}\mathbf{x}_i, \quad i \neq h. \tag{8}$$

The vertex point at the end of reflection is calculated as

$$\mathbf{x}_r = \bar{\mathbf{x}} + \alpha \cdot \left(\bar{\mathbf{x}} - \mathbf{x}_h\right), \tag{9}$$

where α is a positive constant called *reflection coefficient* that is usually considered unitary. Each positive value of the reflection coefficient in equation (9) ensures that $\mathbf{x}_r$ is situated on the opposite side of $\bar{\mathbf{x}}$ than $\mathbf{x}_h$. If the function value in this new point $F_r = F(\mathbf{x}_r)$ meets the condition $F_l < F_r < F_s$, then $\mathbf{x}_h$ is replaced by the value $\mathbf{x}_r$ and the process is repeated with this new simplex.

Conversely, if the value of the function F_r at the end of the reflection is lower than the lowest value of the function $F_l = F(\mathbf{x}_l)$, then there is a possibility that we can still reduce the function by moving further along the same direction. We are looking for a more accurate point $\mathbf{x}_e$ with the expansion method using the relationship

$$\mathbf{x}_e = \bar{\mathbf{x}} + \beta \cdot \left(\mathbf{x}_r - \bar{\mathbf{x}}\right), \tag{10}$$

where β is the expansion coefficient that often takes the value of 2 (or 1.95). If the value of the function F_e is smaller than the value at the end of the reflection step, then replace $\mathbf{x}_h$ with the value $\mathbf{x}_e$ and repeat the process with a new simplex. Otherwise, if the reflection leads to the function value equal or greater than F_r, then create a new simplex by replacing $\mathbf{x}_h$ with the value $\mathbf{x}_r$ and continue the process further.

If the reflection process leads to the point $\mathbf{x}_r$ so that $F_r < F_h$ is true, then replace $\mathbf{x}_h$ with the value $\mathbf{x}_r$ and perform the contraction operation. Conversely, if $F_r \geq F_h$, then perform contraction without any replacement using the relation

$$\mathbf{x}_c = \bar{\mathbf{x}} + \gamma \cdot \left(\mathbf{x}_h - \bar{\mathbf{x}}\right), \tag{11}$$

where γ is the contraction coefficient ($0 < \gamma < 1$). The coefficient usually takes the value of 0.5. If $F_c = F(\mathbf{x}_c)$ is greater than F_h, all points are replaced with a new group of points

$$\mathbf{x}_i = \mathbf{x}_i + \frac{1}{2}\left(\mathbf{x}_l - \mathbf{x}_i\right); \quad i = 0, 1, \ldots, n \tag{12}$$

and the process starts from the beginning with the new simplex. Or simply replace $\mathbf{x}_h$ with $\mathbf{x}_c$ and the process with this simplex will start from the beginning.

The operation in (12) causes that the distance between the old simplex points and the point with the lowest value of the function is reduced by half, and therefore it is sometimes referred to as the *contraction* operation. As a convergence criterion to terminate the algorithm, Nelder and Mead [10] suggested the following equation

$$\left\{\frac{1}{1+n}\cdot\sum_{i=0}^{n}\left(F_i-\overline{F}\right)^2\right\}^{1/2}<e, \tag{13}$$

where e is the pre-defined condition of convergence, and $\overline{F}=F\left(\overline{\mathbf{x}}\right)$.

4 The computational model for analysis

The entire computational model can be in terms of algorithmization divided into two tasks:

- compilation of the parametric FEM model for the ADINA calculation software and related files for evaluation of the results [11],
- creating an objective function in MATLAB [4, 12].

The high time requirement is assumed based on the character of the optimization task, and therefore it is necessary to create models with minimal computational demands [13, 14]. Therefore, when creating a FEM model, the double symmetry of construction and a few simplifications were used. Simplifications consist in the rigid connection of the pressure plate with the hydraulic cylinder, which effects the closing force $F = 2$ MN. This design node would require a contact solution in case of requiring an exact solutions task [15].

Another simplification used is the assumption of the uniform pressure load at the point of contact of the pressure plate with the mold container. The elongated piston rod has taken away degrees of freedom at its free end in the direction of its axis. In the analysis, a linear material model with the following properties was used:

- Young's modulus of elasticity $E = 2.1\cdot10^{11}$ Pa,
- Poisson's number $\mu = 0.3$,
- density of the material $\rho = 7850$ kg/m^3.

Quadratic tetrahedron type elements were used in the FEM model. The total number of equations exceeded 350 000. An efficient iterative solver was used to solve this system of linear equations. Duration of one FEM analysis lasted maximum of 3 minutes (including generating the model). The efficiently assembled model has made it possible to perform an optimization task from a single start point in the short time, as well as mapping the whole area of acceptable solutions to identify local extremes of the objective function [16].

The target function algorithm consists of the following steps:

- evaluation of the restrictive conditions of the equation (4) – in the event of their failure penalizing the objective function and the completion of the evaluation of the objective function [17],
- if the conditions of equation (4) are met, then the input file for ADINA is generated with current values of the optimization variables [11],
- generating the FEM model and launching the task solution,
- evaluation of results – determination of weight, maximum equivalent stresses according to von Mises theory, maximum plate deflection and evaluation of fulfillment of the conditions according to relations (2) and (3) [18, 19],
- in the event of non-fulfillment, the target function is penalized, otherwise the target function value is determined based on the calculated weight of the structure.

The shape of the target function according to equation (5) and its evaluation procedure is suggested for the following reason - if the geometric constraints for the area of permissible solutions (4) are not met, the FEM analysis is not required.

5 Results of the optimization process

The optimization process had to be carried out in two steps due to the large number of local extremes in the investigated area. Therefore, the authors decided to map out a set of optimal solutions and find the best starting step within the permissible interval. Figures 3 and 4 show the map of values for the optimization variables at the start point and at optimum.

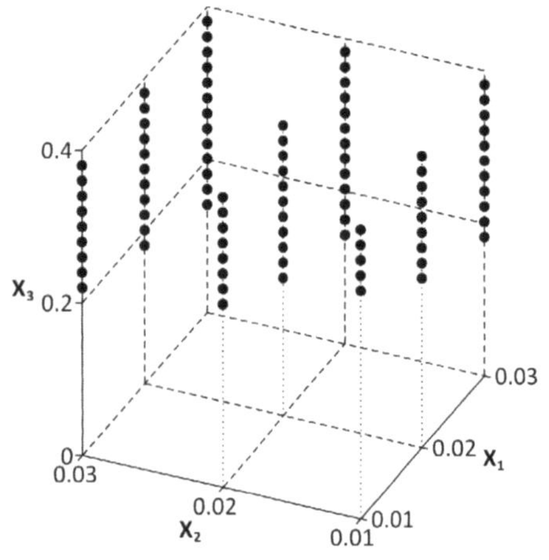

Fig. 3. Map of starting points

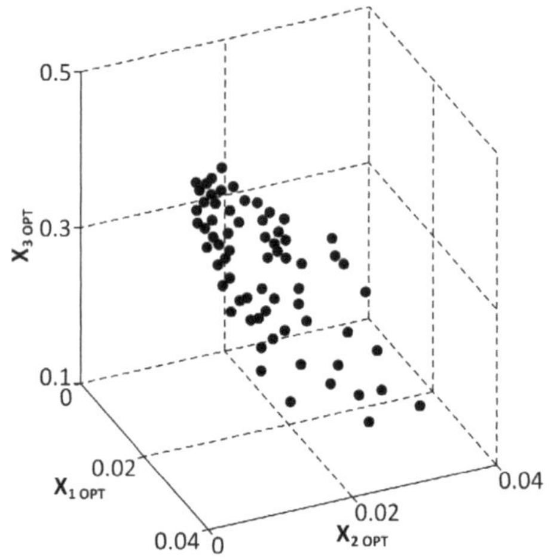

Fig. 4. Map of local optimums

Figure 5 shows a view of the mapped optimum weight of the plate for the different starting points. Based on the analyzed optimization process, the most appropriate starting point was $\mathbf{x} = [0.01\ \ 0.01\ \ 0.36]$ [m] and the subsequently obtained results were globally optimal for the solved area

$$\mathbf{x}_{OPT} = \begin{bmatrix} 0.014 & 0.010 & 0.364 \end{bmatrix} \text{ [m]}$$

with parameters - weight $m_{OPT} = 1298.5$ kg, maximum stress $\sigma_{OPT} = 64.6$ MPa and maximum deflection $U_{max\_OPT} = 0.00049$ m.

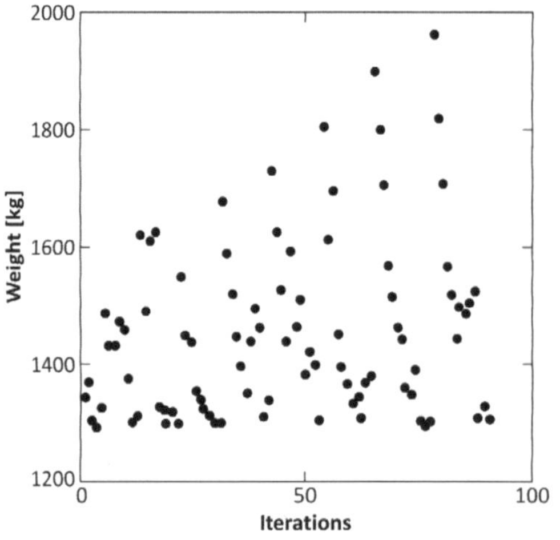

Fig. 5. Map of the optimization results for different starting points

An overview of the optimization process history for the selected starting point is presented in Figs 6 and 7.

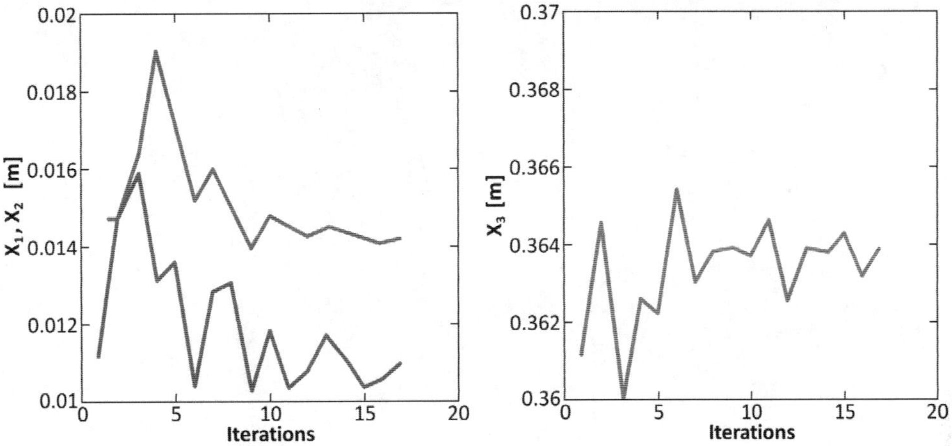

Fig. 6. The convergence process of the optimization variables X_1, X_2 and X_3

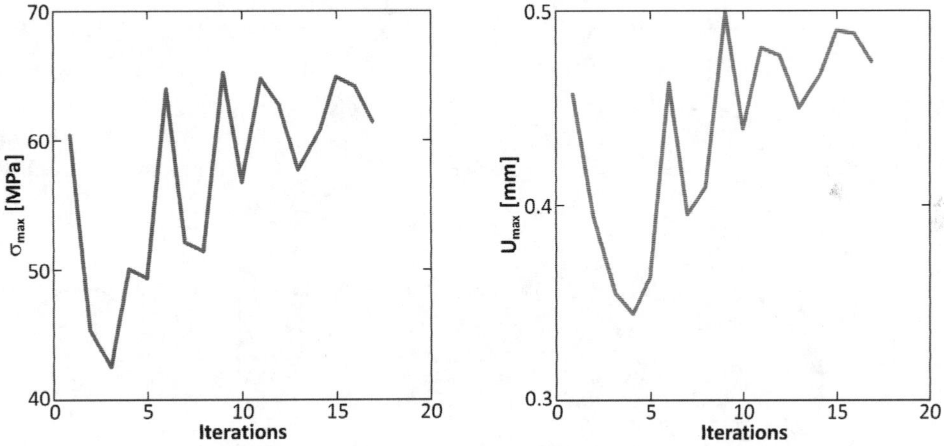

Fig. 7. The convergence process of maximum stress and deflection

The equivalent stress according to the von Mises theory for the selected starting point is presented in Fig. 8.

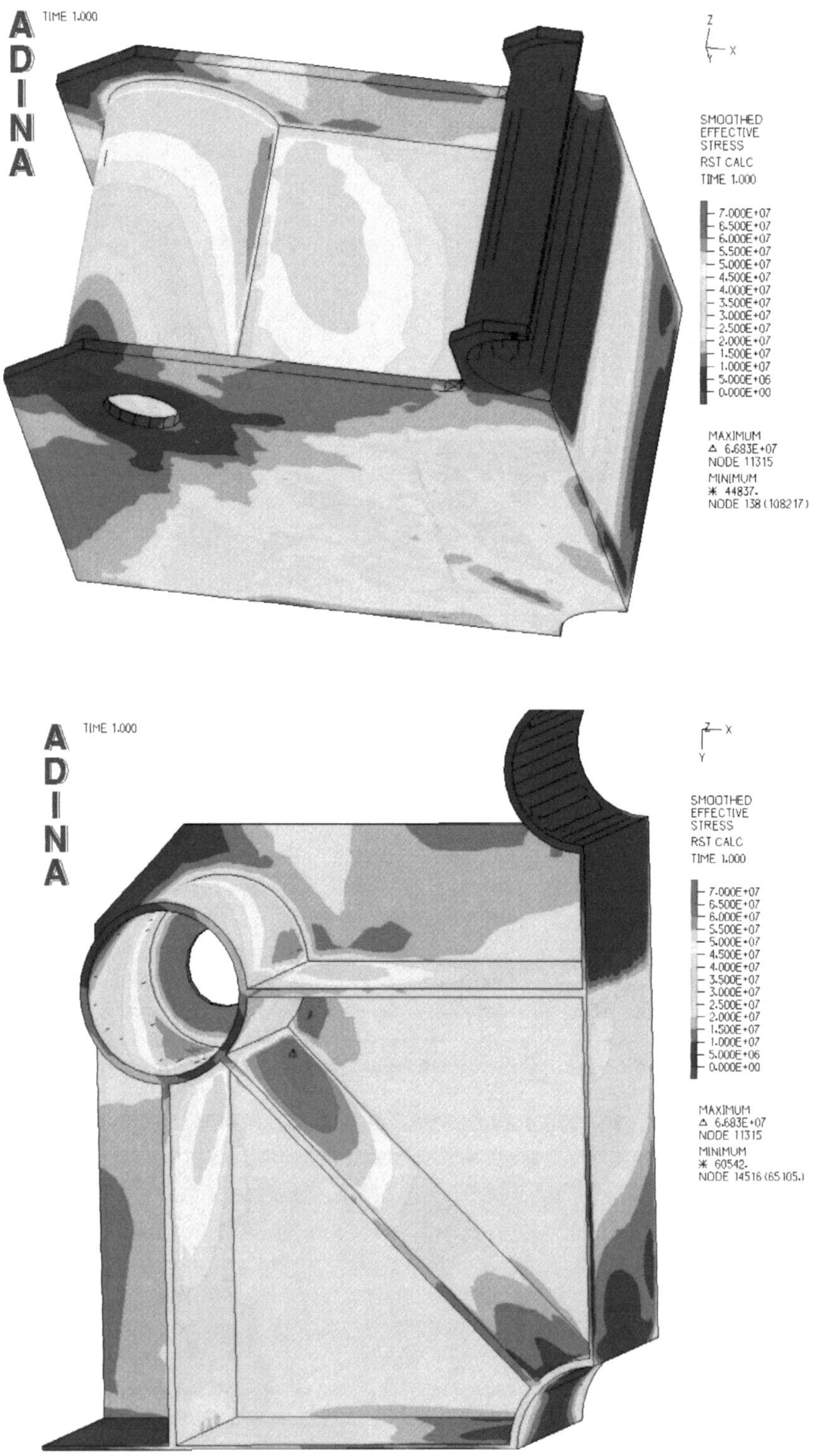

Fig. 8. The equivalent stress according to von Mises theory [Pa]

Conclusions

The process of optimizing the press sandwich plate as an alternative design solution resulting in a reduction of its weight is presented in the article. In the course of solving the problem it has been shown that the starting point significantly affects the final result. This has prompted the authors to mapping the influence of starting values of optimization parameters to the weight and limiting conditions.

The global optimum from the set of local optimums calculated in the investigated area of interest was obtained on the basis of a series of controlled analyzes. The multisoftware approach using ADINA and MATLAB software and the parametric model were used to perform FE analyzes. In addition to the presented calculation methodology for the optimization of extremely stressed sandwich plates, the aim of the paper was to draw attention to the inexplicability in choosing the starting values of the optimization parameters.

This work was supported by KEGA 037ŽU-4/2018 "Implementation of effective computational tools into modern information and communication technologies with the aim of increasing competencies of graduates of technical studies".

References

1. K. J. Bathe, *Finite Element Procedures*. (Prentice Hall, New Jersey, 1982)
2. L. Jakubovičová, M. Sága, *Computational analysis of contact stress distribution in the case of mutual stewing of roller bearing rings*. Novel Trends in Production Devices and Systems, Applied Mechanics and Materials **474**, 363-368 (2014)
3. M. Dudziak, A. Kołodziej, G. Domek, K. Talaśka, *Multi-angularity - identification of parameters and compatibility conditions of the axisymmetric connection with form deviations*. Procedia Engineering **177**, 431-438 (2017)
4. M. Sága, M. Vaško, N. Čuboňová, W. Piekarska, *Optimisation Algorithms in Mechanical Engineering Applications*. (Pearson Education Limited, 291 p., 2016)
5. A. Sapietová, V. Dekýš, M. Sapieta, P. Pecháč, *Application of computational and design approaches to improve carrier stability*. Procedia Engineering **96**, 410-418 (2014)
6. I. Kuric, V. Bulej, M. Sága, P. Pokorný, *Development of simulation software for mobile robot path planning within multilayer map system based on metric and topological maps*. International Journal of Advanced Robotic Systems **14** (6) (2017)
7. *MATLAB*. [online] https://www.mathworks.com/products/matlab.html
8. I. Pobočíková, Z. Sedliačková, M. Michalková, *Application of four probability distributions for wind speed modeling*. Procedia Engineering **192**, 713-718 (2017)
9. W. Spendley, G. R. Hext, F. R. Himsworth, *Sequential Application of Simplex Designs in Optimization and Evolutionary Operations*. Technometrics **4**, 441-461 (1962)
10. J. A. Nelder, R. Mead, *A Simplex Method for Function Minimization*. Computer J. **7**, 308-313 (1965)
11. ADINA, *Theory and Modelling Guide*. [help manual] (Watertown, 2010)
12. L. Žuľová, R. Grega, J. Krajňák, G. Fedorko, V. Molnár, *Optimization of noisiness of mechanical system by using a pneumatic tuner during a failure of piston machine*. Engineering Failure Analysis **79**, 845-851 (2017)

13. J. Flizikowski, M. Macko, J. Czerniak, A. Mrozinski, *Implementation of genetic algorithms into development of mechatronic multi-edge's grinder design*. ASME 2011 International Mechanical Engineering Congress and Exposition, Denver, Colorado, USA, November 11–17, 2011, 1227-1235 (2011)
14. P. Pecháč, M. Sága, *Memetic algorithm with normalized RBF ANN for approximation of objective function and sedondary RBF ANN for error mapping*. Procedia Engineering **177**, 540-547 (2017)
15. W. Piekarska, M. Kubiak, Z. Saternus, *Numerical simulation of deformations in T-joint welded by the laser beam*. Archives of Metallurgy and Materials **58** (4) 1391-1396 (2013)
16. M. Bara, T. Kmita, J. Korzekwa, *Microstructure and properties of composite coatings obtained on aluminium alloys*. Archives of Metallurgy and Materials **61** (3), 1107-1112 (2016)
17. J. Vavro, J. Vavro Jr., P. Kováčiková, R. Bezdedová, *The optimisation of the materials properties for the passenger cars in dependence on defect distribution at the dynamic loading*. Procedia Engineering **136**, 114-119 (2016)
18. A. Vaško, L. Hurtalová, M. Uhríčik, E. Tillová, *Fatigue of nodular cast iron at high frequency loading*. Materialwissenschaft und Werkstofftechnik **47** (5-6), 436-443 (2016)
19. R. Halama, A. Markopoulos, F. Fojtik, M. Fusek, Z. Poruba, J. Famfulík, *Effect of stress amplitude on uniaxial ratcheting of aluminum alloy 2124-T851*. Materialwissenschaft und Werkstofftechnik **48** (8), 814-819 (2017)

The importance of proper evidence and identification of metallurgical materials for effective management of the company

Vladimíra Schindlerová[1,*], *Ivana* Šajdlerová[1]

[1]VŠB – Technical University of Ostrava, Faculty of Mechanical Engineering, 17. listopadu 2172/15, 708 33, Ostrava - Poruba, Czech Republic

Abstract. An important characteristic for efficient management of production systems is the ability of a product, component or material to be tracked. That is, to be assigned with a unique symbol, number, or other code (identifier) that can be traced back both within the production process and to the customer (e. g. when complaining about a defective part). Traceability leads to a cost reduction in eliminating the risks associated with the difficult identification of the material or parts, their handling in pre-production, the manufacturing process, or the storage and sale of finished products to customers. In case of problems, it makes it easier to implement the necessary measures and reduces the time to remedy the situation either within the company or even outside. Individual companies within the Czech Republic usually solve the identification and traceability independently. The paper deals with the results of the analysis of the current state of record keeping and identification of metallurgical materials in selected companies, and presents a proposal for improvement of the current situation in a specific company, especially in the field of work with remaining material.

Keywords: Identification, traceability, barcodes, designation of the material

1 Introduction

Undoubtedly, the company management is aware of the fact that the resulting effect of economic activities has a significant impact on customers´ satisfaction with their products. Efforts to meet the assumed requirements regarding price, quality, availability, delivery speed and reliability are paramount. However, the organization's own management and setting up of individual processes and work with them are also fundamentally involved in the achieved results. Transport, storage and handling employ up to 25% of the company workers, occupy 55% of the area and up to 87% of the time which the material spends in the plant. These activities sometimes account for 15 to 70% of the total cost of the product and also affect the quality of the product. 3 to 5% of the material may be damaged by incorrect shipping, handling and storage [1]. Storage and handling in logistics are an important competitive factor.

---

* Corresponding author: vladimira.schindlerova@vsb.cz

Reviewers: *Josef* Chladil, *Jozef* Pilc

2 Methodological base

Identification is the first step that needs to be done to recognize the particular elements of the system before deciding how they will be further handled.

Identification is a very broad term which is used in various contexts, e.g. in mathematics, citizenship, investigation, media, science and engineering, tracking and inventory management, social sciences. In this sense, identification can be defined as the act of finding out what something (or someone) is. This article deals with the identification connected with tracking of objects (remaining metallurgical materials) in an industrial company.

The remaining material management belongs to the field of inventory management and is certainly one of the areas where there will always be room for further improvements. [2] It can represent a significant source of inefficiency, i.e. wasting, in a wide range of its forms – supplies, surplus material and components, unnecessary manipulation, waiting etc. The main forms of wasting in logistics are described in a number of publications worldwide [1,3,4,5,6]. Insufficient or inappropriate identification and traceability can significantly contribute to losses and inefficiency of both the manufacturing process and the duration of non-technological operations as well as the manufacturing system as a whole. Some sources are presented e.g. in [7] and the practice confirms that the proportion of non-technological times (i.e. times which do not add any value to the product), servicing or overhead times in mechanical assembly manufacturing is much higher than the proportion of technological times (20%). Inappropriate work with remaining material or with useless supplies of material is the area where a hidden potential to save financial resources of the company can be found.

For any object identification (material, semi-finished products, goods, people etc.), it is essential to have the objects marked properly first of all. We can say, traceability begins with marking of particular parts. Marking and traceability are very important in all fields of business. They are used in aerospace, automotive, military, energy, medical and food industry, metallurgy etc.

We can meet with various marking technologies (see table 1) in industrial practice. The nature of the manufacturing, its repeatability, materials used, manufacturing processes as well as required durability of marking influence the choice of technology used. An important aspect when selecting the technology used is the possibility of using standards for marking and its automation. The costs related to the implementation of the selected technology and, last but not least, the management of the organization, which must provide or release the required resources, also play an essential role.

Identification and Data Capture can be understood as automatic objects identifying, collecting data about them, and entering them into computer systems, without human involvement.

Automatic Identification and Data Capture (AIDC) can be used in different fields of business. There are the enabling AIDC technologies as bar codes – linear and 2D, radio frequency (RFID), optical character recognition (OCR), magnetic stripe, smart card, biometrics, voice recognition or real-time location service (RTLS). The beginning of today's successful AIDC technologies is connected with the names of Bernard Silver and N. Joe Woodland and their patent application "Classifying Apparatus and Method" from 1949 [8], (granted 1952), which introduces the first machine-readable bar code, called the Bull's Eye Code. Currently, there are a huge number of standards for AIDC technologies, which can be implemented in a variety of areas [9].

Table 1. Various marking technologies

Technology	Description	Marking: Permanent	Temporary	Machines, integrated devices	Manual	Steel	Standardization marking type
Dot-peen	Work by electromechanically striking a carbide or diamond stylus assembly against the surface of a part to be marked.	✓		✓		✓	✓
Scribing	Uses a hardened tip (a diamond or carbide) to mark the surface of components.	✓		✓		✓	✓
Laser	A high frequency beam generated from a laser source. This beam is then amplified and directed towards a part to be marked.	✓		✓		✓	✓
Electrochemical	By means of a chemical reaction of the electrolyte and the electric current, the mark is applied over the template to the surface. Marking by oxidation or etching.	✓		✓	✓	✓	✓
Hand Stamps	Marking tools made of high quality steel with a hardness of 60 HRC, designed for marking materials up to a hardness of 40 HRC.	✓			✓	✓	
Industrial markers	Ball markers, varnish markers, rigid markers, special markers, water removable markers for different types of materials. They are not suitable for direct contact with food.	✓	✓		✓	✓	
Burner plate	Marking the product with a suitably modified burner plate (e.g. brass).	✓			✓		✓
Electroerosive pencil	The pencil works on the principle of an electric arc. Tungsten electrode.	✓			✓	✓	
Etching pen	An indelible description of almost all metal surfaces. Mark resistant to scratches, abrasion, chemicals, etc.	✓			✓	✓	
Engraving Pen	Pneumatic or electric hand held engraving pen is simply used like a pen and relies on the quality of the handwriting. Marking based on micro-impact.	✓			✓	✓	
Non-contact marking with color point	The ink is splashed under pressure from the spray gun to the desired location.	✓	✓	✓		✓	✓

2.1 Identification of remaining material

A lot of companies have developed standards how to deal with remaining material. They have defined parameters which help them determine whether the waste (scrap) can no longer be used, or the remaining material is worth storing for further manufacturing. Preconditions for further use of remaining material include correct identification, traceability and proper storage. Unless it is possible to find out in a reasonable time what material, in what quantity and quality is available and whether we are able to prepare it (or not) for the manufacturing process quickly and without unnecessary costs, it is advisable to consider the possibility of getting new material and scrapping the residues automatically. In this respect, it is also important to take into account the environmental aspects regarding both the energy intensity for the manufacturing of metallurgical materials and the sustainability of resources for the future.

$$\sum_1^n C_{SRM} < \sum_1^n C_{NM} - \sum_1^n P_{SC} \qquad (1)$$

C_{SRM} – remaining material cost; C_{NM} – new material cost; P_{SC} – scrap price

To decide on the use of remaining materials, we have to ensure that the costs related to the storage and management of the residues are lower than the costs to acquire new material reduced by an amount of the financial resources gained from selling the scrap.
The basic procedure for working with remaining material is shown in the flowchart (Fig. 1).

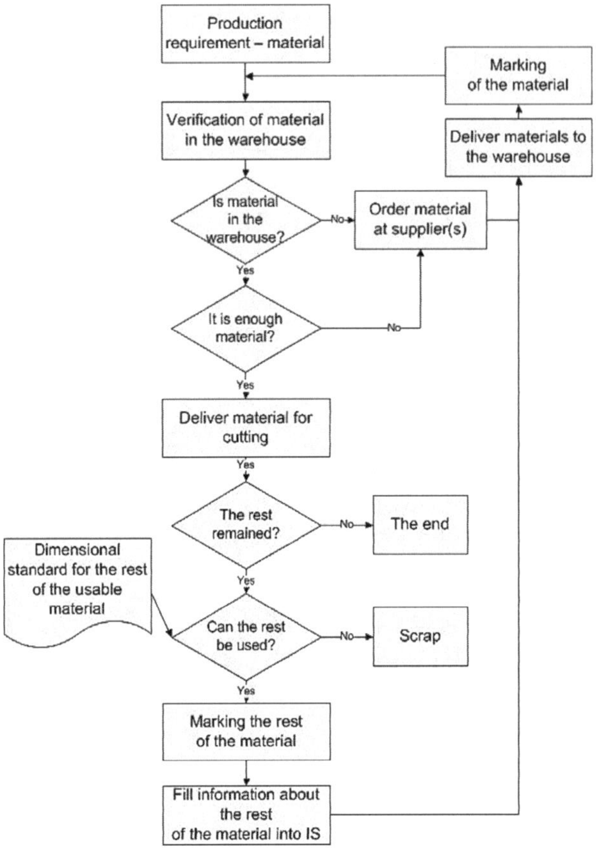

Fig. 1. Flowchart

3 A Case study of evidence and identification of metallurgical materials

The presented study deals with a company which is engaged mainly in the production of flame-cut shapes and the sale of thick metal plates on the domestic and foreign markets, and especially with the area of processing the remaining steel material in this company.

3.1 Survey on the use of automatic identification in several domestic companies

The study itself was preceded by a survey on the use of automatic identification in several domestic companies. The survey was carried out on a sample of 6 companies with 30 – 270 employees which focus mainly on metalworking (e.g. processing of metal plates, profiles, pipes, wires etc.). The survey shows that there is no unified system of automatic identification in domestic industrial enterprises which could be bought as a finished product and implemented easily. It also shows that domestic industrial enterprises want to and make an effort to introduce automatic identification of material, supplies and semi-finished products [10].

Other findings include:

- Organizations in the supply chain use different forms of identification (customers demand their own type of marking), it is impossible to use unified marking.
- In most of the monitored companies, dispatch notes were used, mainly for tracking the production process. Apparently, it is based on the experience of domestic industrial enterprises with the given system and on the fact that this system has proved in the long term.
- Some manufacturing processes, such as burning, grinding, blasting, heat treatments, have a common feature: conventional marking doesn´t last long.
- Companies which have been using the AI system actively for a longer period (5-6 years) accept it positively. The positives also initiate the introduction of other ideas affecting the production efficiency and the manufacturing system (e.g. the signal semaphores).
- In companies with established AI system, readers and printers are located at each workplace. Therefore, new labels can be printed out either after every operation or as needed.
- The importance of high-quality connection and cooperation between AI and IS in the enterprise is crucial.
- In case the own manufacturing system is designed unsuitably (illogically) and is unnecessarily complicated, errors occur far more often because of its complexity.
- It is the human factor that has been identified as the common denominator of errors in the manufacturing system, whether it was reluctance, ignorance or negligence of employees.
- The stored remaining material can be very diverse mainly in piece or custom-made production, which is specific due to a large amount of orders of one or a few pieces. Appropriate identification of residues and their proper storage considering easy manipulation are essential.

3.2 Evidence and identification of metallurgical materials in monitored companies

Based on the results of an inventory check at the end of 2016, almost 1450 tons of steel materials were found in remnants from previous production in store of the monitored company. This amount was estimated at around 19 million CZK which represented approximately 24% of the total inventory of the enterprise. The material was stored on an area of about 300 m^2.

The supplied input material (standardized size sheet) is usually used for the manufacturing of more orders (products) of various sizes and shapes. Part of the material is then scrapped, usable remaining material goes to the stock of remains for further potential production.

The material of many of the stored residues (about 26% of all residues) couldn´t be identified correctly, only the dimensions were obvious. Tests for determining the material in question weren´t allowed by the company´s management due to their cost. Until the decision of the company´s management to keep records of the residues (Excel) and their re-planning into new production, the remnants had been only growing. Then, there was a rapid slowdown in their accumulation. The goal of the company´s management is to reach the 20% limit of the current situation of the stored remaining material.

The management decided to invest in a new information system (K2) which also includes the AI module. All the data is now entered into the IS which is shared by all departments and employees of the enterprise with appropriate rights.

Proposal for the introduction of AI for the identification of remaining material (residues).

The proposal for the introduction of AI for the identification of remaining material had to respect the fact that the company's products are intended for various customers and create inputs for further processing. Permanent marking or branding weren't allowed in this case. Therefore, matrix data and laser marking of the material can't be applied.

After considering the advantages and disadvantages of AI using barcodes for the company (Table 2), the management decided to implement this type of marking.

Table 2. Various marking technologies

Advantages	Disadvantages
- Maturity of Barcodes marking - Small risks associated with Barcodes marking - Simplicity of the method - Low cost of marking itself (label + printing) - Marking speed (label printing) - Speed of follow-up record keeping (code reading) - Lucidity of inventory	- Initial higher purchase costs - Initial more demanding logical planning (when to identify, location of the readers and who will identify) - Training of employees

The choice of suitable labels and their testing to find out whether they will adhere to the steel under given conditions were the most important steps.

10 labels were selected to be tested which varied in material they were produced from and used glue. Their shape, size, etc. weren't emphasised. The aim was to test which labels adhere to the rough surface of the sheet in operating and storing conditions (in a shop floor). The price ranged from 0,5 to 4 CZK / 1 piece. The tested labels were made of paper, metal and plastic.

A sample of sheet with a maximum amount of dust, mill scales and other impurities was selected for testing.

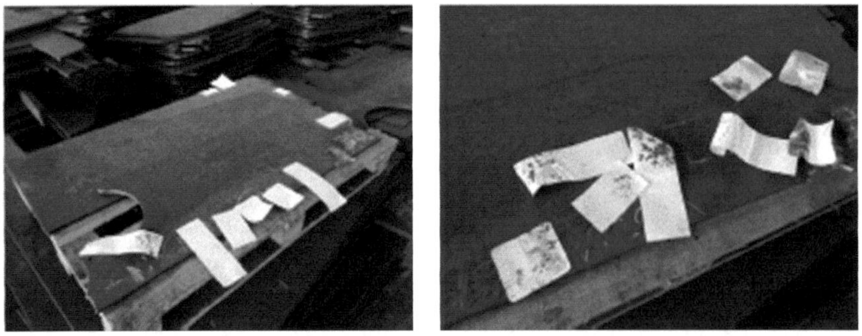

Fig. 2. Tested samples

None of the tested labels adhered to the material without its prior cleaning from dust and dirt. After the material had been cleaned from dust and basic impurities, the labels (concerning the intended use) adhered sufficiently even to the scaled material. They could be torn off the material, leaving the mill scales on the label.

In the following step, the size of the label was decided. Essential was mainly the length of the label. Its overlap had to be sufficient to enable the identification of the material even when stacking the material on top of each other. To avoid larger initial investment, the company chose a conventional paper label. The selected label fulfilled the basic requirements, i.e. sufficient adhesion to metal sheet, low cost per item (0,67 CZK / piece without VAT), availability of labels (are sold commonly), wide range of printers for the chosen label. The disadvantage of the chosen identification method was that it couldn´t be used for heat-treated items (metal labels).

It was suggested to divide the backing material of the label into two parts. When applied to the material, only one part is removed and the other remains backed. Thanks to this, the label is tougher and twists less.

The input costs for the implementation depend on the selected AI and the number of particular components. In this case, the costs were approximately 300 000 CZK excluding VAT (printer, 9 sensors, 9 computers, portable terminal, programming, labels, marking of material) and only for the equipment which is to be used for remaining material management.

Effective handling with remaining material also involves appropriate storage of this material. Proper marking of the material does not mean that it can be traced easily. In this case, it is advisable to carry out the ABC analysis [6] and, based on the results, spatially arrange the storage of the material in the room available, e.g. using the SLP design methodology [11,12]. E.g. it is possible to create sectors for sheets of 30 mm or more thickness where the materials can be stored on suitable pallets. Sheets of smaller thicknesses (the company uses them to a small extent) and remnants of circular shape could be placed on shelves. In this case, it is recommended to implement the elements of the 5S methodology [6] which is suitable not only for the production itself but also in all sections of the manufacturing system.

It has been proved that the height of costs on AIDC and spatial layout of the storage space is much lower than the means obtained from the sale of the unidentified material residues in the form of scrap (about 5 million CZK excluding VAT). Savings obtained from the use of remaining material for further production represent another approximately 10 million CZK. There is only a one-off cost saving. Order and easily identifiable remaining material in stock provide additional day-to-day savings related to time waiting for material or inappropriate production owing to misplaced material etc.

4 Conclusion

The survey of selected companies confirms that domestic industrial enterprises want to and make an effort to introduce automatic identification of material, supplies and semi-finished products.

The achieved results demonstrate that the methods of Automatic Identification and Data Capture can also be successfully implemented into the field of storage of rest (remaining) materials and it is very useful for production spheres which do not have the character of serial or mass production.

The benefit of introducing AI into the company has been proved. The proper work with the remaining material has brought positive results with an impact on the cost, time and efficiency of the production system.

One-off cost savings can vary in double-digit percentage range of the total inventory depending on the initial state of inventory management and remaining material in enterprises. Day-to-day cost savings can vary in single-digit percentage range.

The transition from a familiar system to a new one is conditioned by the supportive approach of the company management and also by its patience. For the proper functioning of the newly established system, it is essential to prepare the employees and convince them of the positives of the changes, qualify them appropriately and motivate them so that they could accept the changes and adapt to them.

The implementation in practice has shown that the management support is crucial for successful implementation changings to practice.

References

1. J. Košturiak, Z. Frolík, *Lean and Innovative Enterprise*. Praha: Alfa Publishing, (2006)
2. V. Schindlerová, I. Šajdlerová, *Influence tool wear in material flow*. Adv. In Sci. And Tech.- Res. J. **11**, 161-165 (2017)
3. J. K. Liker. *The Toyota Way: 14 Management Principles from the World's Greatest Manufacturer*. New York: McGraw-Hill, (2004)
4. J. P. Womack, D.T. Jones, *Lean Thinking: Banish Waste and Create Wealth in Your Corporation*. Simon & Schuster, (2003)
5. S. Shingo, *The Shingo Production Management System: Improving Process Functions*. P. Cambridge, MA: Productivity Press. (1992)
6. J. Bhamu, K. S. Sangwan, *Lean manufacturing: literature review and research issues*. Int. J. of Op. & Prod. Man., **34**, 876-940, (2014)
7. M. Leščišin, V. Líbal, A. Šperlich, *Organization and management of production*. Bratislava: Alfa, 85. (1985)
8. *The History of automatic identification ID systems*. (AIDC 100, Rev A, July 9, 2010) Available: http://www.aidc100.org/index.htm
9. *Standards catalogue: 35.040.50 - Automatic identification and data capture techniques*. (ISO, 2018) Available: https://www.iso.org/ics/35.040.50/x/p/1/u/0/w/0/d/0
10. A. Chmiel, *Record Keeping and Identification of Metallurgical Materials: Diploma Thesis*. VSB – TUO, Faculty of Mechanical Engineering, Department of Mechanical Technology, (2017)
11. R. Muther, K.Haganäs, *Systematic handling analysis (SHA)*. Management and Industrial Research Publications, Kansas City, (1975)
12. R. Muther, *Systematic layout planning (S.L.P.)*. Boston:Cahners Books (1972)

The strength of traditional and self-pierced riveted joints

Anna Rudawska[1*], *Izabela* Miturska[2], *Dana* Stančeková[1], *Jacek* Mucha[3]

[1]Lublin University of Technology, Faculty of Mechanical Engineering, ul. Nadbystrzycka 36, 20-618 Lublin, Poland
[2]University of Žilina, Faculty of Mechanical Engineering, Univerzitna 1, 010 26, Žilina, Slovak Republic,
[3]Rzeszow University of Technology, Faculty of Mechanical Engineering and Aeronautics, al. Powstańców Warszawy 8, 35-959 Rzeszów, Poland

Abstract. The objective of this study is to compare the strength of riveted joints fabricated by traditional riveting (with pre-drilled holes) and self-piercing riveting (SPR) for different types of joints. Riveted joints were produced using steel and aluminum alloy rivets and two types of sheet material: 235JR steel sheet and EN AW 6060 aluminum alloy sheet with the following dimensions: length l = 100 ± 1 mm, width b = 50 ± 1 mm and thickness g = 2 mm. For all tested types of riveted joints (pre-drilled and SPR), 5 sets of joints were fabricated, each set containing 6 samples. The sets of joints differed with respect to the number of rivets (1, 2, 3, 4 and 6 rivets), joint type (single-, three- and four-riveted joints) and lap length. For all tested joints, the highest load capacity was obtained for self-pierced riveted joints, while the lowest – for pre-drilled joints with aluminum alloy rivets. In addition, it was found that the shear strength of self-pierced riveted joints is higher than that of aluminum and steel blind rivets.

Keywords: riveted joints, self-piercing riveting SPR, blind rivets, steel sheet, aluminum alloy sheet

1 Introduction

Assembly joints are an inherent part of human life and activity. In assembly joints individual structural components are assembled and can thus support loads. Depending on whether structural components are joined by a special fastener or not, it is possible to distinguish indirect and direct joints. In indirect joints, rivets or screws can be used as joining elements. In direct joints, no fasteners are used [1-3].

Rivet joints are indirect permanent joints. Riveting consists in joining shapes with thin sheet metals and brackets. This technique is also widely used to join leather and rubber with metal elements, e.g. in saddlery. This type of joints is based on the use of rivets as fasteners. The industry offers a wide variety of general purpose rivets of standard shape and

* Corresponding author: a.rudawska@pollub.pl
Reviewers: *Sorin* Grozav, *Štefan* Michna

size. Depending on the rivet shank shape, we can distinguish solid rivets, hollow rivets, tubular rivets, rivet nuts, blind rivets [4].

Solid rivets are used in metal structures such as bridges, cranes, boilers and towers. Hollow and tubular rivets are used in precision engineering for small structures and in the electrical industry to join brittle, soft or non-uniform materials.

The main advantages of riveted joints include [1,5-7]:

- vibration damping,
- no residual stresses,
- the possibility of joining different materials (metal-metal, metal-nonmetal),
- no interference in the structure of materials being joined (in cold riveting).

The disadvantages of riveted joints include [3,8-10]:

- permanent fastening of structures,
- weakening of joined materials due to hole drilling,
- high labor consumption,
- increased structure weight,
- difficulty in producing tight joints,
- relatively high costs of fabricating joints.

Over the years the technique of producing riveted joints has evolved as a result of modifying rivet geometry and material. This also led to changes in the production technology for riveted joints and the design of new machines and devices for riveting processes. The traditional process for producing riveted joints consists in piercing the materials being joined, placing a fastener in the hole and clenching the rivet. However, this method of producing joints means longer production time and causes difficulties with respect to obtaining the required accuracy and properties of joints. As a result of research on riveting processes, an innovative method for producing riveted joints has been developed, which is known as self-piercing riveting (SPR) [11-16].

Pierce riveting consists in joining materials under load. This leads to considerable savings in assembly departments owing to shorter operational time and improved quality of produced joints. Because material is cold-formed, structural components such as sheet metals and profiles are joined via force- and form-fit mechanisms. This type of joints does not require the pre-drilling of holes. One can distinguish several riveting techniques with or without material piercing [3,11,16-18]:

- clinch rivet (CR) – this joining method consists in clinching rivets by stamping and involves the use of an additional solid fastener. A characteristic of this technique is that no sheet of material is pierced, which allows for joining elements with chemically-protected surfaces (e.g. against corrosion);
- self-piercing riveting (SPR) – this technique enables joining thin-walled elements made of strip and sheet materials using specially designed tubular rivets. In this method, the punch presses the rivet through the top sheet of material. As a result of the rivet geometry and applied punch load, the lower sheet of material is pressed onto the die. In effect, the rivet pierces the top sheet. After that, the lower sheet and the rivet undergo upsetting, which ensures that the deformed lower sheet of material adheres closely to the die surface. As a result, the sheets closely adhere to each other, which is essential as the joint quality depends on geometric match of the die shape to the rivet size and type.
- solid self-piercing riveting (SSPR) – this joining technique is similar to SPR, the difference lies in the punch geometry – in SSPR the rivet acts like a shearing punch. The punch presses on the rivet head leading to piercing of the sheet materials being joined. The excess material is removed through an opening in the lower die – the internal edge of this opening acts like a cutting edge and helps

pierce the lower sheet of material. Finally, a hole is cut in the sheet materials and the rivet is securely locked.

The objective of this paper it to compare the strength of SPR joints and that of joints with pre-drilled holes, for different types of joints.

2 Methods

2.1 Description of joined materials

Riveted joints were fabricated using EN AW-6060 aluminum alloy. This alloy has a relatively low tensile strength, so it is suitable for producing various types of sections. The other joined material was a sheet of non-alloy structural steel of S235JR grade. Due to its good mechanical properties, this steel grade is used for structural components that are exposed to high static and dynamic stresses. In addition, this steel has good machining properties, which allows for producing finished parts of complex shape by machining. Its chemical composition allows for producing high-quality welded joints. The tests were conducted on sheet materials described by the following dimensions: length l = 100 ± 1 mm, width b = 50 ± 1 mm and thickness g = 2 mm.

2.2 Riveted joints

Experiments involved fabricating single-lap riveted joints by two methods: riveting with steel and aluminum alloy blind rivets and self-piercing riveting (SPR) with steel rivets. For every tested joint type 5 series of joints were fabricated, each series containing 6 samples. The series of joints differed in the number of rivets, joint type and lap length. The geometry of these joints is shown in Fig. 1.

The lap length, rivet spacing and edge distance were designed in compliance with the guidelines given in the literature on the subject, as shown in Table1.

Table 1. Dimension calculation [19]

Design assumptions	Calculations, mm	Selected value, mm
Rivet diameter d: d = (1-2)g	(1-2)4=4-8	5.0
Pitch t: t = (3+5)d	(3-5)5=15-25	16.0
Spacing between rivet rows a: a = (0.6-0.8)t	(0.6-0.8)16=9.5-12.8	12.5
Spacing between end rivets and edge e_1 (in loading direction): $1.5 \leq e_1 \leq 2d$	$1.5 \leq e_1 \leq 2d=$ $7.5 \leq e_1 \leq 10$	9.5
Spacing between end rivets and edge e_2 (perpendicular to loading direction): $1.5 \leq e_2 \leq 2d$	$1.5 \leq e_2 \leq 2d=$ $7.5 \leq e_2 \leq 10$	9.0
Rivet hole diameter d_0: $d_0 = d + (0.1\text{-}0.2)$	5 +(0.1-0.2)= 5.1-5.2	5.2

To make sure that the above assumptions are applicable to finished products, the dimensional tolerance was set to ± 1,5 mm for *t, a* and to 1 mm for *e1, e2*.

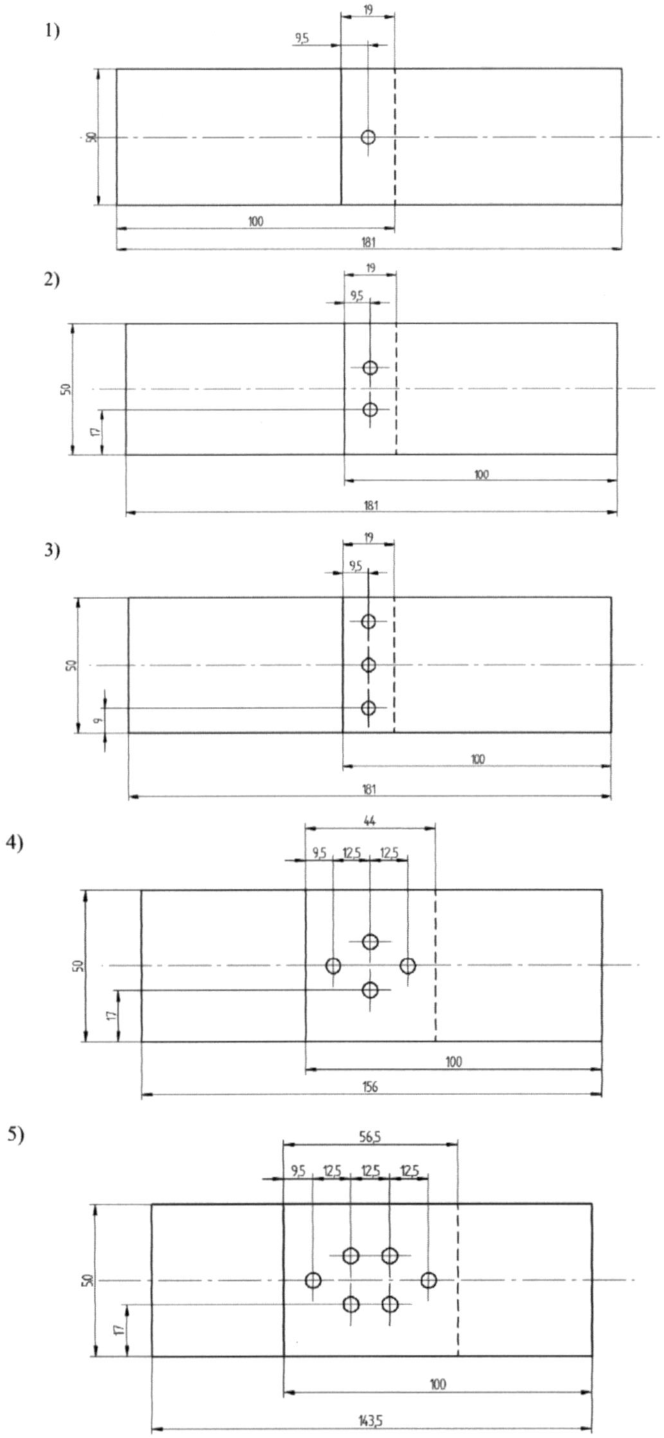

Fig. 1. Top view of fabricated joints: 1) joint with one rivet, 2) single-riveted joint with two rivets, 3) single-riveted joint with three rivets, 4) three-riveted joint with four joints, 5) four-riveted joint with six rivets

2.3 Preparation of joints

Surface preparation is a vital part of the joining process because it has a significant effect on the strength properties of finished joints. To prevent as many structural defects as possible, all surface edges of steel sheets were ground with an abrasive disc to remove all grooves and surface irregularities produced during the cutting process. Further surface treatment of the sheet materials differed in terms of the applied joining method.

2.3.1 Pre-drilled riveted joints

To make pre-drilled (traditional) riveted joints, a template was used (Fig. 2) as a guide for placing the rivet holes in specified points. After that, two drilling operations were performed in the points marked on the sheet: pre-drilling using a drill with a diameter of d = 3 mm and actual drilling using a drill with a diameter d = 5.2 mm.

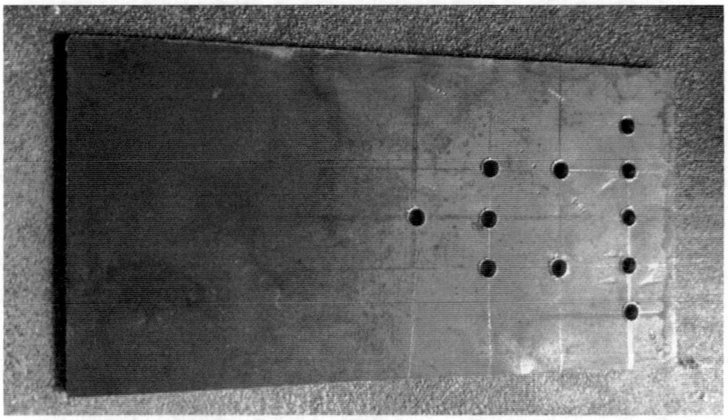

Fig. 2. Template used a guide for placing rivet holes on sheet materials

Following the drilling operations, the samples were ground with 80-grit abrasive paper in order to remove surface irregularities and burrs produced during the drilling operations. After that, the samples were degreased. Next, the sheet materials were subjected to riveting. Two types of rivets were used. EN AW-6060 sheet joints were made using rivets made of aluminum alloy (aluminum – 3.5% magnesium alloy) with a carbon steel mandrel in compliance with DIN EN ISO 15977. In turn, steel sheet riveted joints were fabricated with carbon steel rivets according to DIN EN ISO 15979. The geometry of the two rivet types is shown in Fig. 3.

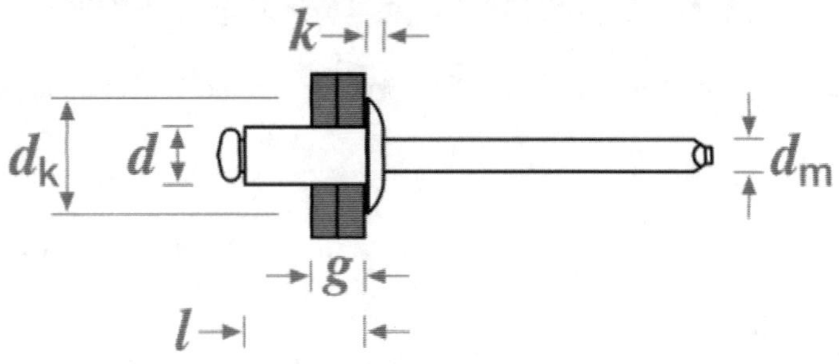

Fig. 3. Schematic design of a blind rivet

The key to the figure and the dimensions of rivets used in the experiments:

- d - nominal diameter, d = 5 mm,
- g - thickness of elements to be joined, g = 4 mm,
- dk - rivet flange diameter, dk = 9 mm,
- dm - nominal mandrel diameter, dm = 2.64 mm,
- l - rivet shank length, l = 8 mm,
- k - rivet flange thickness, k = 1.6 mm.

The rivets were clenched with a bimanual riveting machine. The use of this machine ensured the stability of clenching, suitable clamping load and correct forming of the rivet head. After clenching, rivet joints were fabricated, as shown in Fig. 4.

a)

b)

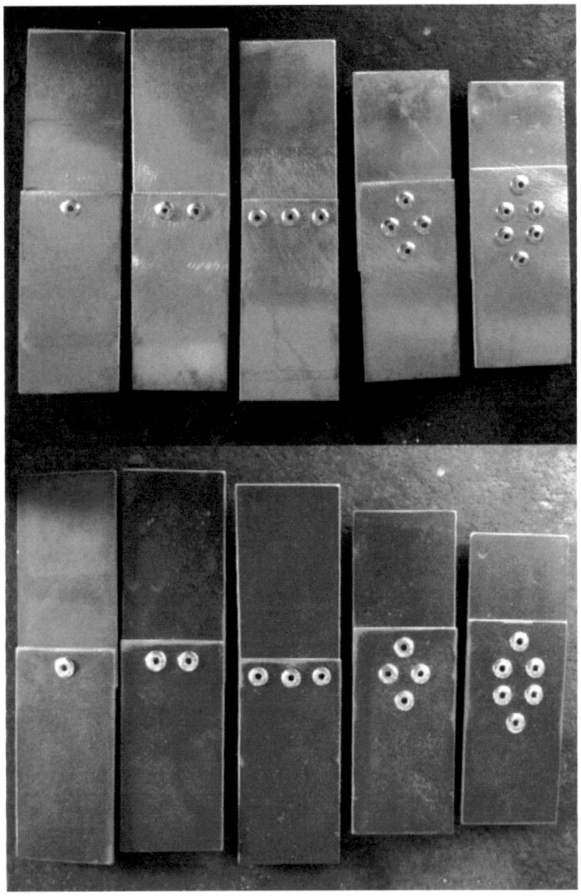

Fig. 4. Pre-drilled riveted joints with blind rivets: a) aluminium alloy, b) steel

2.3.2 SPR joints

Following the grinding of their edges, the steel sheets were ground all over in order to remove grooves and inclusions on the contact surfaces. After that, the sheet samples were degreased and wiped with a dry cloth.

In the self-piercing riveting process, rivets with a 5 mm length and 5 mm diameter were used. These rivets had a shank of type P and a head of type SK made of hardened steel coated with ALMAC. Their hardness is classified as H2 (HV = 410 ± 30). In the catalogue, this rivet type is described as: rivet d = 5 mm, P-SK type.

The tests were performed on a die marked as FM 085 2123. The geometry and dimensions of both the rivet and the die are protected data (know-how) of Bollhoff Fastenings Ltd. To clench the rivets, the punch load was set to 50 kN. The riveting process was performed with a machine provided with a standard C-type frame and a belt feeder for rivets. Fig. 5 shows the joints produced by self-piercing riveting.

a) b)

Fig. 5. SPR joints - view from the side of die (a) and punch (b)

2.4 Strength tests

The failure load for individual riveted joints was measured with the Zwick/Roell Z150 testing machine. The use of this machine enabled the measurement of fundamental parameters and, at the same time, ensured high measuring accuracy.

3 Results of strength tests

3.1 Load capacity of riveted joints

Fig. 6 compares the load capacity of all pre-drilled riveted joints. The analysis of the data in Fig. 6 reveals that the pre-drilled steel riveted joints (with steel blind rivets) have a higher load capacity than riveted joints of aluminum alloy steel sheets (with aluminum alloy blind rivets). This trend was observed for all tested types of joints. It was also observed that in order to equalize the load carrying capacity of the joints with aluminum alloy rivets with that of steel rivets, the number of aluminum alloy rivets must be over twice as high as that of steel rivets.

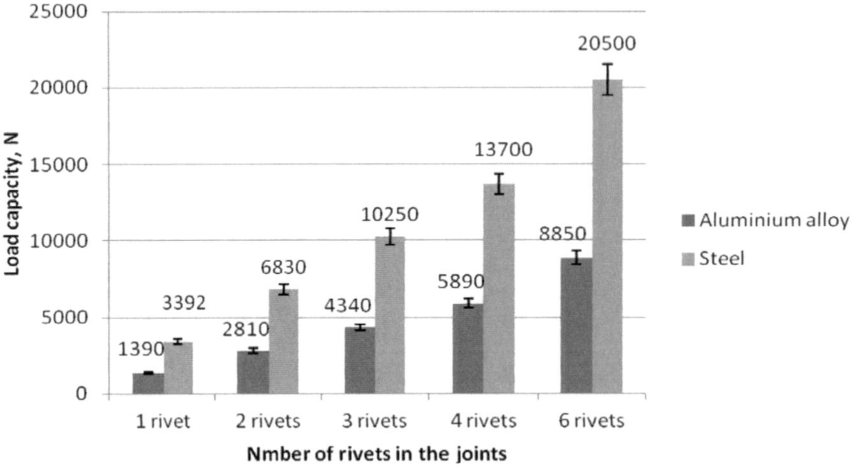

Fig. 6. Load capacity of pre-drilled riveted joints for aluminum alloy and steel sheets

Fig. 7 shows the results of load capacity of riveted joints for steel sheets in a function of the rivet type.

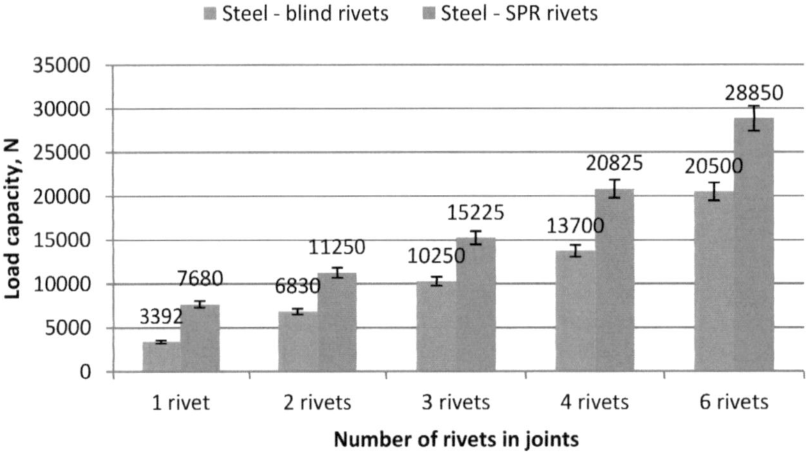

Fig. 7. Load capacity of steel sheet riveted joints versus rivet type and number

Analyzing the data given in Fig. 7 it can be observed that the load capacity of all SPR joints is higher than that of joints with blind rivets made of steel. The difference in load capacity of individual joints ranges from 4288 N to 8350 N. The load capacity is the lowest for single-rivet joints and it increases with increasing the number of rivets in the joints.

3.2 Comparison of the shear strength of riveted joints

The shear strength of tested riveted joints was determined in compliance with Equation (1) [19], and the results are given in Fig. 8.

$$\tau = \frac{F}{\frac{\pi d_o^2}{4} \cdot m \cdot n} \leq k_t \quad (1)$$

where:
k_t - the maximum shear stress;
F - the external force;
d_0 - the diameter of a rivet hole (clenched rivet);
m - the number of shear sections in one rivet; for single-shear rivets – m = 1,
n - is the number of rivets (in lap joints – all rivets).

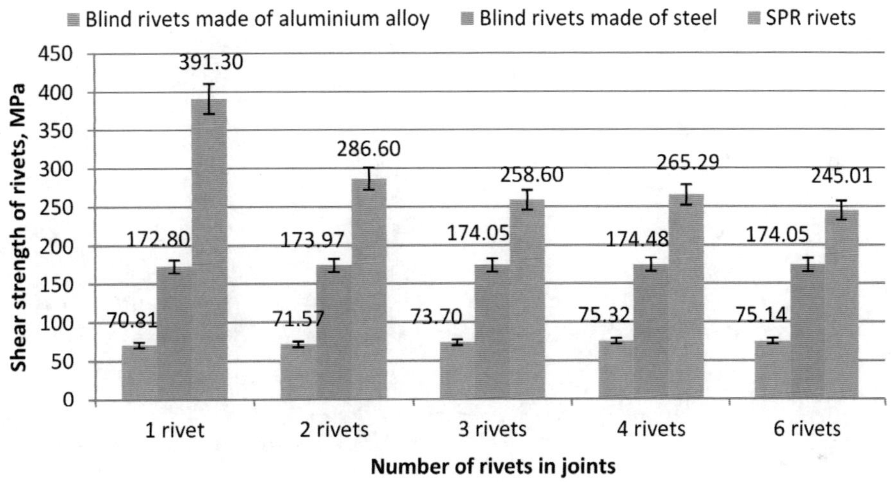

Fig. 8. Shear strength of tested riveted joints

Based on the results given in Fig. 8, the following can be observed:

- the shear strength of SPR rivets is higher than that of joints with blind rivets made of aluminum and steel;
- the shear strength of joints with blind rivets is proportionate to the number of applied rivets, which means that the load force is uniformly distributed over all rivets;
- the shear strength of SPR joints is not proportionate to the number of rivets, in particular the shear strength of single-rivet joints seems to be overestimated, which may result from the deformation of sheets and partial tensile loading of the rivet.

3.2 Comparison of maximum stresses in joined sheet materials

Maximum stresses in the sheet materials being joined were calculated from the tensile strength condition (2) [19], taking account of weakening of the section of sheet materials due to rivet holes, the most prone to damage being the section that crosses through the first rivet row.

$$\sigma_r = \frac{F}{g(b_g - n_1 \cdot d_o)} \leq k_r \quad (2)$$

where:
b - the width of sheet [mm];

g - the thickness of sheet [mm];
n_1 is the number of rivets in the section;
k_r - for a sheet material or plates.

The results of maximum stresses in joined steel sheets versus the type of riveted joint are given in Fig. 9.

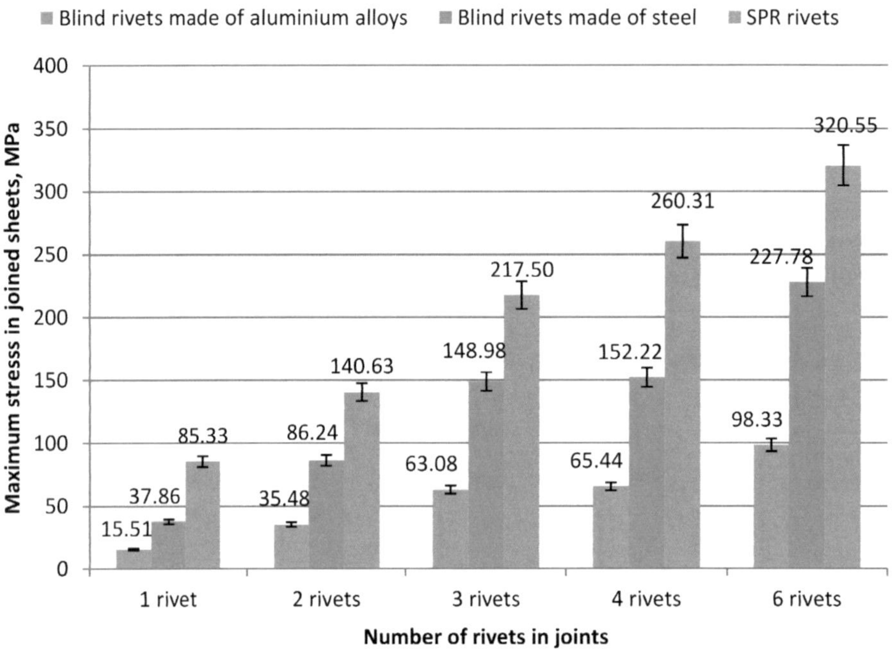

Fig. 9. Maximum stresses in joined sheets in tested rivet joints

Examining the results given in Fig. 9 it can be observed that the stress increases in the steel sheets being joined with increasing the number of rivets, irrespective of the rivet type. The yield stress of riveted joints for steel sheets was exceeded in joints with 4 and 6 steel SPR rivets (R_{eH} = 235 MPa). As for the riveted joints with aluminum alloy rivets, the yield stress of the joined materials was not exceeded (R_{eH} = 150 MPa).

4 Summary and conclusions

The study investigated joints fabricated by traditional riveting and piercing riveting. Pre-drilled riveted joints were fabricated with blind rivets for two reasons: firstly, the design of clenched blind rivets is similar to that of rivets used in pierced riveted joints, and, secondary, the clenching of blind rivets is simple to perform. Given these features, blind rivets are a very popular type of rivets used in assembly processes. Two series of pre-drilled joints: joints with aluminum alloy rivets and steel rivet joints were made, in order to estimate which of them can support higher loads. Self-pierced riveted joints were fabricated using steel rivets, which made it possible to compare the strength properties of these joints with those of pre-drilled joints with steel rivets.

The analysis of the strength test results led to the following conclusions:

- for all tested riveted joints and joint types the highest forces were obtained for self-pierced riveted joints while the lowest – for pre-drilled riveted joints with aluminum alloy rivets;
- the mean shear strength increases with the increasing the number of rivets located in a line perpendicular to the axis of the tensile force – the highest shear strength was obtained for the joints with 3 rivets for all tested types of riveting;
- when the number of rivets was increased from 3 to 4, this led to a 43% decrease in the shear strength of every joint type due to an increase in the lap surface;
- compared to 3-rivet joints, the shear strength of 6-rivet joints is lower by 33% for pre-drilled riveted joints and by 38% for SPR joints, which results from a change of the lap surface;
- SPR joints have higher shear strength than blind rivet joints made of aluminum and steel alike;
- the strength of blind rivet joints is proportionate to the number of applied rivets, which means that the load force is distributed uniformly over all rivets;
- the stress in joined sheet materials increases with increasing the number of rivets, irrespective of the fastener type;
- in the case of riveted joints for steel sheets, the yield point of the joined materials (R_{eH} = 235 MPa) is exceeded in joints with 4 and 6 steel SPR rivets;
- the elongation at maximum tensile force increases with the increasing the number of rivets in pre-drilled riveted joints due to an increase in the maximum force;
- as for the standard deviation of load capacity, the deviation is smaller for pre-drilled riveted joints than that obtained for self-pierced riveted joints, which indicates a higher repeatability of this measure.

Based on the above conclusions, it has been found that the use of the innovative technique of self-piercing riveting combined with the use of SPR rivets leads to higher strength properties of the product and high repeatability of the joining process compared to traditional riveting techniques.

References

1. G. Di Lorenzo, R. Landolfo, *Shear experimental response of new connecting systems for cold-formed structures*. J. Constr. Steel Res. **60**, 561-579 (2004).
2. N. Baurova, A. Anoprienko, Y. Romanova, *Providing dismountable rivet bonded joints through the use of hot-melt adhesives*. MATEC Web of Conferences, International Conference on Modern Trends in Manufacturing Technologies and Equipment (ICMTMTE 2017)**129**, 01004 (2017)
3. J.P. Varia, *The suitability of clinching as a joining method for high-strength structural steel.* J. Mat. Process. Technol. **13**, 242-249 (2003)
4. SH. Cheraghi, *Effect of variations in the riveting process on the quality of riveted joints*. Int. J. Adv. Manufact. Technol. **39**, 1144-1155 (2008)
5. R. Haque, Y. Durandet, *Strength prediction of self-pierce riveted joint in cross-tension and lap-shear*. Mat. Design **108**, 666-678 (2016)
6. C.P. Fung, J. Smart, *An experimental and numerical analysis of riveted single lap joints*. Proc. Inst. Mech. Eng., Part G, J. Aeros. Eng. **208**, 79-90 (1996)
7. L.F.M. Silva, J.P.M. Gonçalves, F.M.F. Oliveira, P.M.S.T. de Castro, *Multiple-site damage in riveted lap-joints: experimental simulation and finite element prediction.* Int. J. Fat. **22**, 319-338 (2000)

8. D. Li, L. Han, M. Shergold, M. Thornton, G. Wiliams, *Influence of Rivet Tip Geometry on the Joint Quality and Mechanical Strengths of Self-Piercing Riveted Aluminium Joints*. Mat. Sci. For. **765**, 746-750 (2013)
9. Y.W. Ma, M. Lou, Y.B. Li, Z.Q. Lin, *Effect of rivet and die on self-piercing rivetability of AA6061-T6 and mild*. J. Mat. Process. Technol. **251**, 282-294 (2018)
10. R. Haque, N.S. Williams, S.E. Blacket, Y. Durandet, *A simple but effective model for characterizing SPR joints in steel sheet*. J. Mat. Process. Technol. **223**, 225-231 (2015)
11. R. Haque, Y. Durandet, *Investigation of self-pierce riveting (SPR) process data and specific joining events*. J. Manufact. Proc. **30**, 148-160 (2017)
12. X. Liu, Y.Ch. Lim, Y. Li, W. Tang, Y. Ma, Z. Feng, J. Ni, *Effects of process parameters on friction self-piercing riveting of dissimilar materials*. J. Mat. Process. Technol. **237**, 19-30 (2016)
13. J. Mucha, *Some aspects of designing process self piercing riveting*. Arch. Mach. Technol. Autom. 29, 91-101 (2009)
14. N.-H. Hoang, R. Porcaro, M. Langseth, A.-G. Hanssen, *Self-piercing riveting connections using aluminium rivets*. Int. J. Sol. Struct. **47**, 427-439 (2010)
15. N.-H. Hoang, A.-G. Hanssen, M. Langseth, R. Porcaro, *Structural behaviour of aluminium self-piercing riveted joints: An experimental and numerical investigation.* Int. J. Sol. Struct **49**, 3211-3223 (2012)
16. G. Di Franco, L. Fratini, A. Pasta, *Influence of the distance between rivets in self-piercing riveting bonded joints made of carbon fiber panels and AA2024 blanks.* Mat. Design **35**, 342-349 (2012)
17. J.F.C. Moraes, H.M. Rao, J.B. Jordon, M.E. Barkey, *High cycle fatigue mechanisms of aluminum self-piercing riveted joints*. Fat. Fract. Eng. Mat. Struct. **41,** 57-70 (2018)
18. R. Haque, *Quality of self-piercing riveting (SPR) joints from cross-sectional perspective: A review.* Arch. Civ. Mech. Eng. **18**, 83-93 (2018)
19. Module 10. Design of Permanent Joints. Version 2ME, IIT Kharagpur. Lesson 2. Design of Rivet Joints. Available at.: http://nptel.ac.in/courses/112105125/pdf/mod10les2.pdf (access on May 2018)

Stiffness of 316L stainless steel support structures proposed for the SLM process

Jiří Hajnyš[1,*], *Marek* Pagáč[1], *Tomáš* Zlámal[1], *Jana* Petrů[1] , *Lukáš* Kousal[1],

[1]VSB – Technical University of Ostrava, Ostrava, Czech Republic

Abstract. The subject of the article is to compare and evaluate the stiffness of the proposed structures relative to the section of the individual rods. The study is applicable to the additive technology of metal production by the 316L (1.4404) by SLM method. In the SLM process the design supports serves as the transporter of the heat generated in the melting bath to the substrate, the secondary function of the structures is for support the powder layer. The article focuses on the supportive function and weight ratio. A total of 4 structures were designed in iteration with three different rod diameters. During construction, the suitability of structures for the SLM method was determined and then the samples were subjected to a mechanical tensile test and the mechanical properties evaluated. In conclusion the most appropriate structure for the supportive function was identified.

Keywords: Stiffness, support, structures, SLM, lattice

1 Introduction

With ever-expanding 3D metal printing technology, it is devoted more and more attention to supporting structures in terms of material savings in their production. Especially in the production of overhangs with an angle of less than 45 °, in the SLM (Selective Laser Melting) process, the area is supported by the metallic powder only, resulting in the sticking of the surrounding particle powder to the component and the significant deterioration of the resulting surface of the material. This phenomenon is due to the poor heat transfer capacity of the powder and the accumulation of too much energy in the area around the overhang. For parts with overhang created without supporting structures additionally occur due to sudden heating of the laser and subsequent rapid solidification of the internal stress and deformation. The aim is to eliminate these undesirable effects. One of the ways to achieve this is to effectively orient the component in the building chamber so that the overhang angle during the construction is in a horizontal position. This option is not always feasible, and there is another option to reduce unwanted phenomena, namely to design a suitable supporting structure that will adequately dissipate heat, securely hold the part in the desired position, and be easily removable. Since the support structures are of no other significance than the above-mentioned reasons, it is a waste of material, when the components are disposed of and

* Corresponding author: jiri.hajnys@vsb.cz

Reviewers: *Nadežda* Čuboňová, *Peter* Košťál

unused. The entire production process is costly and therefore it is necessary to design structures where the minimum material is used and consumed as little energy as possible.

M.F Ashby [1] states in his study that the main structural properties of the lattice of structures significantly influence choice of material, topology and relative density of the grid. Other properties that can affect properties include powder morphology, laser power setting, layer thickness, and scanning speed. Gumruk and Mines [2] examined the lattice of the BCC (Body Centerd Cubic) and designed a theoretical model for the pressure test, resulting in the finding that the effects of block nodes and shear became very important for large d / L (small cell size) rather than small aspect range (d / L <0.1). The structure of BCC was dealt with by Hasan et al. [3], where he found that the subsequent heat treatment of titanium alloys improved the microstructural properties of scattered impurities. Another variation of BCC grids examined by Smith and Guan [4] focused on the BCCz (Body Centerd Cubic with vertical pillars) grid and found that stiffness and yield strength lattices improved markedly when the ratio of one cell was reduced over the others. Another specific structure for the SLM process is the gyroid structure and its variations. Among the first to examine this structure were Yan and Hao [5], their finding was that the effective yield strength dropped with increasing cell size. Variations of the gyriod, called double gyroid lattices, were dealt with by Maskery et al. [6], the conclusion was that more smaller cells could withstand more burdens and fewer cracks, and that heat treatment eliminates diagonal cuts and can accommodate more impact energy. The most important conclusion of the study, however, was that the double gyroid can absorb 3 times more energy than the BCC lattice.

The aim of the experiment is to determine the optimal shape of the structure's lattice in relation to the material used and the stiffness of the structure. The results will be useful for further application in the formation of overhangs and subsequent supporting structures. A similar experiment was carried out by the team around A. Hussein [7], where they found low volume lattice supports are faster to build and efficient for material saving and supports having a larger gap involving greater deformation and manufacturing failure. Our experiment takes this finding into account and works with other shapes and volumes of lattice structures. The main motivation for the experiment was the cost savings of printing the lattice of structures in the SLM process and testing the mechanical properties of the structures in academic conditions.

2 Experimental details (settings)

For the purpose of the experiment, four different lattice structures in iterations with three different beam diameters were designed using different softwares. For modeling basic cube and handles was used Solid Edge and for the creating lattice structures was used nTopology software. The first test structure was a basic SC (Simple Cubic) structure that was used more or less for comparative purposes, as it follows from its very nature that it is not suitable for SLM printing because of horizontal beams that are not self-supporting. The structure is formed by moving the base cube along its elongated edges. The second structure was BCC (Body Centered Cubic) consisting of a cube with particles in the vertices and particles in the center of the cube. The structure was already investigated earlier by Ushijima et. al. [8], but the different diameters of individual beams have not been verified. This arrangement of bars appears to be suitable for printing according to the shape, due to the self-supporting bars at an angle of 45 °. Another test structure (Tetragon Vertex) has already been more complicated in the arrangement of beams, this is a quadrangle where the rods from the center point to the corners of the regular tetrahedron and form an angle of 109 ° and the joining of the edges of the tetrahedron results in an eighth shape wall. The lattice walls are made up of four general triangles and are a so-called Platonic body. The last tested structure was also a tetragon, but centered on the center, which created an interesting structure suitable for printing. All lattice structures were applied to a 17 x 17 x 17 mm cube pattern, for which a clamp holder was formed for mechanical properties testing. The individual beam diameters were 0.5; 0.7; 1 mm. Used material stainless steel 316L (1.4404). Table 1 shows the structures in the modeling software and their size of the individual beams.

Table 1. Used structures and their parameters

<table>
<tr><th colspan="4">Structure 1 - Simple Cubic</th></tr>
<tr><td rowspan="4">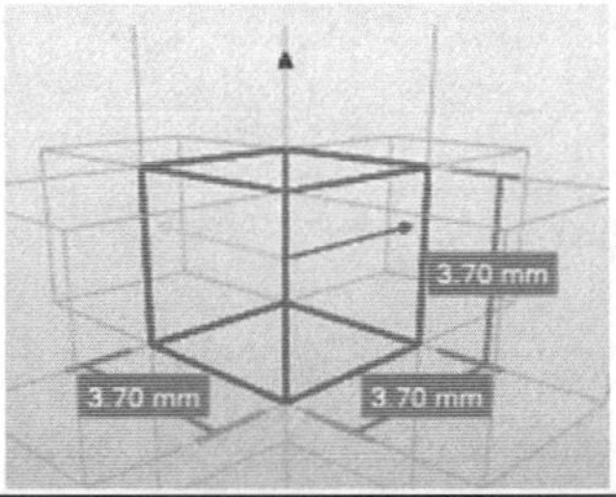</td><td rowspan="4"></td><td>Beam diameter</td><td>Percent volume</td></tr>
<tr><td>0.5 mm</td><td>5.68 %</td></tr>
<tr><td>0.7 mm</td><td>10.85 %</td></tr>
<tr><td>1 mm</td><td>21.39 %</td></tr>
<tr><th colspan="4">Structure 2 - Body Centered Cubic</th></tr>
<tr><td rowspan="4">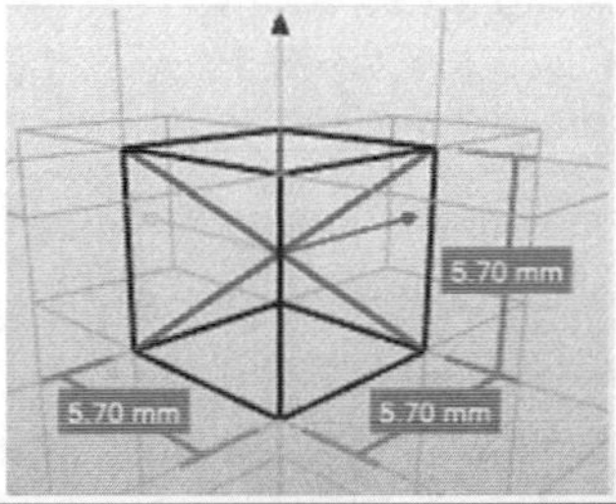</td><td rowspan="4">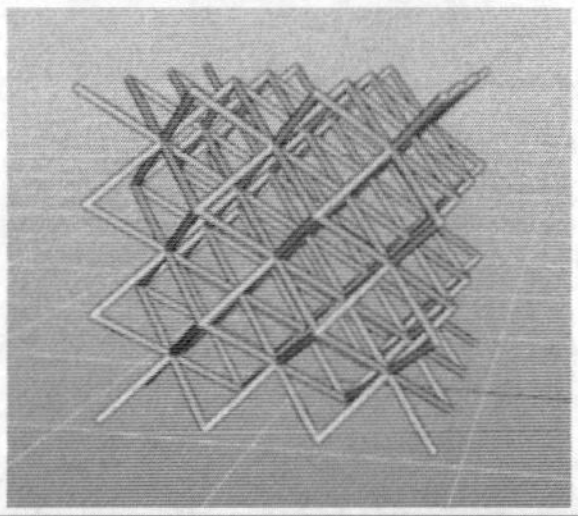</td><td>Beam diameter</td><td>Percent volume</td></tr>
<tr><td>0.5 mm</td><td>3.40 %</td></tr>
<tr><td>0.7 mm</td><td>6.49 %</td></tr>
<tr><td>1 mm</td><td>12.99 %</td></tr>
<tr><th colspan="4">Structure 3 - Tetragon Vertex</th></tr>
<tr><td rowspan="4">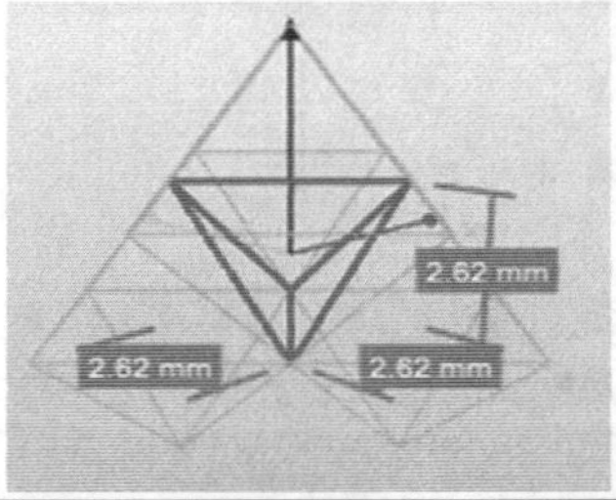</td><td rowspan="4">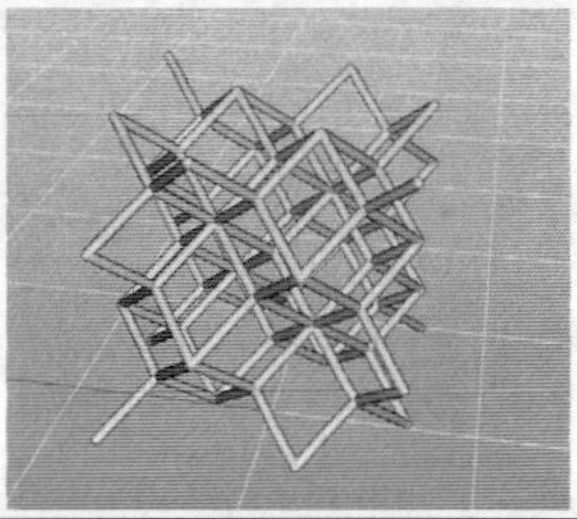</td><td>Beam diameter</td><td>Percent volume</td></tr>
<tr><td>0.5 mm</td><td>4.84 %</td></tr>
<tr><td>0.7 mm</td><td>9.30 %</td></tr>
<tr><td>1 mm</td><td>18.44 %</td></tr>
<tr><th colspan="4">Structure 4 - Tetragon Edge</th></tr>
<tr><td rowspan="4">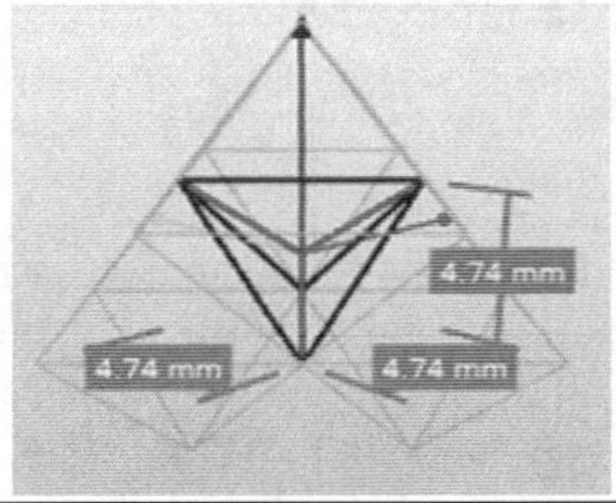</td><td rowspan="4">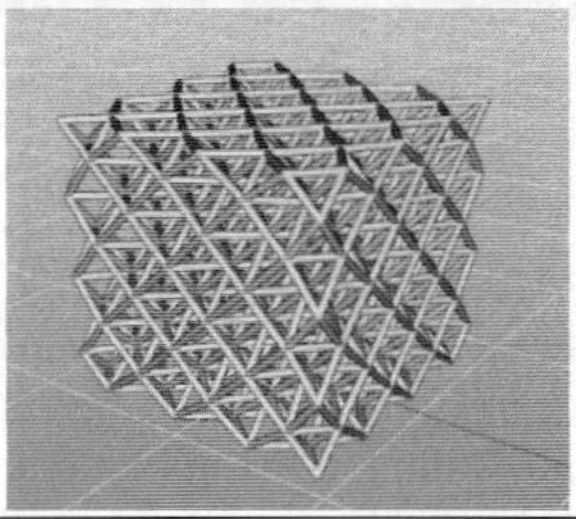</td><td>Beam diameter</td><td>Percent volume</td></tr>
<tr><td>0.5 mm</td><td>12.33 %</td></tr>
<tr><td>0.7 mm</td><td>23.35 %</td></tr>
<tr><td>1 mm</td><td>45.06 %</td></tr>
</table>

All structures were made under the same set conditions (laser performance, scanning speed, hatch distance) and the same strategy of scanning was used, namely "Meander" because of its universal use. The samples produced were then cut off by a belt saw from the substrate and no thermal finishing was applied.

3 Results

Visual inspection was determined as the first evaluation criterion for printed structures. The inspection has been carried out since the very beginning of the press, where no complications have been observed and the build has not been interrupted once. But on cut off samples some fabrication defects can be observed. There was a phenomenon called "warping", which can be caused in two ways. The first reason for this is the heat stress caused by rapid metal solidification, where the thermal stress exceeds the strength of the material and plastic deformation occurs. This phenomenon occurs on the upper layer, which lies below the powder metal and according to Zhao et al. [9], a bending angle is produced towards the laser beam, Zhao found that the thermal residual stress is centered on both edges of the scan path, and the low scanning speed can lead to high cooling rates, resulting in higher thermal residual stresses. Zhang et al. [10] recommends that the substrate preheat temperature be adjusted and remelted to reduce it. Another option is to change the laser scanning strategy. The second reason for the emergence of warping when a part of the curve begins to overlap is the lack of support that does not support the previous layers, which is probably the reason for our defect. Warping occurred only with two structures, namely BCC and Tetragon Vertex with a diameter of beam 0.5 mm, see Fig. 1.

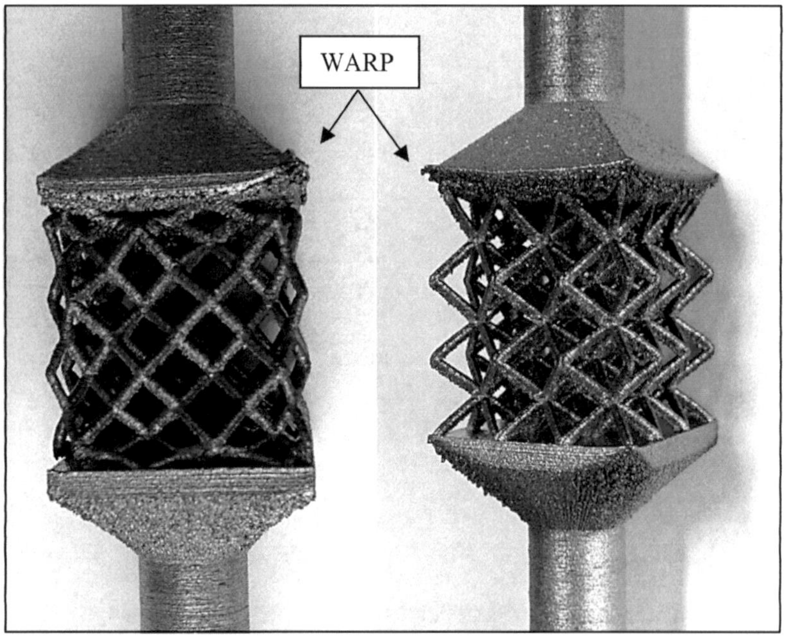

Fig. 1. Warp defect

Another defect that has been observed is "dross formation" This defect results in increased surface roughness and change of overhang geometry, moreover it is a difficult predictable defect and it is difficult to understand the underlying mechanism of formation of dross defect during overhanging. Kovalev and Gurin [11] in their study state that dross formation occurs when the laser scans the already formed layer and the thermal conduction speed is too high, unlike when the laser scans the powder layer and the heat conduction speed is much lower than the fixed created layer. This is often the case when overpowering occurs when the absorbed energy is higher and this results in the melt pool being enlarged and, due to the effects of gravitational and capillary forces, plunging deeper into the powder. The defect manifested itself on the first tested structure of SC, see Fig. 2.

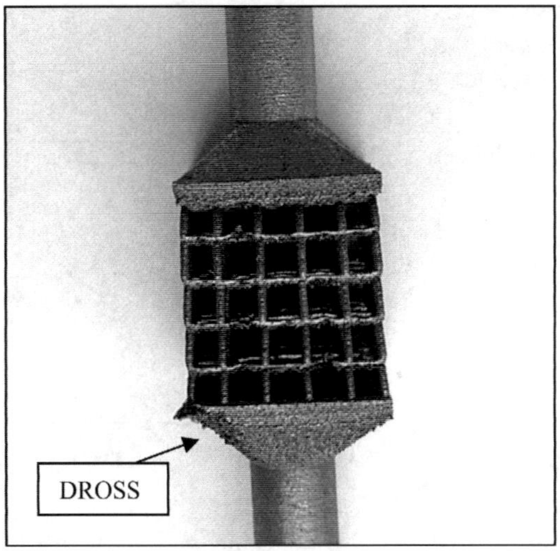

Fig. 2. Dross formation defect

For the next evaluation criterion, the maximum load of the structures that they can withstand without a visible violation was investigated. A standard tensile test was chosen at a load speed of 5 mm/min. Testing was carried out on a desk, two-column, computer-controlled testometric M500-50CT at a room temperature of 21 ° C. Graphs were obtained from the obtained results, where the individual member thicknesses were compared.

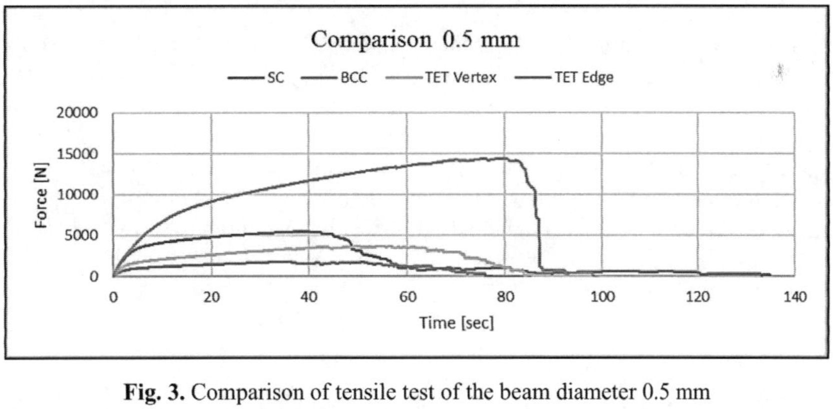

Fig. 3. Comparison of tensile test of the beam diameter 0.5 mm

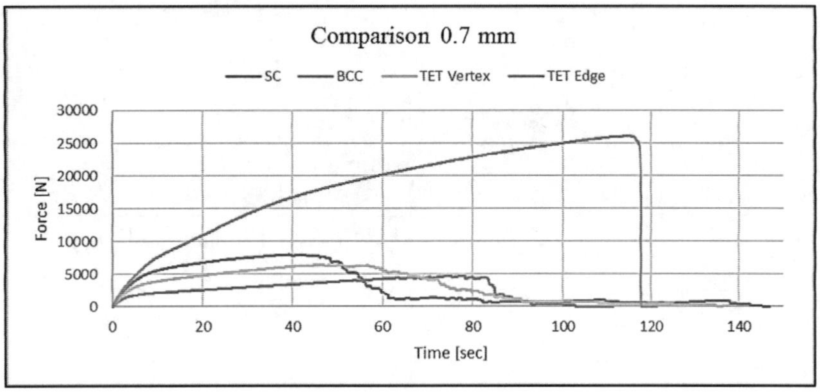

Fig. 4. Comparison of tensile test of the beam diameter 0.7 mm

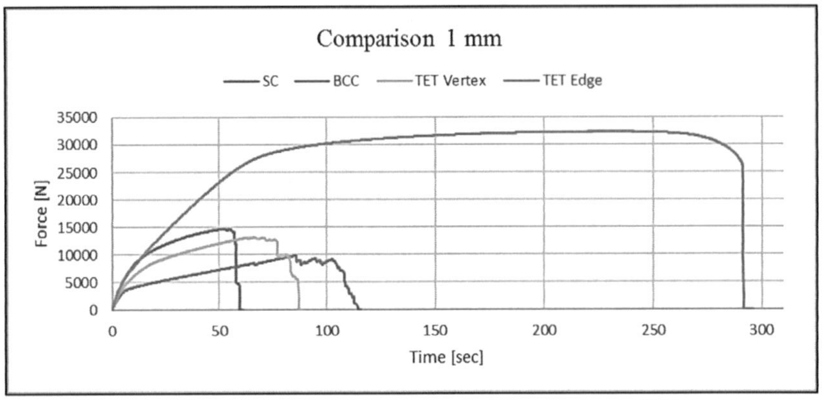

Fig. 5. Comparison of tensile test of the beam diameter 1 mm

It can be seen from the graphs (Fig. 3, 4, 5) that with the increasing thickness of the rods the necessary force is increased to break the samples. The largest load on all three beam diameters it takes most by the Tetragon Edge, but it also occupies the largest percentage volume of all structures tested, so it is a logical result. Interestingly, the results of other structures, predominantly the Tetragon Vertex and Body Centered Cubic structures, hold a similar percentage of the volume, yet the Tetragon Vertex structure, which held almost twice as much load, was significantly more robust. The Simple Cubic Structure has also achieved good results due to the position of the rod on the Z axis. The latest interesting finding is that the smaller the diameter of the rod, the longer the relative elongation and thus the time that the structure breaks.

The last and most important assessment criterion was the relative stiffness of the structures, or the Young modulus of elasticity. The Young's modulus of elasticity is the constant characteristic of the substance and indicates its strength and deformation potential. The higher the value of the module, the higher the voltage required to achieve a certain deformation, always gets positive values. To determine the modulus of elasticity from our experiment, the relative elongation and load forces from the Testometric instrument and the calculation according to formula (1) were used. Calculated values are shown in Table 2.

$$E = \frac{F}{S} : \Delta l \qquad (1)$$

Where:

E – Modulus of elasticity [GPa]
F – Load force [N]
S – Sample area [mm]
Δl – Elongation [mm]

Table 2. Calculated modulus of elasticity

	Beam diameter [mm]		
	0.5	0.7	1
	Modulus of elasticity of structures [GPa]		
Structure 1	1.68	2.28	9.01
Structure 2	0.84	1.69	3.5
Structure 3	5.63	9.07	4.64
Structure 4	1.49	2.02	6.12

The table shows that the greatest values of the modulus of elasticity are achieved by Structure 3 (TET Vertex), which additionally occupies the smallest percentage of the material volume, on the other hand, the worst values of the modulus of elasticity reach Structure 2 (BCC). Of all the data, one can observe the dependence that the larger the diameter of the beam, the overall stiffness is better.

4 Conclusion

Additive production permits the creation of complicated shapes of components that cannot be manufactured by other methods or special preparations are needed. One of the great advantages of additive production is the formation of a lattice of structures that can replace the solid material inside the body and thereby reduce overall weight and material consumption. Lattice structures can also be used for so-called support when they support surfaces. The shapes, sizes and possibilities of fabrication of structures are many, this article aimed to create 4 structures with different diameter of the connecting members. The model structures were printed using the SLM (Selective Laser Melting) method on the Renishaw AM400 under constant conditions. The suitability of the use of the individual structures was assessed using established criteria where the most important criterion was the stiffness of the structure, depending on the percentage of material consumed. The stiffness was determined according to the tensile test, the data was processed into charts and tables.

From the results obtained, several partial conclusions can be drawn:

1) When printing the structures of BCC and Tetragon Vertex with a diameter of 0.5 mm, there was a warping effect and when printing the structure of SC there was dross formation defect

2) Depending on the percentage of material consumed, the Tetragon Vertex Structure can withstand the greatest load to the breaking of the beams

3) The smaller the diameter of the beam, the higher the relative elongation

4) The highest rigidity was seen in the Tetragon Vertex structure (0.7 mm) and the worst structure BCC (0.5 mm)

Overall, the most appropriate structure, depending on the consumed material, is structure no. 3 Tetragon Vertex with a diameter of 0.7 mm and a percentage volume of 9.3%. This structure should be further investigated and further tests and experiments should be carried out.

Article has been done in connection with projects Education system for personal resource of development and research in field of modern trend of surface engineering - surface integrity, reg. no. CZ.1.07/2.3.00/20.0037 financed by Structural Founds of Europe Union and from the means of state budget of the Czech Republic and by project Students Grant Competition SP2018/150 and SP2018/136 financed by the Ministry of Education, Youth and Sports and Faculty of Mechanical Engineering VŠB-TUO.

References

1. M.F. Ashby, *The properties of foams and lattices*. The Royal Society: Mathematical physical and engineering sciences, **68**, p17, (2006)
2. R. Gumruk, R.A.W. Mines, *Compressive behaviour of stainless steel micro-lattice structures*. International Journal of Mechanical Sciences, **68**, 125-139, (2013)

3. R. Hasan, R. Mines, P. Fox, *Characterization of selective laser melted Ti-6Al-4 V micro lattice struts.* Procedia Engineering, **10**, 536-541, (2011)
4. M. Smith, Z. Guan, W.J. Cantwell, *Finite element modelling of the compressive response of lattice structures manufactured using the selective laser melting technique.* International Journal of Mechanical Science, **67**, 28-41, (2013)
5. C. Yan, L. Hao, A. Hussein, P. Young, D. Raymont, *Evaluations of cellular lattice structures manufactured using selective laser melting.* International Journal of Machine Tools and Manufacture, **62**, 32-38, (2012)
6. Maskery, N.T. Aboulkhair, A.O. Aremu, C.J. Tuck, I.A. Ashcroft., *Compressive failure modes and energy absorption in additively manufactured double gyroid lattices.* Additive Manufacturing, **17**, 24-29, (2017)
7. Hussein, L. Hao, Ch. Yan, R. Everson, P. Young. *Advanced lattice support structures for metal additive manufacturing.* Journal of Materials Processing Technology, **213**, 1019-1026, (2013)
8. K. Ushijima, W.J. Cantwell, R.A.W. Mines, S. Tsopanos, M. Smith, *An investigation into the compressive properties of stainless steel micro-lattice structures.* Journal of Sandwich Structures and Materials, **13**, 303-329, (2010)
9. X.M. Zhao, X. Lin, J. Chen, L. Xue, W.D. Huang, *The effect of hot isostatic pressing on crack healing, microstructure, mechanical properties of Rene88DT superalloy prepared by laser solid forming.* Materials Science & Engineering, **504**, 129-134, (2009)
10. B.C. Zhang, L. Dembinski, C. Coddet, *The study of the laser parameters and environment variables effect on mechanical properties of high compact parts elaborated by selective laser melting 316L powder.* Materials Science & Engineering, **584**, 21-31, (2013)
11. O.B. Kovalev, A.M. Gurin, *Multivortex convection of metal in molten pool with dispersed impurity induced by laser radiation.* International Journal of Heat and Mass Transfer, **68**, 269-277, (2014)

Comparison of the mechanical properties and microstructural evolution in the HAZ of HSLA DOMEX 700MC welded by gas metal arc welding and electron beam welding

Peter Kopas[1,*], *Milan* Sága[1], *Michal* Jambor[2], *František* Nový[2], *Libor* Trško[3], *Lenka* Jakubovičová[1]

[1]University of Zilina, Faculty of Mechanical Engineering, Department of Applied Mechanics, Univerzitná 1, 010 26, Zilina, Slovak Republic
[2]University of Zilina, Faculty of Mechanical Engineering, Department of Materials Engineering, Univerzitná 1, 010 26, Zilina, Slovak Republic
[3]Research Centre of the University of Žilina, Univerzitná 1, 010 26, Zilina, Slovak Republic

Abstract. High strength low alloyed steels are nowadays an important group of engineering materials, with numerous industrial applications. Properly chosen chemical composition together with the thermomechanical processing allow to reach high mechanical properties, while the carbon content is maintained low. Resulting microstructure is therefore sensitive to the amount of the heat introduced to the metal during welding. During welding procedure, softening phenomenon can occur in the heat affected zones and it could lead to the degradation of the mechanical properties of the final joint. Nowadays, welding technologies based on the high energy electron beam are able to produce weld joints with very narrow heat affected zone. In this study, authors compare mechanical properties and the microstructural evolution in the heat affected zone of the GMAW and EB welded joint of the DOMEX 700MC HSLA steel. The carried out experiments showed that both evaluated technologies are able to produce joints with appropriate mechanical properties. Welded joints made by the EBW technique showed slightly better mechanical properties of the whole joint, as a result of the narrower HAZ.

Keywords: HSLA, DOMEX 700MC, gas metal arc welding (GMAW), electron beam welding (EBW)

1 Introduction

In the last decade, the high strength low alloyed (HSLA) steels became the standard material used in the engineering constructions. This category of materials offers the higher mechanical properties than standard grades of weldable structural steels, what allow to reach significant reductions of the total component's mass, as the cross-section of components can be reduced. Improved mechanical properties of the high strength low alloyed steels thanks to the carefully

* Corresponding author: peter.kopas@fstroj.uniza.sk

Reviewers: *Wieslawa* Piekarska, *Milan* Naď

designed chemical composition together with the thermomechanical processing, resulting in the fine-grained microstructure of the final products [1-5].

It the numerous engineering applications, the welding is the most widespread joining technology. The high strength low alloyed steels owe for their properties fine-grained microstructure, and the heat introduced to the material during welding procedure can significantly affect microstructure in the heat affected zone a degrade resulting mechanical properties of the weld joint [6-9]. To obtain appropriate mechanical properties of the weld joint of the high strength low alloyed steels welding process has to be carefully designed, to reduce the amount of the heat introduced to the materials and prevents to the softening in the heat affected zone [10]. Modern welding techniques, which use high energy beams, allows to minimize the amount of heat introduced to the joint and obtaining very narrow heat affected zone. On this bases, electron beam and laser beam welding techniques are very interesting options for the welding of high strength low alloyed steels [11-13].

This study compares mechanical properties and the microstructural evolution in the heat affected zone of weld joint of high strength low alloyed steel DOMEX 700MC, obtained by the gas metal arc welding (GMAW) and electron beam welding techniques (EBW).

2 Materials and methods

In the presented study, high strength low alloyed DOMEX 700MC steel of 10 mm thickness was used as an experimental material. The DOMEX 700MC steel belongs to the group of HSLA steels with ferritic structure. The chemical composition and the mechanical properties of experimental material are shown in Tables 1 and Table 2. High values of ultimate tensile strength and yield strength of DOMEX 700MC steel are achieved by the combination of properly chosen chemical composition and thermomechanical processing, what results in very fine ferritic microstructure with uniformly dispersed fine cementite particles (Fig.1).

Table 1. Basic mechanical properties guaranteed by the supplier

	Ultimate tensile strength [MPa]	0.2 offset Yield strength [MPa]	Tensile elongation [%]
DOMEX 700MC	750 - 950	min. 700	min. 12

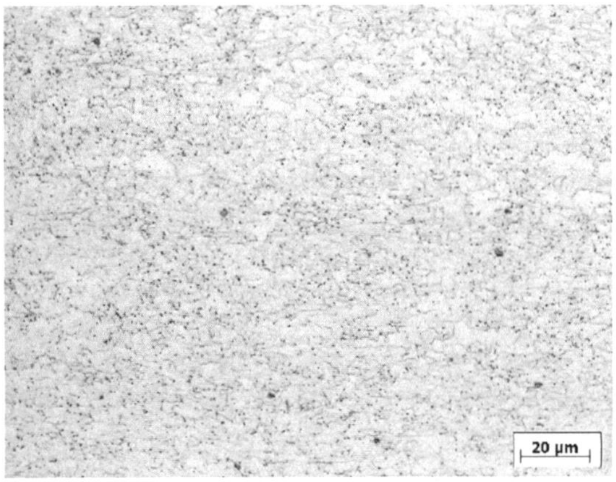

Fig. 1. Microstructure of the DOMEX 700MC HSLA steel

Table 2. The chemical composition of the DOMEX 700MC guaranteed by the supplier

	C	Si	Mn	P	S	Al	Nb	V	Ti
DOMEX 700MC	max. 0.12	max. 0.10	max. 2.10	max. 0.025	max. 0.01	min. 0.015	max. 0.09	max. 0.2	max. 0.15

Two different welding methods were used for the welding of the HSLA DOMEX 700C steel. The GMAW and EB welding. As a preparation for the GMAW procedure, metal sheets were mechanically cleaned before welding and degreased by the acetone. The gap 2 mm was maintained between the sheets and their edges were cut at angle of 50°. The OK AristoRod 69 wire with diameter 1.2 mm was used as a filler metal. For the optimization of GMAW process, several welds were prepared with the welding parameters in the range: voltage 23-30V, current 145-300A and the welding speed 3-8 mm/s. The welding parameters were chosen to obtain a stable welding process with the high quality of final welds, without any macroscopic geometrical and internal defects. As a result of the optimizing process, the following welding parameters were used for the preparation of the testing specimens. For root layer: voltage 23.3V, current 164A and welding speed 4.8 mm/s and for the top layer: voltage 28.2V, current 230A and the welding speed 5 mm/s. Similarly, before the EB welding, sheets were again mechanically cleaned and degreased in acetone and after optimization of the welding process, the following parameters were chosen for the preparation of final welds: acceleration voltage 55 kV, current 115 mA and the welding speed 11 mm/s.

Mechanical properties of the weld joints were obtained by the standard tensile tests. Tree specimens of each type of weld were used, and the average values were calculated. For the evaluation of the properties in the individual areas of weld joint (weld metal - WM and heat affected zone - HAZ), microhardness measurements were performed. Evolution of the microstructure, mainly in the HAZ was examined by optical microscopy.

3 Results and discussion

Macro and microscopic observations of the welds revealed significant differences. The width of the HAZ and the weld metal zone (WM) were significantly larger for the GMAW process (Fig.2). In both welds, two different regions in HAZ were identified. The evolution of the microstructure in the HAZ depends on the thermal cycle to which was exposed during the welding process. Closer to the weld metal (WM), coarse-grain HAZ was observed (CGHAZ). These areas are during the welding completely transformed to the austenite, then coarse and during subsequent cooling transform back to the ferrite. The thickness of this zone was considerably larger for the GMAW as a result of the larger heat input during GMAW process, which result in the longer exposures of that zone in the austenite region. CGHAZ in the EB welds was significantly narrower, but the grain size in that area was larger (Fig. 3a, b) than in the GMAW welds. This phenomenon can be explained that during the EB welding process, the high energy beam input a large amount of heat energy in a very short time, thus the heat has no sufficient time to dissipate in the metal and the maximum reached temperature in the CGHAZ is higher than for the conventional GMAW process. With increased distance from the weld metal, the microstructure changed and the fine-grained HAZ (FGHAZ) was observed in both welds (Fig. 3c, d). During the thermal cycle, these zones were heated to the austenitic area for a very short period of time, thus no coarsening could occur. The thickness of this zone was larger for the GMAW weld, especially in the root area. Between the FGHAZ and base metal, the transition area was observed. The temperature reached in this area during the welding procedure was very close to the AC_1 temperature, thus the resulting microstructure is only partially recrystallized and the phenomenon as a polygonization and recovery took place in this area (Fig. 3e, f).

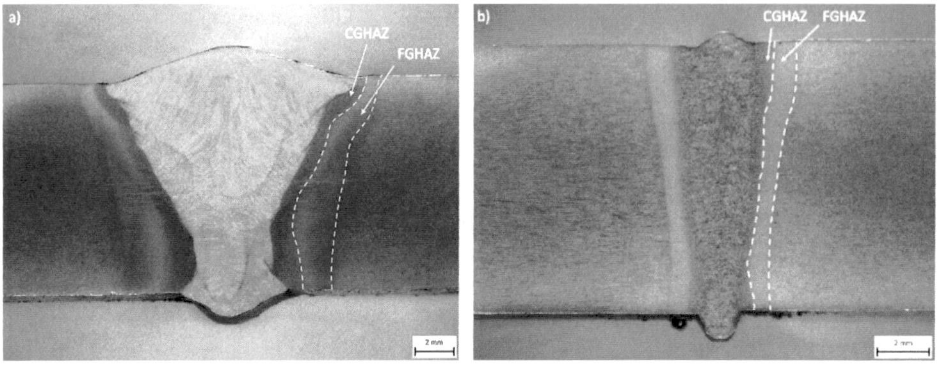

Fig. 2. Macroscopic view of the examined weld joints, a) GMAW, b) EBW

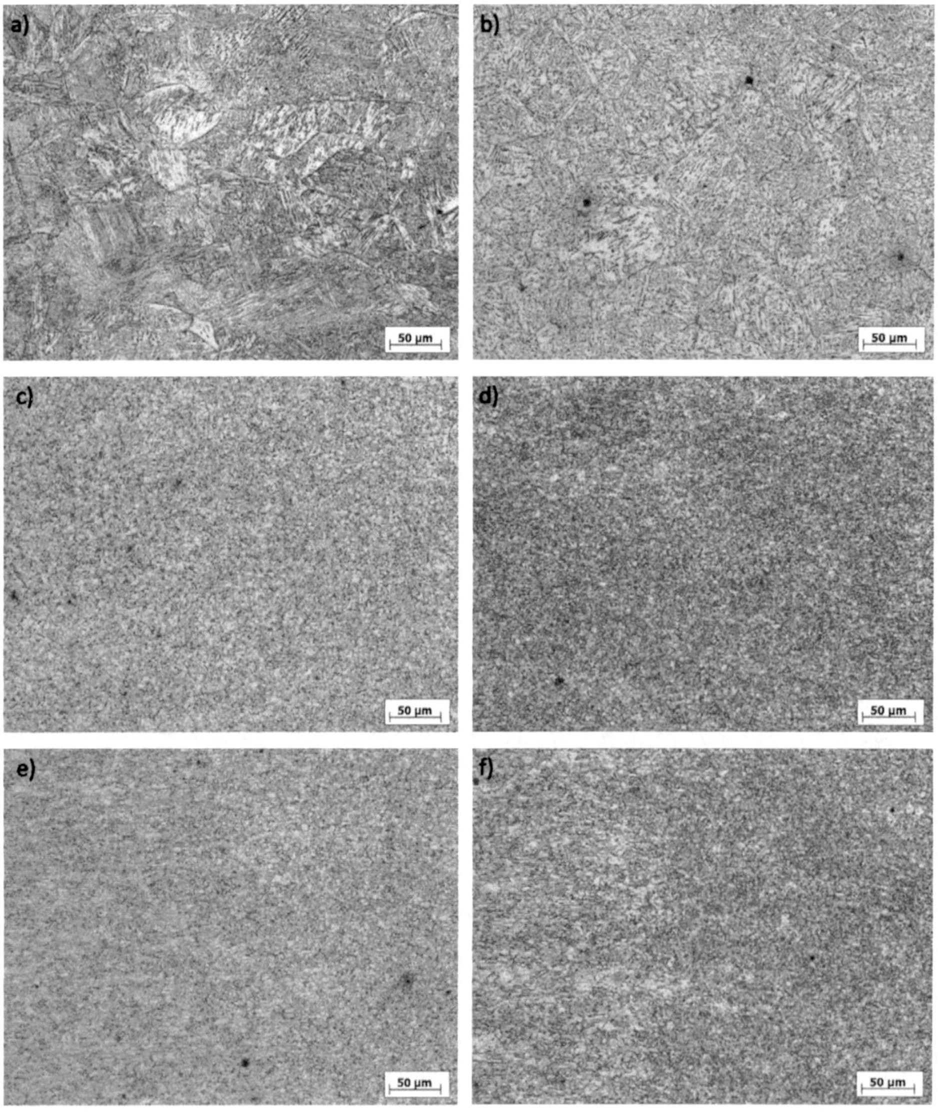

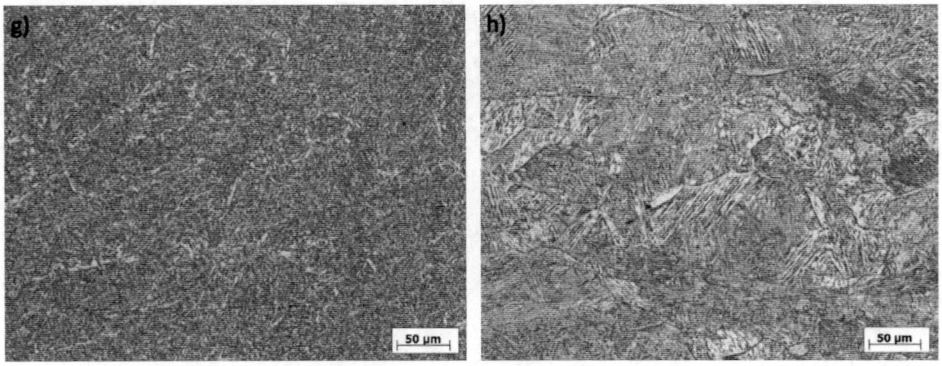

Fig. 3. Microstructures of the HAZ
Figures a, c and e represents CGHAZ, FGHAZ, resp. transition zone of GMAW joint.
Figures b, d and f represent CGHAZ, FGHAZ and transition zone of EBW joint.

Microhardness measurement examined mechanical properties of all areas of the weld joints. The results of the microhardness measurements are shown in the Fig.4. Based on the results, in the GMAW weld, the most critical part is the HAZ, especially transition region between the FGHAZ and the base metal (BM). Average values of microhardness in the BM was 276 HV1 and for the WM values 261 HV1 were recorded. In the HAZ, the highest average values were recorded in the CGHAZ (252 HV1), then in the FGHAZ (235 HV1) and the lowest values were recorded in the transition region between the FGHAZ and the BM (210 HV1).

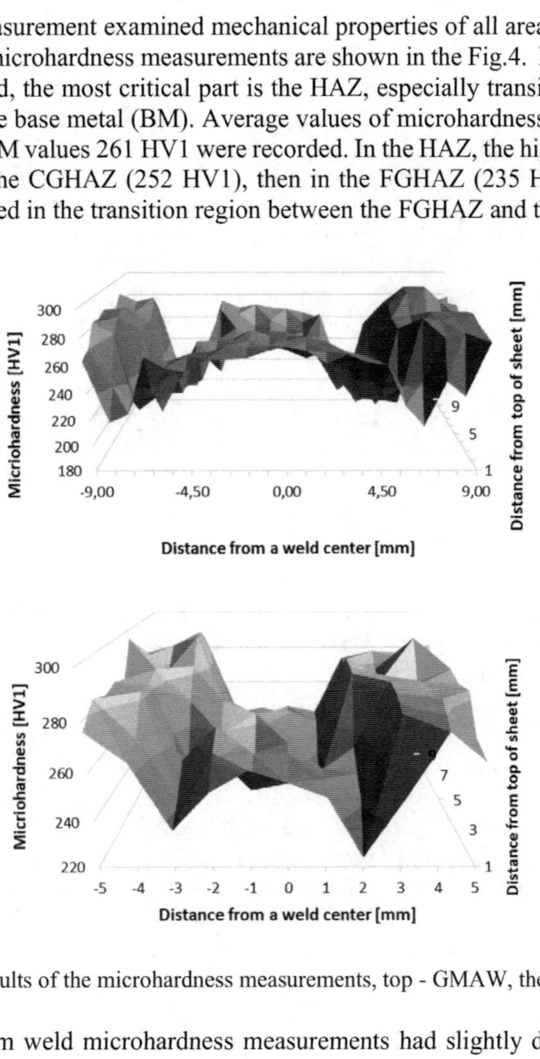

Fig. 4. The results of the microhardness measurements, top - GMAW, the bottom – EBW

The electron beam weld microhardness measurements had slightly different results. The mainly, in the EBW process, there was not used any filler metal, thus the WM zone is a remelted parent material, and this corresponds with the measured values of microhardness (256 HV1). Microhardness values in the HAZ vary for different subzones, but the differences were significantly smaller than in the HAZ of GMAW weld. The average value in CGHAZ was 252 HV1, in the FGHAZ 243 and finally, in the transition region, the average value 236 HV1 was recorded. These observations are in consistency with other authors [14].

Results of the tensile tests are shown in Table 3. Specimens welded by the GMAW were broken in the HAZ, what suggests that the softening in the heat affected zone recorded by the microhardness measurements has a significant effect on the resulting properties of the weld joint. Even that the specimens were broken in the softened regions of HAZ, the measured values of yield strength and ultimate tensile strength still fulfil the requirements for the mechanical properties of the DOMEX 700MC. For the specimen welded by the electron beam, the different behaviour was observed. However, the softening in the HAZ was observed also in the EB weld, this was not so significant and the specimens were broken in the weld metal zone. This suggests that use of EB welding results in the production of very narrow HAZ, in which the level of softening is lower than in the weld made by conventional GMAW. Results of the tensile test showed higher values of ultimate tensile strength, yield strength and tensile elongation for the weld made by electron beam welding compared with the GMAW. The most significant difference is in the values of YS, which were about 57 MPa higher for the EB weld.

Table 3. Results of the tensile tests

	UTS [MPa]	0.2 offset YS [MPa]	Tensile elongation [%]
DOMEX 700MC GMAW	783.5	691.4	9.1
DOMEX 700MC EBW	793.3	748.8	12.2

Both examined welding technologies were able to produce joints with sufficient mechanical properties. The phenomenon of softening in the HAZ was recorded in both, GMAW and EBW welds, but was more significant in the weld joint made by conventional GMAW. The EBW produces markedly narrower HAZ, and the resulting properties of the joints made by EBW were higher. It is however questionable if the very high operating cost of the using EBW procedures are worth of the small advantages in the slightly higher YS introduced by the EBW process.

4 Conclusions

Based on the carried out tests and analysis, the following can be concluded:

- Both used method, GMAW and EBW are able to produce high-quality weld joint, with appropriate mechanical properties.
- In the HAZ, several subzones (CGHAZ, FGHAZ and transition region) were identified in both examined welds. In the EBW joint, the narrower HAZ was recorded, same as the thickness of individual subzones was lower.
- The softening of HAZ was observed in both tested joints, but it should be pointed out that the level of softening was significantly lower in EBW joint.
- Based on the result of the tensile test, in the GMAW joint, the most critical zone is the transition zone between the FGHAZ and BM, as the all tensile specimens were broken in that area. This is with consistency of the microhardness measurements, where the lowest values of microhardness were recorded in the transition area.
- Even in the EBW joint, softening of HAZ was recorded, this was not so significant. Tensile test specimens were broken in the WM area.
- Tensile tests showed that the weld joints made by EBW exhibits about 57 MPa higher yield strength and about 3% higher tensile elongation.

This work was supported by the Slovak Research and Development Agency under the contract No. APVV-14-0096 and No. APVV-16-0276. Also this research was supported by Scientific Grant Agency of the Ministry of Education, Science, Research and Sport of the Slovak Republic under the contract VEGA No. 1/0951/17.

References

1. V. Sharma, A. S. Shahi, *Quenched and tempered steels welded with micro-alloyed based ferritic fillers*. Journal of Materials Processing Technology, **253**, 2-16 (2018)
2. S. Błacha, M. S. Węglowski, S. Dymek, M. Kopyściański, *Microstructural and Mechanical Characterization of Electron Beam Welded Joints of High Strength S960QL and Weldox 1300 Steel Grades*. Archives of Metallurgy and Materials, **62,** Issue 2A, 627-634 (2017)
3. K. Borko, B. Hadzima, M. Neslušan-Jacková, *Corrosion resistance of Domex 700 steel after combined surface treatment in chloride environment*. Procedia Engineering, **192**, 58-63 (2017)
4. D. Arsic, V. Lazic, R. Nikolic, S. Aleksandrovic, B. Hadzima, M. Djordjevic, *Optimal welding technology of high strength steel S690QL*. Materials Engineering, **22**, 33-47 (2015)
5. R. Ulewicz, P. Szataniak, *Fatigue cracks of Strenx steel*. Materials Today: Proceedings, **3**, No. 4, 1195-1198 (2016)
6. L. Jakubovičová, B. Ftorek, V. Baniari, A. Sapietová, T. Potoček, M. Vaško, *Engineering Design of a test device*. Procedia Engineering, **177**, 520-525 (2017)
7. Handrik, M., Vaško, M., *Design of a computational model for multi-body contact of deformable bodies in the flowing fluid*. Matec Web of Conferences, 157 (2018)
8. X. Qiang, X. Jiang, F. S. Bijlaard, H. Kolstein, *Mechanical properties and design recommendation of very high strength steel S960 in fire*. Engineering Structures, **112**, 60-70 (2016)
9. R. Oyyaravelu, P. Kuppan, N. Arivazhagan, *Metallurgical and mechanical properties of laser welded high strength low alloy steel*. Journal of Advanced Research, **7**, 463-472 (2016)
10. M. Mazur, R. Ulewicz, F. Nový, P. Szataniak, *The structure and mechanical properties of Domex 700 MC steel*. Communications - Scientific Letters of the University of Zilina, **15**, No. 4, 54-57 (2013)
11. R. Halama, A. Markopoilos, F. Fojtik, M. Fusek, Z. Poruba, J. Famfulík, *Effect of stress amplitude on uniaxial ratcheting of aluminum alloy 2124-T851*. Materialwissenschaft und Werkstofftechnik, **48** (8), 814-819 (2017)
12. M. Kubiak, T. Domanski, V. Dekýš, A. Sapietová, *Measurement of strain during tension test of welded joint using multi-camera 3D correlation system*. Procedia Engineering, **177**, 107-113 (2017)
13. P. Kováčiková, R. Bezdedová, J. Jr.Vavro, J. Vavro, *Comparison of numerical analysis of stress-strain states of cast iron with vermicular graphite shape and globular graphite shape*. Procedia Engineering, **136**, 28-32 (2016)
14. A. Dzyubyk, V. Palash, I. Khomych, S. Hrynus, *Optimization of welding modes for high strength low-alloy DOMEX 700 steel*. Ukrainian Journal of Mechanical Engineering and Materials Science, **3**, No. 2, 47-54 (2017)

MATEC Web of Conferences **244**, 01010 (2018) https://doi.org/10.1051/matecconf/201824401010
ITEP'18

Dynamic simulation for optimisation solution of manufacturing processes

Vladimíra Schindlerová[1,*], *Ivana* Šajdlerová[1], *Dominika* Lehocká[2]

[1]VŠB - Technical University of Ostrava, Faculty of Mechanical Engineering, 17. listopadu 2172/15, 708 33, Ostrava - Poruba, Czech Republic

[2]Technical University of Kosice, Faculty of Manufacturing Technologies with a seat in Presov, Bayerova 1, 080 01 Presov, Slovak Republic

Abstract. One way how to study the real behaviour of industrial processes or systems in practice is to use computer simulations. We can simulate different conditions and find optimal parameters without increased risk. The right application of these parameters in practice can produce the desired results. The advantage is not only the safe verification of various variants of the simulated parameters, but also the possibility of their use in different areas of industrial practice. This article deals with an example of the use of simulation in the production of the selected automobile cooling system component. The simulation model was created to design the correct number of Kanban circuits to shorten production lead time and to reduce inter-operational supplies. The suitability of using computer simulations to optimize the production processes and systems in practice can be confirmed based on comparison of the results from the computer simulation with results achieved in practice.

Keywords: Dynamic Simulation, Optimisation, Manufacturing Processes, Kanban

1 Introduction

Evaluating the quality of delivered products or provided services by customers depends on the degree of fulfilment of the expected requirements whether it is the ability to deliver products or services corresponding to the customers' ideas as for the appearance and design, correct amount, agreed delivery times or required quality at reasonable prices. The current market environment forces the companies to respond to changes and customer demands faster and in a more targeted way. It is necessary to search for suitable solutions in all business processes, minimizing the risk of incorrect decisions. This environment brings large business opportunities on one hand and possible risks on the other. The relatively big and stable interest of the management of the companies is reflected in the modernization of management of process systems, development of new high-quality products and minimization of costs incurred. The simulation systems which can assist in the decision-making processes associated with the design of manufacturing processes and systems are one of the possible solutions [1].

* Corresponding author: vladimira.schindlerova@vsb.cz

Reviewers: *Peter* Demeč, *Milan* Edl

2 Methodological base – Kanban, dynamic simulation

The production pace is determined by the market requirements and it is the task of companies to meet these requirements as much as possible. Currently, various planning and business process management methods are used for the efficient use of available manufacturing resources [1,2,3]. The question is, which methods or techniques are the most suitable solution. One of many applicable logistics solutions for smooth flow and shortening of the sequence of individual production processes is the KANBAN method (Japanese term for a label or card).

2.1 Kanban

The decentralized organization production system Kanban aims to deliver the material into the manufacturing process at the lowest turnover capital within a certain time frame. When integrating the Kanban system, we have to plan with respect to consistent sequence of the product on the final assembly line, schematic layout of the machinery, standardization of processes, change of production etc. [2]

2.2 Dynamic simulation

The simulation is also referred to as a modern method of analysing complex business processes which include the elements of coincidental and dynamic behaviour. It is a very successful method of analysing business processes. The applications in practice have shown that companies which have used the optimisation for their business processes have achieved financial, material and personnel savings. [3,4]

Simulation tools

In practice, we can encounter various simulation software, examples of some of which are presented in Tab. 1.

Table 1. Setting Word's margins

MATLAB	For mathematical purposes, technical fields and disciplines of economics – MathWorks company [5].
ARENA	General simulation software for industrial applications and business process re-engineering, graphical animation system, principles of hierarchical modelling, designated for any industries – Rockwell Automation company [6].
PROMODEL	Simulation product for prediction in various fields and industries – industry, army, aviation, health service – Promodel Corporation [7].
SIMPROCESS	Simulation of company processes, it integrates process mapping, discrete simulation and activity – based costing – CACI International Inc. [8].
SIMSCRIPT	Programming language for simulation of projects; simulation and modelling of applications with interactive graphical user interface, presentations and 2-D animation of graphics – CACI International Inc. [8].
WITNESS 14, WITNESS Horizon	Set of tools for modelling and simulation of any company processes in various fields and industries – Lanner Group Ltd. [9].
SIMUL8	Simulation software which can be used by anyone for visualization, evaluation, and enhancement of processes; commercial simulation software –SIMUL8 Corporation [10].

Computer simulations can be used in numerous practical situations, for example at optimisation of complex production systems, including logistic ones, scheduling of production, forecasting of prices on financial markets, as process tools for engineering simulation, municipal simulation models and traffic engineering, simulation of behaviour of structures under various conditions, meteorological forecasts, project management, risk management, etc.

Use of simulation software in company practice has numerous advantages:

- it is possible to simulate without an intervention into real operation;
- it is also possible to experiment with situations which could not be tested in real conditions;
- simulation offers more comprehensive view of the resolved problem and leads to better understanding of the real system;
- simulation offers broader overview of company processes;
- it is possible to resolve even very complex systems that cannot be resolved by analytical methods;
- several different variants of solution can be verified relatively quickly;
- it is possible to investigate behaviour of the system in real, accelerated or decelerated time;
- it is possible to find or to verify real functioning of the system in contrast to suppositions and guesses of individuals. [1].

Difficult interpretation of the results of some simulation tasks is a disadvantage of the simulation, as well as the fact that they are often highly time consuming or expensive and require knowledge and experience of the workers with creation of models and knowledge of work with the simulation software.

Below, we will show an example of successful simulation in practice implemented with the use of the WITNESS simulation software. The WITNESS simulation tool is a tool used for simulating and optimising manufacturing, logistics and servicing systems. It is produced by Lanner Group Ltd. The advantage of the system is its use which is based on interactive simulation of systems of certain events introduced in the organization using physical elements depending on the timeline. [8]

3 Example of practical application

The company needed to design a number of Kanban circuits to shorten production lead time and reduce in-process inventory. [10]

The aim of the enterprise was the application of the Kanban system using a computer simulation in a particular case of automobile cooling system production which points out to a 90 % reduction in stock inventory of semi-finished products, helps to a fast transmission of information and increase of flexibility of all workplaces.

Analysis of the current state revealed the following facts:

- the company works in three-shift operation; one shift lasts 480 minutes;
- 1750 pcs of final products are produced daily (automobile cooling system components);
- components A and B are parts of the product;
- the company working breaks in total represent 45 minutes of working time per work;
- the time of one working cycle of the workplace includes the time necessary to operate the machine, move the finished product to the next operation and insert a new product into the machine;
- batches of 50 pcs are degreased (260 sec = 4. 3333 min);

- after degreasing, there are again 1750 pieces of the A and B manufacturing segments in each buffer. Subsequently, one after the other according to the type, they are taken from the buffers and inserted into the CNC machine;
- conveyor belts move at 5m/min;
- the assembly is the further step in the technological process. The machined input components A and B are transformed into the resulting products;
- current production time for 1750 pcs = 2486. 3 min.

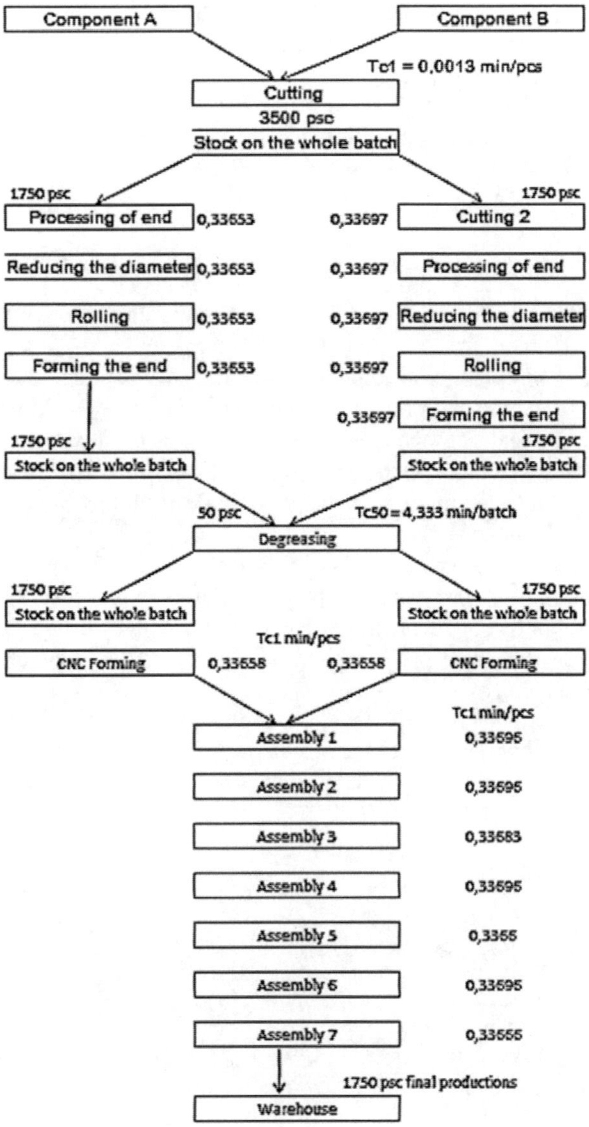

Fig. 1. Current state (scheme of the automobile cooling circuit production process)

The procedure for the simulation of the production model of the automobile cooling circuit was determined in the following steps:

- assembly of the original variant using simulation elements;
- determination of process times, including service times, among particular workplaces;
- defining the logistics flow of the material and its amount;
- assigning workers to single-purpose machines;
- starting the simulation run (simulation is done in minutes).

The simulation defined by the Witness software works on the principle of monitoring the used elements and their acting in the simulation model which is created by the basic used elements on the basis of a real manufacturing system.

Diagram of the model was plotted using the elements from the WITNESS library (Fig. 2).

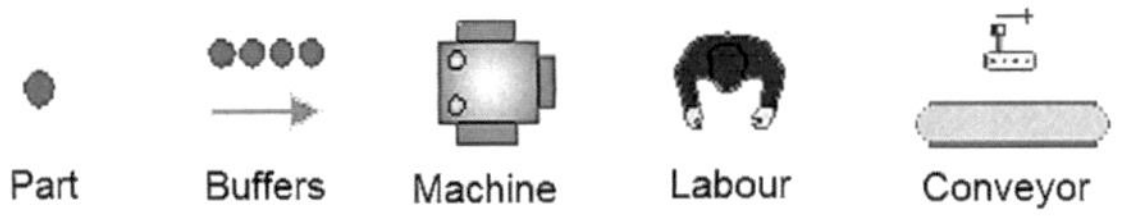

Fig. 2. Elements used for creation of the simulation model. [8]

- Element "Part"- represents the A and B components passing through the manufacturing model.
- Element "Buffers"- represents the containers or rather the storage places for unfinished production or warehouses with finally finished products. They are very important for the Kanban system application since they determine the size of inter-operational supplies.
- Element "Machine" – represents the single-purpose machines in the simulation which are used in the manufacturing process of the automobile cooling system. The machines are simulated using real parameters.
- Element "Labour", represents a worker who has the task of operating the production machine and storing the components in prepared buffers with semi-finished products.
- Element "Conveyor" - conveyor. In particular, it is a conveyor belt for transporting the material in the last stage of production, which involves the assembly of the A and B components in one unit.

Once all known parameters have been set, the model is run in the simulation program to find out the current total production time of the product which is 2486,3 min for the required 1750 pcs. Figure 3 shows the model of the current state of the automobile cooling system production.

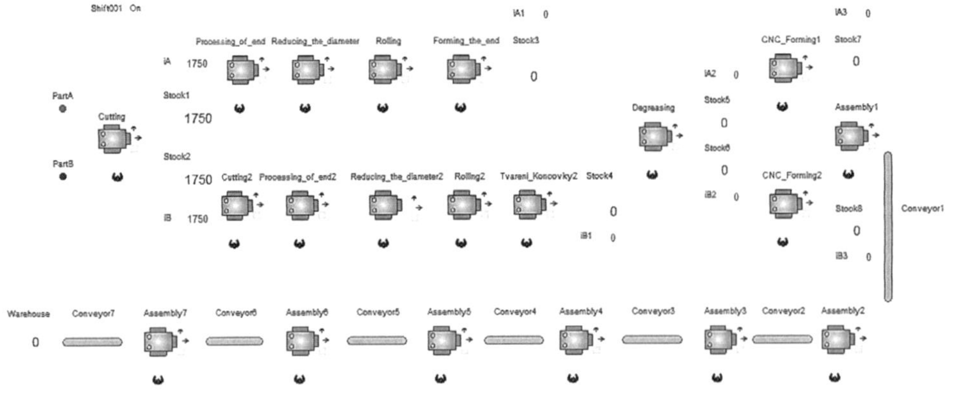

Fig. 3. Diagram of the current state of WITNESS

Results obtained by the simulation

As a part of the simulation, variants were created for forming the Kanban circuits with regard to the increase of the company efficiency in terms of production times and operational stocks which entail increased costs. The proposed variants of the Kanban circuits were elaborated in the Witness simulation program and subsequently, they were evaluated and compared with the model of the current state in the company. The results of particular variants are presented in Table 2.

Table 2. Simulation results of particular Kanban circuits

	Original state	Variant A	Variant B	Variant C
Graphic display of proposed Kanban circuits	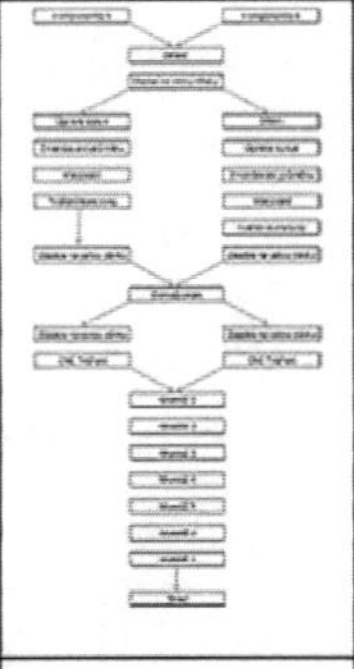	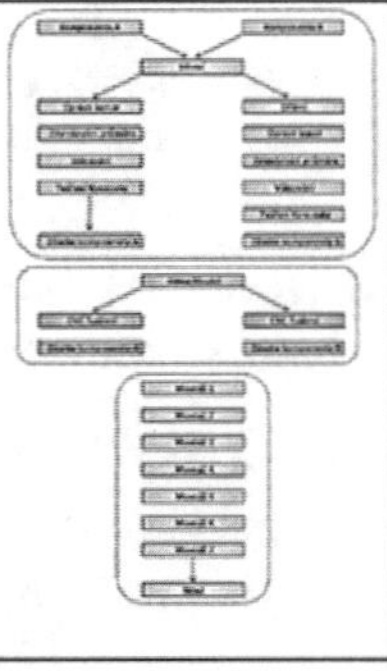	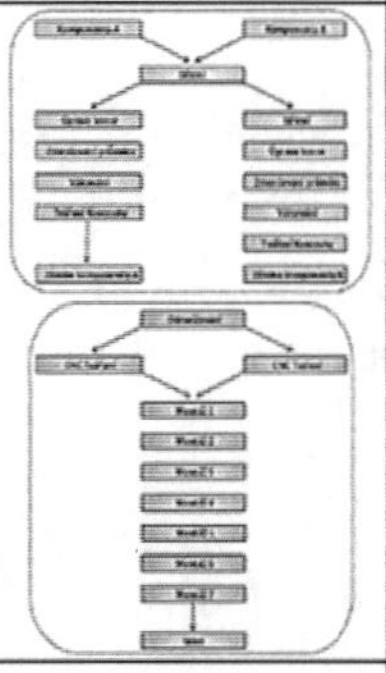	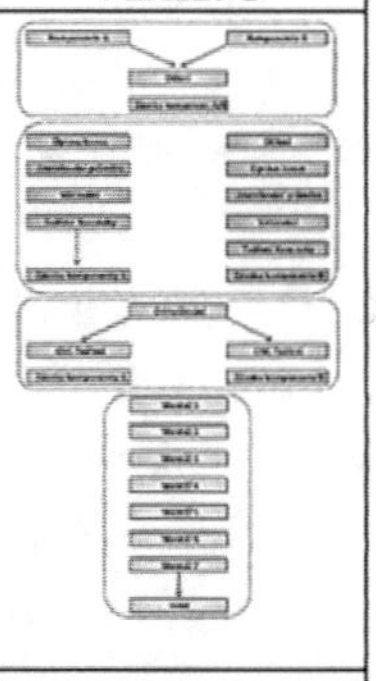
Number of Kanban circuits	0	3	2	4
Size of inter-operational supplies	3500 pcs	150 pcs	100 pcs	200 pcs
Production time for 1750 pcs	2486,3 min	889,2 min	903 min	889,2 min

The optimal choice of the most suitable variant always depends on what becomes the priority. In terms of the production time, variants A and C are comparable. In case of the size of inter-operational supplies, variant B seems to be optimal. The choice of the most suitable variant depends on the priorities of the manufacturing company.

4 Conclusion

The paper deals with the issues of the use of simulation of company processes in practice. Simulations are one of the possibilities how to verify the envisaged changes in the production process without any bigger risk, or how to set optimal conditions of the system. We simulated on a practical example from design a number of Kanban circuits to shorten production lead time and reduce in-process inventory.

Simulations include a huge advantage when compared with real systems, they enable experiments and have a significant degree of variability and repeatability. Dynamic simulations have a great potential as for the extent and possibilities of use. They can not only be used in the design and optimization of production processes but they can also offer an interesting solution for planning and production management due to their properties. Simulations are one of the ways to verify the proposed changes in the production process or to set optimal system conditions without any bigger risk.

The case study was focused on the simulation of KANBAN circuits for the production of components for automobile cooling systems. It has been proved that the practical use of the simulation model based on a real production process enables to verify the possibility of

KANBAN circuit settings and thus highly reliably plan the optimal use of production equipment and increase the throughput of the manufacturing system.

By using this model, which allows tracking the inputs as well as the degree of completion of individual products, the company can interactively use the proposed production planning model.

The model is designed to be able to correspond with computer outputs so that each product could be assigned to his technological processing, created in several input variants into the manufacturing section and so that the most suitable variant could be chosen.

References

1. R. E. Shannon, J.J. Swain, D. Goldsman, R.C. Crain, J. R. Wilson, *Introduction to Simulation*. Proceedings of the 1992 Winter Simulation Conference, Virginia USA, 65-73 (1992)
2. I. Mašín, M. Vytlačil, (1996). *Way to higher productivity: strategy based on industrial engineering*. Liberec: Institute of Industrial Engineering, (1996)
3. A.M. Law, *Simulation modeling and analysis*. 4th ed., New York: McGraw-Hill, 2007
4. V. Schindlerová, I. Šajdlerová, P. Mohyla, *Optimization of metallurgical processes using dynamic simulation*. METAL 2016, Conference Proceedings, TANGER Ltd, 2013-2018 (2016)
5. *MathWorks - MATLAB and Simulink for Technical Computing*, (© 1994-2018 The MathWorks, Inc.) Available: http://www.mathworks.com/
6. *Arena Simulation (Software by Rockwell Automation: Products*, (© 2018 Rockwell Automation) Available: http://www.arenasimulation.com/
7. *ProModel - Technology Enabled Predictive Analytics Simulation* (© 2018 ProModel Corporation) Available: http://www.promodel.com/
8. *CACI Profile (online)*. (© 2018 CACI International) Available: http://www.caci.com
9. *WITNESS: Witness*. (Witness © 2018). Available: http://witness.cz
10. *SIMUL8 Simulation Software - For Visual Process Simulation Modeling*. (© 2018 SIMUL8 Corporation.) Available: http://www.simul8.com
11. L. Žilt, *Record Keeping and Identification of Metallurgical Materials: Diploma Thesis*. VSB – TUO, Faculty of Mechanical Engineering, Department of Mechanical Technology, (2016)

Static analysis of transmission tower beam structure

Lenka Jakubovičová[1,*], *Alžbeta* Sapietová[1], *Jan* Moravec[1]

[1]University of Žilina, Faculty of Mechanical Engineering, Department of Applied Mechanics, Univerzitná 8215/1, Žilina, Slovakia

Abstract. The main topic of this paper is a static analysis of transmission tower beam structure fixed to the earth. It contains facts necessary for a need to create and apply tower beam structures. Beam elements were used to design a model and a static analysis was performed in finite element program Ansys Workbench. Three types of analyses were applied. The modal analysis was applied to verify the model compatibility. The static analysis was applied to find maximal deformations and the bending moment caused by overhead conductors and self-weight. Eigenvalue buckling was applied to state a loss of stability. Finally, stability and suitability of the whole construction was considered.

Keywords: modal analysis, Eigenvalue buckling, FEM, Ansys Workbench

1 Introduction

The historical development itself required an origin and a beginning of application of power line pylons for electric power transmission [1]. The first transmission of electricity over long distances was performed by the scientist Stephen Gray and his friend Granville Wheeler in the 18th. century. A distance of transmission was 765 feet, which is around 230 metres. In 1891 the first transmission network of three-phase alternating current took place during the international electricity exhibition in Frankfurt. Miller together with his colleague M.O. Dolivo-Dobrovolskij and other professionals were able to construct a network of the length of 110 miles (around 177 km) from town Lauffen to Frankfurt.

Current was transmitted by means of copper wires in a three-phase alternating current system with the efficiency around 75 %. At the beginning of the 20th century, first 110 kV and later 220kV facilities were constructed. Hand in hand with increasing needs for transmission of energy there was an increase in needs for static and mechanical stability of transmission structures.

Transmission structures have various constructions and application follows needs for their operation, placement in a ground, a type of tension and transmitted volume of electrical energy. Tower bodies usually have quadrangular structures with a constant or linearly varying cross section. In the submitted paper we analyze a standard high voltage transmission tower concreted in the base [2].

*Corresponding author: lenka.jakubovicova@fstroj.uniza.sk

Reviewers: *Wieslawa* Piekarska, *Robert* Grega

2 Finite element model

2.1 Cross Sections

A model of the high voltage transmission tower is a beam structure. Each beam is of a different profile (tab. 1, tab. 2). In general, in the structure there are used seven different profiles of the shape "L" (fig.1) and "U" (fig.2). A principle to design a model of the transmission tower consists in a fact that individual beams are presented as lines in the structure (fig. 4) and they form a beam structure together. Each line represents a concrete beam to which a profile is assigned and, thus, also properties belonging to the profile.

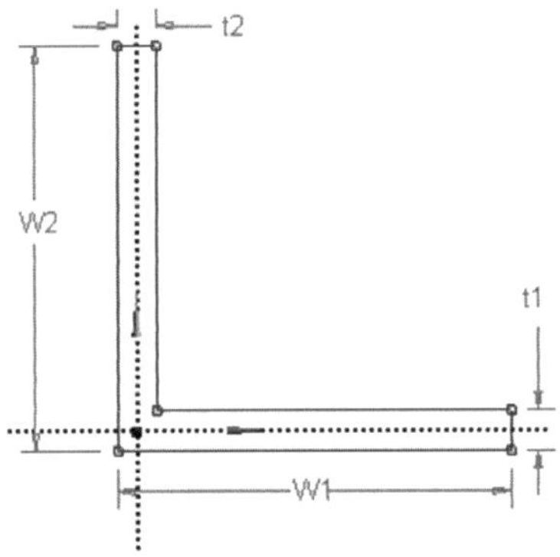

Fig. 1 Dimensions of L profile

Table 1 - Table of dimensions of applied L profiles

Profile	Dimension of a profile quote in [mm]	
	W1=W2	**t1=t2**
R150k15	150	15
R130k12	130	12
R120k10	120	10
U60k5	60	5
U50k4	50	4
U35k4	35	4

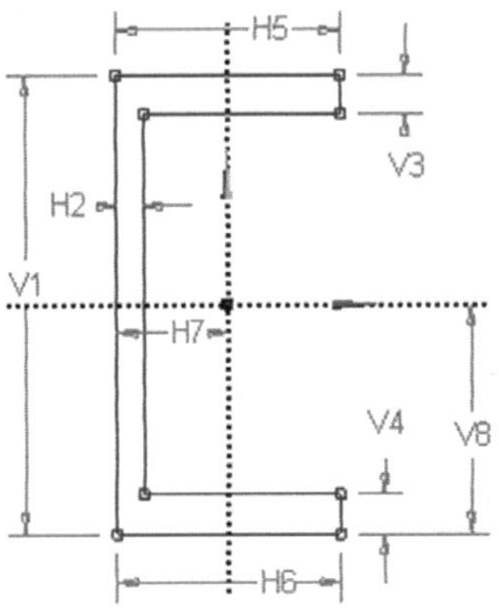

Fig. 2 Dimensions of profile U100

Table 2 Dimensions of profile U100

Profile	Dimension of a profile quote in [mm]							
	H2	**H5**	**H6**	**H7**	**V1**	**V3**	**V4**	**V8**
U100	6	50	50	25	100	8.5	8.5	50

2.2 Boundary conditions

Electric conductors are attached to consoles of the high voltage transmission tower and they act with their own weight in the point of attachment.

The load of the transmission tower will represent weight of 6 electric conductors attached to consoles together with self-weight of the whole construction. Points of attachment will be in the same heights and, thus, we may assume that we speak of a symmetric field from the both sides [3]. Towers will be spaced a =100m from the both sides. For power transmission will be applied AlFe wire marked 42/25 according to STN 024310 and its parameters together with other parameters necessary for load calculations were calculated according to the publication Mechanics of outdoor power lines[4].

The load calculation was considered according to a condition of the symmetric field in fig.3.

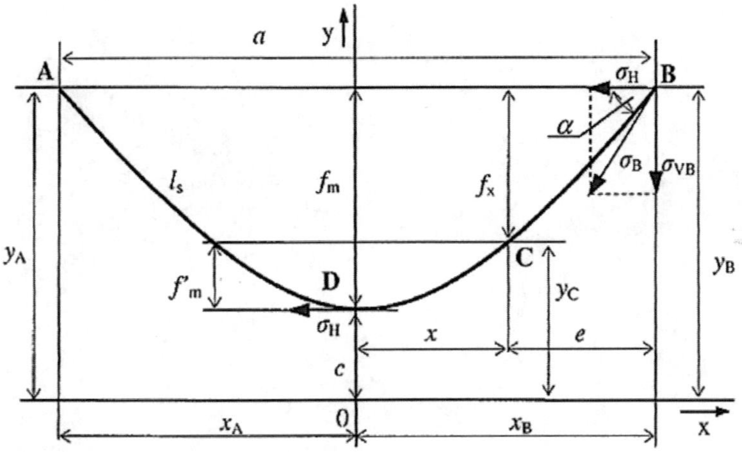

Fig. 3 Symmetric catenary

The parameter of parabola

$$c = \frac{\sigma_H}{\gamma \cdot z} = \frac{177.0315 \cdot 10^6}{0.046702 \cdot 10^6 \cdot 5.9} = 642.5\ m. \tag{1}$$

The maximum deflection of the catenary

$$f_m = c \cdot \left(\cosh\left(\frac{a_v}{2c}\right) - 1\right) = 642.5 \cdot \left(\cosh\left(\frac{100}{2 \cdot 642.5}\right) - 1\right) = 1.9465\ m. \tag{2}$$

Total mechanical stress in the suspension point B from one wire

$$\sigma_B = \sigma_H \cdot \cosh\left(\frac{a_v}{2 \cdot c}\right) = 177.0315 \cdot \cosh\left(\frac{100}{2 \cdot 642.5}\right) = 177.567\ MPa. \tag{3}$$

Total force in the suspension point B from one wire

$$F_B = \sigma_B \cdot S_l = 177.567 \cdot 10^6 \cdot 67.07 \cdot 10^{-6} = 11909.48576\ N. \tag{4}$$

A vertical force component in the suspension point B

$$F_{VB} = \sigma_{VB} \cdot S_l = 13.8 \cdot 10^6 \cdot 67.07 \cdot 10^{-6} = 925.566\,N. \tag{5}$$

Total force in the suspension point F_B = 11909.5 N will be applied as a force on 6 suspension points of geometry at the same angle. Three forces will be oriented positively and three negatively, fig. 5.

To obtain the correct orientation of powers it is necessary to distribute force into vector components in Ansys Workbench, fig. 4.

The angle at which the forces act:

$$\sin\alpha = \frac{F_{VB}}{F_B} = 0.07774198696 \;=>\; \alpha = 4.459°. \tag{6}$$

Force in the suspension point in the horizontal direction:

$$F_{HB} = F_{VB} \cdot \cos\alpha = 11869.564\,N. \tag{7}$$

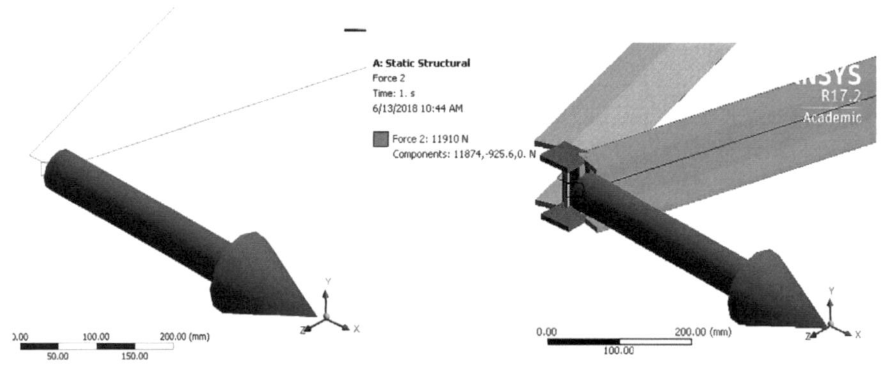

Fig. 4 Force components

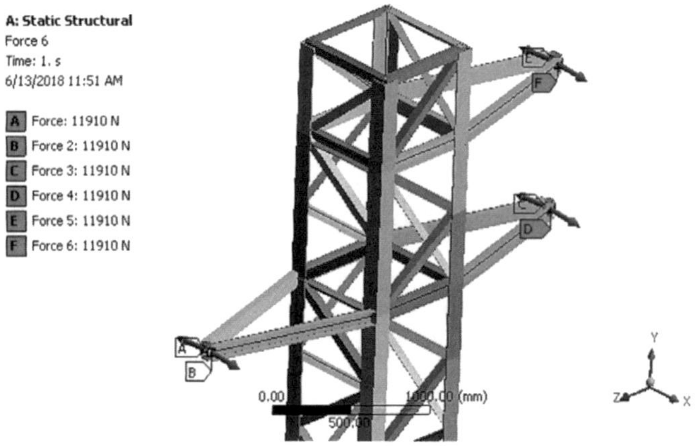

Fig. 5 Load by forces from wires

In figure 6 all degrees of freedom are removed from elements marked in the blue colour. That represents the condition of a concreted section of the transmission tower.

 https://doi.org/10.1051/matecconf/201824401011

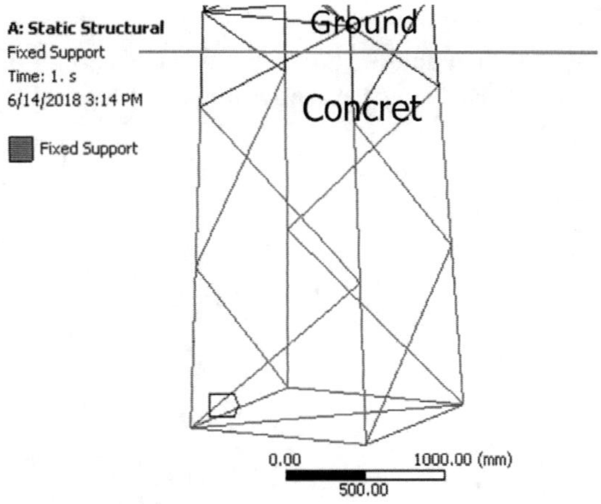

Fig. 6 Boundary condition "Fixed Support"

3 Analysis

The modal analysis was applied to control compatibility of the whole model. The first six own frequencies were equal or close to zero and that mean six degrees of freedom for a body [5, 6]. In a further analysis of the transmission tower we need to pay our attention to a static analysis and a buckling analysis, as we deal with a slim construction [7, 8].

3.1 Finite-element analysis

Using this analysis we are able to verify stiffness of construction affected by the load caused by suspension wires and standard gravity force. In figure 7 there is shown total deformation and figure 8 contains Total Bending Moment.

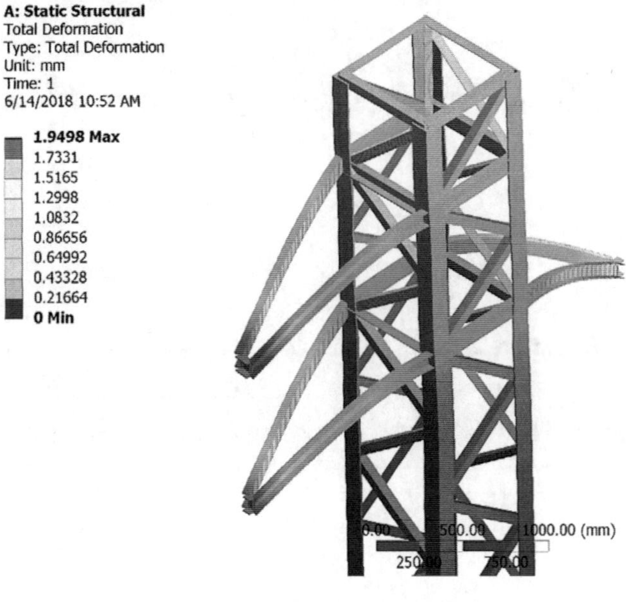

Fig. 7 Total deformation caused by suspension wires

The maximal value of total deformation according to figure 7 was measured as 1.95 mm. The value of the deformation points to a slight impact of load on a console holder and also on the whole construction.

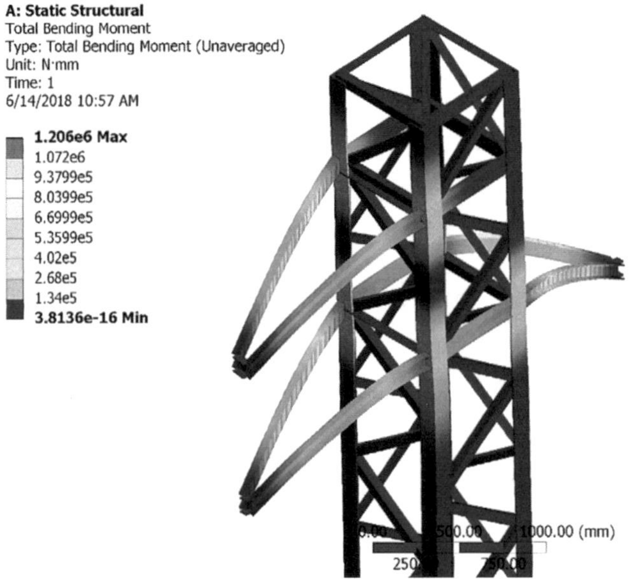

Fig. 8 Total bending moment

Stress in bending in the point with a maximum bending moment will be calculated according to the relation (8).

$$\sigma_{OMAX} = \frac{M_O}{W_O} = \frac{1206}{0.000041402287} = 29.13\ MPa. \qquad (8)$$

For steel 11503, the maximum allowable stress is σ_{DOV}=245 MPa.

Stress condition together with safety coefficient which includes also ice and icing on wires:

$$\sigma_{OMAX} \cdot k \leq \sigma_{DOV} \qquad (9)$$

$$29.13 \cdot 4 = 116.52\ MPa\ < 245\ MPa \Rightarrow\!\!\!\!> allowed. \qquad (10)$$

3.2 Eigenvalue Buckling

The third analysis will be an eigenvalue buckling analysis of the transmission tower, as it is a slim construction. This type of analysis provides us information regarding identification of maximum values of load in which a damage of the analyzed structure may occur under the influence of a stability loss [9, 10]. In Table 3 there are stated the lowest values for a stability loss for a given tower. The first value 7.2388 provides us the information that if we increase the amount of load 7.2388 times, the first case of a stability loss occurs. If we increase load by 11.875 times, the second case of stability occurs.

Table 3 Results of Eigenvalue Buckling analysis

A real load multiplier
7.2388
11.875

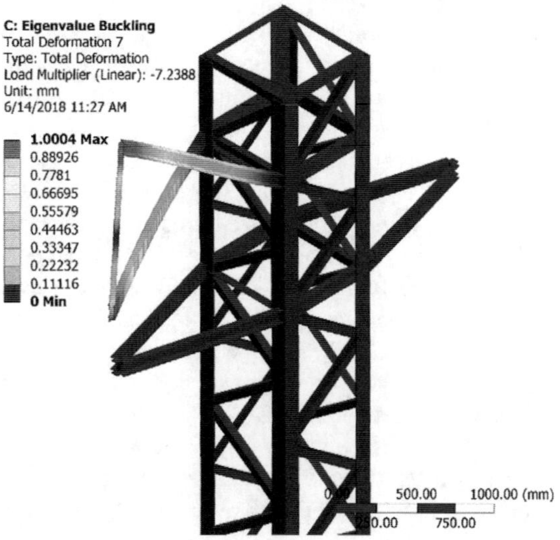

Fig. 9 Total deformation of the first value of a stability loss

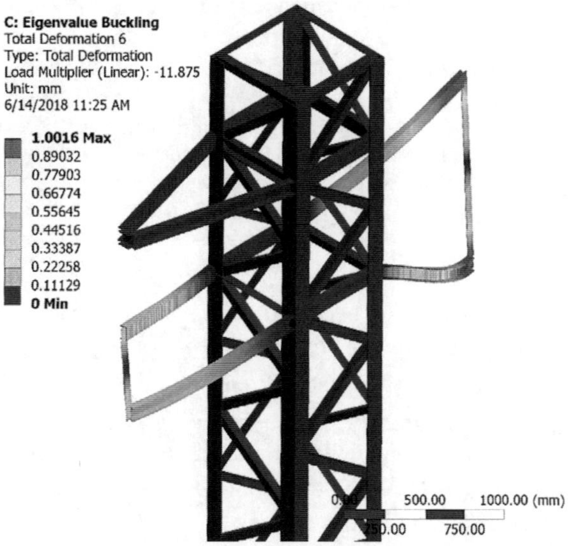

Fig. 10 Total deformation of the second value of a stability loss

Results of the analysis approve that the construction is stable enough by a given load by wires and self-weight. Load acting on the power tower must increase almost sevenfold to cause the first case of a stability loss. From measured results it is also possible to analyze the weakest element of the whole construction which is the console holding wires.

Summary

The aim of the work was to prepare a model and to perform a statistic analysis of the transmission tower beam structure that is fixed to the earth. We made three types of different analyses. The first one, a modal analysis, was performed after formation of geometry with the aim to verify a compatibility of the model. The second analysis was a static analysis. Its aim was to find out, respectively verify areas with the highest value of deformation and a bending moment to prevent a stability loss of the whole construction or an element of the power tower (a console for mounting of suspension wires). The third analysis, Eigenvalue buckling, was performed after a meshing and boundary conditions for the model with the aim to identify possible conditions of stability losses.

Finite-element mesh was created and boundary conditions defined. Concreting of the power tower up to the height of 2 metres was represented by a removal of all degrees of freedom to elements that were in concrete. To remove degrees of freedom we used the function "Fixed Support". A size and orientation of load under an influence of suspension wires were calculated according to a presumption of symmetric overhead conductors. The resultant force acting on the suspension point F_{VB}=11905.6 N was applied on 6 suspension points at the angle of α = 4.459°. The final deformation was measured at the value of 1.95 mm. A very low value of deformation proves the fact that a construction will manage mounting of conductors without any problems. Another evidence of the suitable construction is also a verification by means of strength conditions. The maximum stress is σ_{OMAX} = 29.13 MPa after multiplication by a safety coefficient k = 4, i.e., 29.13·4 = 116.52 MPa is still less that the allowed load σ_{DOV} = 245 MPa for steel used in a construction of transmission power towers.

Results obtained by the finite-element program Ansys Workbench can be considered as approximate. By multiplication by safety coefficient we obtain deformations and bending moments under critical values. An advantage of the process of a solution is its speed and flexibility. In case of a change of applied profiles, we do not need to define the whole geometry from the ground, we just need to allocate a new profile to a selected body.

This work was supported by the Slovak Research and Development Agency under the contract No. APVV-14-0096 and by grant KEGA 015ŽU-4/2017 - Digital technologies for machine building study programs as a part of strategy "Internet of Things".

References

1. Flizikowski, J., Macko, M.: *Method of estimation of efficiency of quasi-cutting of recycled opto-telecommunication pipes*. Polimery/Polymers, XLVI, 53-59 (2001)
2. K. J. Bathe, *Finite Element Procedures*. Prentice Hall, New Jersey (1982)
3. J. Kortis, L. Daniel, M. Handrik, *The Numerical Analysis of the Joint of the Steel Beam to the Timber Girder*. Procedia Engineering, **91**, 160-164, (2014)
4. A. Otčenášová, *Mechanika vonkajších silových vedení*. Žilina: University of Zilina - EDIS – publisher ŽU.. **236** (2010)
5. R. Grega, J. Krajňák, L. Žuľová, G. Fedorko, V. Molnár, *Failure analysis of driveshaft of truck body caused by vibrations*. Engineering Failure Analysis, **79**, 208-215 (2017)
6. A. Vasko, M. Vasko, J. Belan, E. Tillova, *Comparison of fatigue properties of nodular cast iron at low and high frequency cyclic loading*. Procedia Engineering, **177**, 576-581, (2017)

7. W. Piekarska, M. Kubiak, M. Žmindák, *Issues in numerical modeling of phase transformations in welded joint.* XXI Polish-Slovak Scientific conference machine modeling and simulations MMS 2016, Procedia Engineering, **177**, 141-148 (2017)
8. P. Kovacikova, R. Bezdedova, J. Jr.Vavro, J. Vavro, *Comparison of numerical analysis of stress-strain states of cast iron with vermicular graphite shape and globular graphite shape.* 20 th International conference machine modeling and simulations MMS 2015, Procedia Engineering, 136, 28-32 (2016)
9. M. Handrik, M. Vasko, P. Kopas, V. Mozer, *The linear and nonlinear stability loss of structures due to thermal load.* Procedia Engineering, **136**, 359-364, (2016),
10. L. Zuľova, R. Grega, J. Krajňák, G. Fedorko, V. Molnár, *Optimization of noisiness of mechanical system by using a pneumatic tuner during a failure of piston machine*, **79**, 845-851, (2017)

Problems of blackening of steel parts in technical practice

Sylvia Kuśmierczak[1*], *Martin* Makovský[1]

[1]FME, Institute of technology and Materials, Pasteurova 3334/7, 400 96, Usti nad Labem, Czech Republic

Abstract. Problems of degradation are mainly solved in finished products, in a number of cases it is manifested during production. A great influence on the degradation processes of the products has the surface quality and surface finish. The article deals with the problem of blackening of steel, which has been treated with alkaline blackening. The throttle shaft made of automatic steel will be analyzed. Degradation was manifested by a change in surface quality after alkaline blackening. The methods of light and electron microscopy in combination with complex analysis of production flow and technology conditions were used for analysis. Based on the results of analyzes it was found that the degradation processes are closely related to the quality of preparation of the surface. Owing to the incorrect pretreatment of the surface, oxides occurred locally. At the sites of oxides, after alkaline blackening, the surface layer was peeled off, resulting in deterioration in the quality of the product.

Keywords: Degradation, Blackening, Steel, EDX analysis

1 Introduction

Degradation is primarily known as a manifestation of environmental influences or degradation of material properties for already finished products. However, degradation may already occur in the component manufacturing process. Significant influence on degradation processes has the effect of surface quality and surface treatment.

The surface properties depend on the surface structure, the chemical, physical, mechanical and geometric properties of the surface layer [1-3]. The free surface structure forms the surface of a large set of grains, which forms the interface between the solid phase and the surroundings. A characteristic feature of the surface is a step change in the symmetry of forces acting on the atomic nuclei of the crystal lattice. Within the volume forces are compensated in different directions, while on surface they are not compensated [4,5].

Under chemical properties we mean changes in chemical composition or oxidation states at the interfacial interface. On the surface, in contact with the liquid medium, a combination of anodic oxidation (metal dissolution) and cathodic reduction (oxygen reduction or hydrogen excretion) can take place. The dissolution of the metals can take place in two ways.

* Corresponding author: sylvia.kusmierczak@ujep.cz

Reviewers: *Karol* Vasilko, *Dana* Stančeková

The first method is active dissolution where the surface is not coated with any protective film. The second option is passive dissolution, when insoluble compounds are formed on the surface [4].

Many methods are used to improve surface quality. They may be mechanical, chemical, electrochemical, thermal and chemical-thermal treatments. The purpose of the surface treatments is to obtain a higher corrosion resistance of the product, to achieve the desired appearance or to improve the mechanical properties of the surface. It is mainly used as a finishing operation [6-8].

All surface treatments, other than mechanical, are based on the formation of a layer on the base material that meets the required properties.

This article focuses on the blackening of steel parts. Alkaline blackening is a coating that is the most common way to dye steel products, and above all to ensure improved appearance of the part being treated. Its purpose, combined with preservatives, is also an increase in corrosion resistance. The result of alkaline blackening is the formation of a conversion oxide layer of about 1 μm thick. The principle of alkaline oxidation blackening consists in iron oxidation by immersion of ferrous metal products into a concentrated hot aqueous solution of a mixture of sodium hydroxide with a suitable oxidant and other ingredients added to the bath. These are inorganic or organic chemicals, called accelerators, inhibitors, wetting agents, complexion agents and others [9-11].

The process itself takes place in several steps (Fig. 1). First, degreasing parts and rinses will result in the removal of dirt from the surface of the component (dirt, oil residues after pre-treatment). If any residual grease remains on the surface of the component, stains will form on the surface. Subsequently, alkaline blackening is performed in the first bath, followed by cascade washing. In most cases, two-stage blackening is performed to ensure a uniform layer across the entire surface of the part, so the last two steps are done again. After the second blackening, the parts are rinsed with hot water to ensure that the scrubbing residue does not enter the preservation bath. After rinsing, the components are preserved to ensure corrosion resistance.

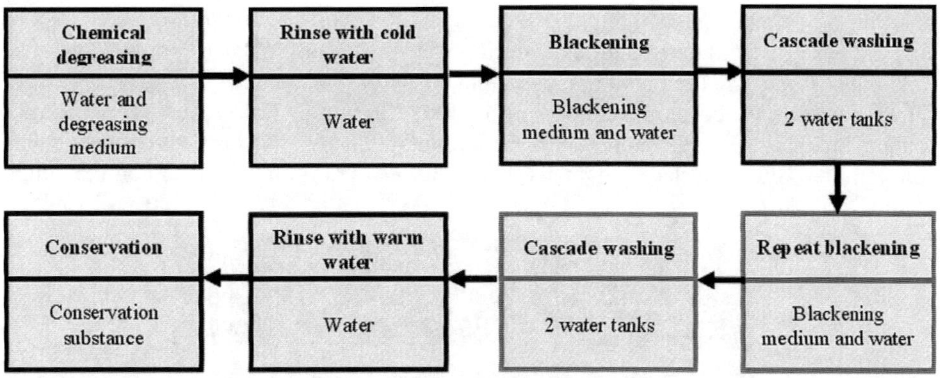

Fig. 1. Blackening process

The quality of the resulting surface layer depends on several factors. One of them is the quality of pre-fabrication of the component. It is very important that the products are well degreased and that there are no impurities on them to prevent the formation of the surface layer. Another factor is the temperature of the bathing bath. The calcination takes place at a boiling point of the sodium hydroxide solution, which is approximately 140 °C. At this temperature there is a considerable evaporation of water. By reducing the water content the concentration of the bath changes and thus the boiling point changes. Therefore, it is necessary to constantly monitor the temperature of the scrubbing bath and to check the quality of the preparation, e.g. adding water or adding individual components of the scrubbing agent (NaOH and oxidizers) [10].

As with any process, even with alkaline blackening, defects occur. Some can be repaired, for example, by re-blacking, but some can be not been repaired. It may be small black layer thickness, uneven layer thickness, brown spots on the surface or corrosion.

The aim of this paper is to analyze the causes of the specific defects that arise on surface-treated components and this is alkaline blackening. The component being analyzed will be a throttle shaft that affects the amount of intake air into the engine.

2 Description of the problem

The parts were made of standard automatic steel 11SMnPb30 according to CSN EN 10027-1. For the production of these components are used chip machining technologies mainly turning, milling and grinding. Part of these products are further processed by surface treatment - alkaline blackening. Each of the technologies used, the manual operations carried out between the various machine operations (e.g. dimension checks), not least storage, bring a number of risk factors. These factors are reflected in defective products. These may be defects in size, shape, appearance, or corrosion. The defect, which is dealt with in this article is manifested by the occurrence of colour spots on the surface of the component and its cavities. There are more manifestations of this defect. The first signs are colour spots on the cylindrical surface or in the threaded hole. Another manifestation is the presence of impurities or sediments in holes components.

Factors that affect quality can be divided into internal and external. Among the internal factors we can include the chemical composition of the material (content of harmful, alloying elements), the material structure (grain size, pores size), inclusions (type, distribution, size and shape), internal stresses in the entire crossection and surface layers, material (roughness, surface hardening, etc.). Among the external factors we can include the influence of environment (temperature, humidity during storage, transport and production), used technology (type, used parameters), used tool (material, geometry, quality, wear), cutting fluid (composition, age, temperature) Washing and drying (temperature, bath concentration, soaking time, drying temperature, drying time), blackening (grease residues on components, concentration of dirt, surface roughness, blackout, bath temperature), manipulation, human factor.

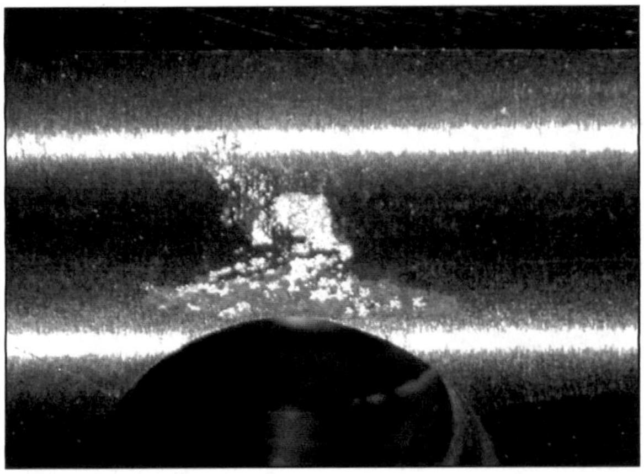

Fig. 2. Stains above the hole

Fig. 3. Deposits in the hole

3 Analysis of causes of part degradation

Stain and sediment analysis was performed visually and using the Tescan Vega 3 electron microscope.

Stains appeared near the flap holes. These are irregular shapes that have a partially white (Fig. 4a) and a partially red colour (Fig. 4b).

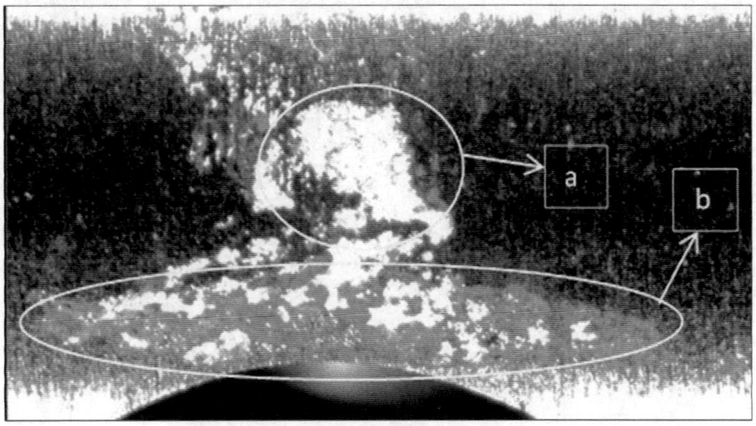

Fig. 4. Stain at hole.

From an electron microscope observation, it is clear that on the surface of the material the oxide is present (Figure 5). This shows on the presence of oxides on the surface of the component before the blackening process. Oxides on the surface were probably due to improper pre-treatment of the surface before the blackening process.

Due to the presence of oxides on the surface of the component, the protective layer, formed during the blackening process, has not been bonded to the base material. Subsequently, the black layer was peeled off. The surface of the material is highly cracked after peeling.

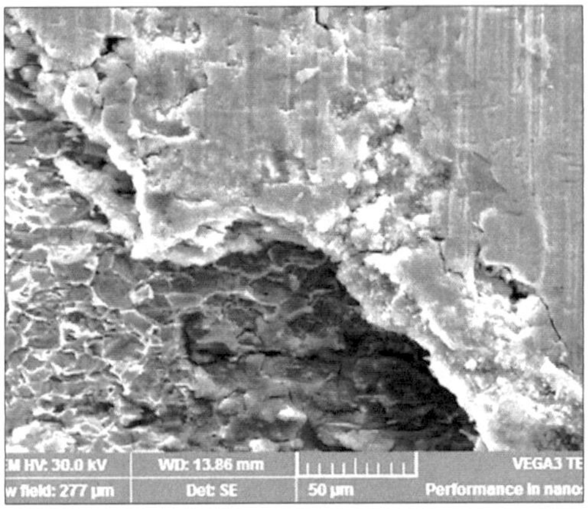

Fig. 5. Layer of oxides, REM

A surface was conducted EDX analysis to determine the orientation chemical composition in and outside areas the presence of oxides. The results (Tab. 1) show that in the areas where the surface layer has peeled, the oxygen content is much lower. In addition, there are other differences in the chemical composition of selected analysed areas.

The greatest differences between the results of surface EDX analysis in the non-oxide region (Figure 6a) and the oxide regions (Fig. 6b) are in Fe, O and C. The increased oxygen content is due to the formation of iron oxides on the surface of the component. Increased content C (approx. 4 x) may be due to imperfect degreasing. Interesting is also Na content. Although blackening occurs in sodium hydroxide solution, and the rinses are carried out in sodium hydroxide solutions, no Na content was found.

Table 1. Results of area EDX analysis [wt%]

Area	Fe	C	Mn	S	Mo	Pb	Bi	Si	O	Cu	Na	Ca
a	74.86	13.76	13.49	0.09	6.52	0.23	0.47	0.12	0.00	0.03	0.00	0.02
b	34.45	15.50	17.85	0.28	4.31	0.99	1.41	0.00	34.57	0.11	0.00	0.02

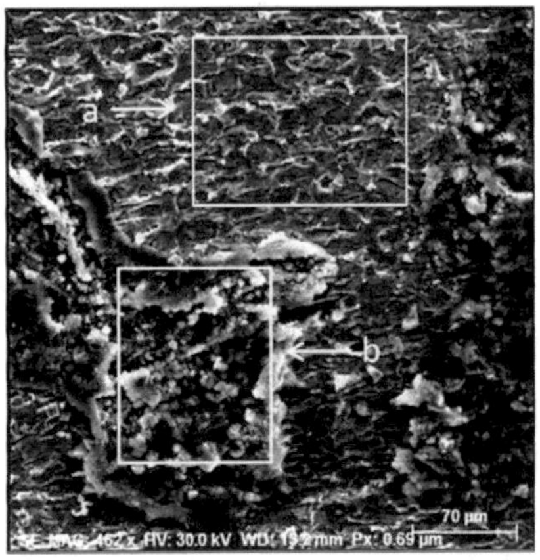

Fig. 6. Analyzed areas, area EDX analysis

In addition, analyzes were carried out to determine the causes of the formation of sediments (Fig. 3) in the holes of the part. It was supposed to be the sediment of the foam produced during the blackening process. Incorrectly set conditions in the scrubbing bath can result in foam formation. This foam was then blown into the holes in the component and did not eliminate the washing of parts. Subsequently, in the drying process, the foam crystallized (Fig. 7). This assumption confirmed electron microscope observation. Detailed observations show a crystalline sediment structure. Furthermore, it can be seen that crystallization proceeded at different speeds (Fig. 8).

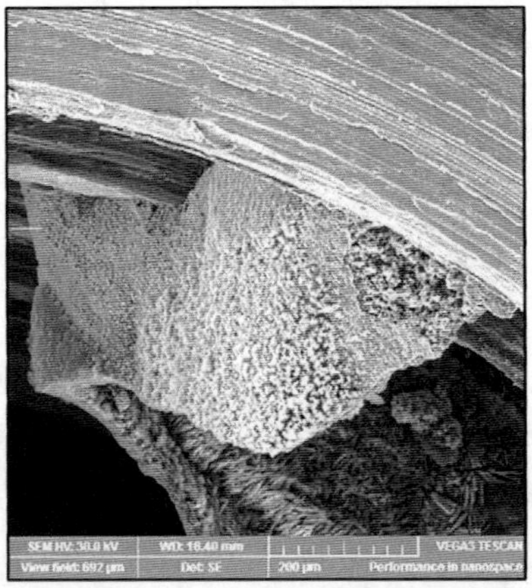

Fig. 7. Sediments in the hole, REM

Fig. 8. Sediments, detail, REM

Surface EDX sediment analysis also confirmed that it is a scouring bath residue (Tab. 2). It is predominantly a mixture based on iron oxides.

Table 2. Results of area EDX analysis [wt%]

Sample	Fe	C	Mn	S	Mo	Pb	Bi	Si	O	Cu	Na	Ca
Sediments	25.94	9.35	4.08	0.00	3.70	0.11	0.20	0.00	56.22	0.00	0.20	0.00

3 Conclusions

The aim of the paper was to show the problem of blackening of steel parts in technical practice, which has to solve the occurrence of defects on the components, which have been treated with alkaline blackening. Specifically, it was a throttle shaft, where it was important to analyze the cause of the defects and remove this cause.

On the basis of an assessment of internal and external factors, it was assumed that degradation could in the first case have been caused by incorrect pre-treatment of the surface before the blackening process. In the second case the suspected impurities in the holes of the parts.

Based on the results of electron microscopy, it can be stated that in the first case the products were not sufficiently degreased and subsequently passivated. This has resulted in the formation of oxides during storage and transport. Oxides at the surface of the component caused a weakening of the adhesion of the subsequent surface treatment. Due to the reduced adhesion, the surface layer was peeled off. One way to reduce this defect is by introducing the pickling of the components before the blackening process. In the second case the cause is the uncorrected prepared blackening process. By blackening form the foam on the surface of the tank. This foam was then blown into the holes in the components. The washing process did not remove it and the foam crystallized.

This paper confirms that in the surface treatment technologies, which consist of several stages, it is important to follow the prescribed conditions and procedures for each of the intermediate steps. This is the only way to assure the required quality and durability of the manufactured part.

Authors are grateful for the support of grant No. CZ.1.05/4.1.00/11.0260 EDIMARE and SGS FME JEPU in Ústí nad Labem for 2018.

References

1. D. Stančeková, M. Šajgalík, J. Petru, N. Náprstková, P. Balaz, *Implementation of coating for failure elimination of dial gauges.* METAL 2015, **24**, (2015)
2. V. Kucera, D. Vojtech, *Influence of the heat treatment on corrosion behaviour and mechanical proper-ties of the AA 7075 alloy.* Manufacturing Technology, **17** (5), 747-752 (2017)
3. N. Naprstkova, J. Cais, P. Kraus, M. Tavodova, *Using of electron microscopy for tool wear evaluation.* Engineering for Rural Development, **16**, pp. 647-652 (2017)
4. V. Kraus, *Povrchy a jejich úpravy*, Plzeň, (2009)
5. M. Muller, *Evaluation of single-lap adhesive bond quality by means of electron microscopy methods.* Manufacturing Technology, **16** (5), 1056-1062 (2016)
6. C. H. Simmons, N. Phelps, *Chapter 27 - Surface Finish and Corrosion of Metals.* Manual of Engineering Drawing. Oxford, 229-235, (2012)
7. B. Kratochvíl, V. Švorčík, D. Vojtěch, *Úvod do studia materiálů*, Praha (2005)
8. K. Raz, J. Hora, P. Pavlata, *Unconventional materials usage in design of vehicle bodies.* Manufacturing Technology, **17** (5), 823-827 (2017)
9. R. W. Farrell, *Blackening of ferrous metals.* Metal Finishing, **105**, 10, 390-396, (2007)
10. J. Kudláček, *Alkalické černění oceli*, MM Průmyslové spektrum, **66**, (2008)
11. Z. Kolska et col. *Surface treatment of materials for variable applications and surface properties and characterization* Manufacturing Technology, **16** (5), 949-955, (2016)

Possibilities of using interlayers during diffusion welding of Ti Gr2 and AISI 316L

Jaromir Moravec[1,*], *Iva* Novakova[1], *Tomasz* Kik[2]

[1]Technical University of Liberec. Studentská 2, 461 17 Liberec 1. Czech Republic
[2]Silesian University of Technology, Welding Department, Konarskiego 18A, Gliwice, Poland

Abstract. Joining of materials with different physical and mechanical properties is always very problematic. As one of the possibilities, there can be used method of joint creation in solid state – e.g. by the diffusion welding. With this method it is possible to joint very different materials – even metals with non-metals. Despite the unquestionable advantages of this method, there is necessary to take into account also its limits. Among of them, there is creation of brittle intermetallic phases on the boundary of materials. This occurs e.g. at creation heterogeneous joints between titanium and highly-alloyed austenitic steel. The major aim of this paper is to describe how it is possible to eliminate brittle intermetallic phases created on boundary between Ti Grade 2 and AISI 316L steel by using metal interlayers.

Keywords: Diffusion Welding, Interlayers, Titanium Grade 2, AISI 316L steel, Heterogeneous Welds

1 Introduction

Diffusion welding is fairly quite a new technology in the research field. It has numerous applications in distinct areas such as aerospace, nuclear, automotive, electronics and sensor industries. This has led to increasing attention for the researcher to develop new and advanced materials.

Diffusion, as every transport effect, can be described by two aspects. Phenomenological diffusion theory deals with the overall balance of transport process in solid substance undertaken in the given conditions. This theory takes into account only macroscopic perspective. Microscopic perspective is considered in the atomic diffusion theory. This theory respects internal composition of material including crystal lattice interferences. In the case of non-existing motive power on atoms, their jumps are random. However these jumps occur in specific crystallographic equivalent directions.

The diffusion process can be divided into two types: self-diffusion and hetero-diffusion. Self-diffusion takes place in pure metal. During self-diffusion, atoms migrate randomly throughout the crystal lattice. Due to the random movement of atoms in the grid, concentration gradient arises in crystal lattice but mass of pure metal isn´t changed. Concentration gradient can be expressed by Fick's laws when the Fick's first law (eq. 1)

* Corresponding author: jaromir.moravec@tul.cz

Reviewers: *August* Sládek, *Ivo* Hlavatý

indicates a variation of concentration in the x-direction per unit of time. The diffusion flux J [$mol \cdot m^{-2} \cdot s^{-1}$] of atoms per unit time and unit area in the direction of the x-axis is proportional to the concentration gradient. Mathematically, it is expressed as:

$$J = -D \cdot \left(\frac{\partial C}{\partial x} \right) \tag{1}$$

Where D [$m^2 \cdot s^{-1}$] is diffusion coefficient, C [$mol \cdot m^{-3}$] is concentration of elements, x [m] is diffusion direction and t [s] is time. In most practical cases, diffusion processes are non-stationary ones, because diffusion processes are changing over time. It can be deduced as relation called Fick's second law (eq. 2). In mathematical form, it is represented as:

$$\frac{\partial C}{\partial t} = \frac{\partial}{\partial x} \left(D \cdot \frac{\partial C}{\partial x} \right) = D \cdot \frac{\partial^2 C}{\partial x^2} \tag{2}$$

Hetero-diffusion occurs between two phases of the material. It is more complex phenomenon and possible only if atom has sufficient amount of energy to displace from one position to another one. When the atom moves in the grid, it creates a hole or vacancy in its node.[1,2]

2 Possibilities to joint Ti Gr2 and AISI 316L without interlayers

There are several methods how to create heterogeneous joints between titanium and highly-alloyed stainless steel. As the 1st possibility there is fusion welding, because both materials have similar magnitudes of melting temperatures. For own jointing can be used e.g. TIG method (however with precise shielding of weld pool), eventually electron-beam welding. However, in both cases takes place immixture of both materials at creation different intermetallic compounds (IMCs) in the weld metal. Mostly there are phases as TiFe and $TiFe_2$ that are typically of their high hardness and brittleness [3]. So as the aim of the welding process there is tendency to apply processes, which can eliminate quantity of areas, where are these IMCs created. Diffusion welding seems to be quite suitable possibility welding technology, because area of mutual diffusion of elements varies here just within the few tens of micrometers.

Nevertheless, even there is very narrow diffusion area, IMCs are also created within it and hence there is important increase of hardness values as can be seen from Fig. 1 (measurement of hardness HV0.02). Due to that, ductility of created joint is very low. So as a solution, there is utilization of interlayers from elements, which don´t create hard and brittle IMCs with titanium. Their purpose is to create with both materials diffusion joint, but also to prevent mutual diffusion of Ti and Fe at the same time.

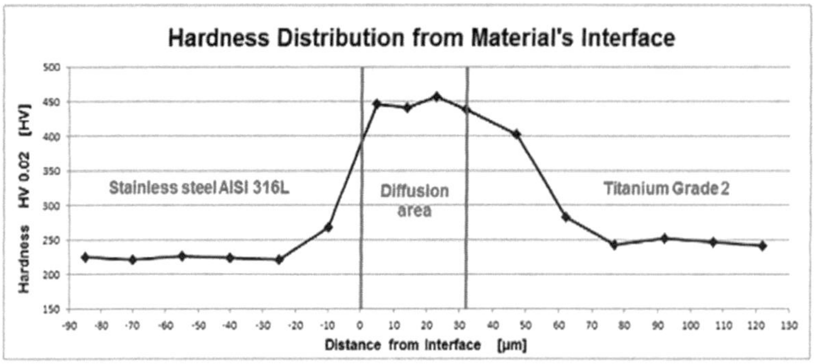

Fig. 1 Measurement of micro-hardness HV 0.02 over the diffusion area [3]

3 Possibilities to joint Ti Gr2 and AISI 316L by the interlayers

Based upon the performed recherché there was determined that for diffusion bonding of titanium and highly alloyed austenitic steels it is possible to use Ni, Ag or eventually Cu interlayer. Together with titanium, these elements create compounds that are tough enough and thus suitable to remove abnormal brittleness in the joint area. In all cases monitored in recherché was for the diffusion welding used thermal-mechanical simulator Gleeble and that is why the same device was also used for the experimental verification of the individual interlayers properties.

As the first interlayer used for the bonding there was used the Nickel one. From the research paper of Chatterjee and Kundu [4], it is investigated that the diffusion bonding between commercially pure Ti Gr 2 and AISI 304 steel with the help of Ni interlayer reveals good result on final welded joint in view of ductility. They used bonding temperature within range from 850 up to 950°C, but applied pressure wasn´t mentioned in the article. Also Fang-Li Wang [5] and his colleagues used diffusion bonding between commercially pure Ti Gr 2 and AISI 304 steel with the help of Ni interlayer. They used impulse pressuring in range from 8 up to 20 MPa with duration of 60 up to 150 sec at temperature 850°C.

During the experimental verification of Ni interlayer application suitability was determined that maximal available pressure at temperature 870°C is 3.2 MPa. Then there is very rapid deformation on the side of pure titanium, which is unacceptable from the industrial utilization point of view. At temperature 900°C is maximal available pressure lower than 1 MPa and at temperature 850°C must be pressure below 6.5 MPa.

At diffusion welding in vacuum and using device Gleeble 3500 there was determined that at temperature 870°C and pressure 3 MPa is not bonding between Ni interlayer on one side and Ti + AISI 316L on the other side sufficient enough - even after keeping holding time 50 min at used temperature. Magnitude of deformation for these samples after 50 min achieved only 0.182 mm. Moreover, there was experimentally determined that diffusion bonding of sufficient quality with Ni interlayer is created not below temperature 1050°C, pressure 4.5 MPa and holding time 20 min. Nevertheless, this was possible to be verified only for steel AISI 316L, which can resist such temperature and pressure without any problems. The total deformation of sample under these parameters didn´t exceed 0.2 mm.

By utilization of the impulse pressuring diffusion welding acc. to Fang-Li Wang [5] parameters it is also possible to achieve diffusion bonding of high quality. However, it is necessary to take into account deformation over 15% on the side of pure titanium, which is more than impracticable from the light of industrial utilization.

As the second interlayer used for the bonding Ti Gr2 and AISI 316L there was used the Ag interlayer. From the research paper of Yongqiang Deng and his colleagues [6] it is reported that using of Ag interlayer exhibits good result in light of micro-hardness distribution of bonded joint. Yongqiang Deng used device Gleeble and experiments were performed at temperatures 825, 850, 870°C and pressure 6 MPa. At experimental verification of Ag interlayer application suitability, there was carried out experiment at temperature 850°C, because at temperature 875°C and pressure 6 MPa would be again presented a very intense deformation of Ti Gr2.

Fig. 2 reveals SEM image with EDS scanning line of sample. Several distinct layers from left to right side (Ti/Ti s.s/TiAg/Ag/SS) were observed. Titanium solid solution (abb. Ti s.s) 19.55 μm, Ti-Ag IMC (TiAg) 28.33 μm and Ag interlayer 82.62 μm thicknesses were measured in the joint interface. From EDS analysis is evident that during holding time on temperature there occurs very intense diffusion between Ti and Ag. However, on the side of AISI 316L is diffusion minimal. Yongqiang Deng states that acc. to Ag-Fe/Cr/Ni binary diagrams [7], Ag cannot form IMCs with the main elements in SS. That is true, but there should be also state that solubility of Ag in Fe, Cr and Ni in solid state is almost on zero value. So it means that diffusion processes probably will not take place, which is also confirmed by result of EDS analysis. Acc. to binary diagram Ag-Fe is solubility of Ag in iron at temperature 850°C lower than 0.01 wt%. Thus the bonding between surfaces is created only due to adhesion.

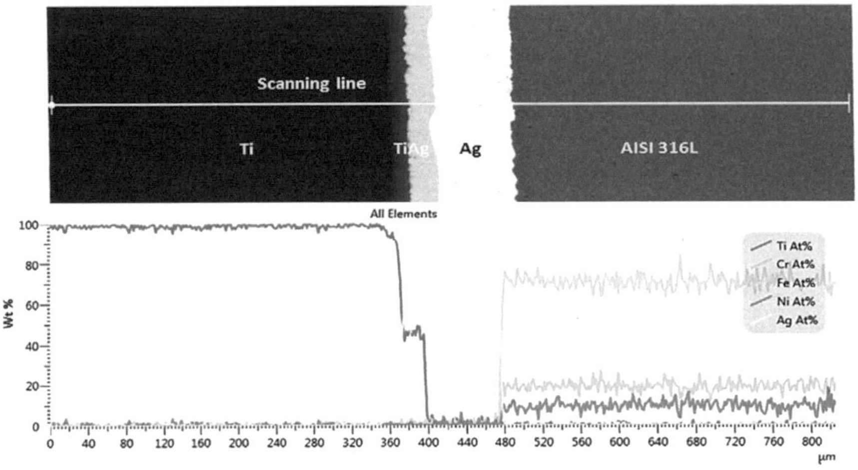

Fig. 2 SEM image and corresponding EDS scanning line of results for the bonded joint with Ag interlayer (Gleeble - 850°C; 6 MPa; 20 min)

To verify such presumption, there was performed experiment where samples Ti-Ag-SS were firstly diffusively bonded at device Gleegle 3500 at temperature 850°C, pressure 6 MPa and holding time on temperature 20 min. After cooling and unloading from Gleeble, samples were inserted into vacuum furnace for holding time of 10 hrs at temperature 850°C. The aim of experiment was to verify, whether there would be at least partial diffusion of Ag into material AISI 316L. Moreover, there was effort to determine diffusion rate of Ag into titanium. Results of experiments are shown in Fig. 3. It is obvious that high holding time at temperature 850°C resulted in fractures on the TiAg-Ag interface. They are (despite the high ductility of Ag interlayer) probably caused by differences between thermal expansion coefficients for titanium and silver. Furthermore, there was proved that even under 10-times extension of holding time, there isn´t diffusion of Ag interlayer into steel AISI 316L. Thus there´s no need to find or optimize diffusion welding process parameters for Ag interlayer.

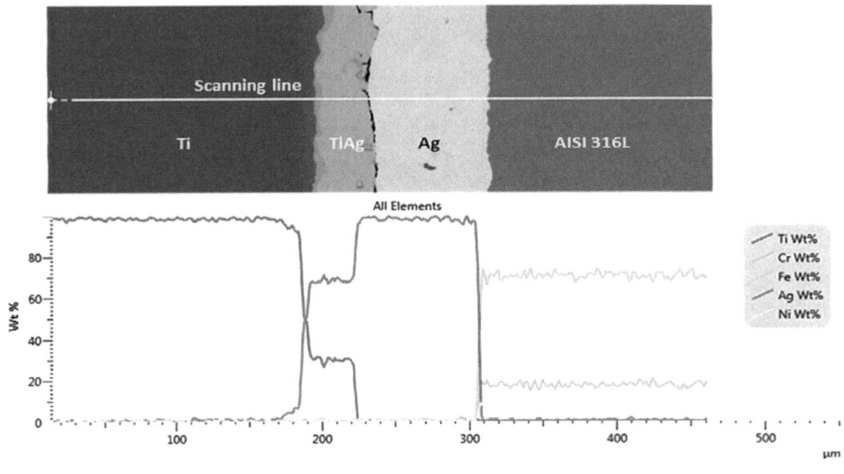

Fig. 3 SEM image with EDS scanning line scanning of results for the bonded joint with Ag interlayer (Gleeble - 850°C; 6 MPa; 20 min) and 10 hrs in vacuum furnace at temperature 850°C

 https://doi.org/10.1051/matecconf/201824401013

As another recommended type of interlayer for combination Ti-SS there is Cu interlayer. This problems is solved e.g. by S. Kundu and his colleagues at bonding the commercially pure Ti Gr 2 and AISI 304 steel. Cu interlayer reveals good result on final welded joint in view of ductility [8]. Therefore, it is anticipated that Cu can be estimated as a potential candidate to be used as interlayer to improve the joint ductility. S. Kundu performed the diffusion welding with Cu interlayer for temperatures 850 up to 950°C, pressure 3 MPa and holding time 90 min. Based upon the experiments, he states that at temperatures 900°C and higher, Cu interlayer can´t stop diffusion of Fe, Cr and Ni into Ti and reversely Ti into SS. That is why there are created brittle IMC´s in diffusion area, which have a crucial influence on increase of hardness and brittleness in the joint area. Because of that, there was decided to carry out experiment with Cu interlayer at temperature 850°C, pressure 3 MPa and holding times 30, 60 and 90 min.

There was only a partial diffusion and thus insufficient joint strength at temperature 850°C and holding time 30 min. This result is shown in Fig. 4. There was joint failure in the Cu interlayer, when part of interlayer remained on the SS part (Fig. 4 – left) and part on the titanium part.

Fig. 4 Failure of Cu interlayer in sample SS-Cu-Ti after diffusion welding (850°C; 3 MPa; 30 min)

As it can be seen from Fig. 5 (acquired by the optical microscope Olympus DSX 500) there was not achieve full diffusion bonding even after extension of holding time to 90 min. Nevertheless, from the measurement of micro-hardness HV0.1 it is clear that in the weld joint are not created hard and brittle IMC´s and total hardness HV0.1 didn´t exceed magnitude 265 HV. To create full diffusion welding through the whole cross-section, there is necessary to extend diffusion time or increase diffusion activation energy by higher temperature. Based upon these facts, there was decided to apply another holding time at temperature 850°C in the vacuum furnace for sample diffusively welded by device Gleeble (850°C, 3 MPa, 90 min). To be specific, holding time at temperature 850°C was now 5 hrs.

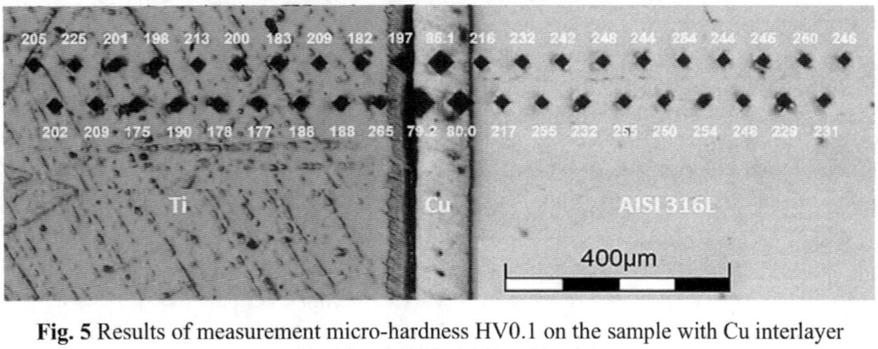

Fig. 5 Results of measurement micro-hardness HV0.1 on the sample with Cu interlayer (Gleeble - 850°C; 3 MPa; 90 min)

From the result shown in Fig. 6, it is obvious that total diffusion time 6.5 hrs at temperature 850°C is sufficient to create quality diffusion bonding through the whole cross-section both on the side of Ti and on the side of SS. However, used diffusion time was too long, because there is diffusion of Ti through the whole thickness of Cu interlayer 0.125 mm into SS and simultaneously there is diffusion of Fe, Cr and Ni into Ti. Due to this fact there were created brittle IMC's, which not only increased micro-hardness HV0.1 up to 418 HV (see Fig. 7), but also caused fractures creation on the boundary with steel AISI 316L.

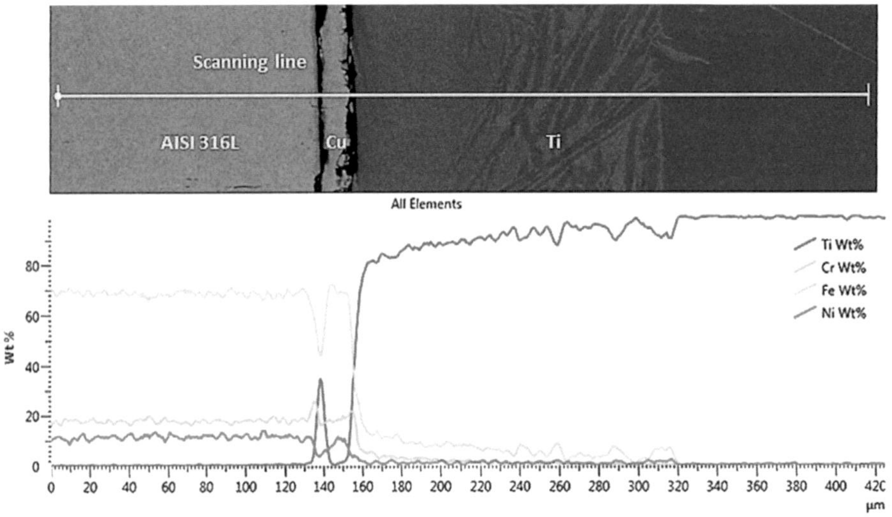

Fig. 6 SEM image and relevant EDS scanning line of results for the bonded joint with Cu interlayer (Gleeble - 850°C; 3 MPa; 20 min, vacuum furnace 850°C; 300 min)

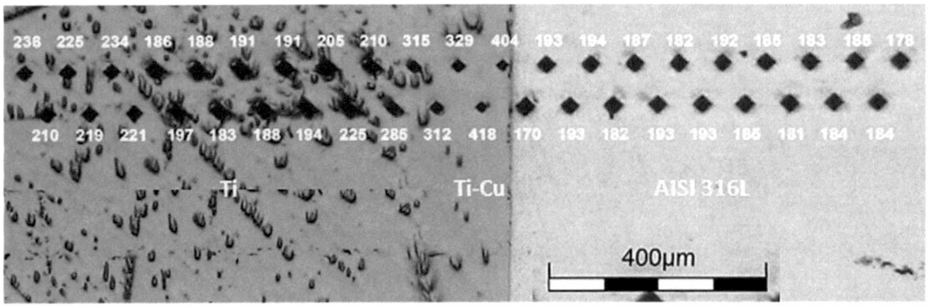

Fig. 7 Results of measurement micro-hardness HV0.1 on the sample with Cu interlayer (Gleeble - 850°C; 3 MPa; 20 min, vacuum furnace 850°C; 300 min)

4 Conclusion

Diffusion welding is quite progressive technology at production heterogeneous welds created in solid state. On the other hand, this technology has also its limitations - mainly at welding materials, which create brittle IMCs on the materials boundary during diffusion. As example of such joint there is joint Ti-Fe – eventually with other elements as Cr, Al, etc.

One of the possibilities how to eliminate joint brittleness at diffusion welding there is usage of the metal interlayers. These interlayers have several objectives. 1st of them is to create transient layer, which prevent connection of elements that create hard and brittle IMCs during diffusion. Therewithal it should be valid for these interlayers that they have at least partial solubility in solid state with the welding materials and also that they do not create own brittle IMCs with these materials and their major alloying elements.

In the paper were described possibilities to use 3 types of metal interlayers (Ni, Ag, Cu) at diffusion welding of pure titanium Gr 2 and austenitic highly alloyed steel AISI 316L. In light of requirements pose on theses interlayers it seems that Ni interlayer is the most suitable one. Nickel has very high diffusion resistance in comparison to majority of basic alloying elements. Thus there is diffusion only in very narrow surface layer of Ni and due to the low diffusion rate, there aren't generally created intermetallic phase Ti_2Ni, which creates up to 40 wt% of Ti in Ni. On the other hand, nickel interlayer needs higher magnitudes of activation energy given by higher temperature that should be at least 1000°C for diffusion. However, titanium Gr 2 reveals very high amount of deformation already at 900°C and pressure 1 MPa. That is why it would be probably better to use this interlayer in alloyed alloys of Ti having higher strength under high temperatures.

The 2nd tested interlayer was Ag one. This interlayer creates with Ti solid solution whose hardness doesn't exceed pure Ti hardness. However, the problem about utilization of such Ag interlayer is on the side of AISI 316L, because solubility of Ag with Fe, Cr and Ni is almost on zero value at temperatures up to 1000°C. That is why on the boundary of these materials isn't created any bonding or just adhesive bonding.

The last tested interlayer was Cu one. Copper has very good solubility in solid state both with Ti and also with major alloying elements of steel AISI 316L. Nevertheless, problem here is to determine proper holding time on temperature so that there is sufficient diffusion through the whole cross-section of bonded sample and at the same time to prevent diffusion of Ti, Fe, Cr and Ni through the whole Cu interlayer thickness. However, in this case can be possible to apply suitable technological precautions as can be e.g. to increase Cu interlayer thickness at least to range from 0.25 up to 0.3 mm.

This publication was written at the Technical University of Liberec, Faculty of Mechanical Engineering with help of the institutional support KSP 117/2200.

References

1. N. F. Kazakov, *Diffusion Bonding of Materials.* Moscow: Mashinostroenie Publishers (1985)
2. W. D. Callister, jr., *Materials Science and Engineering: an introduction.* USA: John Wiley & Sons, Inc. (2007)
3. J. Moravec, M. Dikovits, M. C. Beal, I. Novakova, R. Chandezon, J. Sobotka, *Selection of the proper diffusion welding parameters for the heterogeneous joint Ti grade 2/AISI 316L.* Manufacturing Technology, Vol. **17**, Issue 2, pp. 231-237 (2017)
4. S. Kundu, S. Chatterjee, *Effect of bonding temperature on interface microstructure and properties of titanium 304stainless steel diffusion bonded joints with Ni interlayer.* Materials Science and Technology, Vol. **22**, Issue 10, October 2006, pp. 1201-1207 (2006)
5. F. L. Wang, G. M. Sheng, Y. Q. Deng, *Impulse pressuring diffusion bonding of titanium to 304 stainless steel using pure Ni interlayer.* Rare Metals, **Vol. 35**, Issue 4, 1 April 2016, pp. 331-336 (2016)
6. Y. Deng, G. Sheng, Ch. Xu, *Evaluation of the microstructure and mechanical properties of diffusion bonded joints of titanium to stainless steel with a pure silver interlayer.* Materials and Design, Vol. **46**, April 2013, pp. 84-87, (2013)

7. Ag-Fe binary diagram, http://www.crct.polymtl.ca/fact/phase_diagram.php?file=Ag-Fe.jpg&dir=FTlite
8. S. Kundu, M. Ghosh, A. Laik, K. Bhanumurthy, G.B. Kale, S. Chatterjee, *Diffusion bonding of commercially pure titanium to 304 stainless steel using copper interlayer.* Materials Science and Engineering A, Vol. **407**, Issue 1-2, 25 October 2005, pp. 154-160 (2005)
9. Fe-Cu binary phase diagram, https://www.metallab.net/chemsoc/alloys.php?id=26

MATEC Web of Conferences **244**, 01014 (2018) https://doi.org/10.1051/matecconf/201824401014
ITEP'18

The contribution to the modal analysis of the control system

Alžbeta Sapietová[1]**, *Filip* Chovanculiak[1], *Vladimír* Dekýš[1], *Lukáš* Gajdoš[2]

[1]University of Žilina, Faculty of Mechanical Engineering, Univerzitná 1, 010 26, Žilina, Slovakia
[2] Schaffler Slovakia, Dr G Schaefflera 1, Kysucké Nové Mesto, Slovak Republic

Abstract This article deals with the analysis of the suitability of using selected solution methodologies while taking into account the different degrees of simplification used. Three ways of determining the first frequency of the mechanism were presented on the lever mechanism of the control circuit. The solution is carried out analytically, then via a calculation model in the program on the dynamics of systems elements MSC ADAMS, and lastly, an experimental measurement was carried out.

Keywords: Dynamics, Custom Frequency, Experiment, Analytical Computation, Match

1 Introduction

Understanding of the physical processes is essential to the proper proposal of the construction. It is therefore natural to come up with design besides relevant calculations or experiments. Those shall, based on knowledge of the designer, provide information on the anticipated features of the proposed object.

These calculations are carried out analytically or by using computational models in specialized computer programs [1]. In both of these approaches, there is some degree of simplification, for example, of the shape of the object, its material parameters, boundary or initial conditions [2]. When solving these analytically, these simplifications may be based on the requirement of the task being solvable. Computing programs, because they are complex, usually require the entry of a large number of input parameters, which are not always known. These parameters are often estimated or entered with their implicit values when one starts the computing task. It is therefore natural to assume that analytical and computer models can "live their own lives" and then provide corresponding matching or different results.

If a construction prototype is designed, it is possible to perform experimental measurements. The immediate results of the measurements and their processing shall take into account the specifics of experimental procedures. Then it can also be said that the experiment "lives its own life". In addition, the experiment may also bring about phenomena which the methodology for measuring does not take into account [3,4,5].

If the results of the calculations and the experiment match, then we encounter this type of a claim: "The results of the calculations were confirmed by an experiment, or the results of

* Corresponding author: alzbeta.sapietova@fstroj.uniza.sk
Reviewers: *Jozef* Bocko, *Josef* Soukup

the experiment were confirmed by calculation". However, various approaches in computational and experimental solutions can mean getting different results. It is preferable to talk about match or no match of results, and not about their confirmation. This issue is dealt with, for example, [6].

2 Modal analysis

Modal analysis is a tool that allows understanding the vibrations of structures, or parts of them. It uses the decomposition of the complex process of vibration into basic, so-called modal contributions. Each contribution is characterised by modal frequency, shape of the vibration [7,3], and damping. The knowledge of these data helps understand the transmission characteristics of the mechanical system as well, when we assess the responsiveness of the system to an alarm, for example, by analysing the selected element by vibrating it [8,9]. The response in this case may be unexpectedly high, which can happen, for example, in the creation of resonance in the system. It can be approximately zero in the analysed frequency, if the point of measuring the response is the nodal (zero) point of the dominant shape of the vibration for this natural frequency. In another case, evoking of an unwanted non-linear response may occur for instance.

Finding the presence of phenomena with large and often unacceptable amplitudes of vibration in the real system and looking for the cause of these phenomena is often the reason for doing such an analysis. Designated natural frequencies are then confronted with the results from practice by using computer modal analysis. This procedure can be considered the initial step for verifying the hypothesis about creating resonance in the system at its natural frequency [10].

In the article, a computation modal analysis in the selected mechanism will be carried out in the programme for dealing with the dynamics of object systems of the MSC. ADAMS. The results of this solution are then compared with the analytical method using Lagrange equations of the second type and the experiment. Finally, a comparison of the results obtained is made.

3 Model of control system

A 3D model of the analysed mechanism was imported from the NX CAD to the MSC. ADAMS in the parasolid format. The accuracy of the position of each of the members has been verified using the body markers [11].

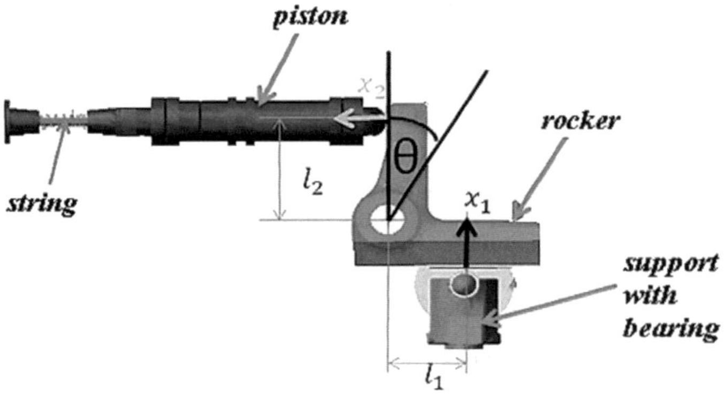

Fig. 1. The model of the analysed control system

This mechanism scans the changes of position of the support by bearing, and through the lever, it transforms this position to the position of the piston. It is a system 1DOF in which the lever arm θ angle was chosen as an independent variable. The mechanism is composed of 12 members. For the sake of clarity, its representation was reduced (Fig. 1).

3.1 Solution with MSC ADAMS

A contact [12-14] was defined between the piston and the lever, and Coulomb's formulation was used for frictional force [15]. For springs used in the model, data on stiffness, damping and preload was entered [16].

Due to use of non-linearities, the linear model could not be used, or rather, it was not possible to use the Adams/Linear module. Therefore, the pulse stimulus was simulated by a very short force (pulse replacement) and the time dependence of the center of mass of piston was determined. It was a model with 1 DOF. Based on the stated dependency, the natural frequency of damped vibration was determined by the Fourier transformation of the generated signal. Fig. 2 shows the time course of response and its Fourier transformation. In this way, the frequency of 285 Hz was determined.

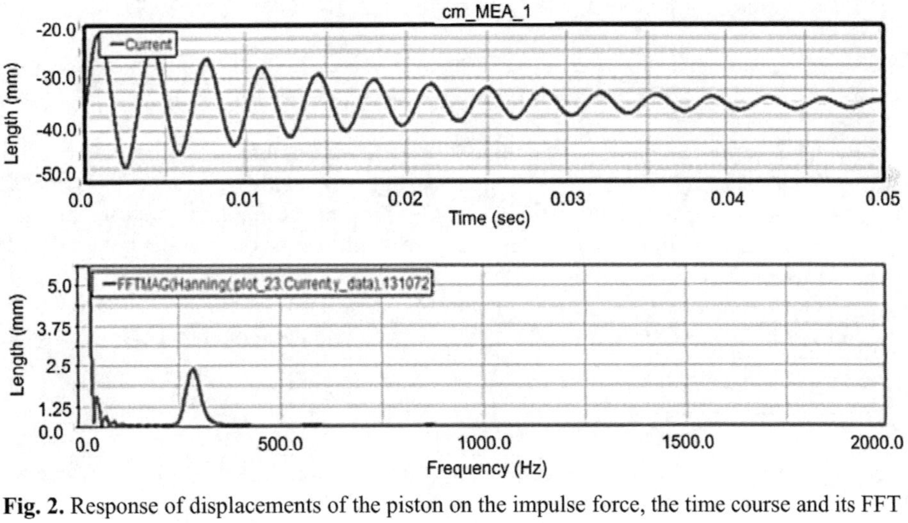

Fig. 2. Response of displacements of the piston on the impulse force, the time course and its FFT

3.2 Solution with Lagrangian formulation

The simplified estimation using Lagrange equations of the second kind was based on the following simplifications: it was a model with 1 DOF; no spring bias and damping was considered; there was no consideration of the contact between the piston and the lever or the presence of frictional force in this bond. The model with conservative forces was used using the following equation:

$$\frac{d}{dt}\left(\frac{\partial E_k}{\partial \dot{\theta}}\right) - \frac{\partial E_k}{\partial \theta} + \frac{\partial E_p}{\partial \theta} = 0,$$

where E_k is the kinetic energy; E_p is position energy; θ is the generalized coordinate (the angle of the rocker arm); generalized angular velocity and t is time, [17-19].

In this way, the frequency of 284 Hz was determined.

3.3 Experimental solution

The module of the control mechanism was mounted on the jig, and a miniature 3 axis accelerometer B&K4520 was installed on the lever. The mechanism was stimulated by impulse hammer IH101-500 by striking the piston, and the impulse hammer and accelerometer were mounted on a four-channel USB module NI9234. Data was collected in NI SignalExpres and processed using the frequency response function (FRF). The output of the measurement is shown in Fig. 3, where a moderate peak around the desired frequency is marked.

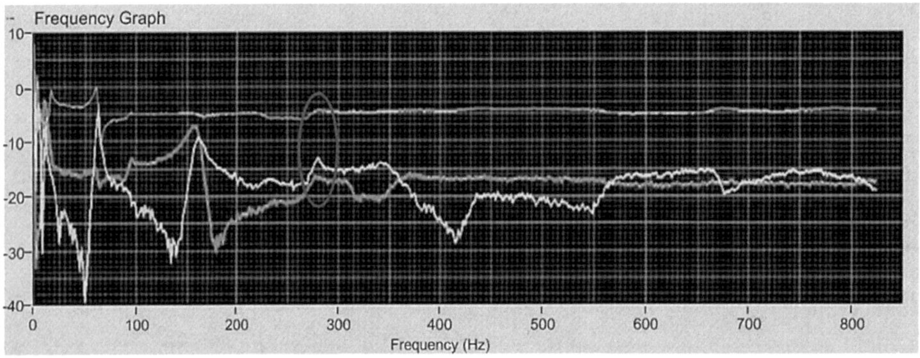

Fig. 3. FRF function for all three measuring channels on a lever. A moderate peak is marked on the expected resonant frequency, more precisely on 275 Hz.

Although there is a readable peak around 275 Hz in the amplitude spectrum on Fig.3, but phase characteristics did not show sufficient change around this frequency which would generally result from a change of approximately 3.14 rad before and after the resonance peak.

At first glance, everything is fine. However, the results of the experiment point to the fact that the above peak is less pronounced, demonstrable only intuitively. This can be caused by making a measurement on a real-life control system module that "lives its own life" independent of our simplifications used in the previous computational models.

4 Conclusion

Three ways of determining the first custom frequency of the mechanism were presented on the lever mechanism of the control circuit. The individual ways used a different degree of simplification.

In the Lagrangian formulation for the analytical solution absorption was neglected, and contact and presence of non-conservative forces was not taken into account.

In a computer model in MSC ADAMS a more comprehensive approach was used, when also damping and spring preload data was entered; the contact between the lever and the piston and Coulomb friction model were taken into account at the point of contact. The results for the first natural frequency were, despite the simplifications, identical.

The result of the experimental measurement provided a result with an estimated insignificant resonance frequency deviating by about 5% from the values obtained by the calculation.

This article aimed at presenting an approach to solving the estimation of natural frequency of a non-linear system in MSC ADAMS. In our opinion, on the basis of the simplifications used, we would like to once again reiterate disagreement with the claims such as: the calculation was confirmed by the experiment or the experiment was validated by the calculation.

This paper was supported by VEGA 1/0795/16 and KEGA 017ŽU-4/2017.

References

1. P. Orsansky, B. Ftorek, P. Durcansky, *Mathematical Model of a Closed Hot Air Engine Cycle Using MATLAB (R)/Simulink (R),* AIP Conference Proceedings, **1608**, 173-176 (2014)
2. Z. Poruba, J. Szweda, J. Podešva, *Boundary Conditions and Constraint Equations for Simulation of Energy Exchange*. Applied Mechanics and Materials, **315**, 929-935 (2013)
3. J. Vavro, J. Vavro, Jr., P. Kovacikova, R. Bezdedova, *The optimisation of the materials properties for the passenger cars in dependence on defect distribution at the dynamic loading,* Procedia Engineering, **136**, 114-119, (2016)
4. D. Stancekova, M. Sajgalik, J. Petru, N. Naprstkova, P. Balaz, *Implementation of coating for failure elimination of dial gauges.* METAL 2015 Conference Procedings of the 24th International Conference on Metallurgy and Materials, 1162-1169, (2015)
5. D. Barta, M. Mruzek, *Importance of real operating parameters for design of public transport vehicles drive.* Scientific journals of the maritime university of Szczecin-zeszyty naukowe akademii morsikiej w Szczecinie, **39**(111), 5-10, (2014)
6. M. Richardson, *Modal analysis versus finite-element analysis.* Sound & vibration, **39**(9), 5-9, (2005)
7. S. Miláček, *Modální analýza mechanických kmitů.* 2.vyd., Vydavatelství ČVUT, 154 s. (1991)
8. M. Sága, L. Jakubovičová, *Simulation of vertical vehicle non-stationary random vibrations considering various speeds.* Scientific Journal of Silesian University of Technology – Series Transport, **84**, 113-118 (2014)
9. T. Harcarik, J. Bocko, K. Maslakova, *Frequency analysis of acoustic signal using the Fast Fourier Transformation in MATLAB*, Procedia Engineering, **48**, 199-204, (2012)
10. M. Nad, L. Cicmancova, *Effect of Shape Parameters on the Modal Properties of Stepped Ultrasonic Concentrator.* Applied Mechanics and Materials, **309**, 43-49, (2013)
11. M. Macko, K. Tyszczuk, G. Śmigielski, J. Flizikowski, A. Mroziński, *The use of CAD applications in the design of shredders for polymers*. MATEC, **157**, 02027, (2018)
12. M. Dudziak, G. Domek, A. Kołodziej, K. Talaśka, *Contact Problems Between the Hub and the Shaft with a Three-angular Shape of Cross-section for Different Angular Positions*. Procedia Engineering, **96**, 50-58, (2014)
13. J. Podešva, J. Szweda, Z. Poruba, *The Substructures Technique Including Contact Problem.* Applied Mechanics and Materials, **315**, 884-888, (2013)
14. M. Zmindak, P. Pastorek, P. Falat, M. Sapieta, *Hardening monitoring and contact stress analysis of test samples*, Proceedings of the institution of mechanical engineers part f-journal of rail and rapid transit, **230**(1), 253-257, (2016)
15. L. Jakubovičová, M. Sága, M. Vaško, *Numerical study of influence of mutual slewig of roller bearing rings on the principal stresses at contact area.* Scientific journal of Silesian University of Technology – Series Transport, **84,** 83-91 (2014)
16. P. Kassay, J. Homisin, P. Copan, M. *Urbansky, Verification of torsional oscillating mechanical system dynamic calculation results.* Scientific journal of silesian university of technology-series transport, **84**, 29-34, (2014)

17. J. Vavro, Jr., J. Vavro, P. Kovacikova, R. Bezdedova, *Kinematic and dynamic analysis of the manipulator for removal of rough tyres,* Procedia Engineering, **136**, 120-124, (2016)
18. J. Famfulik, J. Mikova, M. Richtar, R. Halama, *Random vector approach to the calculation of the number of railway vehicles to hold in reserve,* Tehnicki vjesnik-technical gazette, **24**(1), 7-14, (2017)
19. P. Kopas, M. Blatnický, M. Sága, M. Vaško, *Identification of mechanical properties of weld joints of AlMgSi07.F25 aluminium alloy*. Metalurgija **56** (1-2), 99-102 (2017)

Measurement of industrial robot pose repeatability

Ivan Kuric[1], *Vladimír* Tlach[1,*], *Zuzana* Ságová[1], *Miroslav* Císar[1], *Igor* Gritsuk[2]

[1]University of Žilina, Faculty of Mechanical Engineering, Department of Automation and Production Systems, Univerzitná 1, 010 26 Žilina, Slovak Republic
[2]Kherson State Maritime Academy, Department of Operation of Ship Power Plants, 73000 Kherson, 20 Ushakova avenue, Ukraine

Abstract. The article deals with measurement of pose repeatability of industrial robot Fanuc LR Mate 200iC. Laser interferometer Renishaw XL-80 and digital indicator Mitutoyo Absolute Digimatic ID-F are used for the measurement. The measurement methodology is based on so-called ISO cube placed in the most used part of robot workspace. Individual steps of the solution, the analysis of measurement data and results are presented in the article.

Keywords: Industrial robot, Pose repeatability, Laser interferometer, Digital indicator

1 Introduction

The primary role of industrial robots was to relieve man from physically demanding and monotonous work and from working in harmful and hazardous environment. With development of production facilities and increasing production flexibility, the requirements for industrial robots are also increasing and new areas of deployment are emerging. [1, 1] One of the trends is the use of industrial robots for tasks that have so far been carried out by dedicated devices, or for performing non-production tasks like operating CNC machine tools by an industrial robot. At the same time, automation is constantly increasing the number of industrial robots. Based on a press release issued by IFK (International Federation of Robotics), an estimated global growth of more than three million industrial robots is expected in 2020 [3].

The choice of an industrial robot for a particular application is determined by its basic criteria such as the payload, the number of degrees of freedom, the robot's reach based on the size of the workspace, and the like. [4] The features of an industrial robot can be viewed in more detail through its performance criteria. One of the most frequently mentioned performance criteria listed by industrial robot manufacturers in the respective datasheets is the repeatability value. More specifically, this is a pose repeatability, which represents the ability of the robot to repeatedly return to the same position from the same direction. [5]

* Corresponding author: vladimir.tlach@fstroj.uniza.sk
Reviewers: *Sorin* Grozav, *Radu* Cotetiu

Pose repeatability and many other performance characteristics are defined in the International Standard ISO 9283. This standard defines the importance of individual performance criteria, the method of their calculation, together with the recommended procedure and the measurement conditions. [6] Another document related to this issue is the ISO/TR 13309 technical report, published by the International Standards Organization in 1995. Its aim is to provide an overview of metrological methods and devices for measuring performance criteria in accordance with ISO 9283 standard.

The measurement of performance criteria of industrial robots is a complex task requiring different measuring devices and measurement methods. This issue is addressed by a number of scientific papers. [7] The most frequently measured performance criteria include accuracy and repeatability. Measurements are carried out with the use of various measuring devices such as laser tracker [8, 9], indicator [10, 11], camera systems [12, 13], and the like. Increasingly used are measuring devices primarily designated to assess the condition of CNC machine tools. We are talking about the laser interferometer and the ballbar measuring device [5].

The article deals with measurement of pose repeatability of industrial robot Fanuc LR Mate 200iC with the use of the Renishaw XL-80 laser interferometer and the Mitutoyo Absolute Digimatic ID-F digital indicator.

2 Pose repeatability of industrial robot positioning

Pose repeatability (RP) stands for the ability of the TCP (Tool Centre Point) of the robot to always return to the same (programmed) position from the same direction. One-way acquisition of the required position minimizes the effects of backlash in the individual joints of the industrial robot and their effect on the measurement result. Bidirectional or multidirectional repeatability may be several times greater than one-way repeatability. [5] For graphic display (Fig. 1.a), the pose repeatability is the radius of the sphere whose centre is located in the so-called barycenter. The barycenter (G) represents the point whose coordinates in the space correspond to mean values calculated from the repeatedly acquired TCP positions of the given robot.

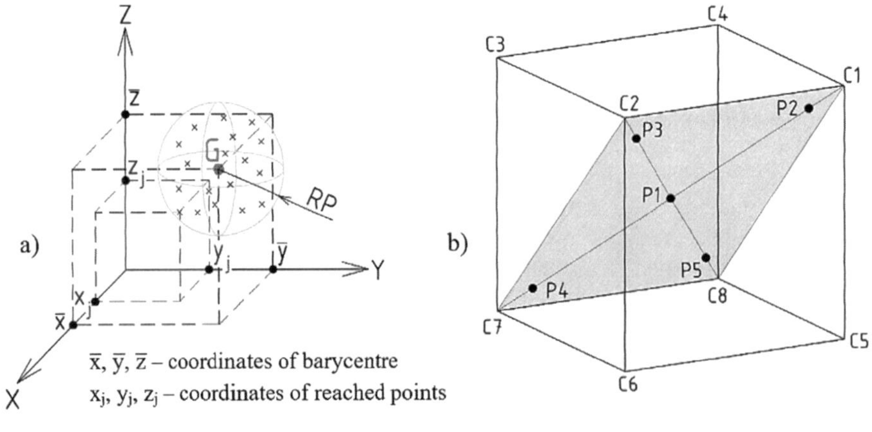

Fig. 1. Graphical displaying of pose repeatability and ISO cube

The ISO 9283 standard for measuring pose repeatability cites the procedure based on the so-called ISO cube. This is an imaginary cube placed in the robot's workspace, at the point where its most frequent use is assumed. The ISO cube should have the maximum admissible volume and, in the case of six-axis robots, an inclined plane with five measuring points P1-P5 (Fig. 1.b). The measurement process itself consists of at least thirty cycles during which the TCP point of the robot moves in a prescribed manner between the five measurement points. 13Each position of the TCP point is acquired by a one-way accession, precisely because of the minimization of effects of the backlash on robot joints.

3 Measurement methodology

The article presents a procedure for measuring pose repeatability on the industrial robot Fanuc LR Mate 200iC. It is robot with serial kinematics, located in the laboratory of the Department of Automation and Production Systems, of the University of Žilina. The robot is part of an automated assembly workplace in which it performs tasks related to the final assembly and handling of the assembled product. Based on standard recommendations, an imaginary ISO cube with a 470mm long edge was placed in the robot workspace (Fig. 2). Individual cube edges are parallel to axes of the robot's World Coordinate System (WCS). The size and location of the ISO cube was chosen to maximally cover the area of the robot's workspace associated with its work at the workplace.

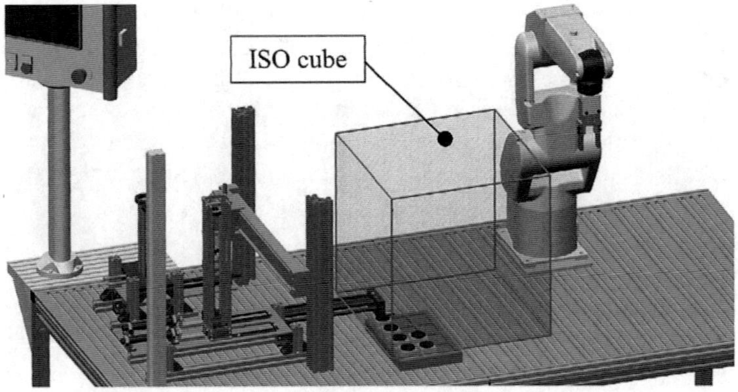

Fig. 2. Location of ISO cube in the robot workspace

3.1 Laser Interferometer Renishaw XL-80

The laser interferometer Renishaw XL-80 is mainly used for measuring machine tools. The system consists of the XL-80 laser unit, the XC-80 compensation unit for sensing environmental conditions (temperature, pressure, humidity) and measuring optics. Depending on the selected measurement optics, several types of measurements can be implemented with this device. Linear measurement was used to measure pose repeatability of the industrial robot. The principle of linear measurement lies in moving the reflector (moving mirror) along the transmitted laser beam. As the measurement runs alongside the line along the laser beam, it is not possible to measure five points located on the slanting plane as defined by ISO 9283.

The proposed method consists of three separate measurements along X, Y and Z axes. The starting point of all three measurements is P0, located at the vertex of the ISO cube (Fig. 3a). For each of the measurements, five measuring points (P1 - P5), spaced along edges of the ISO cube, are selected. To these points, the already mentioned point P0 and the point P6 are added. Measurement is not performed at those points, but they are used by the Renishaw LaserXL measuring program to detect a recurring measurement cycle. Thanks to this, the entire measurement process may be automated. During the measurement, the TCP point of the robot moves between points P0 and P6, with the stop and acquisition of the measuring position in both directions. During this type of measurement, measured values make it possible to observe, among other things, the effect of backlash on robot joints.

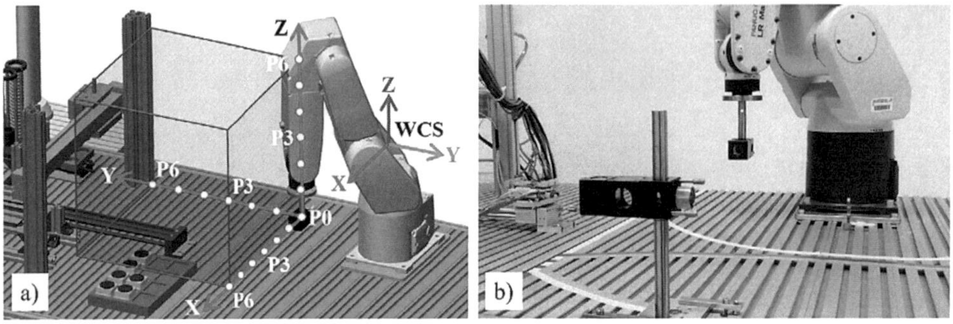

Fig. 3. Location of measuring points in ISO cube and realization of measure with laser interferometer

Each of P1 - P5 points was measured 30 times in the positive direction of the motion (P0 - P6) and 30 times in the negative direction of the motion (P6 - P0). During the measurement, the end of robot´s arm was loaded with a weight of 0.75kg, which contained the measuring optical apparatus and the accessories for its clamping onto the end of robot´s arm. Since the purpose of the workplace of the robot's location (Fig. 3b) is to make presentations and to teach specialist courses; measurements are made at the TCP point velocity value of 2m/s (50% of the robot's maximum velocity). The selected velocity is the velocity used at the given workplace.

3.2 Digital indicator Mitutoyo Absolute Digimatic ID-F

Using an indicator to measure industrial robot performance criteria is a relatively simple and not too costly solution. For pose repeatability measurement, digital indicator Mitutoyo Absolute Digimatic ID-F with a resolution of 0.001mm was selected. Repeatability of the Fanuc LR Mate 200iC robot is stated at ±0.02mm by its manufacturer. To make the repeatability measurement at the given robot possible, a measuring device of at least one order of precision greater must be chosen for the purpose.

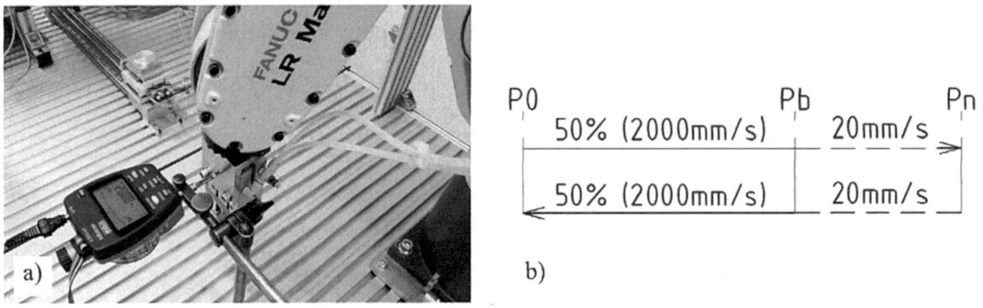

Fig. 4. Measurement methodology with digital indicator

The measurement methodology has been chosen to approximate laser interferometer measurement as closely as possible. However, each of measured axes must consist of five separate measurements to cover points P1 to P5, because the measuring range of the selected indicator does not cover all measured points. The indicator contact is applied to the flange during the measurement by means of which the gripper is clamped on the end of robot´s arm (Fig. 4a). This is the gripper used during work performed by the robot at the assembly workplace. The flange with the gripping head represents a 0.71kg load on the end of robot´s arm. The measurements were made at 50% of the velocity for reasons already mentioned above. The work does not take place at maximum velocity, but even at this velocity, the contact spot of the indicator would be impacted with great force, possibly damaging the measuring system. Therefore, an auxiliary point Pb was placed on the path between P0 and the measuring point Pq, where the velocity decreases to 20mm/s (Fig. 4b).

4 Evaluation of measurement

The AP & RP program was used for the purpose of processing and evaluating measured data presented in this article (Fig. 5), created at our department as a diploma thesis. It is designed for the calculation and processing of pose accuracy and pose repeatability measurements when those are taken with a laser interferometer and a digital indicator. In case of the laser interferometer, measured data are recorded by Renishaw LaserXL program, the output of which is the *.rtl file. This file is loaded into the AP & RP for processing. When measurements are taken with a digital indicator, the above-mentioned program serves not only the purpose of processing measured values but also for the purpose of their receipt via serial communication. In addition, it possible to store the sorted measured values in two file types - Microsoft Excel and a text file.

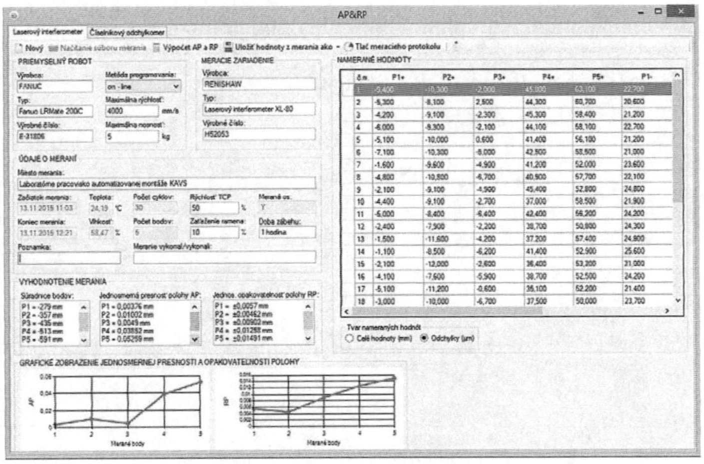

Fig. 5. User interface of AP&RP program

The calculation of pose repeatability both measuring devices is done by the relation (1), based on the sample standard deviation S_{Pq}.

$$RP_{Pq} = \pm 3S_{Pq} = \pm 3\sqrt{\frac{1}{n-1}\sum_{j=1}^{n}\left(Pq_j - \overline{Pq}\right)^2} \tag{1}$$

Where Pq is point of measurement.

4.1 Evaluation of measurement by laser interferometer

The pose repeatability was calculated in the AP&RP program, with the mean of measured values for X, Y and Z axes given in Table 1. The Fanuc LR Mate 200iC manufacturer states the robot's repeatability value ± 0.02mm. This is a certain average repeatability with which the robot should acquire the programmed position in any area of the workspace. If this value is compared with values in Table 1, it is possible to say that calculated repeatability values confirm the value reported by the manufacturer. They are even better.

Table 1. The average value of pose repeatability

Axis of measuring	**Axis X**	**Axis Y**	**Axis Z**
The average value of pose repeatability	±0,016mm	±0,009mm	±0,012mm

Fig. 6, Fig. 7 and Fig. 8 is a graphical processing of measured values for both the positive and negative direction of the robot's TCP point motion. Based on individual charts, and a more detailed view of measurement results can be obtained, and few conclusions can be made.

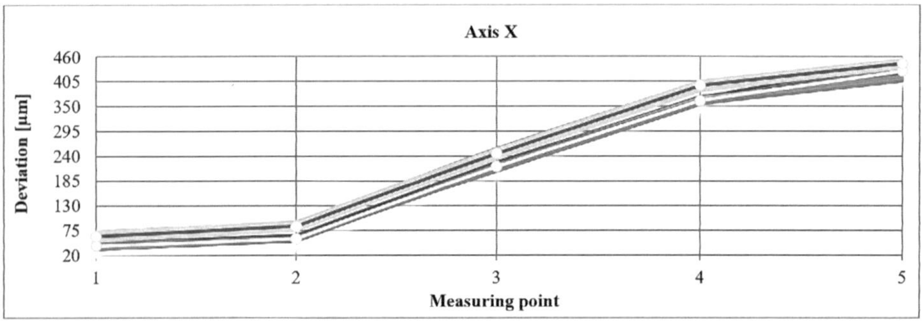

Fig. 6. Graphical processing of X axis measurement

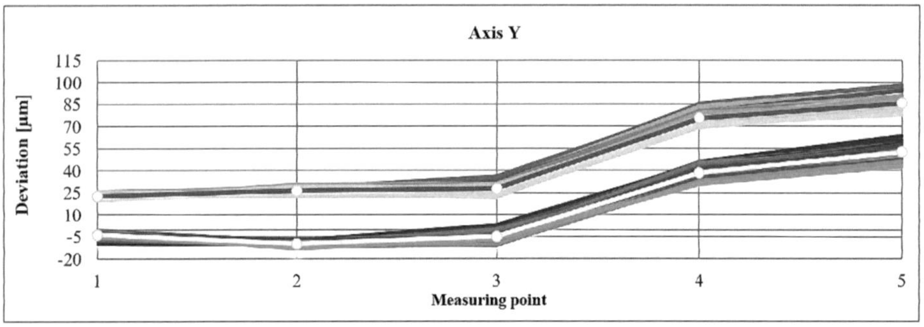

Fig. 7. Graphical processing of Y axis measurement

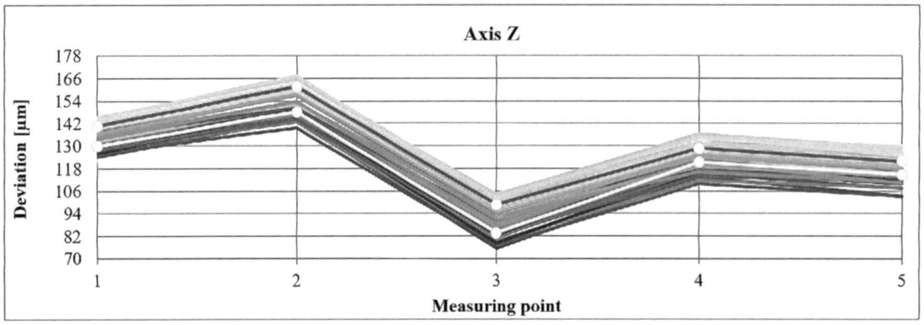

Fig. 8. Graphical processing of Z axis measurement

In case of X and Y (Fig. 6, Fig. 7) axis measurements, the increasing distance of the TCP point of the robot from the beginning of measurement translates into the increasing value of pose repeatability. This trend is likely a result of serial kinematics, where the greater the extension of the robot's arm, the greater the effect of bending forces impacting its end of arm.

This statement is also supported by chart results for the Z axis (Fig. 8), where an upward trend of such significance is not observable.

Since a bidirectional measurement method has been chosen, backlash manifestation in robot's joints can be observed in the graphic processing of measured values. The most significant manifestation of backlash is observable in measurements in direction of the Y axis (Fig. 7). The graph consists of two separate value bundles, the lower of which consists of values obtained in the positive direction of motion (P0 - P6) and the upper bundle represents values obtained in the negative direction of motion (P6 - P0). The difference between mean values of both bundles is caused by the backlash of robot joints. Obtained measurements do not facilitate determination of the backlash value for specific robot joints. This is due to the fact that the motion of the robot arm during the measurement process consists of simultaneous motion of several controlled robot axes. Also, a different expression of the backlash in measured axes X, Y and Z is due to a different range of movements of robot joints.

At the same time, it should be noted that all presented results refer only to the part of the robot workspace in the selected ISO cube and under selected measurement conditions. In other workspace areas, these claims may not apply.

4.2 Evaluation of Measurement by Digital Deviation Meter

The chosen measurement method, in which the indicator contact is applied to the flange of the gripper, does not allow for measurements of the Z axis in the positive direction of motion (measurements with the starting point P0). Conversely, if the negative Z-axis measurement is used with the starting point at P6, the shortness of the magnetic stand used to hold the indicator allows measurements to be made only at points P1 and P2. For this reason, experiments were performed only in the positive direction of X and Y axis movement.

Figures Fig. 9 and Fig. 10 show a graphic comparison of pose repeatability values when measured by the laser interferometer and the digital indicator. Since the indicator was re-set after measuring each of points, it is not possible to evaluate the average pose repeatability on both measured axes. The comparison shows that in both measured axes better repeatability is achieved when digital indicator is employed. This result may be caused by a decrease in the robot's speed at point Pb before reaching the measuring point.

In addition, measured values show the same trend of increase in the value of pose repeatability, depending on the robot's arm extension, as is the case of laser interferometer measurement.

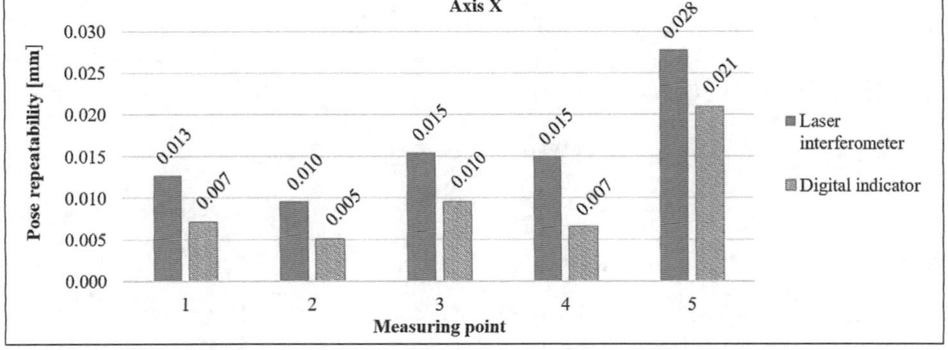

Fig. 9 Comparison of pose repeatability values when measured by the laser interferometer and the digital indicator for X axis

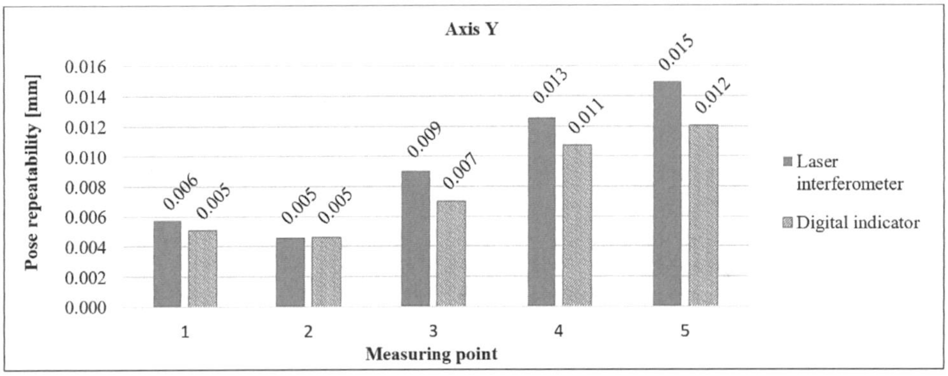

Fig. 10 Comparison of pose repeatability values when measured by the laser interferometer and the digital indicator for Y axis

5 Conclusion

This article presents pose repeatability measurement of the industrial robot Fanuc LR Mate 200iC. The measurement methodology is based on the so-called ISO cube located in the workspace of the robot, in the area of its most common use. Individual experiments are carried out using the laser interferometer and the digital indicator.

Measurement results, obtained by both of measuring devices, were processed in the AP&RP program, developed at the Department of Automation and Production Systems as a diploma thesis. This program was designed to simplify the processing of measured data obtained by laser interferometer and digital indicator.

By evaluating results of pose repeatability, it can be said that the repeatability of the robot at the selected area of its workspace corresponds to the repeatability value stated by the manufacturer (± 0.02mm). At the same time, the trend of repeatability increase was observed, depending on the robot's arm extension, which may be attributed to the bending force acting on the end of robot´s arm. Furthermore, reference is made to the fact that better results are achieved in digital indicator measurements, but those are most likely related to the chosen measurement path. Although the measurement procedure was chosen so that measurement with the digital indicator would be as close as possible to the laser interferometer measurement, certain differences could not have been avoided. Specifically, slowing down of the robot's TCP point prior to position acquisition. It might have been indeed this reduction in velocity that probably caused the difference between two measurements. In order to verify the theory, however, it is necessary to repeat measurements in the future by unifying measurement procedures of two measuring devices.

This work was supported by the Slovak Research and Development Agency under the contract No. APVV-16-0283.

References

1. I. Kuric, V. Bulej, M. Sága, P. Pokorný, *Development of simulation software for mobile robot path planning within multilayer map system based on metric and topological maps.* International journal of advanced robotic systems, **14**, (2017)
2. B. Mičieta, M. Edl, M. Krajčovič, Ľ. Dulina, P. Bubeník, L. Ďurica, V. Biňasová, *Delegate MASs for Coordination and Control of One-Directional AGV Systems: A*

Proof-of-Concept. International journal of advanced manufacturing technology, **94**, 415-431 (2018)

3. C. Heer, *Robots double worldwide by 2020.* International Federation of Robotics. [online] Available on Internet: https://ifr.org/ifr-press-releases/news/robots-double-worldwide-by-2020
4. A. Sapietová, M. Sága, I. Kuric, Š. Václav, *Application of optimization algorithms for robot systems designing.* International Journal of Advanced Robotic Systems, **15** (2018)
5. M. Slamani, A. Nubiola, I. A. Bonev. *Assessment of the positioning performance of an industrial robot.* Industrial Robot: An International Journal, **39**, 57-68 (2012)
6. A. Nubiola, I. A. Bonev. *Absolute calibration of an ABB IRB 1600 robot using a laser tracker.* Robotics and Computer-Integrated Manufacturing, **29**, 236-245 (2013)
7. W. S. Newman, C. E. Birkhimer, R. J. Horning, A. T. Wilkey. *Calibration of a Motoman P8 robot based on laser tracking.* Robotics and Automation, 2000. Proceedings. ICRA'00. IEEE International Conference on. **4**, 3597-3602 (2000)
8. A. Şirinterlikçi, M. Tiryakioğlu, A. Bird, A. Harris, K. Kweder. *Repeatability and Accuracy of an Industrial Robot: Laboratory Experience for a Design of Experiments Course.* Technology Interface Journal, **9** (2009)
9. J. Brethé, D. Lefebvre. *Risk ellipsoids and granularity ration for industrial robot.* International Journal of Factory Automation Robotics and Soft computing, **2**, 93-101 (2007)
10. J. Józwik, D. Ostrowski, P. Jarosz, D. Mika. *Industrial robot repeatability testing with high speed camera Phantom v2511.* Advances in Science and Technology Research Journal, **10**, 86-96 (2016)
11. M. Abderrahim, A. Khamis, S. Garrido, L. Moreno. *Accuracy and Calibration Issues of Industrial Manipulators.* Industrial Robotics- Programming, Simulation and Applications, 131-146 (2007)
12. A. Nubiola, M. Slamani, I. A. Bonev. *A new method for measuring a large set of poses with a single telescoping ballbar.* Precision Engineering, **37**, 451-460 (2013)
13. STN ISO 9283 Manipulating Industrial Robots – Performance criteria and related test methods (1995)

Comparation of nonuniform and uniform Monte - Carlo Searching

Marián Handrik,[1,1*], *Jana* Handriková[1], *Milan* Vaško[1], *Filip* Dorčiak[1]

[1]University of Žilina, Faculty of mechanical engineering, 010 26 Žilina, Slovak Republic

Abstract. Nonuniform Monte-Carlo method is often used for optimization and solution of function mapping. This method has some disadvantages. New genetic algorithm, based on uniform Monte-Carlo is proposed by authors reduce disadvantage of nonuniform Monte-Carlo method. Both of these methods are based on random number generation and therefore the solutions are approximate. Statistical evaluation of solutions is used for comparison.

Keywords: uniform and nonuniform,Monte-Carlo method, optimization, solution function mapping, statistical evaluation.

1 Introduction

In solving the real problems of technical work, we often encounter the concept of uncertainty in the task a simple example of uncertainty in engineering practice can be manufacturing tolerances of machine parts, unknown quantity of material in the bins, and so on. Uncertainties of this nature in contact, modal-spectral and other types of tasks have a significant impact on the value of the obtained results. From the point of view of designing machine components with required service life or security level, it is therefore necessary to find such combination of indeterminate parameters of the solved task in which the value of the investigated function acquires extreme values [1-4]. Often, however, the uncertainties are bounded by their minimum and maximum values and in this case, the task changes to the interval analysis of the problem [5-9]. The basic and the easiest method of solving this type of task is the Monte-Carlo method, where we randomly generate points from the uncertainty interval and perform an analysis of the investigated problem for these random combinations of uncertainty, and finally we determine the combination of unknown parameters at which the investigated function has reached the extreme [10-20].

2 Nonuniform Monte - Carlo method

The basic requirement for point generation by the Monte-Carlo method is the assumption of generating points with equable or normal configuration of probability density for point coordinates. This assumption is quite often used in practice. Usage of this method while

*Corresponding author: marian.handrik@fstroj.uniza.sk

Reviewers: *Tomasz* Domański, *Michal* Šofér

generating of points in the solution area and depending on the unknown parameters is time-efficient and easy in case of algorithmic approach. On the other hand, if this method is explored in more detail we will find several shortages:

- under the condition of equable distribution of probability density of individual coordinates, we can also obtain the set of points shown in Fig. 1 and point distribution histogram for axis x, similarly for axis y. Histograms indicate an equable distribution of probability density, whereby, points are grouped in two sub-areas;
- creating of areas with a large number of clustered points and white spots without the presence of points;
- increasing the quantity of points used in mapping of the solution may not increase the accuracy of the solution or ensure the accuracy of the solution obtained for the subsequent generation and usage of additional points;
- low probability of generating points near combinations of infimum and supremum in mapped uncertainty area. It also often happens that during solving of practical problems, the solution is at the point of combining the infimum and supremum intervals of the mapped area, viz. there is no local or global extreme in the mapped area but the value at the border gets an extreme value;
- the need for a relatively high number of points for area mapping. This fact is not directly related to the Monte-Carlo method but with the scope of the analyzed task, so it is necessary to take it into account when choosing the number of mapping points. If we have the established condition that we want to get the solution with probability P = 0,1 in the uncertainty interval $I \in \langle 0,1 \rangle$ for one-dimensional task based on the probability theory, we need $n = \frac{1}{P}$ points to map the area. Then, for mapping the area which includes the number of m uncertainties parameters with the same probability, we need $n = \left(\frac{1}{p}\right)^m$ points for mapping of the area.

To remedy the above shortcomings, several modifications of the Monte-Carlo algorithm have been created. The most common of them is the modification of the Hypercube method, its principle lies in dividing the mapped area into subareas where the Monte-Carlo classical method will be used. Such subdivision should provide a minimum number of mapping points for each subarea.

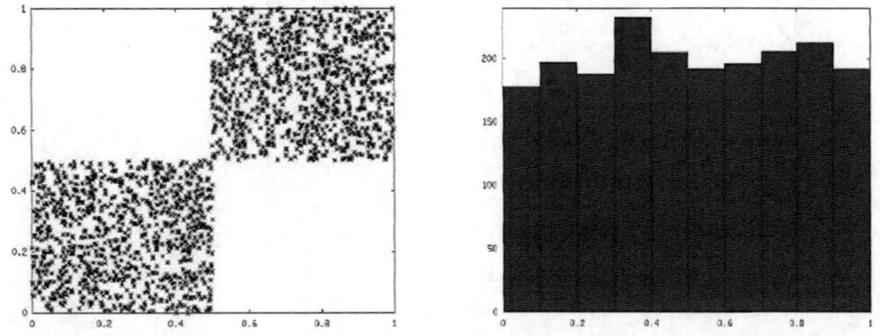

Fig 1. Uneven distribution of points in two sub-areas and histogram of the coordinates of the x points

3 Genetic algorithms

This type of algorithms is characterized by a specific procedure for generating points for mapping of solution. With these algorithms, a set of offspring - points will be generated and, on a base of the evolutionary condition, the offspring or a group of offspring, which fits better to the evolutionary condition, will be ranked among the parents. The variety of algorithms for generation of offspring and evolutionary conditions creates a space for creating a large number of computational procedures for determining parents - area mapping points in the category of generic algorithms Fig. 2.

If we set the main goal while creating of a new genetic algorithm, such as eliminating of disadvantages of the Monte-Carlo method under points 2-4 and at the same time under condition of randomness of the generated offspring points. The algorithm will not generate the same points – parents in case of relaunching. The group of genetic algorithms, generated in this way, can be named as uniform Monte - Carlo. Such gradual filling of white spaces simultaneously removes the disadvantage of the nonuniform Monte-Carlo method referred under the point 3 and guarantees the accuracy of the solution depending on the number of generated points.

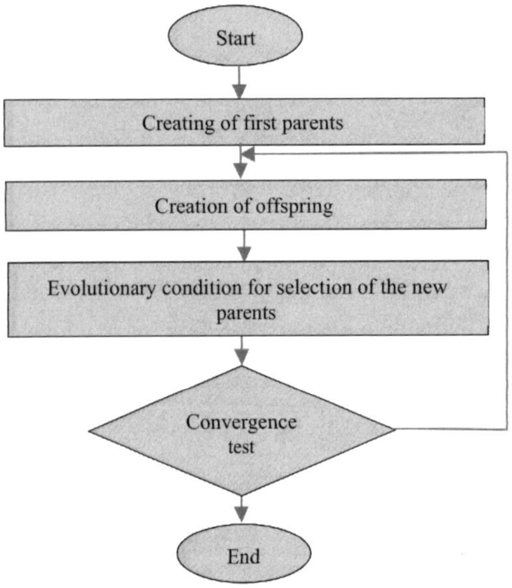

Fig 2. Developmental flowchart of evolutional algorithms

4 Uniform Monte – Carlo method

Algorithm of uniform Monte - Carlo searching Fig. 3 can be divided into the following basic steps:

- generation of a combination of supremum and infimum. When creating an algorithm, we must realize that it is necessary to create a combination of two values in a *m* dimension field, the total number of combinations is then given by a relation $n=2^m$ Individual combinations can be obtained by creating a vector $v=[0,1,...,n-1]$, which is then converted into a binary system. By the transformation, we will obtain *m* sizeable combination of nulls and base values, after changing the nulls and base values (1) to the infimum and supremum each number then represent desired combinations of infimum and supremum. This algorithm is universal in terms of the analyzed task dimension. At first glance, program implementation of the algorithm is complicated with respect to binary transformation, but most of the programming languages have this feature as a standard one;
- when generating the offspring, we will use the Monte-Carlo classical method, i.e. generating of random numbers. During generating of points, the condition of equable distribution of points for one coordinate of generated *m* dimensional field of random points.

- subsequently, we measure the distance between offspring and parents;
- finding of the offspring with the greatest distance to the parents;
- end algorithm test, here are a few possible steps presented. Finishing if we find a determined number of parents in advance. Ending of parents search if the largest distance between parents and offspring falls below a predetermined value, the first descendant reaches this value. Ending of parents searching if the largest distance for a group of newly-generated parents will be less than the prescribed value. It is not possible to find any white space for generating a new parent.

The gradual thickening and filling of white spots is shown in Fig. 4, the first 100 generated points are shown by *, following 100 points are displayed by the +, and the last 100 by the x. The randomness of points distribution for repeated generation of points in uniform Monte - Carlo algorithm for mapping the area is Fig. 4.

The basic disadvantage of the uniform Monte - Carlo algorithm comparing to the nonuniform Monte-Carlo method is that the algorithm is time-consuming. The time consuming of the algorithm is caused due to a large number of necessary measurements of points distance. We will document this time consuming on the example used in the following section, where we compare the results obtained with the nonuniform Monte-Carlo method and the uniform Monte - Carlo algorithm. At the first analysis, we will require 100 points generated in the area of interest. While generating of new offspring, we will always generate 50 new points. We will get 4 points in case of generating of the infimum and supremum combination. Then, when searching for the fifth point, we will measure the distance between 4 parents and 50 offspring, i. e. 200 times in total. When searching for the sixth point, we need to implement the distance measurements 250 times. Overall, we need to measure the distance 247200 times to determine 100 points. On the other hand, this algorithm can be easily parallelized by dividing the offspring vector into an appropriate number of vectors according to the number of calculating fibers.

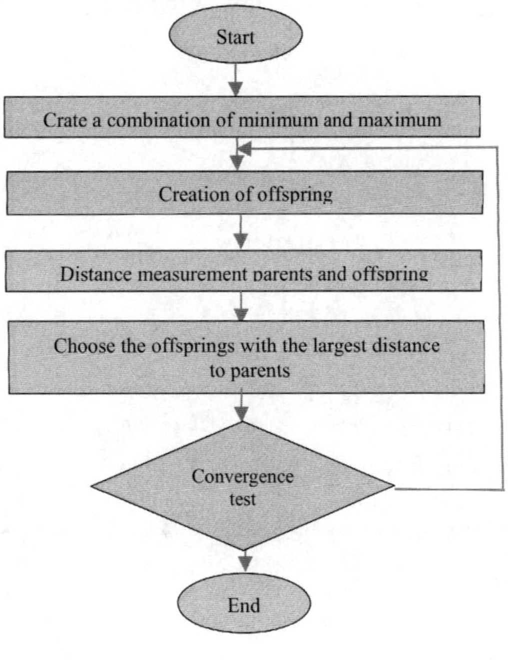

Fig 3. Developmental flowchart of uniform Monte – Carlo method

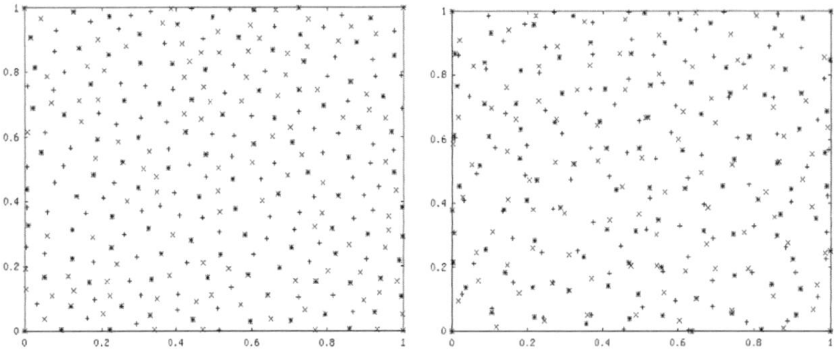

Fig 4. Gradual thickening of the area with points by generating additional points and random layout for re-generation

5 Numeric experiments evaluation the nonuniform Monte – Carlo method and uniform Monte – Carlo method

We use two examples to experimentally evaluate these two approaches for mapping the uncertainty area. In the first example, we compile the function (1) Fig. 5, which will have a maximum value in the field of mapping at the point $x_{opt} = [0.25\ 0.25]$, i. e.

$$y = -x_1^2 - x_2^2 + 0.5x_1 + 0.5x_2 + 5 \tag{1}$$

In the second example, the maximum of function, which was mentioned above (2), will be displaced out of the mapping area Fig. 5, the function will have the following shape and the maximum value will be in the point $x_{opt} = [-0.25\ 0.25]$, i. e.

$$y = -x_1^2 - x_2^2 - 0.5x_1 + 0.5x_2 + 5 \tag{2}$$

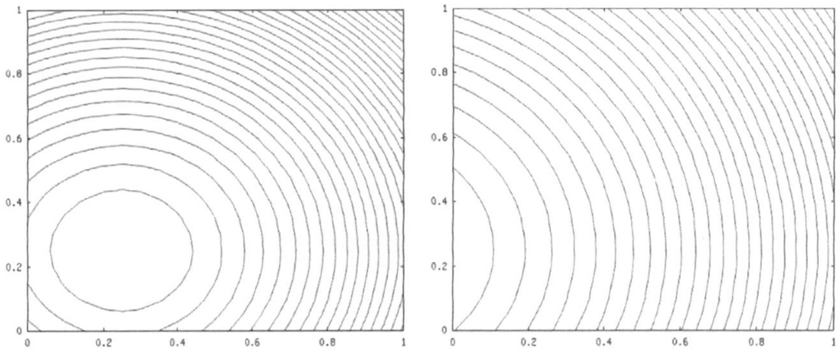

Fig 4. Layers of the function used in the example 1 and example 2

Next, it is necessary to set the criteria, which will help us to evaluate the algorithms used by us. When setting the first criterion, we will evaluate the accuracy of the points finding (3) with the extreme of functions given by the relation (1) a (2), x_{NMC}- - the point of extreme, which was found by usage of the nonuniform Monte - Carlo method, x_{UMC}- the point of extreme, which was found by usage of the uniform Monte - Carlo method. When evaluating the accuracy, we will measure the distance of such a point from the point of the extreme function determined analytically:

$$\Delta_{NMCvUMC} = \left\| x_{opt} - x_{NMCvUMC} \right\|^2 \tag{3}$$

If we base on the evaluation of accuracy of finding an extreme function point in this way, we can compare the algorithms and tell which of them has higher chances to approach to the point with a true extreme of function, while searching the extreme point of function.

The second criterion will be used to determine the value of the integral (4) of the functions given by the relations (1) and (2) in the mapped area according to the following formula:

$$V=\int y \cdot \mathrm{ds} = \frac{S}{n}\sum_{k=1}^{n} y_k \qquad (4)$$

In this case, we determine the integral value by a rectangular rule for the function with two variables. The criterion, defined in this way, gives us some information about the quality of the whole area mapping. Taking into consideration, that both nonuniform Monte-Carlo and uniform Monte - Carlo methods work with random numbers, and with each attempt to find the values, mentioned above, according to relations (3) and (4), we will find another point or other value of the integral, we can say that it is not possible to compare the results obtained during individual analyzes easily. We can set a comparative value, which consists of a standard deviation of the calculated values of the extreme function location and the integral value.

Because of the relevance of the statistical processing of results, we will repeat each analysis 100 times for a set number of points (100, 200, 300, 500, 1000) in the mapped area. The mapped area will be measured as $\langle 0\ 1\rangle \times \langle 0\ 1\rangle$ and the number of new generated offspring for determining the new parent of the uniform Monte - Carlo algorithm will be 50. The analysis results for the first example where the investigated function has the shape (1) with the function extreme located in the mapping area are shown in the Tab. 1. Similarly, for the second example, the function being investigated (2), where the function's extreme is located outside of the mapping area Tab. 2.

Table 1. Standard deviation values for the first example

Number of test points	Nonuniform Monte-Carlo		Uniform Monte – Carlo		Comparison of methods	
	$\sqrt{D(\Delta_{\mathrm{NMC}})}$	$\sqrt{D(V_{\mathrm{NMC}})}$	$\sqrt{D(\Delta_{\mathrm{UMC}})}$	$\sqrt{D(V_{\mathrm{UMC}})}$	$\frac{\sqrt{D(\Delta_{\mathrm{NMC}})}}{\sqrt{D(\Delta_{\mathrm{UMC}})}}$	$\frac{\sqrt{D(V_{\mathrm{NMC}})}}{\sqrt{D(V_{\mathrm{UMC}})}}$
100	0.023482	0.023220	0.014664	0.0071718	1.6013	3.2377
200	0.016538	0.015291	0.010730	0.0053820	1.5413	2.8411
300	0.015365	0.010797	0.0099421	0.0041930	1.5454	2.5750
500	0.010629	0.010873	0.0075251	0.0036659	1.4125	2.9660
1000	0.0079789	0.0080184	0.0045661	0.0021895	1.7474	3.6622

Table 2. Standard deviation values for the second example

Number of test points	Nonuniform Monte-Carlo		Uniform Monte – Carlo		Comparison of methods	
	$\sqrt{D(\Delta_{\mathrm{NMC}})}$	$\sqrt{D(V_{\mathrm{NMC}})}$	$\sqrt{D(\Delta_{\mathrm{UMC}})}$	$\sqrt{D(V_{\mathrm{UMC}})}$	$\frac{\sqrt{D(\Delta_{\mathrm{NMC}})}}{\sqrt{D(\Delta_{\mathrm{UMC}})}}$	$\frac{\sqrt{D(V_{\mathrm{NMC}})}}{\sqrt{D(V_{\mathrm{UMC}})}}$
100	0.030873	0.045354	0.014732	0.015343	2.0956	2.9560
200	0.019580	0.034757	0.0075412	0.0086804	2.5964	4.0041
300	0.014893	0.027139	0.0072452	0.008213	2.0556	3.3044
500	0.010964	0.021610	0.0051722	0.0066126	2.1198	3.2680
1000	0.0067649	0.014710	0.0033634	0.0045439	2.0113	3.2373

6 Conclusion

If we base on the results shown in Tab. 1, we can state that the uniform Mote - Carlo method has better assumptions for finding an extreme point of function comparing to the nonuniform Monte-Carlo method. If the function has an extreme in the mapped area, the ratio of the standard deviation of the extreme point finding accuracy for the nonuniform Monte-Carlo method to the standard deviation of the extreme point finding accuracy by using the uniform Monte - Carlo algorithm is in the interval $\langle 1.41 \;\; 1.75 \rangle$.

We can claim the similar fact for finding a point with a function's extreme, if the maximum is located outside of the mapping area Tab. 2. When searching for a point with a maximum value, the standard deviation for the uniform Monte - Carlo method is smaller comparingly to the nonuniform Monte-Carlo method in the interval $\langle 2.01 \;\; 2.60 \rangle$.

When comparing the results of determining the integral value of the investigated function by the relation (4) for the function with the extreme in the mapping area (1), the ratio of the standard deviations for the nonuniform Monte-Carlo method and the uniform Monte - Carlo algorithm is in the interval$\langle 2.57 \;\; 3.66 \rangle$. We have also obtained the similar results for determining the value of the integral for the function with the maximum function value at the boundary of the mapped area given by the relation (2). During comparing the standard value deviations of the integral in the nonuniform Monte Carlo method and the uniform Monte - Carlo algorithm, we can see their ratio in the interval$\langle 2.95 \;\; 4.00 \rangle$.

According to these results, we can say that the WSS algorithm is more useful than the nonuniform Monte-Carlo method in case of looking for the extreme of function as well as mapping of the whole area. Uniform Monte - Carlo algorithm enables us to get more equable distribution of points on the mapped area. If we compare it with a totally random distribution of points in case of the nonuniform Monte-Carlo method, we will see that the uniform Monte - Carlo algorithm has more significant effect on determining the value of the integral of a given function, then, on the mapping the value of the function on the investigated area, whence on the finding of the extreme point of function.

A significant shortcoming of the uniform Monte - Carlo method is a time consuming. On the other hand, this disadvantage should not be an obstacle in case of the practical solution of the tasks of technical practice. For this type of tasks, analysis of the desired function may last from a few seconds to a few hours. In the case of time-consuming FEM analysis of target function, the time-consuming of the uniform Monte - Carlo algorithm is comparable to the duration of one FEM analysis of the target function and, overall, is not so significant. The assumption of finding a point, which is closer to the extreme of the function, is significantly higher for the uniform Monte - Carlo algorithm.

The contribution was created on the basis of the financial support of the grant project APVV 14-0096 a KEGA 037ŽU-4/2018.

References

1. P. Kumar, P. Luo, F. Gaspar, C. Oosterlee, *A multigrid multilevel Monte Carlo method for transport in the Darcy-Stokes system*. Journal Of Computational Physics, **371,** 382 – 408, (2018)
2. L. Jakubovicova, M. Saga, *Computational analysis of contact stress distribution in the case of mutual stewing of roller bearing rings.* Novel Trends in production devices and systems, **474**, 363-368 (2014)
3. J. Zapomel, V. Dekys, P. Ferfecki, A. Sapietova, M. Saga, M. Zmindak, *Identification of Material Damping of a Carbon Composite Bar and Study of Its Effect on Attenuation of Its Transient Lateral Vibrations.* International Journal of applied mechanics, **7** (2015)

4. P. Orsansky, B. Ftorek, P. Durcansky, *Mathematical model of a closed hot air engine cycle using MATLAB Simulink.* AIP conference proceedings **1608**, 173-176 (2014)
5. V. Dekys, P. Kopas, M. Sapieta, O. Stevka, *A detection of deformation mechanisms using infrared thermography and acoustic emission.* Novel Trends in production devices and systems, **474**, 315-320 (2014)
6. R. Halama, P. Macura, L. Pecenka, F. Fojtik, M. Sofer, *Experimental Analysis of Residual Stresses in Backup Roll and FE solution.* Experimentalni analyza napeti - experimental stress analysis, 85 – 90 (2011)
7. R. Grega, J. Krajnak, L. Zulova, G. Fedorko, V. Molnar, *Failure analysis of driveshaft of truck body caused by vibrations.* Engineering Failure Analysis **79**, 208-215 (2017)
8. G. Domek, M. Dudziak, A. Kolodziej, *Timing belt gear design for mechatronics system.* Procedia Engineering, **96**, 39-43, (2014)
9. T. Domanski, W. Piekarska, M. Kubiak, Z. Saternus, *Determination of the final micro-structure during processing carbon steel hardening.* Procedia Engineering, **136**, 77-81 (2016)
10. M. Macko, J. Flizikowski, Z. Szczepanski, K. Tyszczuk, G. Smigielski, A. Mrozinski, J. Czerniak, A. Tomporowski, *CAD/CAE Applications in Mill's Design and Investigation.* Lecture Notes in Mechanical Engineering, 343 – 351, (2017)
11. J. Flizikowski, M. Macko, J. Czerniak, A. Mroziński, *Implementation of genetic algorithms into development of mechatronic multi-edge's grinder design,*Proceedings of the asme international mechanical engineering congress and exposition, **7**, 1227-1235, (2012)
12. M. Łazarska, T. Woźniak, Z. Ranachowski, A. Trafarski, G. Domek, *Analysis of acoustic emission signals at austempering of steels using neural networks,* Metals and Materials International, **23**, 426-433, (2017)
13. R. Halama, A. Markopoulos, F. Fojtik, M. Fusek, Z. Poruba, J. Famfulík, *Effect of stress amplitude on uniaxial ratcheting of aluminum alloy 2124-T851.* Materialwissenschaft und Werkstofftechnik, **48**, 814-819, (2017)
14. V. P. Mateichyk, V. P. Volkov, P. B. Komov, I. V. Gritsuk, et al., *Special features of vehicle condition monitoring using onboard diagnostics systems,* Project management, systems analysis and logistics: Scientifc journal, K.: NTU, **13**, 126-138, (2014)
15. Y. Turygin, P. Bozek, I. Abramov, Y. Nikitin, *reliability determination and diagnostics of a mechatronic system,* Advances in science and technology-research journal, **12**, 274 – 290, (2014)
16. E. Kalentev, S. Vaclav, P. Bozek, V. Tarasov, A. Korshunov, alexander, *numerical analysis of the stress-strain state of a rope strand with linear contact under tension and torsion loading conditions,* Advances in science and technology-research journal, **11**, 231 – 239, (2017)
17. P. Krawiec, G. Domek, Ł. Warguła, K. Waluś, J. Adamiec, *The application of the optical system ATOS II for rapid prototyping methods of non-classical models of cogbelt pulleys.* MATEC Web of Conferences, **157**, (2018)
18. P. Krawiec, *Numerical Analysis of Geometrical Characteristics of Machine Elements Obtained Through CMM Scanning,* Progress in Industrial Mathematics at ECMI 2008, **15** , 933–938 (2010)
19. T. Lack, J. Gerlici, *a modified strip method to speed up the calculation of normal stress between wheel and rail,* Experimental stress analysis 51, **486**, 359 – 370, (2014)
20. J. Palieskova, M. Pajtasova, A. Feriancova, A. D. Ondrusova, K. Holcova, J. Vavro, J., Jr., S. C. Mojumdar, thermal properties of fillers based on organoclays in the polymeric materials, Journal of thermal analysis and calorimetry, **199**, 939-943, (2015)

Application possibilities of the low-temperature repairs on creep-resistance turbine components from material GX23CrMoV12-1

Jaromir Moravec[1], *Iva* Novakova[1,*], *Jiri* Sobotka[1], *Tomasz* Kik[2]

[1] Technical University of Liberec. Studentská 2, 461 17 Liberec 1. Czech Republic
[2] Silesian University of Technology, Welding Department, Konarskiego 18A, Gliwice, Poland

Abstract. Papers deals with the problems of repairs at new and also already loaded components of steam and gas turbines. Generally, it is about welding repairs at valve and turbine chambers, transmitters, vane carriers, tripping valve body and so on. There are both technological defects caused by casting (e.g. shrinkage cavities, misruns, porosity and cracks) and defects caused by operation (thermal-stress fatigue, incorrect method of operation, etc.). Many of the mentioned defects take effect not until the final technological method - mostly at machining. However, in this production phase it is not possible to get form the supplier new production elements due to economical and also delivery time reasons. All of that results in delay of production and penalty. As the major aim of this paper, there was to find such repairs techniques which can ensure serviceability of mentioned devices under the standard regime. In these cases is needed to use much lower tempering temperature than under the standard production, mainly because of necessity to eliminate deformation caused by heating of part that is not possible to be subsequently removed by machining.

Keywords: Welding, Heat Treatment, GX23CrMoV12-1, Repairing, Strengthening

1 Introduction

Different types of defects can negatively influence behavior of creep resistance materials especially at production and running of casted turbine components. These defects are generally caused either by used production technologies or due to the subsequent operational loading. Nevertheless, these production defects are very difficult to be removed because these materials and products of them are typical of e.g. their highly-alloying, complex shape or abnormal dimensions of casted parts. [1,2] Some of these defects (the visual ones) are repaired directly in the foundries with subsequent specified heat treatment of the whole part in furnace (for martensitic steels tempering at temperatures 710 up to 740°C). Moreover, ultrasonic and eddy current NDT is used during visual inspection of every part. [3,4] However, there are defects that can´t be detected during visual inspection. These defects are generally detected at subsequent technological operations, mainly during/after machining when the given part already has the final assembly dimensions. Nevertheless, generally

* Corresponding author: iva.novakova@tul.cz

Reviewers: *Jozef* Meško, *Daniela* Kalincová

during machining can be detected (or even made) other defects as e.g. are cracks in the zones of local stress waves. Beside the mentioned types of defects can be at energy devices detected also defects caused by operational conditions (parameters and purity of steam, blackouts, starting rates and so on), eventually by infringement of the specified operational parameters by operator (e.g. unauthorized use or manipulation). [5,6]

There are generally used covered electrodes (111 acc. to ISO 4063) to repair such kind of defects. With such approach, there is needed to carry out welding with pre-heating and the whole part (or just area to be repaired) is heat treated after welding. If the part dimensions are not abnormal, heat treatment can be done in the furnace. If the part dimensions are abnormal, there is necessary to perform pre-heating and subsequent heat treatment of part by means of the electric heating elements (so-called muffs). However, in this case can´t be used the common processing process because of high pre-heating temperatures and subsequent heat treatment. Repairs are supposed to take into account fact that there is necessary to eliminate deformation of machined surfaces and plastic stress accumulation, cause to achieve the same operational service life. Also creation of undesirable structural changes must be avoid and eliminate also. Last, but not least, there is thermal degradation of basic materials at components, which have been operated for a long-time.

2 Experimental part

In the own experiment was used martensitic steel GX23CrMoV12-1. Under the normal and elevated temperature, it is stainless steel alloyed by chromium (Cr), nickel (Ni) and manganese (Mn). Among the most important strengthening mechanisms, which take effect during high-temperature explosion of such type of heat-resisting and modified steels, there are precipitation strengthening and solid solution substitutional strengthening.

Microstructural stability of this steel is positively influenced by low coarse-rate of $M_{23}C_6$ particles. It is due to fact that during creep loading, there is growing and subsequent coarsing of these particles, whereas coarse-rate increases with the increasing Cr content in steels. Another creep resistance increase is done by nitrogen alloying and also due to vanadium nitride precipitation within the grain.

Thus solid solution strengthening is subsequently achieved by substitution, whereas the main influence has the increasing molybdenum content in the solid solution. It can be also substitutional strengthened by the same mechanism as using tungsten (W), as it is already used at creep resistance materials – e.g. steel P92. Moreover, there was proved that there is no reason to increase the molybdenum equivalent content in steel over the solubility limit. Higher Mo and W content in steels results in secondary phases precipitation which are rich in Mo or W - e.g. M_6X and $(Fe, Cr)_2(Mo, W)$. So finally there is lowering of solid solution strengthening. [7,8]

Steels GX23CrMoV12-1 is suitable to be used up to temperature 600°C, whereas its ultimate strength under creep achieves 59 MPa for this temperature and temperature exposition for 10^5 Hrs. They are often used as steam turbines components as can be e.g. blades, rotors or compressors in USC power plants. Such steels offer a very good ratio between strength properties and resistance against oxidation under high temperatures - all of it under relatively low cost.

In Fig. 1 is shown the structure of delivered material GX23CrMoV12-1. It is tempered martensitic material with the initial hardness 259 HV for loading that corresponds to method HV10 acc. to ČSN EN ISO 6507-1. In Tab. 1 is shown the chemical composition of GX23CrMoV12-1 steel that was determined by spectrometer Q4 Tasman. In Tab. 2 are summarized mechanical properties of this tested material.

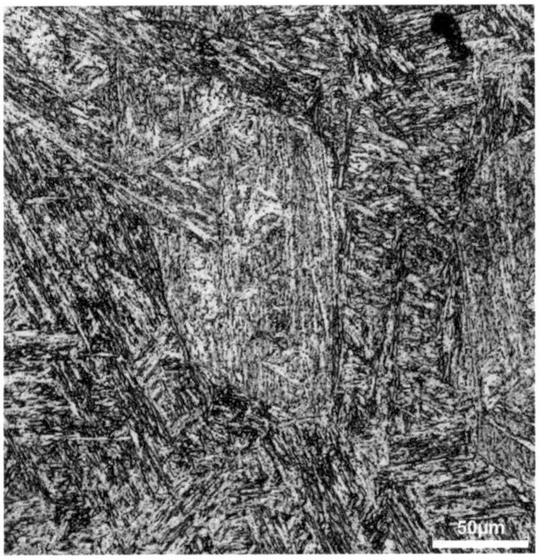

Fig. 1. Structure of tested steel GX23CrMoV12-1

Table 1. Chemical composition of GX23CrMoV12-1determined by spectrometer Q4 Tasman

Chemical composition of GX23CrMoV12-1 [wt.%]								
C	**Si**	**Mn**	**Cr**	**Mo**	**Ni**	**V**	**N**	**Nb**
0,238	0,360	0,609	11,82	1,148	1,151	0,301	0,054	0,007

Table 2. Measured mechanical properties of delivered steel GX23CrMoV12-11

Yield strength		**Ultimate strength**		**Total ductility A_5 at 20°C [%]**	**Impact energy KV2 at 20°C [J]**
$R_{p0,2}$ at 20°C [MPa]	**$R_{p0,2}$ at 600°C [MPa]**	**R_m at 20°C [MPa]**	**R_m at 600°C [MPa]**		
590	312	821	364	19	29

Before the classification of welding procedures and creation WPS, there was necessary to create the preliminary repairing welding procedure pWPS that is needed to be verified in light of both welding and subsequent thermal treatment of welds. Experimental verification is done by samples, which match to repairs on real part. In this case was as sample used multi-layer weld made into groove machined into workpiece having shape of plate with dimensions as following: 400 x 150 x 50 mm. Groove created along the whole length of plate had depth 20 mm, lower width of groove 15 mm, transient radii R5 and upper width of groove 25 mm. Groove is schematically shown in Fig. 2.

Groove was fulfilled totally by 21 runs in 7 layers over each other, whereas each layer contained 3 runs side by side. Welding was done by method 111 acc. to ISO 4063 by electrodes BÖHLER FOX 20 MVW (E CrMoV12 B 4 2 H5). Root layer was welded by electrodes with diameter 2.4 mm, filling layers by electrodes with diameter 3.2 mm and also 4.0 mm and top layer by electrodes with diameter 5.0 mm. Repairing by welding was performed by pre-heating of part on minimal temperature 420°C and interpass temperature 480°C. During the whole time of repairing was temperature hold slightly over the temperature M_s. Pre-heating was carried out by means of resistance heating elements (muffs) connected to device Weldotherm 1, whereas the whole welding sample was wrapped in the insulating cotton Sibral. After welding there was a controlled cooling of sample on temperature 100°C to enable the complete martensitic transformation in the repair area. It was followed by heat treatment after holding time 2 Hrs.

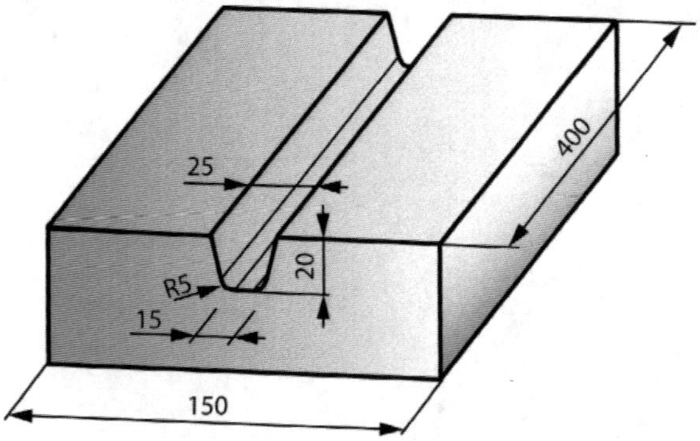

Fig. 2. Geometry of groove used for the multiple weld

In the case of standard welding procedures for steel GX23CrMoV12-1 is welding followed by heat treatment that consists of heating on temperature 730°C and holding time on this temperature from 4 up 8 Hrs. However, in the case of repairs there can´t be used so high temperature for thermal treatment due to reasons mentioned in Chap. 1. That is why there was for repairs firstly proposed to apply tempering temperature of 550°C. To be specific, there was heating rate 100°C/Hr on tempering temperature 550°C, holding time 6 Hrs and subsequent cooling by cooling rate 50°C/Hr on temperature 300°C. Finally, sample cooled down spontaneously in wrap after that.

Properties of sample that was prepared and thermal treated as was described above were evaluated by means of hardness measurement acc. to Vickers HV10 on device Qness Q30A. From the thermal treated part were taken samples for subsequent metallographic evaluation and hardness measurement. These samples were processed by the commonly used metallographic procedure. The etchant Villela Bain was used to highlight structure. Hardness was measured in the parent material, HAZ and weld metal. Totally there were 303 incisions located in 7 longitudinal and 3 transverse lines in such manner that there was evaluated hardness in every of 7 welding layers. Incisions were done acc. to standard ČSN EN ISO 9015-1 for loading HV10 with axial distance 1 mm. Measured values of hardness are summarized in Fig. 3. For the weld joint, magnitude of hardness in the weld metal and in HAZ should not exceed over 350 HV.

From the results is clear that tempering at temperature 550°C for 6 Hrs is completely insufficient. Hardnesses in weld metal and HAZ significantly exceeded the maximal allowed magnitude of 350 HV. That is why there was decided to extend the holding time at tempering temperature 550°C up to 10 Hrs. However, measured results revealed that also in this case is applied thermal treatment insufficient in light of weld metal and HAZ hardness values. There was certain decrease of hardness values indeed, but it was not enough. Thus holding time was extended on 45 Hrs because of that. Results measured under this condition are shown in Fig. 4. It flows from them that after this thermal treatment there was sufficient decrease of hardness values in weld metal, but hardness values in HAZ still exceed the maximal allowed hardness 350 HV. Due to fact that another extension of tempering time seemed to be useless, there was decided to increase tempering temperature on 580°C and keep the same tempering time – so 45 Hrs. In this case, measured results proved a strong decrease of hardness values in weld metal (i.e. under 350 HV).

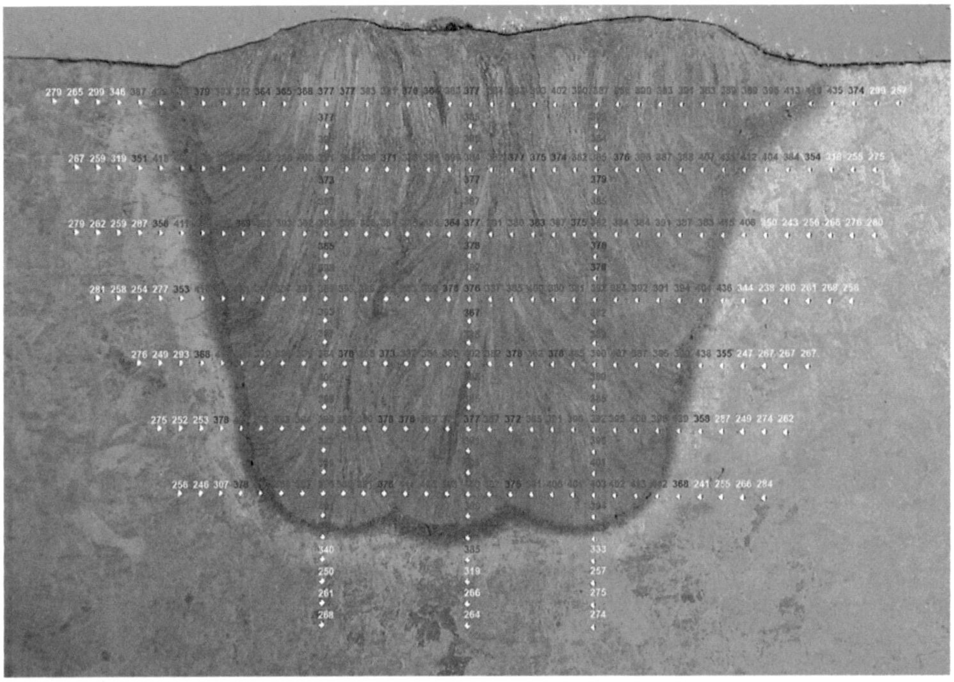

Fig. 3. Hardness values for multiple weld after welding and heat treatment at 550°C with holding time 6 Hrs

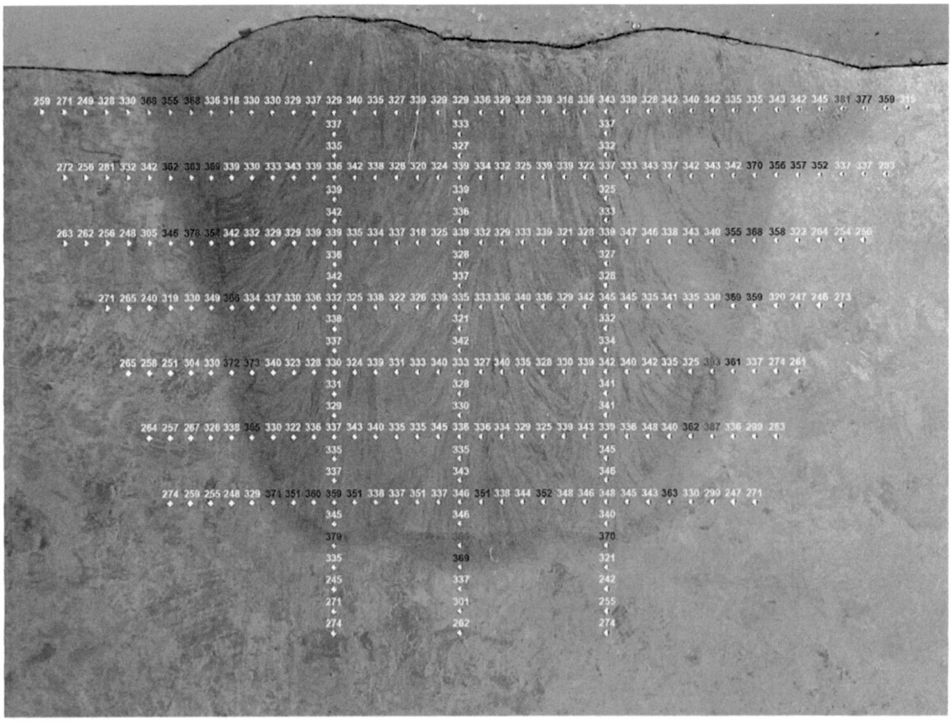

Fig. 4. Hardness values for multiple weld after welding and heat treatment at 550°C with holding time 45 Hrs

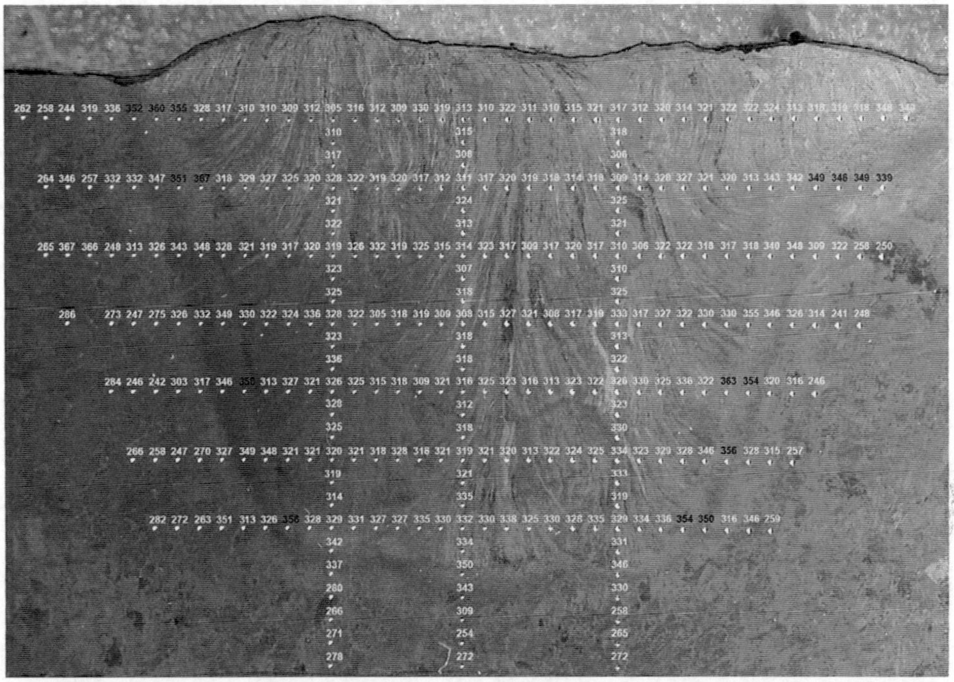

Fig. 5. Hardness values for multiple weld after welding and heat treatment at 580°C with holding time 45 Hrs

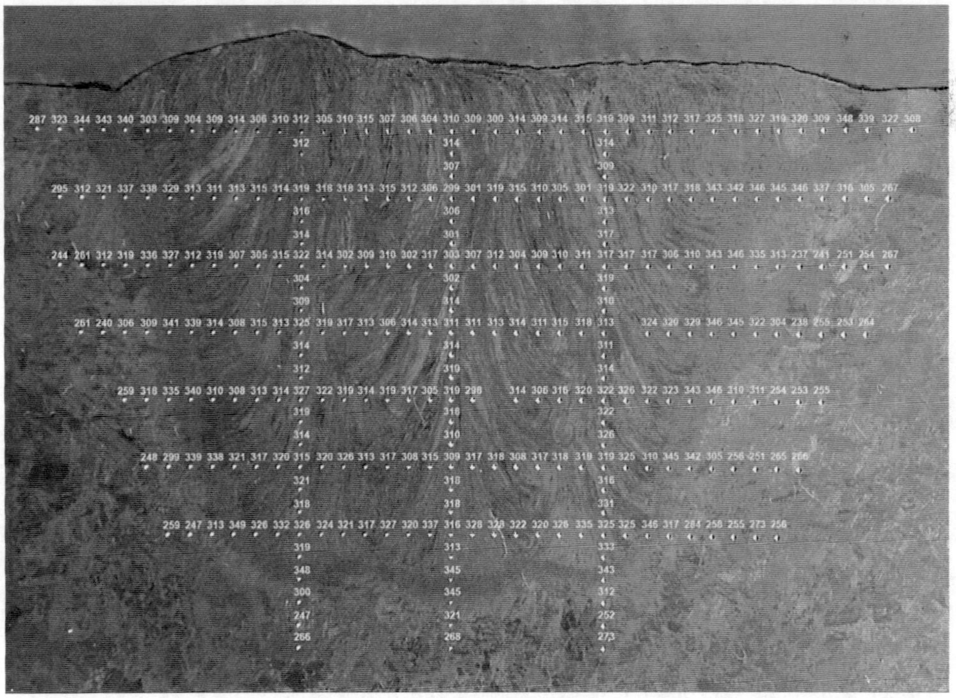

Fig. 6. Hardness values for multiple weld after welding and heat treatment at 595°C with holding time 45 Hrs

Nevertheless, hardness values in certain areas of HAZ still exceeded the required magnitude. So it means that also this approach is not fully suitable for considered repairing. So finally there was decided to increase tempering temperature once more – now up to 595°C. In this case (see fig. 6) were for weld metal and the complete HAZ measured hardness values lower than 350 HV. In the weld metal they varied within 300 HV and 337 HV and in HAZ within 327 HV up to 348 HV.

Acc. to standard ČSN ISO 148-1, these measured results should be confirmed by Charpy impact test. Positioning of testing samples in weld joint and notch orientation was performed acc. to ČSN EN ISO 9016. Samples were prepared in such manner to be tested both in HAZ (VHT 0/3) and in weld metal (VWT 0/3 and VWT 5/3). Own test was carried out on the Charpy impact tester LabTest CHK 450 J and always for 3 samples. Measured values of impact energy KV2 and impact toughness are summarized in Tab. 3.

Table 3. Measured values of KV2 and KCV2

No. of sample	VHT 0/3		VWT 0/3		VWT 5/3	
	KV2 [J]	KCV2 [J.cm^{-2}]	KV2 [J]	KCV2 [J.cm^{-2}]	KV2 [J]	KCV2 [J.cm^{-2}]
1	22	27.5	38	47.5	41	51.3
2	29	36	46	57.5	33	41.3
3	26	32.5	34	42.5	37	46.3
Arithmetic mean	25	32	39	49	37	46.3
Minimal value	22	27.5	34	42.5	33	41.3

3 Conclusion

At welding 9 – 12% Cr steels is very necessary to use not only pre-heating, but also subsequent thermal treatment at relatively high temperatures. This ensures both decreasing of residual stresses after welding and mainly carbon diffusion in martensite and thus also partial decreasing of material strength and hardness values at simultaneous increase of ductility and impact toughness. Moreover, dynamic of these processes is increasing with increasing tempering temperature.

Nevertheless, it is not possible to use tempering temperatures high enough at repairs of real parts. It is partly due to intensive oxidation on the machined parts surfaces, but mainly due to creation of deformation, which can´t be removed by subsequent machining. Thus as a result, there is tendency to lower tempering temperatures as low as possible.

There was observed tempering of weld metal (created by electrodes BÖHLER FOX 20 MVW) for tested material GX23CrMoV12-1 and decrease of hardness values already under tempering temperature 550°C and relevant holding time 45 Hrs. Despite the partial decrease of hardness values in HAZ, their magnitudes still varied within the range of 355 HV up to 393 HV. Lower tempering temperature couldn´t decrease hardness values fast enough, as it is in the weld metal. In HAZ there represent tempering temperature 595°C with holding time 45 Hrs (minimally) limit parameters for subsequent thermal treatment. By application of this thermal treatment it was possible to achieve that in the whole measuring area harness values didn´t exceed over 350 HV. Also magnitudes of impact energy in weld metal were totally sufficient. These impact energies KV2 varied for test VWT 0/3 and VWT 5/3 within 33 J up to 46 J.

Also in HAZ it managed to approximate impact toughness to magnitude 29 J that corresponds to thermally treated parent material. From the Charpy impact test were for 3 samples achieved values as 22 J, 26 J and 29 J. It already means quite good toughness under the surrounding temperature. Because the repairing components will work under temperatures about 600°C, there can be presupposed much higher magnitude of impact toughness and so very small probability of brittle failure occurrence. Also for working temperature start and stop times would be obtained impact toughness sufficient for repairing area.

This publication was written at the Technical University of Liberec, Faculty of Mechanical Engineering with the support of the grant project MPO TRIO 10510 and with the support of the Institutional Endowment for the Long Term Conceptual Development of Research Institutes, as provided by the Ministry of Education, Youth and Sports of the Czech Republic in the year 2018.

References

1. P. Hanus, P., E. Schmídová, *Influence of the Welding Process on the Martensitic and Dual Phase High Strength Steels*, Manufacturing Technology, **Vol. 16**, 2016, pp. 702-707, (2016)
2. S. Kuśmierczak, *Evaluation of Degradation of Heat Stressed Pipelines.* Manufacturing Technology, **Vol. 15**, pp. 1006-1010, (2016)
3. M. Patek, A. Sladek, M. Mician, *Destructive testing of the weld joints on split sleeve for branch connections repairs.* Communications - Scientific Letters of the University of Zilina, **Vol. 17**, Issue 4, pp. 65-69, (2015)
4. J. M. Lee, H.D. Seo, H. Chung, *Efficient welding distortion analysis method for large welded structres*, Journal of Materials Processing Technology, **Vol. 256**, June 2018, pp. 36-50, (2018)
5. J. Novotný, J. Honzíková, V. Pilous, K. Stránský, *Verification for the Causes of the Degradation of Welded Joints in Power Plant.* Manufacturing Technology, **Vol. 16**, pp. 1106-1110, (2016)
6. M. Mičian, M. Patek, A. Sládek, *Concept of repairing branch pipes on high-pressure pipelines by using split sleeve.* Manufacturing Technology, **Vol. 14**(1), pp. 60-66, (2014)
7. J. Koukal, M. Sondel, D. Schwarz, V. Foldyna, *Development and microstructure of advanced creep resistant ferritic steels.* International Journal of Microstructure and Materials Properties, **Vol. 6**, Issue 1-2, May 2011, pp. 122-131, (2011).
8. P. Mohyla, L. Havelka, K. Kouřil, *Mechanical properties testing of P92 welded joints prepared by manual metal arc welding.* Tehnicki Vjesnik, **Vol. 25**, Issue 1, February 2018, pp. 60-63, (2018)

Nanocomposite coatings in polymeric matrices and their effect on friction coefficient

Jan Novotný[1,*], *Irena* Lysoňková[1], *Milan* Sapieta[2]

[1]J.E Purkyně University in Ústí nad Labem, Faculty of Mechanical Engineering, Ústí nad Labem, Czech Republic
[2]University of Žilina, Faculty of Mechanical Engineering, Univerzitna 1, 010 26, Žilina, Slovak Republic

Abstract. This paper deals with possibilities of production of new nanocomposite coatings in polymer matrix on aluminium alloys, namely the formation of a composite coating in a polymer matrix on an aluminium alloy. It is a PMMA coating (polymethyl-methacrylate) with the addition of TiO_2 particles. Working with these particles requires not only safety but also a suitable preparation process to obtain particles of suitable size, their subsequent homogeneous distribution in the coating (particles of this size are influenced by electro statically attractive forces and have a strong tendency to aggravate).The aim is to determine if the coatings will affect the surface's condition and its properties. The focus of the work is on selecting of the appropriate preparation of sample technology, examining the particle distribution in the coating and the effect of the coating on the coefficient of shear friction. Consequently, what can be achieved is the coefficient of shear friction of the surface of the coated part compared to the uncoated surface (at different particle concentrations in the spin) and the distribution of TiO_2 particles on the surface of the sample.

Keywords: Nanocomposite coating, polymer, coefficient of friction, surface analysis.

1 Introduction

The usage of polymeric materials as a matrix for nanocomposite materials has become a frequent case. This is mainly due to their features and availability. It is important to keep in mind the size of the nanoparticles used and their amount in the matrix. The aim of the research is to investigate the possibility of forming PMMA matrix nanocomposite coatings. An influencing factor is particle selection and coating technology. For evaluation, it is important to measure and analyse the resulting coatings. Therefore, shear friction coefficient by tribometric assays and particle size distribution in electron microscopy will be tested.

* Corresponding author: jan.novotny@ujep.cz

Reviewers: *Róbert* Čep, *Mária* Čilliková

2 Used materials

2.1 Characteristics of the underlying material

The base material, which is the material on whose surface we form the coating, is the Al-Si alloy. The base material was obtained from previous research of metal forms designed especially for the production of tires. These metal molds are made of Al-Si alloys.

2.2 Matrix material

Polymethyl methacrylate (PMMA) was selected as the matrix material. It is an amorphous thermoplastic with perfect light transmission. Due to the chemical composition, it is a polar plastic that has excellent mechanical properties. It has a wide range of applications, especially in the automotive and construction industries. The modulus of elasticity reaches up to 3200 MPa. [1-4]

2.3 Reinforcement material

Titanium dioxide (TiO_2) was chosen as the reinforcement material of the nanocomposite coating. It is very suitable for coating for its properties in nanoparticle scale. Its material of high hardness is antistatic and is resistant to mechanical wear. [5-7] It has a high refractive index and is just like PMMA UV resistant. TiO_2 nanopowder from the previous research was used. TiO_2 was milled as a nanopowder by a planetary ball mill. [8-11]

3 Nanocomposite coating preparation

As a matrix for the nanocomposite coating, polymethylmethacrylate (PMMA) was used, which is available to buy in the form of plates. Titanium dioxide (TiO_2) in the form of a milled nanopowder was used as reinforcement.
A solution method was chosen for the preparation. As the PMMA solvent, chloroform ($CHCl_3$) was chosen.

Table 1. Labelating of created samples

Solution dilution - 1:10		Solution dilution - 1:10	
Application Method - dipping		Application Method - by brush	
Sample	Concentation of TiO_2 [%]	Sample	Concentation of TiO_2 [%]
1A	0	1C	0
2A	0.01	2C	0.01
3A	0.1	3C	0.1
4A	1	4C	1
Solution dilution - 1:3		Solution dilution - 1:3	
Application Method - dipping		Application Method - by brush	
Sample	Concentation of TiO_2 [%]	Sample	Concentation of TiO_2 [%]
1B	0	1D	0
2B	0.01	2D	0.01
3B	0.1	3D	0.1
4B	1	4D	1

The final coating solution was already applied to samples of the backing material. Two coating methods were chosen: first being the dipping method, which consists in submerging the prepared backing material into the coating solution. The second method was the brush application method. Brush dipped in the coating solution and such is apply it to the substrate. The resulting coating is allowed to dry at room temperature.

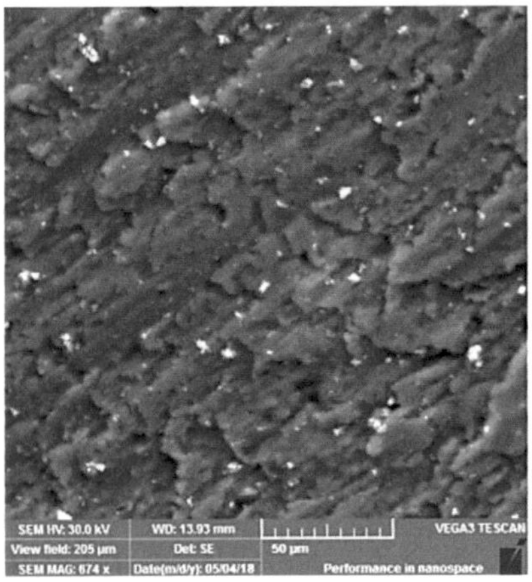

Fig. 1. Sample 4A coated by dipping method in a solution diluted 1:10

Figure of the sample 4A (Figure 1) shows that the higher titanium dioxide concentrations in the coating (1%) by the chosen coating method - by dipping in a solution diluted 1:10, i.e. the same coating and dilution method as in all samples marked as A, it results in a better distribution of the particles in the coating. However, there are still areas with accumulated dust particles in the coating. Care must be taken to ensure that the environment is clean.

4 Research tools - Tribometric assays

It was necessary to analyse the examined coatings for the evaluation. The coefficient of shear friction of the coated surfaces was tested by tribometric assays.

Measurement of the friction coefficient was performed by the universal tribometer Bruker UMT Tribolab. We used a two-axis power sensor with a maximum load of 1000N and 50N.

Input values:

- Tribometer Body - ball ⌀ 10 mm
- Track length $x = 20$ mm
- Velocity $v = 10$ mm·s^{-1}
- Measuring time $t = 60$ s
- Load force $F_z = 5$N

5 Results of the Measurement

The usage of polymeric materials as a matrix for nanocomposite materials has become a frequent case. This is mainly due to their features and availability. It is important to keep in mind the size of the nanoparticles used and their amount in the matrix. The aim of the research is to investigate the possibility of forming PMMA matrix nanocomposite coatings. An influencing factor is particle selection and coating technology. For evaluation, it is important to measure and analyse the resulting coatings. Therefore, shear friction coefficient by tribometric assays and particle size distribution in electron microscopy will be tested.

Table 2. Measured coefficients of friction

Sample	Coefficients of friction	Standard deviation
Sample without coating (SO)	0.1216	0.00898

Table 3. Measured coefficients of friction 1A – 4A

Sample	Concentation of TiO_2 [%]	Coefficients of friction	Standard deviation
1A	0	0.3788	0.03119
2A	0.01	0.3812	0.03796
3A	0.1	0.4075	0.03773
4A	1	0.4076	0.04143

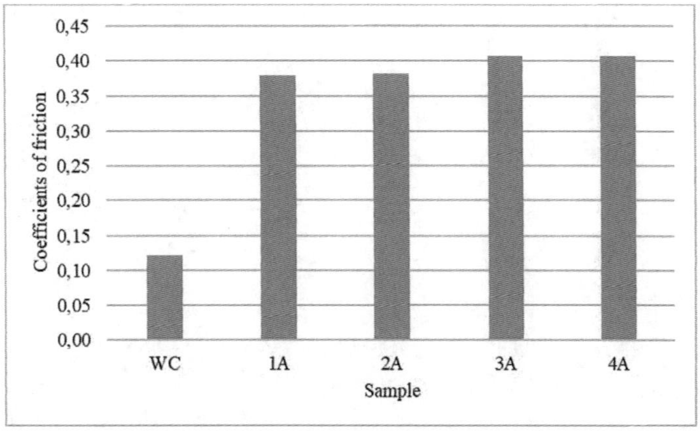

Fig. 2. Comparison of measured coefficients of friction (A)

The friction coefficient values measured in samples 1A to 4A are compared in Figure 2. There is well visible a jump increase in the coefficient of friction coefficient value of coated samples over a non-coated sample. Furthermore, a slight increase in the friction coefficient value can be observed with the increasing concentration of titanium dioxide in the coating.

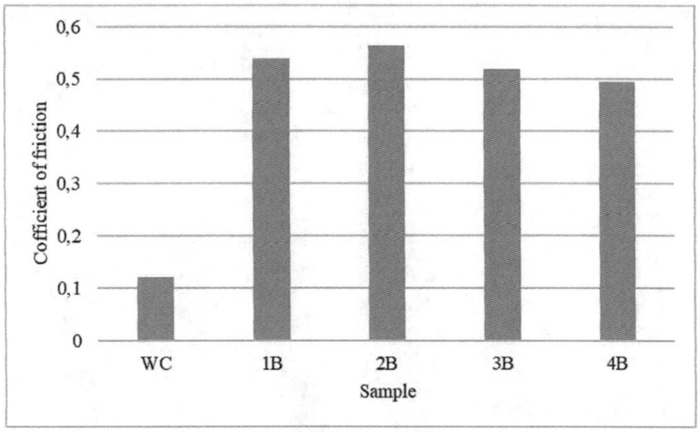

Fig. 3. Comparison of measured coefficients of friction (B)

The friction coefficient values measured in samples 1B to 4B have the highest increase. Its value is almost fivefold compared to the uncoated sample. It can be seen from Figure 3 that the increase is remarkable for all coated samples. However, the rule does not be applied to the increasing coefficient of friction with the increasing percentage of titanium dioxide.

Table 4. Measured coefficients of friction 1C – 4C

Sample	Concentation of TiO_2 [%]	Coefficients of friction	Standard deviation
1C	0	0.3876	0.03998
2C	0.01	0.4147	0.03768
3C	0.1	0.4155	0.04208
4C	1	0.4166	0.05153

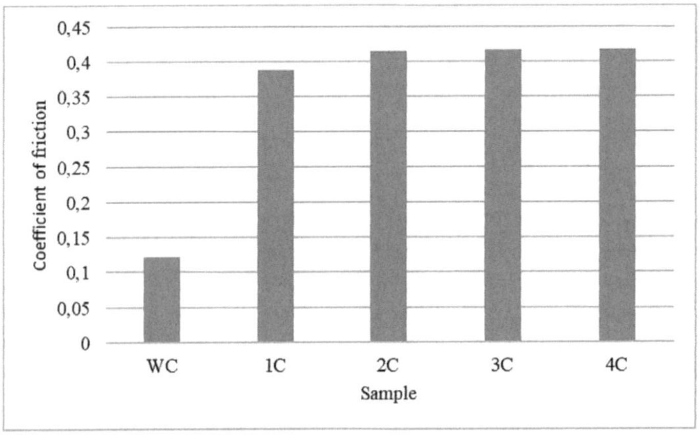

Fig. 4. Comparison of measured coefficients of friction (C)

For samples marked 1C to 4C also applies the jump increase of the coefficient of friction value. Here we can observe a slight increase in the coefficient of friction and the increasing percentage of titanium dioxide in the coating. The highest coefficient of friction is thus on the sample 4C, which is also observable in Figure 4.

Table 5. Measured coefficients of friction 1D – 4D

Sample	Concentation of TiO_2 [%]	Coefficients of friction	Standard deviation
1D	0	0.4733	0.04683
2D	0.01	0.4909	0.04756
3D	0.1	0.4935	0.05142
4D	1	0.4941	0.06319

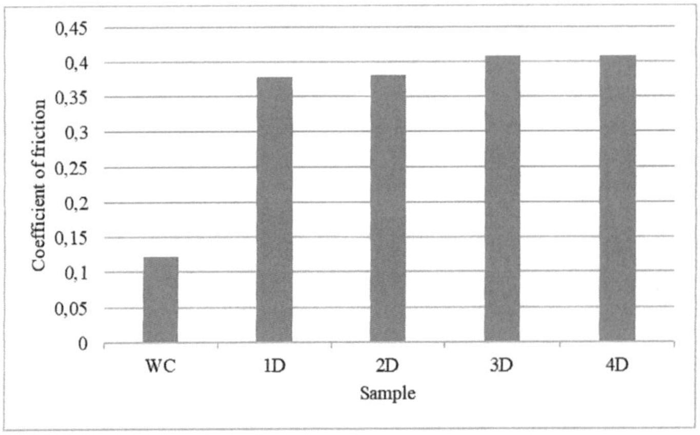

Fig. 5. Comparison of measured coefficients of friction (D)

The friction coefficient values measured in the 1D to 4D samples show again a stepwise increase in the friction coefficient of the coated samples compared to the uncoated sample. It can be seen in Figure 5 that the greatest friction coefficient is for sample 4D is. The increase is approximately four times the coefficient of friction of the uncoated material. It is also a valid phenomenon that increasing the percentage value of titanium dioxide increases the coefficient of shear friction.

6 Conclusions

The aim of the experiment was to evaluate the measured friction coefficients of the coated samples and their comparison with the uncoated sample. It was important to choose the kind of added reinforcement particles in the matrix. Several variants of the concentrations of the added particles were also tested in the coatings, namely titanium dioxide particles. There were prepared two variants of dilution of the coating solution and the technology of its application to the underlying material.

Measurement of the friction coefficient produced an expected and targeted finding that the coated samples exhibited a jump increase in the coefficient of friction versus uncoated samples. On the frame of the three variants of the coated samples, there is a link between the increasing TiO_2 concentration and the friction coefficient value. For all samples A, B, C, D, more than three times the coefficient of friction was measured compared to uncoated samples. A new beneficial finding in general is that the highest coefficient of friction in coated samples was found in samples with the highest TiO_2 concentration, it means with concentration 1% of TiO_2 in polymer matrix.

References

1. L. Běhálek, *Polymery.* [online], https://publi.cz/books/180/18.html
2. J. Krzak, J. Filipiak, C. Pezowicz, A. Baszczuk, M. Miller, M. Kowalski, R. Będziński, *The effect of substrate roughness on the surface structure of TiO2, SiO2, and doped thin films prepared by the sol-gel method.* Acta of Bioengineering and Biomechanics, vol. 11, nr 2, 21-29 (2009)
3. Čuboňová, N. - Císar, M. *Design of camera mount and its application for monitoring machining process.* Advances in Science and Technology, Research Journal, 9(26), 2015, 34-40.
4. J. Petru,T. Zlamal, R. Cep, D. Stancekova, M. Pagac, O. Vortel, *Mechanisms of cutting insert wear and their influence on cutting ability of the tool during machining of special alloys.* Conference: 3rd International Conference on Manufacturing Engineering and Technology for Manufacturing Growth (METMG 2015), Canada 36-40 (2015)
5. R. Horký, *Návrh nanokompozitních povlaků v polymerní matrici a jejich vliv na vlastnosti povrchu.* Univerzita J. E. Purkyně v Ústí nad Labem, Ústí nad Labem, (2018)
6. M. Novák, N. Náprstková, *The influence of cutting conditions on surface roughness during steel X38CrMoV5 grinding.* Key Engineering Materials 581, 247-254 (2014).
7. P. Valášek, R. D'Amato, M. Müller, A. Ruggiero, *Mechanical properties and abrasive wear of white/brown coir epoxy. Composites Part B.* Engineering 146, 88–97 (2018)
8. J. Novotný, M. Jaskevič, J. Cais, T. D. Hung, *The shape and structure of titanium dioxide nanoparticles.* In Manufacturing Technology, Vol. 16, No. 5, pp. 1102-1106 (2016)

9. T. Hanes, P. Hvizdoš, M. Ťavodová, D. Kalincová, J. Hricová, P. Beňo, *Coating surface roughness measurement made on coining dies,* Manufacturing technology : journal for science, research and production. - ISSN 1213-2489. - Vol. 14, no. 3, 309-317 (2014)
10. Stanček, J., Bulej, V. *Design of driving system for scissor lifting mechanism*. Academic journal of manufacturing engineering. **13**(4), 38-43 (2015)
11. J. Zapomĕl, V. Dekýš, P. Ferfecki, A. Sapietová, M. Sága, M. Žmindák, *Identification of material damping of a carbon composite bar and study of its effect on attenuation of its transient lateral vibrations*. Int. Journal of Applied Mechanics **7** (6), (2015)

Effect of incorrect mix ratio on strength of two component adhesive Butt-Joints tested at elevated temperature

Jakub Szabelski[1,1]

[1]Lublin University of Technology, Faculty of Mechanical Engineering, Nadbystrzycka 36, 20-618 Lublin, Poland, j.szabelski@pollub.pl

Abstract. The aim of this study was to determine the impact of the incorrect mix ratio on the strength of joints bonded with a commercial epoxy adhesive compo-sition. The performance of cold-cured as well as accelerated cured butt joints was monitored at elevated temperature conditions. The obtained data was subjected to statistical analysis to show the correlation between joint strength at high temperature and incorrect mix ratio. The degradation of adhesive material properties with increase of hardener ratio in adhesive material was observed, as well as the change in failure type from mixed type to clearly cohesive (for inadequate volume of harde-ner) or adhesive (for excessive amount of hardener). Surprisingly insufficiency of hardener doesn't affect the strength of joint in such manner. General recommend-dation were drawn for the preparation of two-component epoxy adhesives for joints to be used in elevated temperature, particularly when uncertainty regarding the correct resin/hardener mix ratio and future research was planned.

Keywords: adhesive joints, adhesive materials, tensile strength, testing, elevated temperature, incorrect mixing ratio,

1 Introduction

The availability of a wide range of adhesive compositions on the market requires of the engineers basic knowledge regarding the groups and characteristics of common products. The differences between adhesives concern their application technology, cure procedures and, last but not least, the strength properties of the adhesive joint, which is after all a crucial aspect determining the suitability of a given adhesive for the particular application. The classification of adhesives for structural applications proposed in [1] includes detailed groups of adhesives and subgroups classified according to cure mechanisms:

1) Adhesives cured by chemical reaction:
   a) two-component adhesives,
   b) one-component adhesives activated by catalyst or curing agent,
   c) adhesives cured by moisture,
   d) adhesives cured by UV light exposure,

[1] Corresponding author: j.szabelski@poiiub.pl

Reviewers: *Miroslav* Müller, *Petr* Valášek

   e) anaerobic adhesives catalysed by substrate,
   f) adhesives in the form of tapes, films, granules, etc,
2) Adhesives cured by solvent evaporation:
   a) contact adhesives,
   b) adhesives cured by pressure,
   c) reactive adhesives,
   d) resin-based adhesives,
3) Adhesives cured on cooling (hot-melt adhesives).

However in practice, adhesives frequently escape the traditional classification by representing the characteristics of different groups. Epoxy adhesives, for that matter, may form compositions cured by: mixing the resin with the curing agent, or by exposing the one-component composition to heat treatment.

Adhesives that set as a result of a chemical reaction between their components upon mixing share one common feature: they are produced as multi-component products, which cure as a result of a chemical reaction between two or more separate components. It is of critical importance that the constituents of the adhesive mix must be dosed at the correct ratio. The "base material" and the curing agent usually differ in colour so as to facilitate the visual inspection of the correct dosage while mixing. The working life of the mixed composition must be observed, as beyond that time the adhesive cannot be successfully dispensed and bond the adherends. The suggested mix ratio is specified by the manufacturer in the form of fixed ratio or a ratio range, given by volume or by weight. Apparently, it is only by strict adherence to component mixing ratio and specified bonding conditions that we can obtain a quality joint that will exhibit optimal strength properties. However, certain findings, such as presented in the work [2] are contrary to the accepted state, proving that certain adhesives develop full shear strength at a different mix ratio than declared by the manufacturer or suggested by the stoichiometric analysis.

Chemo-setting adhesive mass with an insufficient content of hardening agent per resin unit will not become completely polymerised. On the other hand, excessive addition of hardener will produce a fragile and brittle adhesive composition. Furthermore, the excess of unprocessed hardening agent could result in corrosion of the metallic surfaces of adherends [3]. Other properties of adhesives, such as heat resistance or stiffness, may also become compromised by the incorrect resin/hardener mix ratio. The role of imbalance in the proportion of hardener in the adhesive mix should be further explored, with a view to benefiting adhesive joint strength properties.

With the majority of formerly conducted studies into determining the optimal resin/hardener ratio focusing on experimental character, there have been few researches employing computational methods to resolve the issue. Although stoichiometric analyses could provide data regarding the molecular balance in the reaction (hence the correct concen-tration of reagents), nevertheless the obtained values rather serve as a reference point for the experimental analyses [4, 5]. In the case of epoxy adhesives, a certain range of curing agents, apart from the traditional involvement in the polymerisation as a result of reactions between components, play an additional role and act as catalysts in the polymerisation among epoxy groups, thus reducing the participation of these groups in the reaction with the hardener. As a result, the actual amount of the hardening agent necessary for the reaction to occur is lower than it would seem based on the results from the stoichiometric analysis. For instance, the optimal amount of acid anhydride curing agents is estimated at 85-95% of the stoichiometric relationship [6].

Epoxy adhesives are cured by long-chain fatty acid polyamide, obtained in the reaction of dibasic organic acid, employed in polycondensation with surplus ethylenediamine or diethylenetriamine, as a result of which amides with free amine groups are produced. Thus obtained resin/hardener ratio does not need to be strictly observed, and its measuring may be carried out in the visual method, until reaching a roughly similar amounts to the recommendded ones [7, 8].

The main aim of the conducted research was the analysis of degradation of cylindrical butt adhesive joints resulting from the incorrect resin/hardener mix ratio, not in room- but in elevated temperature conditions. Examined adhesive joints were prepared both: cold-cured as well as accelerated cured at high temperature. The goal was to discover if lack of hardener is as unwanted as its excess or if one of this occurrences is more undesirable than the other. There has been significant research conducted on strength of adhesive joints regarding mix ratio of components [9, 10, 11, 12, 13, 14] however, none of these consider butt joints examined at elevated temperature and cured using various mechanisms.

2 Materials

Regarding the selection of the adhesive composition subjected to testing, it was resolved that, for the sake of its suitability for a range of various applications, the joints will be bonded with the versatile and multi-purpose Loctite Hysol 9492. The producer declares that it can be used in traditional bonding applications, as well as sealing and repair of various materials (*e.g.* to remove pores and surface irregularities from casts and forgings). The adhesive is available in a double-cartridge dispensing system with a static mixer nozzle, mixing the adhesive upon dispensing at a 2:1 resin/hardener mix ratio by volume, 100:50 mix ratio by weight. The adhesive mass is a white opaque paste, which working life once it is mixed is approx. 15 min. The fixture time of the adhesive composition, *i.e.* the time it develops the shear strength of 0.1 MPa, is 75 min. The selected adhesive is quite commonly used for research due to its heat resistance and non-conductivity properties as well as handy dispenser and above all good market accessibility [15, 16].

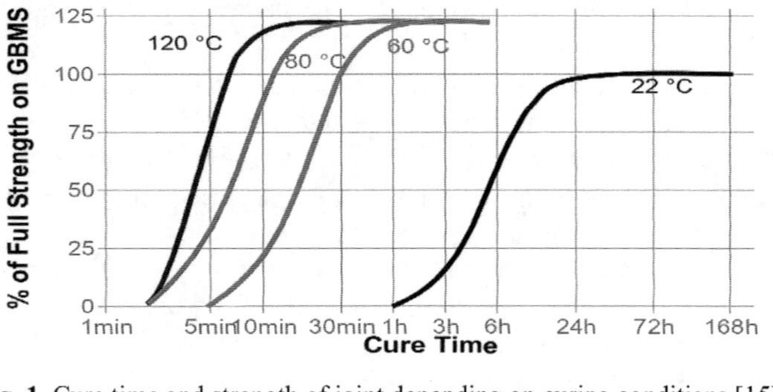

Fig. 1. Cure time and strength of joint depending on curing conditions [15]

Hysol 9492 develops complete cure within 3 days at room temperature, but elevated temperatures may be used to accelerate the cure. Elevated temperature cure allows obtaining 25% higher joint strength than cold-cure (Fig. 1) [15, 16].

3 Research method

The tests conducted to evaluate the axial-tensile strength of samples are fully compliant with the International Standard EN 15870:2009 (Adhesives – Determination of tensile strength of butt joints) [17]. The tests were performed on the test set-up based on the MTS Bionix – Servohydraulic Test System. The set-up furthermore included the environmental chamber, for testing at elevated/reduced temperature in the range of temperatures from 40°C to +250°C.

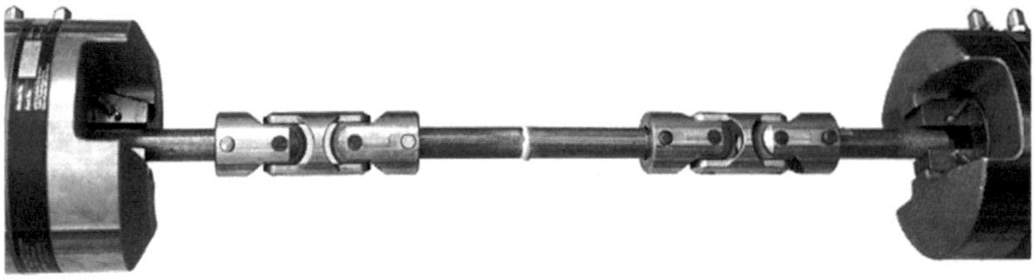

Fig. 2. Articulated grip of the testing machine

The samples were clamped with uniaxial self-centring hydraulic grips for cylindrical specimens (the specimen diameter range of ø14-20.8mm), with self-developed double-articulated grips, compensating for axial errors of clamped specimens, which ensured axial tension loading of the test specimens (Fig. 2). The tests were controlled using dedicated software: MTS Test Works 4 SH.

The cylindrical test samples (diameter d = 20 mm, length L = 100 mm) used in the experiments were made of S235JR steel drawn bar. The butting face of the samples was prepared for bonding according to the standard PN-EN 13887:2005 (Structural Adhesives - Guidelines for surface preparation of metals and plastics prior to adhesive bonding) [18]. The ends of test samples were drilled through to create the holes for attaching the specimens in the articulated grip. The tests were carried out in 80 repetitions, *i.e.* 80 test sample series, whose butting faces were subjected to additional mechanical treatment upon joint failure, with a view to reusing the samples in another test series.

The required surface roughness of all samples for bonding (Ra = 3.4±0.6 μm - which represented an intermediate-quality finish) was achieved by turning. In order to minimise the impact of corrosion processes, the time between mechanical pre-treatment and bonding did not exceed 2 h.

The proper coaxial positioning and alignment of adherends during bonding was reali-sed on a specially-built stand. The self-developed apparatus employed the fixed mass of the top adherend as the source of loading on the joint; thus exerted load amounted to 8kPa.

Prior to bonding, the specimens were subjected to necessary pre-treatment operations of cleaning and degreasing by means of a degreasing and industrial cleaning agent, Loctite 7036. The adhesive mass was dispensed following careful measurement of resin and hardener by means of FAWAG TP2/1 precision balance, which offers accuracy of 0.1 g. The total mass of the adhesive composition produced in a single measurement was 15-20 g. The adhesive components were mixed for 3 min, after which time the obtained mass was subjected to vacuum degassing for additional 5 min. The total time from the degreasing to application of the adhesive did not exceed 15 min (always within working time). The adherends were bonded with the amount of adhesive that guaranteed consistent bond, free from debonding on account of adhesive mass shortage. Upon assembly the adhesive surfaced on the sides of the specimens in the form of a minor flash, which was not in any way removed [17].

The complete process of sample preparation, involving adherend preparation, degreasing, preparing the adhesive mix, joint assembly and post-curing (if necessary) was compliant with the relevant norm, and did not exceed the total time of 4 h. In each series, 4 sets of 9 samples of identical resin/hardener ratio were produced: 2 post-cured in the thermal hardening process, and 2 cured at room temperature [11]. This procedure minimised potential test result errors resulting from discrepancies in the adhesive preparation process. Samples to be post-cured were placed in a cold electric heat chamber, controlled with a Shimaden FP93 digital controller. Post curing was conducted according to the plan of experiment (Table 1). The post curing time was controlled and registered until stabilisation of set temperature level. Both sample variants, post-cured and non-post-cured, were conditioned at constant 25°C and constant relative humidity of 50%.

Table 1. Adhesive butt joint strength testing plan

Resin/hardener ratio	Cure conditions	
1 : 0.30	3 days at 25°C	1 h at 100°C
1 : 0.40		
1 : 0.45		
1 : 0.50 (r)		
1 : 0.55		
1 : 0.70		

Bonded joints were fixed in the testing machine and subjected to tensile testing at the constant speed of 4 mm/min until failure. Each attempt was monitored to register the strength characteristics of each tested joint. The observed failure was described according to ISO 10365:1992 (Adhesives – Designation of main failure patterns) [19].

3 Results

Mean arithmetic values of strength test results, accounting for the coefficient of variation for each test sample series, are shown in Table 2. The letter "*r*" denotes the initial resin/hardener ratio, *i.e.* recommended by the manufacturer. Deviations from the initial ratio are described as a change of resin volume proportion to the recommended resin proportion, and as a real value of the applied ratio. The test results distribution in particular series is represented by the coefficient of variation, *i.e.* the ratio of standard deviation to the mean arithmetic values of test results in a given series.

Table 2. Mean values of elevated-temperature static strength of butt adhesive joints

Resin/hardener ratio	Cure conditions	Static joint strength [MPa]	Coefficient of variation
1 : 0.3	3 days at 25°C	23.2	5%
	1 hour at 100°C	25.7	18%
1 : 0.4	3 days at 25°C	22.7	8%
	1 hour at 100°C	24.4	10%
1 : 0.45	3 days at 25°C	22.7	5%
	1 hour at 100°C	20.7	16%
1 : 0.5 (r)	3 days at 25°C	20.6	6%
	1 hour at 100°C	16.7	12%
1 : 0.55	3 days at 25°C	13.9	11%
	1 hour at 100°C	10.6	9%
1 : 0.7	3 days at 25°C	6.4	7%
	1 hour at 100°C	5.2	13%

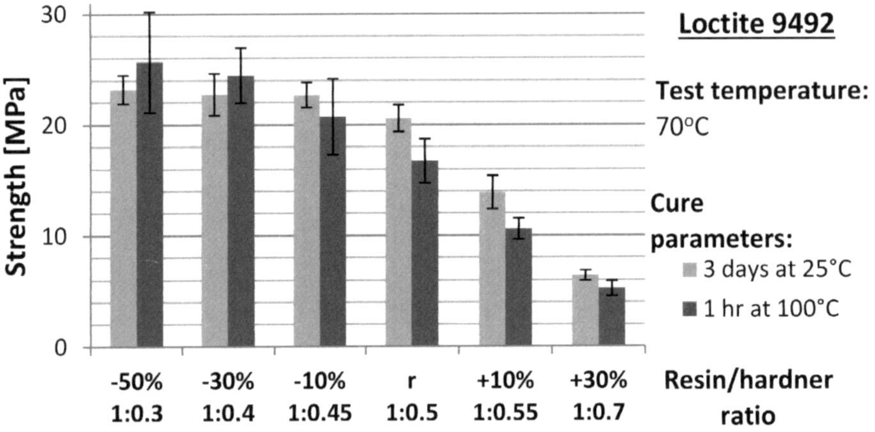

Fig. 3. Collective joint strength test results, with marked statistical analysis results (full line – equality of means, broken line – variance of means)

In order to determine whether the observed differences show statistical significance, the results were subjected to statistical analysis with the implementation of Statistica 10 software (at the standard level of significance α=0.05). The analysed values that exhibited no variance, at the assumed level of significance, are connected with a full line, whereas the broken line connects values that did show the statistically relevant difference (Fig.3.).

As it has been mentioned, the observed failure was described according to the definetions detailed in the International Standard ISO 10365:1992 Adhesives – Designation of main failure patterns [19]. This organised designation has enabled evaluating the effect of such parameters as adhesive components ratio and type of cure on the character of adhesive joint failure. The surfaces of samples were visually inspected to determine the failure mode.

Table 3. Failure modes of adhesive joints bonded with Loctite 9494 at 70°C

Resin/hardener ratio	**Cure conditions**	**Failure mode**	**Fracture image**
1 : 0.3	3 days at 25°C	CF	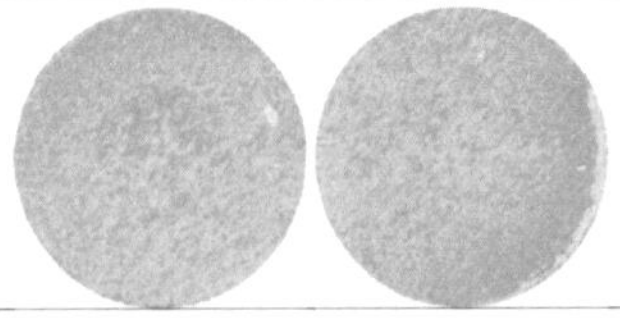
1 : 0.5 (r)		56%AF+CF 44% CF	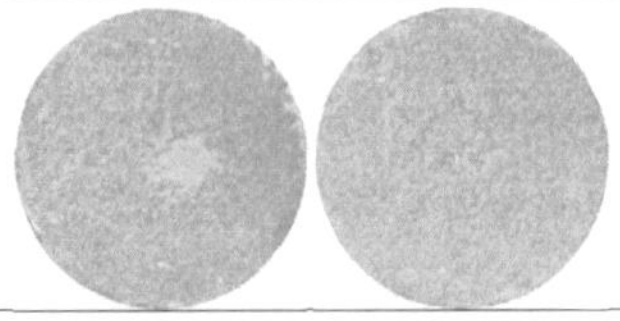
1 : 0.7		AFCP	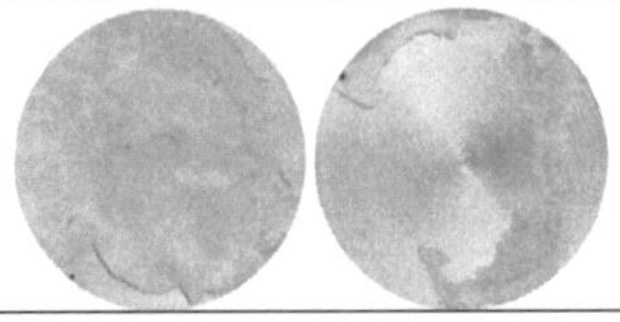

Resin/hardener ratio	Cure conditions	Failure mode	Fracture image
1 : 0.3	1 hour at 100°C	CF	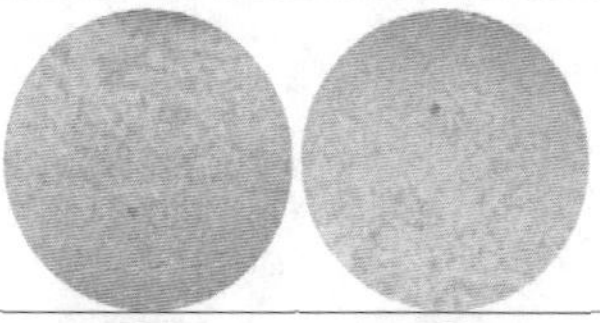
1 : 0.5 (r)		78% AF + CF 22% AF	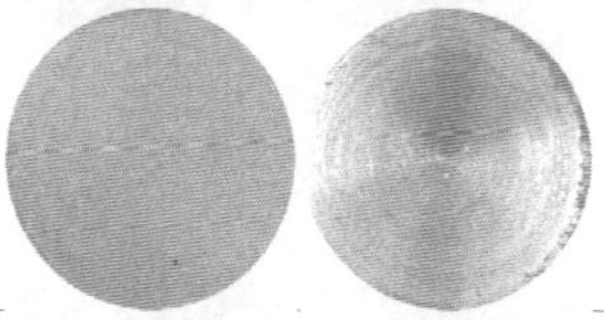
1 : 0.7		AF	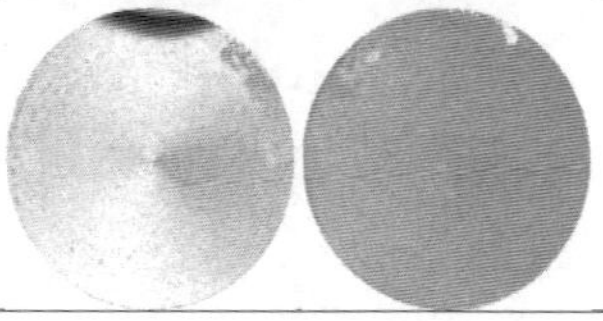

Table 3 presents the analysis of fracture of joints cured only at room temperature and post-cured. The joint performance was tested at elevated temperature conditions.

Particular test specimens were selected according to the rule:

- minimal tested resin/hardener ratio,
- initial ratio for strength tests, described as in actual strength tests,
- maximal tested resin/hardener ratio.

The failure modes, as per the aforementioned norm, are given as: AF – adhesive failure, CF – cohesive failure, AC + AF – combined adhesive-cohesive failure, ACFP – adhesive-cohesive failure with peeling. When the mode of failure observed in a series exhibited features of several types of failure, their percentage participation is given.

4 Discussion

The presented butt adhesive joint strength test results are characterised by high uniformity within particular series, which is highlighted by low values of coefficient of variation, not exceeding 15%. The statistical analysis conducted in the reported study investigated whether the mean values, accounting for the scatter of results, obtained from selected test series are equal.

With regards to testing at elevated temperature, it was observed that the joint strength developed by the post-cured specimens is 19% higher than in the case of non-post-cured specimens (Fig. 4). Manipulating the resin/hardener mix ratio also proved to affect the post-cured joint strength. It was shown that the increase of hardener content in the mix leads to reduced effect of post-curing on the joint strength of test samples. Simultaneously, at elevated temperature, the post-cured joints bonded with an adhesive mix of increasing hardener content (from 1:0.45 ratio onwards) exhibited an increasingly lower strength. It was therefore shown that Loctite 9492 reacts negatively to post-curing at excessive hardener content in the adhesive mix, particularly in joints that operate in high temperature conditions. From the chemical perspective, these curing conditions indicate saturation of bonds and suggest that the curing agent acts like a solvent or plasticiser. This may lead to reducing the thermal and chemical resistance of the joint.

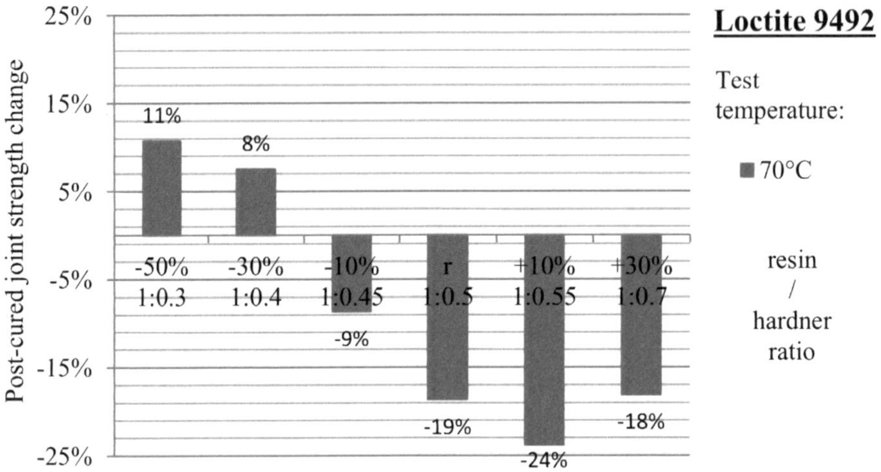

Fig. 4. Relative change of mean static strength of butt adhesive joints between post-cured and non-post-cured joints

The results obtained from the tests appear to indicate that subjecting the joint to thermal treatment while the epoxy adhesive is setting escalates the effect of mix ratio discrepancy on the joint strength. Fig. 5. compares relative strength of the joint with the mean strength of joint at recommended mix ratio (r = 100%) in a given series.

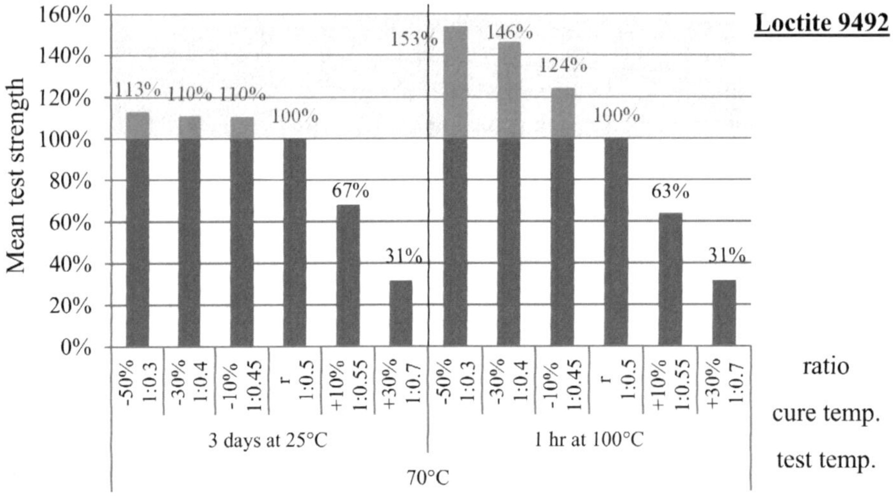

Fig. 5. Relative change of mean static strength of butt adhesive joints for the tested resin/hardener mix ratios

A part of strength parameters recorded for altered mix ratio showed in comparison negligible differences. From the data in the figure it is apparent that gradual decreasing the proportion of hardener in the mix down to 40% of the recommended amount appeared not to affect the strength of non-post-cured joints, regardless of the testing temperature. It may be, however, safely assumed that further decreasing of hardener content below 50% would result in a significant joint strength drop, nevertheless, this case was beyond the scope of this study. In the case of post-treated joints tested at elevated temperatures, it can be seen that a 30% decrease in the hardener content in the mix was translated to a marked increase in the strength

(46-51%: Loctite 9492) compared to the joint bonded at the recommended hardener ratio. In non-post-treated joints, however, the rise in the strength was less significant, amounting to 10%. The joint strength was inhibited by 30% at 10% increased hardener ratio, and by 50-60% when the hardener content in the adhesive mix was increased by 30% (Fig. 5).

Fig. 6. collates the comprehensive overall strength test result, *i.e.* the relative increase in the static strength of post-cured butt adhesive joints bonded at different hardener proportion variants related to non-post-treated joints at identical hardener proportion variants. What the figure appears to indicate is that the relative joint strength change in post-cured samples is particularly beneficial for joints bonded with reduced hardener content (compared to the initial resin/hardener ratio recommended by the manufacturer). The adhesive subjected to thermal treatment develops a strong bond even in the event of hardener deficiency. This is attributed to increased mobility of adhesive particles, which in turn boosts their reaction and chain forming capabilities.

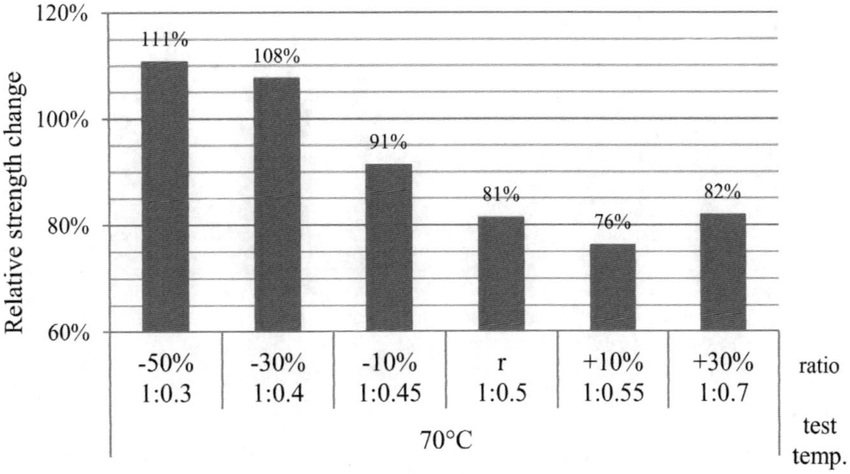

Fig. 6. Relative change of mean static strength of post-cured butt adhesive joints for the tested resin/hardener mix ratios with reference to initial ratio

The potential reason for the increased strength of post-cured joints with hardener deficiency might be the positive effect of mutual polymerisation between resin particles and increased polymerisation as a result of improved chemical functionality of epoxy resin particles, rather than the negative effect of insufficient amount of hardener. On the other hand, increasing the hardener proportion in the mix leads to a situation when the hardener begins to act like a solvent, thus reducing the level of polymerisation. Moreover, in the case of a chemical reaction between the unprocessed hardener and the adherend surface, the adhesive properties of the surface may become inhibited by surface contamination or substrate damage. In all the described defect scenarios, the strength of the assembled joint may be seriously affected.

The analysed adhesive joints exhibited different failure modes. While it was the cohesive failure that dominated in specimens bonded with the adhesive mix deficient in hardener, the gradual increase in the content of hardener translated into occurrence of the adhesive failure mode in tests. Fig. 7. shows a detailed presentation of the observed butt adhesive joint fracture modes together with their mean static strength and the dependence on the type of cure, the testing temperature, and resin/hardener mix ratio variants.

Test samples bonded with Loctite 9492 exhibited a purely cohesive failure mode when the recorded joint strength values were the highest, *i.e.* over 22 MPa – which was measured at the recommended standard mix ratio and at reduced hardener content. The increase in the hardener proportion in the adhesive mass even by 10% led to a change in the joint failure mode: adhesive and cohesive in the substrate, although considerably high strength values were recorded. It appears to indicate that the reduced capacity of the adhesive to form bonds results from the excess of unprocessed hardener in the adhesive mass, which causes partial bonding and inconsistent bondline. Regardless of the curing method, the samples showed approximately similar adhesive-cohesive character of failure in the small range of mix ratio imbalance (±10%). The reduction in the hardener proportion in the adhesive mass positively affected the strength of bonded assemblies, and furthermore led to an increasing share of cohesive failure; in the case of a 50% deficiency in hardener in the mix the failure mode was purely cohesive.

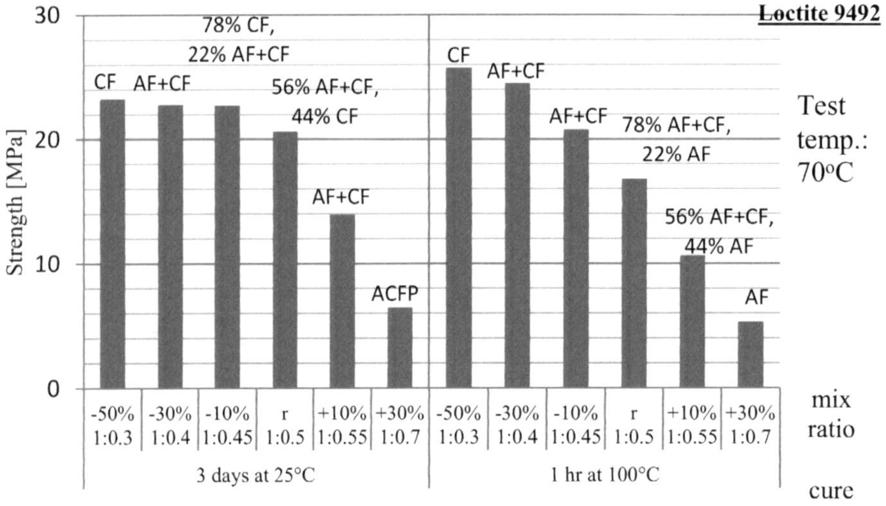

Fig. 7. Mode of joint failure versus mean static strength of butt adhesive joints depending on cure method and resin/hardener mix ratio

On the other hand, when the volume of hardener in the mix increased, the mode of failure of non-post-cured samples showed a gradual tendency towards the homogeneously adhesive-cohesive, and when the hardener excess reached 30%, the adhesive with peel. In the same conditions the post-cured joints showed an increasing tendency to fail adhesively. In both cases, of the post-cured and the non-post-cured joints, further increase in the hardener content diminished the strength of the assembly (Fig. 7.).

4 Conclusions

The results obtained from the tests lead to the conclusion that the resin/adherend mix ratio imbalance affects the strength of post-cured joints. It was found that post curing adhesive butt joints bonded with an adhesive composition with excess of hardener in the mix may thus lead to compromising the strength of joint exposed to heat during its service lifetime. The presented results shed new light on the applicability of thermal post-curing as a universal means to improve the strength properties of adhesive joints. Nevertheless, post-curing had a positive effect on the strength of bonded assemblies in the case of hardener-deficient mix ratio. The experiment detected the evidence that excessive amounts of hardener in the adhesive composition may, however, have a considerably worse effect on the final joint strength parameters. An implication of this observation may be the general recommendation for the preparation of two-component epoxy adhesives, particularly when uncertainty regarding the correct resin/hardener mix ratio. The negative impact of hardener excess on the adhesive properties of the bond material was observed in the change of failure mode of joint specimens with a gradual increase of the hardener proportion in the mix: the initial cohesive or adhesive-cohesive failure gradually shifted towards the adhesive failure mode. The failure mode shift was observed in the reported above technological conditions of joint formation and testing, for a particular adhesive composition. The disparity observed at elevated testing temperature regarding the change in the strength of post-cured joints, induced by the tested resin/hardener mix ratio imbalance as compared with the non-post-cured joints, indicates a higher sensitivity of thermally post-cured samples to incorrect resin/hardener mix ratio.

The obtained results are a basis for continuation of research regarding other material properties of adhesive (hardness, resilience, Young's modulus, etc.) and microstructure of adhesives with non-recommended mix ratio.

References

1. E. M. Petrie, *Handbook of Adhesives & Sealants,* 1st Edition, McGraw-Hill Professional (1999)
2. J. Godzimirski, *Wytrzymałość doraźna konstrukcyjnych połączeń klejowych*, Warsaw: WNT (2002)
3. J. Kuczmaszewski, *Fundamentals of metal-metal adhesive joint design,* Lublin: Lublin University of Technology (2004)
4. P. Czub, Z. Bończa-Tomaszewski, P. Penczek, J. Pielichowski, *Chemia i technologia żywic epoksydowych,* 4th Edition., Warsaw: WNT (2002)
5. J. Bishopp, *Epoxide adhesives: curatives,* in Handbook of Adhesion, D. Packham, Red., Chichester, John Wiley & Sons, Ltd, 150-155 (2005)
6. J. Pielichowski, A. Puszyński, *Chemia Polimerów*, Kraków: TEZA (2004)
7. M. Chanda, S. K. Roy, *Industrial Polymers, Specialty Polymers, and Their Applications,* CRC Press (2009)
8. A. Rudawska, M. Czarnota, *Selected aspects of epoxy adhesive compositions curing process,* JOURNAL OF ADHESION SCIENCE AND TECHNOLOGY, vol. **27**, no. **13**, 1933-1950 (2013)
9. E. O. Ozgul, M. H. Ozkul, *Effects of epoxy, hardener, and diluent types on the hardened state properties of epoxy mortars*, Construction and Building Materials, volume 187, 360-370 (2018)
10. G. Doyle, R. A. Pethrick, *Environmental effects on the ageing of epoxy adhesive joints*, International Journal of Adhesion and Adhesives, vol. **29**, no. **1**, 77-90 (2009)
11. J. Michels, J. S. Cruz, R. Christen, C. Czaderski, M. Motavalli, *Mechanical performance of cold-curing epoxy adhesives after different mixing and curing procedures,* Composites Part B, 434-443 (2016)
12. M. Muller, P. Valasek, *Adhesive Bond Degradation*, Engineering for Rural Development, Jelgava (2010)
13. P.-S. Shin, Z.-J. Wang, D.-J. Kwon, J.-Y. Choi, I. Sung, D.-S. Jin, S.-W. Kang, J.-C. Kim, L. K. DeVries, J.-M. Park, *Optimum mixing ratio of epoxy for glass fiber reinforced composites with high thermal stability*, Composites Part B: Engineering, 132-137 (2015)
14. R. F. Minty, L. Yang, J. L. Thomason, *The influence of hardener-to-epoxy ratio on the interfacial strength in glass fibre reinforced epoxy composites*, Composites Part A: Applied Science and Manufacturing, vol. **112**, 64-70 (2018)

15. Loctite Research, Development & Engineering, *Technical Data Sheet: EA 9492 (Hysol)*, (2014)
16. J. Szabelski, J. Domińczuk, *Analiza Wpływu Obróbki Termicznej Na Wytrzymałość Połączeń Gwintowo-Klejowych*, Innowacje w Zarządzaniu i Inżynierii Produkcji, vol. 1, R. Knosala, Opole, 805-814 (2014)
17. Standard PN-EN 15870:2009E - *Adhesives. Determination of tensile strength of butt joints*
18. Standard PN-EN 13887:2005P - *Structural adhesives. Guidelines for surface preparation of metals and plastics prior to adhesive bonding.*
19. Standard PN-EN ISO 10365:1998P - *Adhesives - Designation of main failure patterns.*

Capabilities of processing and visualization of production facilities diagnostic data

Matej Kandera[1,*], *Miroslav* Císar[1], *Ivan* Zajačko,

[1]University of Žilina, Faculty of Mechanical Engineering, Department of Automation and Production Systems, Univerzitná 1, 010 26 Žilina, Slovakia.

Abstract. One of the main requirements of the safe and sustained workflow of production facilities and machines, is the proper choice and implementation of suitable technical diagnostic method. In the broader sense technical diagnostics includes approaches and methods of technical objects condition determination. This process is based on specific devices parameters inspecting, which values are directly or indirectly caused by certain changes of monitored production facilities condition. In order to reliably detect faults in right time, selected parameters indicating status of the production device should be monitored continuously. At a time when an Internet of Things platforms implemented to the industry business systems is the great temptation from the view of investors and company prestige are these principles common in the process of machine parameters online monitoring and its cross connections within the business realms. This paper describes specific software ThingWorx, which is a great example of sophisticated and universal IoT platform and also describes its application for production facilities diagnostic data collection and visualization.

Keywords: condition monitoring, maintenance, Thing

1 Introduction

High manufacturing quality, reliability, product accuracy and last but not least efficient use of financial resources, are the main parameters which would define a successful engineering company. Since at the time when the great effort is oriented on this parameters, like high quality combined with economic efficiency of manufacturing process, there is limited space for accidental faults of production from the manufacturers side which can immediately cause profit decrease and even under certain circumstances can cause the regression of the whole company. Downtimes of the manufacturing or assembling processes caused by unexpected faults of all types represent the most significant issues mainly in the fully automated production lines, where literally every tenth of second play its role. Especially in automotive industry where even a few seconds lasting downtime can cause significant financial losses. For these and many other reasons, the technical diagnostics and thus proper and effective planning of maintenance of production facilities are of great importance. [1]

On the other hand, diagnostics of the cause of faults or damage and subsequently production machine and facilities maintenance represents a significantly demanding financial

* Corresponding author: matej. kendra@fstroj.uniza.sk
Reviewers: *Nicoale* Ungureanu, *Jozef* Jurko

part within the business costs, mainly considering smaller or starting companies. Not to mention cases, when facility damage or degradation is detected after machine breakdown or when damage became unsustainable. From these and many other reasons, it is required to reveal previously mentioned faults and degradations and its progress as soon as possible to ensure minimum financial repairs impact. In the numerous number of fault cases, choice and application of suitable diagnostic method can be the right solution. Choice of such a system is dependent on a type of faults and parameters that we want to detect and also on the type of the machine or device on which we want to apply such a diagnostic method, whether it will be implemented to some sort of assembly or production line, CNC machines, industrial robots or other comparable devices or technical objects. Selection of the right choice also depends on the quantity of the diagnosed devices and on special requirements, for example "machine to machine communication". [2,3]

Real-time monitoring systems and devices are often implemented for this particular purpose. This systems, also known as condition based monitoring systems are capable of on-line information reflecting of all important machine parameters. Condition monitoring is based on non-destructive and non-invasive methods of measuring of required system characteristics such as frequencies or modal domains measuring and verification for the purpose of detecting whatever deviations. These shall indicate machine failure or degradation of any device parts in advance. In the field of mechanical engineering these monitoring systems have to increase economical profitability and to simplify production facilities maintenance in order to justify their implementation.

This article deals with maintenance, with the aim focused mainly on area of the condition monitoring. There are few platforms chosen as an example of such a software for developing web applications which can process and visualize diagnostic data acquired from diagnosed devices or facilities and subsequently implement this devices, to the Internet of Things. The most advantageous candidate is the use of a large PTC tool called ThingWorx. It is presented as an Internet of Things platform that allows to easily create web applications for the collecting, processing and visualization of data obtained from the sensory and control subsystems of various production facilities. Main purpose is to use the above-mentioned software to capture and monitor data defining the machine state during production and also take advantages of the tools and services that the software provides to the user.

2 Condition based maintenance

In order to understand why the use of condition monitoring is important or relevant, it is necessary to understand what is actually the essence of condition-based maintenance and what is the difference between the concepts of technical diagnostics and machine condition monitoring.

According to the CSN EN 60300-3-11 standard, maintenance is defined as a set of all technical, administrative and managerial steps or actions during the lifecycle of the machine on which the maintenance is aimed to sustain its serviceability or to return it to the maintained state. [4]

The most common types and methods of technical maintenance are listed in the Figure 1. Important for us now is a part of "before detected faults" which contains bigger group of preventive maintenance and specifically we are dealing with condition based maintenance. This method is also known as predictive maintenance, because the potential failure of the machine is predicted by regular condition monitoring of chosen parameters, that indicates the condition of the equipment, which is measured continuously or periodically. Following maintenance is therefore performed at the optimal time. Benefits of the predictive maintenance are obvious in comparison to methods like "run-to-brake" or preventive maintenance, but it requires access to the reliable condition monitoring techniques which can not only define or describe the current state but also provide reasonable prediction for the remaining lifetime. [5,6]

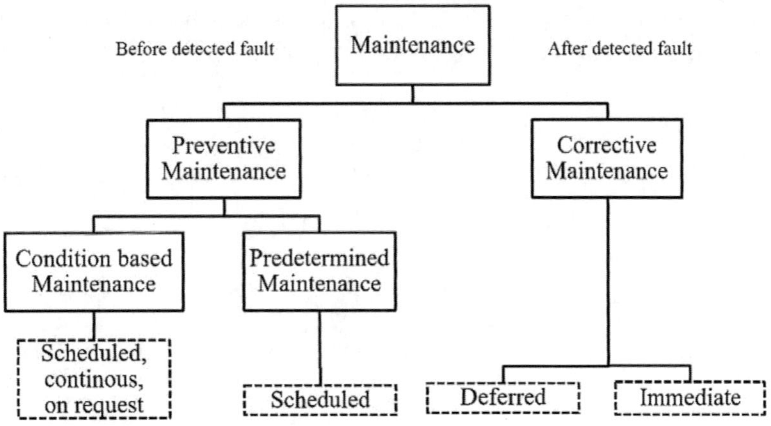

Fig. 1. Maintenance approaches and strategies.

3 Condition monitoring

Condition monitoring does not serve to reveal the reason of fault. It should serve only to detect the fact that there are some undesirable deviations in the monitored values or that some kind of fault has occurred. This process is based on tracking chosen parameter values and their changes and is a major component of predictive maintenance. Main aim of condition monitoring relies on the ability of monitoring present state or condition of devices and on ability of predicting of monitored parameters to the future while the device is in operation. This means that information about internal effects must be obtained externally while the machines are in operation. [6,7]

Machine condition monitoring is an important part of condition-based maintenance (CBM) that is recognized as the most effective maintenance strategy in a wide range of industries. Before preferred "run-to-break" strategy has ensured maximum operational time between shutdowns, but the breakdowns during using this strategy have been occasionally catastrophic, with serious security consequences, production losses, and cost of subsequent repairs. The first response to the deficiencies of this strategy was "preventive maintenance", in which device is repaired at certain intervals, thus ensuring very little possibility of breakdowns between these repairs. However, the disadvantage of such a maintenance strategy is that it is in some cases carried out even though it is still unnecessary, large number of spare parts and, last but not least, a lot of repair time are consumed, which reflects on its financial demands. Condition monitoring can improve Overall Equipment Effectiveness (OEE) with actionable insights and alerts at every level of production lines. [6]

Whole process begins on the production line, where sensors of different types can be installed as required or as are requested by the customer. The most commonly monitored parameters are vibrations, temperature, lubricant degradation or contamination, and electric current. Data from the sensors is processed in the data connectivity gateway to which they are connected either directly by Ethernet, serial ports like RS-232 or are connected via Wi-Fi, Bluetooth or another wireless method, where the data is translated between sensor protocols and processed before upcoming procedure. Subsequently data is processed in cloud services, for example using a specialized IoT platform. Then virtual representation is created, with identical properties like real monitored object which represents digital a reflection of the monitored device and on this twin monitored properties and values are represented. The measured data is in the end visualized by various visual elements in the web based user interface or data can be further exported. Another solution is the setting of limits after which the system send a warning to the specific mail or otherwise notify the responsible person. Structure of such a condition monitoring system is described on the figure 2.

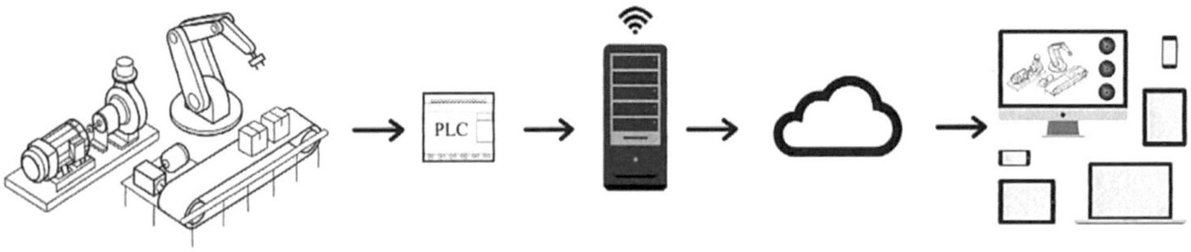

Fig. 2. Structure of condition monitoring system

4 Example of application

There are many ways for the purpose of creating application suitable for condition monitoring. One of them is to use independent programs for each part of process from data acquiring thru data processing to the form in which it is possible to analyse them until the last step which is building and compiling specific program in one of the commonly used programming languages and creating user interface for the purpose of graphical representation of monitored parameters values. [8]

Another way to deploy an application for condition monitoring is the possibility of using one of the big platforms. These can be strictly specialized only on condition monitoring or can offer solutions for predictive maintenance too. In these days is big trend to associate industry with the phenomenon of upcoming Industry 4.0 and because of this reason there cannot be omitted platforms that make it possible to implement Internet of Things solutions. Their tools can be also used to create web-based applications for condition monitoring and event monitoring of the single machine or whole workshop machinery. Among the best rated IoT platforms of the year 2018 are included for example: Amazon Web Services, Microsoft Azure IoT, Google Cloud Platform and one of the best rated platforms is also ThingWorx from PTC. All of these software serves to M2M (machine to machine) connection, respectively for the interconnection of individual devices within the enterprise and offer the possibility of visualization of the measured parameters which are in our case, diagnostic data from the production machines or production lines.

However, significant level of programming skill is required from the user or programmer to implement most of these platforms. ThingWorx is in this aspect more user-friendly as it allows to develop IoT web-based applications without considerable large knowledge of programming. Instead it works on intuitive "drag and drop" method and uses pre-created objects, so called widgets and things.

4.1 ThingWorx

This particular software allows users to build a specific and unique web applications from defining a digital representation of examined object thru defining properties and services to building its final visual interface. The ThingWorx Industrial IoT Platform delivers tools and technologies that allow to rapidly develop and deploy powerful IoT applications and AR experiences too. ThingWorx is a complex of entities which represent real processes, solutions and applications with which it is possible to create or model real solutions of specific tasks in a way based on creation of digital twins. Digital twin in this scope of view represents exact reproduction of certain physical object or real system across its lifecycle using a real-time data for purpose of understanding, learning and reasoning. [9,10]

As a description can indicate, ThingWorx platform is based on creation of the "Things". Thing is the digital representation of real world subjects or processes with its own properties and business logic. Structure of such a "Thing" is described in the figure 3., where are four main characteristics shown.

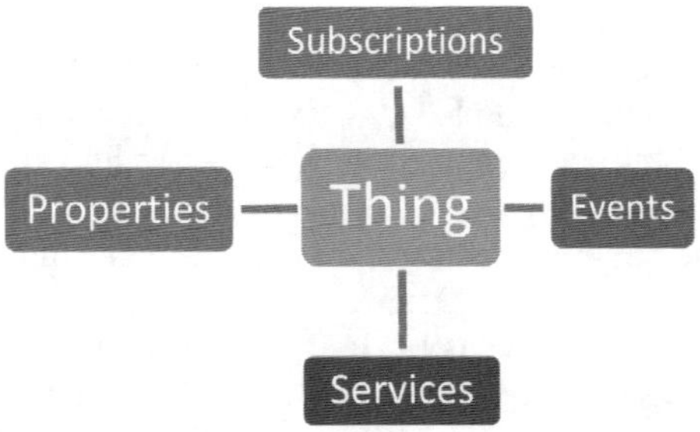

Fig. 3. Structure of the „Thing ".

Main characteristics of the Thing are:
• Properties – also can be thought as parameters or variables. Represents behaviour of actual thing or just modelled process.
• Services – every function which is ThingWorx allowed to execute. They can be defined at the level of creating a ThingTemplate, ThingShape or Thing.
• Events – represents change of the state or in value of some property. Can interrupts ThingWorx Core and subscribe to for purposes of receiving notification about errors or some faults.
• Subscriptions – it is name for each action which is executed as a stimulus when one of the defined events occurs.

Values of all these characteristics are configured by application programmer and are bound to the created thing. [9]

4.2 Application development example

As mentioned in chapter above, the basis of work in this software is to define "Things" and their characteristics. That is why the development of the application in this platform starts with this part. It is not a strict steps sequence ordered during the application development process but there are some main steps that user will not avoid.

One of these steps is defining or creating of the Tags, which serves to organize the work for the purpose of easiest saving and exporting process, categorizing and grouping individual items throughout platform, version control and for the purpose of searching and migrating entities between serves. In general Tag consists of two parts: Vocabulary and term (vocabulary: term) and it is recommended to create tags just before creating other entities in this platform. [9]

It is recommended to take advantage of defining a ThingTemplate during the process of creating individual Things. ThingTemplates are reusable building blocks used to maintain scalability and flexibility of our application development. By using ThingTemplates, it is possible to define a set of similar or the same objects with those of the defined basic properties and services. One ThingTemplate is therefore capable of serving as a multi-item template without the need for repetitive and time-consuming multiple submissions of the same data to each Thing. Within the ThingTemplate creation, there is a defining of properties included, that can be static or dynamic, persistent, and can be logged. Subsequently during creating Thing, the process is followed by the user filling a name that would clearly defines what device is it representing and bind the new thing with the pre-created ThingTemplate and assigns the required tags to it. Entering all of this data is done through a relatively simple tables filling with additional Java editing options. [8,10]

For the purpose of data storing there might be used Value Stream. This tool represents one of the way, quick and easy, to record the values or data with timestamp in Foundation Server. Other possibilities are to use data tables or simply the streams. When creating the application for condition monitoring in ThingWorx, one of the most important parts is the definition of a Data source that represents real-time data values. Data acquired from the sensors on machines or another device, can be securely extracted to the ThingWorx using OPC UA (Open Platform Communications Unified Architecture), or using preferred industrial data connectivity protocol. Data can be using OPC server communicated continuously among PLCs on the shop floor, RTUs (Remote Terminal Units) in the field, Human Machine Interface (HMI) stations, and software applications on PCs desktops. Main benefit of OPC Compliance usage is, that it allows real-time communication even between hardware and software from different vendors. For the purpose of this demonstration, there was an OPC software KEPServerEX 6 used. Other options for ThingWorx connectivity is usage of ThingWorx Edge SDKs, Edge Microserver, REST API or Device Clouds (AWS IoT, Azure IoT Hub, ...) while each of these options has its pros and cons. [9,11]

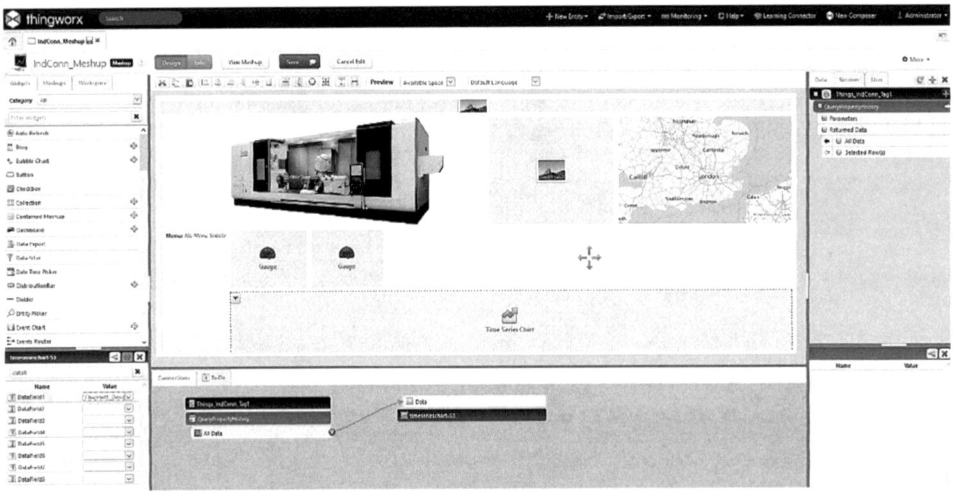

Fig. 4. Mashup builder user interface.

Figure 4 illustrates an example of the Mashup builder interface, the creating module for the application final user interface. Mentioned interface creation is based on a very intuitive "drag-and-drop" system using various widgets such as maps, spreadsheets, charts, various types of buttons and gauges, and various other elements. Platform offers the option to add other required widgets that are not found in the main menu via the marketplace. The data for each widget can be bound from the Data Table again with drag-and-drop method. [9]

Figure 5 shows the appearance of the proposed user environment as seen by the users in their browser.

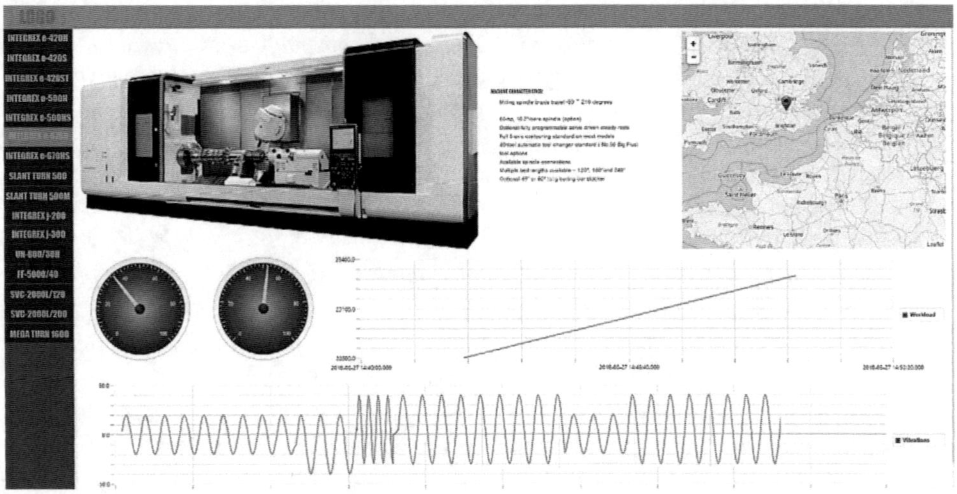

Fig. 5. Example of the final user interface created in "Mashup builder".

5 Conclusion

Machine condition monitoring as an important part of condition based maintenance, which is more and more recognized as the most efficient strategy for carrying out maintenance in industry. The benefits of such monitoring can be summarized in several parts. First of all, it allows to implement predictive maintenance as it needs actual data about the values of the machine parameters in this maintenance method. The condition monitoring also accelerates the detection of defects or wear of machine parts that would otherwise be detected only after detailed measurements or during the machine maintenance. As in the case of human diseases, even in the case of industrial facilities, it is true that the sooner these defects or deviations from the normal values are detected, the less expensive and easier it is to remove them and return the machine to its fully functional state. CM also ensures safe operation and preserving constant machine accuracy without undue deviations and quality of manufactured parts. Another benefit could be real-time online representation and visualization of measured parameter values. Such representation is understandable, intuitive, and applicable even without the necessity for special training.

The platform presented in this article offers a solution for the implementation of condition monitoring and the creation of web based monitoring applications based on digital representations of real objects that have their own properties and business logic. The main advantage of such solution, is that it allows to create condition monitoring applications in a simpler way, in comparison to other platforms as it works mainly on drag-and-drop principle, it offers large amount of ready to use objects and templates, and it does not require knowledge of high-level programming languages.

The outputs of paper was supported by grant VEGA 1/0504/17 "Research a development of methods for multicriterion accuracy diagnostic of CNC machines.

References

1. V. Bulej, J. Uríček, V. Poppeová, R. Zahoranský, M. Rupiková, *Study of the workspace of hybrid mechanism Trivariant.* Applied mechanics and materials. ISSN 1660-9336, **436**, 366-373 (2013)
2. I. Kuric, V. Bulej, M. Sága, P. Pokorný, *Development of simulation software for mobile robot path planning within multilayer map system based on metric and topological maps.* International journal of advanced robotic systems, **14**, (2017)
3. V. Bulej, G. V. Soianovici, V. Poppeova, *Material flow improvement automated assembly lines using Lean logistics.* Annals of DAAAM for 2011 & Proceedings of the 22nd international DAAAM symposium "Intelligent manufacturing & automation: Power of knowledge and creativity". ISBN 978-3-901509-83-4, 253-254 (2011)
4. H. Opocenska, M. Hammer. *Contribution to maintenance issues in company practice.* MM Science Journal, **12**, 748-755 (2015)
5. V. Peťková, *Teória a aplikácia vybraných metód technickej diagnostiky.* 1. publication. Košice : Technická univerzita, 235 p., ISBN 978-80-553-0483-0. (2010)
6. R., B. Randall, *Vibration-based Condition Monitoring: Industrial, Aerospace and Automotive Applications.* Chicester: John Wiley & Sons, Inc., 308 p. ISBN: 978-0-470-74785-8 (2011)
7. I. Kuric, I. Zajačko, M. Císar, *Analytical intelligence tools for multicriterial diagnostics of CNC machines.* Advances in Science and Technology Research Journal, 10.32., (2016)
8. J. Stanček, V. Bulej, *Design of driving system for scissor lifting mechanism.* Academic journal of manufacturing engineering, ISSN 1583-7904, **13**, 38-43 (2015)
9. *ThingWorx Glossary.* [online] Available on Internet: https://developer.thingworx.com /resources/guides/thing worx- glossary/glossary
10. E. Pivarčiová, et al., *Analysis of control and correction options of mobile robot trajectory by an inertial navigation system.* International Journal of Advanced Robotic Systems, 15.1: 1729881418755165., (2018)
11. B. Mičieta, M. Edl, M. Krajčovič, Ľ. Dulina, P. Bubeník, L. Ďurica, V. Biňasová, *Delegate MASs for Coordination and Control of One-Directional AGV Systems: A Proof-of-Concept.* International journal of advanced manufacturing technology, **94**, 415-431. ISSN 0268-3768 (2018)

Influence of roller - burnishing on surface roughness parameters and roundness

Augustín Görög[1*], *Ingrid* Görögová [1], *Dana* Stančeková[2], *Miroslav* Janota[2]

[1]Slovak University of Technology in Bratislava, Institute of Production Technologies, Faculty of Materials Science and Technology in Trnava,, Bottova 25, 917 24 Trnava, Slovak Republic
[2]University of Žilina, Faculty of Mechanical Engineering, Univerzitna 1, 010 26, Žilina, SK,

Abstract. In order to increase productivity, machining times and market competitiveness of machining production systems, it is important to continuously develop existing technological solutions. The paper deals with surface roughness and roundness after roller burnishing outer rotary surface. Roller burnishing was applied on turned surfaces with different surface roughness. Each outer rotary surface was roller burnished twice with the same conditions (pressing force, peripheral speed of machine part, feed of tool). The parameters of the surface roughness Ra, Rq and Rt were measured after each single roller burnishing. The results show that the surface roughness is decreasing after the first roller burnishing cycle (additional roller burnishing had only negligible influence on surface roughness) and the original surface roughness has a significant influence on change of surface roughness after roller burnishing.

Keywords: surface roughness, roundness, roller burnishing, measuring

1 Introduction

Surface geometry machined with traditional processes consists of scallops causing additional finishing operation. Surface roughness bellow 0.1 mm [1] is required for minimizing friction losses and high fatigue strength. To increase production efficiency and these requirements, it is necessary to look either for new tool designs gathering better machining outputs or either for more cheaply actions toward optimization of the machining parameters [2].

Most of the investigations towards improvement of fine machining processes are concerned with the process outputs: surface roughness and surface hardness. It was pointed out by many investigators that an improvement in wear resistance can be very easy achieved by burnishing, but very few actual studies analysed environmental implications and versatility. [3] Roller burnishing (Fig. 1, [4]) is a low-cost surface treatment method of cold working to induce compressive residual stresses in the top surface layer of the burnished material. Stoić at all [3] defined roller burnishing as a fine machining process that is used to improve certain physical and mechanical properties, such as surface roughness, corrosion resistance, friction coefficient, wear, and fatigue resistance. This process is used when the goal is to achieve a high quality surface finish strengthened by the strain hardening

* Corresponding author: augustin.gorog@stuba.sk

Reviewers: *Sylvia* Kuśmierczak, *Vladimír* Vrba

phenomenon [5, 6]. In the RB process a cylindrical roller is pressed across the part being processed. The plastic deformation created by the roller is a displacement of the material that flows from peaks into valleys under pressure, and results in a surface finish with a strain-hardened surface [7, 8].

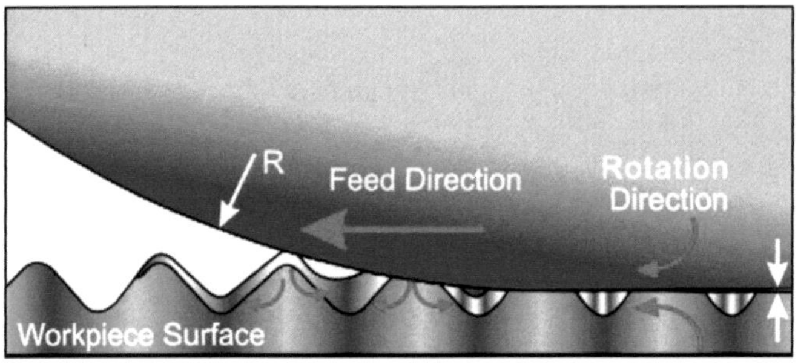

Fig. 1. Scheme of roller burnishing [4]

When roller burnishing is conducted on a lathe, the workpiece is spinning and the burnishing tool is moved across the surface at a constant rate. Many of processes for the manufacturing of high-quality surfaces can be replaced by roller burnishing (e.g. galling, grinding, honing, polishing) [9, 10]. Roller burnishing is commonly applied to cylindrical workpieces on both external and internal surfaces, and its tools are similar to roller bearings [11].

In last few decades roller and ball burnishing have been studied by a number of researchers on various materials. El-Axir [12] found that the burnishing feed and burnishing force feed have the most significant effect on both micro hardness and surface roughness. Luo et al. [13] studied the effect of the burnishing parameters on the burnishing force and material micro hardness via theoretical analysis. The results showed that burnishing depth and feed were the most significant factors. Franzen et al. [14-16] found that the rolling process parameters had a great influence on the tribological properties of burnished surfaces. The optimum hardness and roughness were obtained for a specific regime whose decisive parameters are the applied force and the number of passes of the burnishing tool. El-Tayeb et al. [17-20] investigated the influence of the contact conditions between the roller and workpiece on the quality of the burnished surface as well as tribological behaviour in roller burnishing.

The aim of the article is to experimentally determine the change of surface roughness. Comparison of the original surface roughness of the turned surface with the surface roughness after roller burnishing, respectively after two roller burnishing. On selected surfaces will be monitored change of the roundness, respectively roundness profile.

2 Experimental methods

2.1 Roller burnishing tool

Burnishing tool (Fig. 2), which was used during experimental work, is duo-trolley toll with dynamometer. The basic dimensions of the tool:

- diameter of the forming rollers D = 40 mm,
- clamping holder section s = 35 x 30 mm,
- clamping holder length l = 160 mm,
- total tool length L = 320 mm,
- maximum tool height v = 81 mm,
- maximum tool width h = 35.5 mm.

At press tool on cylindrical surface, indicating gauge evince particular deviation. This deviation depends on force parameter, which pressing the burnishing tool.

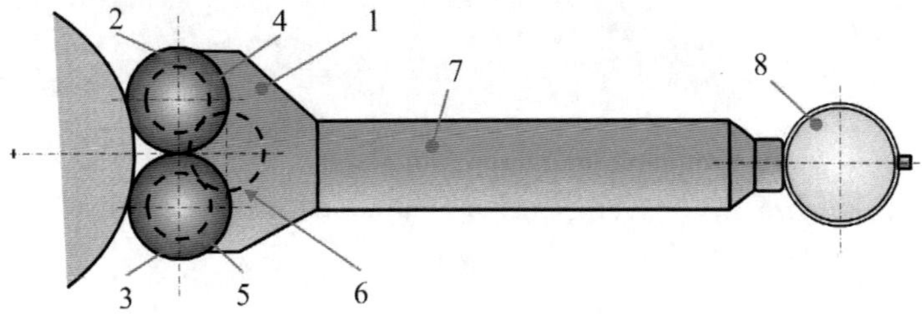

Fig. 2. Burnishing tool: 1 - supporting body of roller head, 2, 3 - rollers, 4, 5, 6 - gear wheel, 7 - holder, 8 - dial indicator

Before using roller burnishing tool is necessary to determine functional dependence of deviation reviewed on indicating gauge on pressing power. This reliance was determined by roller burnishing tool pressing using known pressing force to the cylinder with equivalent diameter and material. Pressing force was in range of 0-5000N. This process use to be called as a device calibration. Obtained relationship of pressing power and deviation is shown in fig. 3.

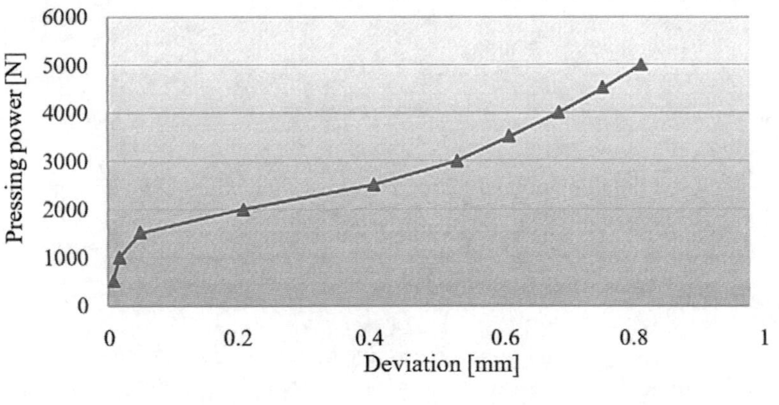

Fig. 3. Obtained relationship of pressing power and deviation

Pressing force was adjusted so, that roller burnishing was pressed step by step into roller burnishing surface and the deviation was reviewed on indication gauge.

2.2 Workpiece material and experiment conditions

In the present work, the S 335 J2G3 steel (diameter 41 mm) was used as workpiece material and the chemical composition of material is given in Table 1. The workpiece was selected for present work because this material is mostly used in engineering industries for shafts, spindles, studs, etc.

Table 1. Chemical composition (in weight %)

Carbon (C)	Manganese (Mn)	Silicon (Si)	Phosphorus (P)	Sulphur (S)	Nitrogenium (N)	Aluminium (Al)
Max. 0.200	Max. 1.600	Max. 0.550	Max. 0.050	Max. 0.045	-	Max. 0.015

There were turned nine surfaces on the workpiece (Fig. 4) with different surface roughness in random order. The average value of arithmetical mean deviation of the profile Ra was from 1.67 μm to 5.34 μm.

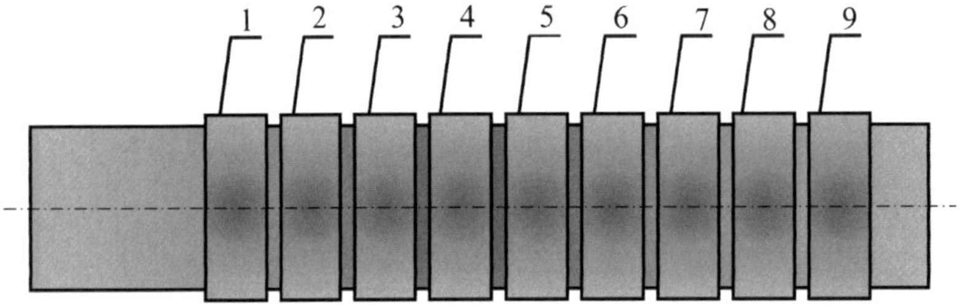

Fig. 4 Shaft used for experiments

All nine surfaces were roller burnished twice under the same conditions:

- Machine: lathe SUI 40,
- Peripheral speed: 72.13 m/min,
- Pressure force: 3000 N,
- Feed: 0.3 mm (feed rate 168 mm/min),
- Cooling: oil.

The experimental conditions used in this work were chosen in accordance with the manufacturer's recommendations of the roller burnishing tool.

3 Attained results

Surface roughness was measured after each roller burnishing (every surface on the three various places) with device type Surtronic 3+ (Rank Taylor Hobson Ltd., UK). The referent profile length l_r = 0.8 mm, observed length l_n = 4 mm. Three parameters were measured, as declaring the STN EN ISO 4287 standard:

- Arithmetical mean deviation of the profile Ra,
- Root mean square deviation of the profile Rq,
- Total height of profile Rt.

For better view, the smallest and the highest measured data of surface roughness are listed below in tab. 2.

Table 2. Measured values of surface roughness

After turning:	After the first roller burnishing:	After the second roll burnishing:
➢ **Ra = 1.25-5.67 μm** ➢ **Rq = 1.58-6.51 μm** ➢ **Rt = 9.4-36.4 μm**	➢ **Ra = 0.65-0.99 μm** ➢ **Rq = 0.81-1.28 μm** ➢ **Rt = 5.6-10.1 μm**	➢ **Ra = 0.70-1.32 μm** ➢ **Rq = 0.88-1.69 μm** ➢ **Rt = 4.9-12.0 μm**

The change of surface roughness (parameters Ra, Rq, Rt) is shown in fig. 5, 6 and 7.

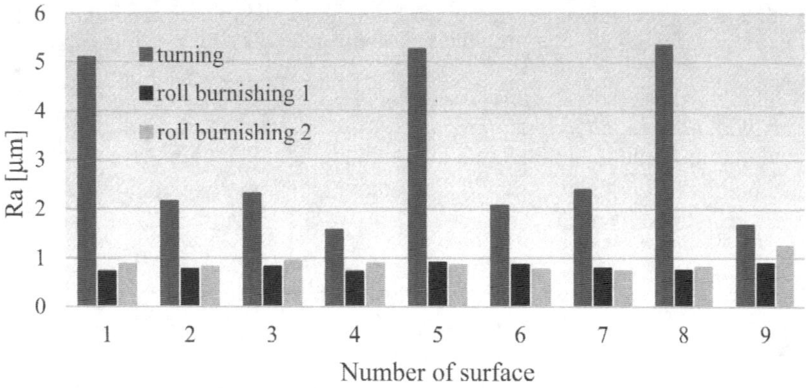

Fig. 5. Surface roughness – parameter Ra

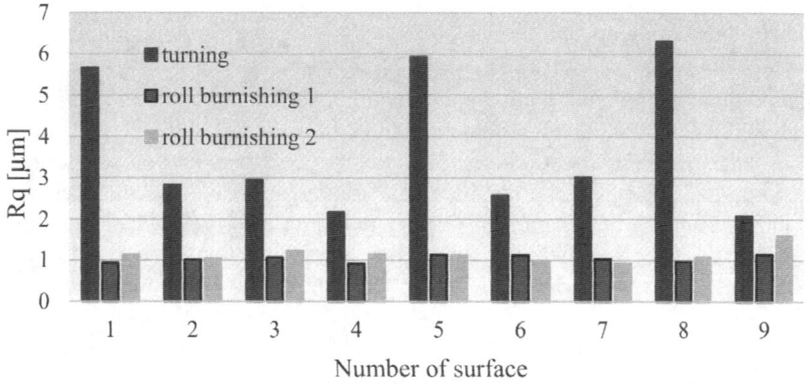

Fig. 6. Surface roughness – parameter Rq

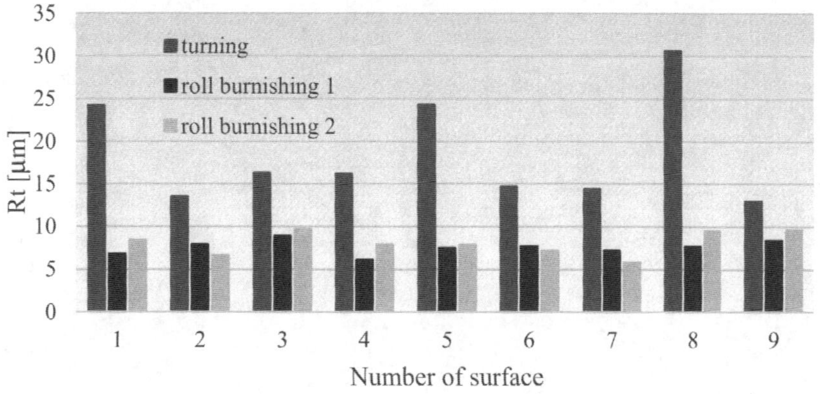

Fig. 7. Surface roughness – parameter Rt

Because the burnishing is essentially a cold forming process, in which the metal near a machined surface is displaced from protrusions to fill the depressions [21-23], there is the fig. 8 for to better illustrate changes of the surface profile.

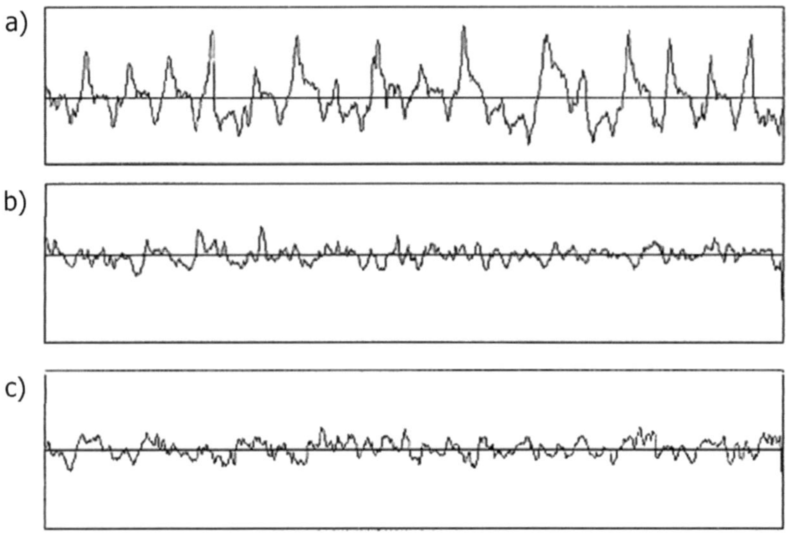

Fig. 8. Example of roughness profile (surface 3): a) after turning, b) after first roller burnishing, c) after second roller burnishing

On selected areas 8, 9 was measured roundness by the MK 300C device with rotating stylus. The MCC method (minimum circumscribed circle) was chosen as the evaluation method. In the MCC method, the profile center is determined by the smallest circle that contains the measured profile. From this center point, an inscribed circle fitted inside the profile is drawn. The out-of-roundness is the maximum inward departure from the circumscribed circle. Tab. 3 presents measured values of roundness on the surfaces 8 and 9.

Table 3. Roundness of surfaces 8 and 9

Number of surface	Roundness [μm]		
	Turning	First roller burnishing	Second roller burnishing
8	34,4	12,6	11,7
9	10,8	11,6	16,9

Figures 9 and 10 present a change of the roundness profile for better illustration because the roundness is only one numerical value.

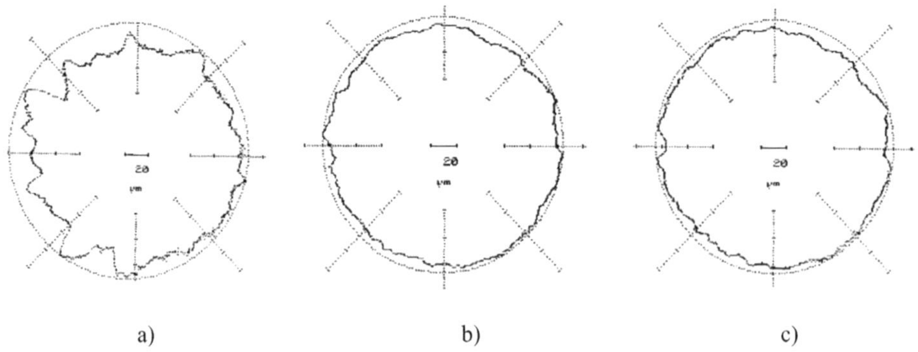

Fig. 9. Roundness profile of surface 8: a) after turning, b) after first roller burnishing, c) after second roller burnishing

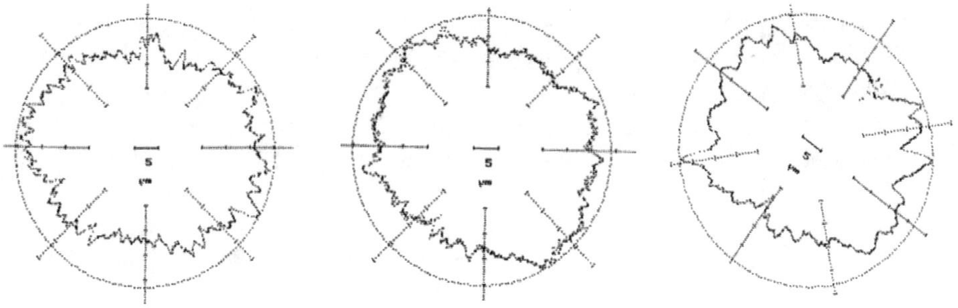

a) b) c)

Fig. 10. Roundness profile of surface 9: a) after turning, b) after first roller burnishing, c) after second roller burnishing

4 Conclusion

For commercial steel the experimental work showed:

- the original surface roughness had not significant influence on surface roughness after roller burnished - the rolled surface has approximately the same surface roughness on all surfaces regardless of the original surface roughness after turning,
- the surface roughness is decreasing after the first roller burnishing cycle (all monitoring parameters - Ra, Rq, Rt),
- if the original surface roughness is high, then the change of surface roughness after roller burnishing is significant. The turning surface with the highest surface roughness (surface No. 8) achieved an improvement of the Ra parameter of 86 percent, of the Rq parameter of 85 percent, of the Rt parameter of 85 percent,
- if the original surface roughness is low, then the change of surface roughness after roller burnishing is not so significant. The turning surface with the lowest surface roughness (surface No. 9) achieved an improvement of the Ra parameter of 47 percent, of the Rq parameter of 46 percent, of the Rt parameter of 35 percent,
- the above information confirm the manufacturer's tool recommendation, the surface roughness before roller burnishing should be Ra = 3.2 - 6.3 μm,
- additional roller burnishing had only negligible influence on surface roughness - on some surfaces have improved, on some surfaces have worsened,
- if the original roundness is high, then it can be expected to improve after roller burnishing. If the roundness of the turned surface is low, cannot be expected to improve it, even it is possible the deterioration of the roundness.

The article was made under support grant project KEGA 022ŽU-4/2017 Implementation of on-line education in the area of precise technologies with an impact on educational process to increase skills and flexibility of students of engineering fields of study.

References

1. P. R. Babu, T. S. Prasad, A.V.S. Raju, A. Jawahar Babu, *Effect of internal roller burnishing on surface roughness and surface hardness of mild steel*, Journal of Scientific & Industrial Research **68**, 29-31 (2009)
2. S. Dolinšek, J. Kopač, *Mechanisms and types of tool Wear; particularities in advanced cutting materials*, Journal of Achievements in Materials and Manufacturing Engineering **19,** 11-18 (2006)

3. A. Stoić, I. Lacković, J.Kopač, I. Samardžić, D. Kozak, *An investigation of machining efficiency of internal roller burnishing*. Journal of Achievements in Materials and Manufacturing Engineering **40,** 188-194 (2010)
4. B. B. Buldum , S. C. Cagan, *The optimization of surface roughness of az91d magnesium alloy using ANOVA in ball burnishing process*. Turkish Journal of Engineering **1,** 25-31 (2017)
5. A. Kułakowska, L. Kukielka, K. Kukielka, R. Patyk, L. Malag, L. Bohdal, *Possibility of Steering of Product Surface Layers Properties in Burnishing Rolling Process*. Applied Mechanics and Materials **474**, 442– 447 (2014)
6. L. Malag, L. Kukielka, K. Kukielka, A. Kułakowska, L. Bohdal, R. Patyk, *Problems Determining of the Mechanical Properties of Metallic Materials from the Tensile Test in the Aspect of Numerical Calculations of the Technological Processes*. Applied Mechanics and Materials **474**, 454–459 (2014)
7. S. P. Dwivedi, S. Sharma, R. K. Mishra, *Effects of roller burnishing process parameters on surface roughness of A356/5%SiC composite using response surface methodology*. Advances in Manufacturing **2**, 303–317 (2014)
8. F. Gharbi, S. Sghaier, H. Hamdi, T. Benameur, *Ductility improvement of aluminum 1050A rolled sheet by a newly designed ball burnishing tool device*. International Journal of Advanced Manufacturing Technology **60**, 87–99 (2012)
9. R. J. Kosarevych, B. P. Rusyn and R. V. Torska, *Modelling of the Propagation of Pitting Corrosion by Point Processes,* Materials Science **51**, 673–681 (2016)
10. B. P. Rusyn, N. P. Anufrieva, N. R. Hrabovska, V. H. Ivanyuk, *Nondestructive Testing of the State of Surfaces Damaged by Corrosion Pitting*. Materials Science **49**, 516–524 (2014)
11. L. N. López de Lacalle, A. Lamikiz, J. A. Sánchez, J. L. Arana, *The effect of ball burnishing on heat-treated steel and Inconel 718 milled surfaces*. International Journal of Advanced Manufacturing Technology **32**, 958–968 (2007)
12. M. H. El-Axir, *An investigation into roller burnishing*. International Journal of Machine Tools and Manufacture **40**, 1603–1617 (2000)
13. H. Luo, J. Liu, L. Wang, Q. Wang, *The effect of burnishing parameters on burnishing force and surface microhardness*. International Journal of Advanced Manufacturing Technology **28**, 707–713 (2006)
14. V. Franzen, M. Trompeter, A. Brosius, A. E. Tekkaya, *Finishing of Thermally Sprayed Tool Coatings for Sheet Metal Forming Operations by Roller Burnishing*. International Journal of Material Forming **3**, 147–150 (2010)
15. Dodok, T., Čuboňová, N., Kuric, I. *Workshop Programming as a Part of Technological Preparation of Production*. Advances in science and technology. Research Journal, 11(1), 2017, 111-116
16. M. Sapieta, A. Sapietová, V. Dekýš, *Comparison of the thermoelastic phenomenon expressions in stainless steels during cyckic loading*. Metalurgija **56**, Issue: **1-2,** 203-206 (2017)
17. N. S. M. El-Tayeb, K. O. Low, P. V. Brevern, J. *Influence of roller burnishing contact width and burnishing orientation on surface quality and tribological behaviour of Aluminium 6061*. Journal of Materials Processing Technology **186**, 272–278 (2007)
18. J. Petru, T. Zlamal, R. Čep, D Stančeková, M. Pagač, O. Vortel, *Mechanism of cutting insert wear and their influence on cutting ability of the tool during machining of special alloys*. METMG 2015 - 3rd international conference, Vancouver, Canada, 36-40 (2015)

19. J. Novotný, J. Cais, N. Náprstková, *Analysis of aluminium alloys AlSi7Mg0.3 and AlMg3 by means of X-Ray Diffraction.* Manufacturing Technology, Vol. **14**, No. **3**, 392-397 (2014)
20. M. Drbul, A. Czan, M. Sajgalik, M. Piesova, K. Stepien, *Influence of normal vectors on the accuracy of product's geometrical specification.* Conference: 12th International Scientific Conference of Young Scientists on Sustainable, Modern and Safe Transport Location: High Tatras. Procedia Engineering, Vol.**192**, 119-123 (2017)
21. R. J. N. Malleswara, K. R. A. Chenna, R. P. V. Rama, *The effect of roller burnishing on surface hardness and surface roughness on mild steel specimens.* International Journal of Applied Engineering Research **1**, 777 (2011)
22. M. Kowalski, P. Karolczak A. Stadniczenko, *Applicability of wavelet analysis in evaluation of roughness and waviness profiles of turned surfaces.* Technological Engineering, Iss. **7**, vol. **2**, 66-69 (2010)
23. V. Poppeova, V. Bulej, R. Zahoransky, J. Uricek, *Parallel mechanism and its application in design of machine tool with numerical control.* Applied Mechanics and Materials, Vol. **282**, 74-79 (2013)

Innovation of hardness tester mechanism and its functionality test

František Bárnik[1,*], *Milan* Sága[1], *Ondrej* Štalmach[1]

[1]University of Žilina, Faculty of Mechanical Engineering, Univerzitná 1, 010 26, Žilina, Slovakia

Abstract. The article is focused on constructional design of device mechanism innovation. In the introduction is describe device and his original mechanism. Next section is focused on constructional design of device mechanism innovation and his describe. Another step is comparison between original mechanism and innovated mechanism. In the conclusion are evaluated theoretical and practical benefits of innovation.

Keywords: hardness, hardness tester, innovation

1 Introduction

Actual engineering practice, which is in the period of automated high-productivity production, is characterized by an increase in quality requirements. Because many factors affect the production process, it is not possible to produce material with ideal properties. The manufactured components exhibit minor or greater deviations from the ideal mechanical properties. Therefore, the function of product reliability and durability is affected not only by the keeping accuracy of the dimensions, the accuracy of the geometric shape and mutual position of the elements of parts, but also keeping the mechanical properties of the material. Constant development of progressive technologies used in mechanical engineering, progress in the design of products and the rapid development of information technology, push manufacturers of measuring technology to constantly upgrade measuring instruments. The aim of the article is to approach the constructional design of device mechanism innovation. The deviace is hardness tester HPO-250 KR/AQ. The hardness tester HPO-250 KR / AQ is a versatile hardness measuring device capable of measuring material hardness by 3 methods: Rockwell, Brinne and Vickers. These are the 3 most commonly used methods of hardness measurement in industry The task was to design constructional changes to the device mechanism so that the device can measure and evaluate the hardness of the material using the computer program by all the hardness measurement methods provided by the device (Rockwell, Brinell, Vickers), without the need for manual exchange of indenters bodies and lenses.

* Corresponding author: frantisek.barnik@fstroj.uniza.sk

Reviewers: *Wieslawa* Piekarska, *Jozef* Bocko

2 Hardness tester HPO-250 KR/AQ

The hardness tester HPO-250 KR/AQ (Fig. 1) is intended for measuring material hardness with the Vickers, Rockwell and partially with the Brinell method using a Ø2.5 mm ball. In automatic mode the process of measuring is controlled by PC. The imprint scanning is carried out by a sensitive CCD camera with high resolution, necessary for precise processing of measured results. The camera is connected with PC by BNC cable. The hardness tester HPO-250KR/AQ works under Windows 7 operating system. The program of device is a new software for collecting and statistic data evaluation of hardness tester, which works according to the Brinell, Vickers and Rockwell method. This product fully cooperates with the network and enables data export to the server for further processing. The database is also accessible from the network enviroment, so that the measured results are also accessible to another users in case of necessity [1,2].

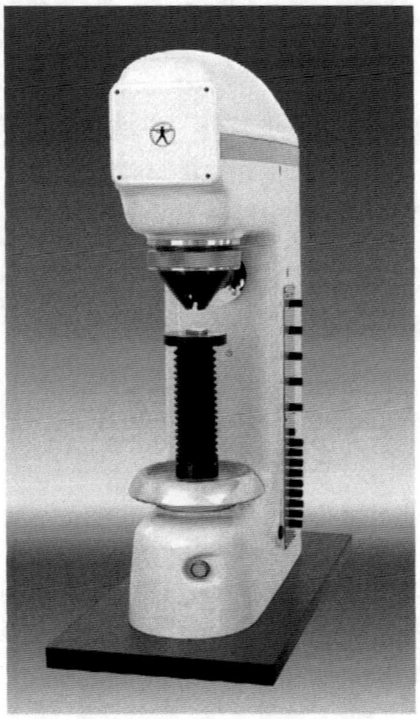

Fig. 1. HPO-250KR/AQ

2.1 Description of the device mechanism

The original mechanism of the hardness tester HPO-250 KR / AQ (Fig. 2). The actual assembly of lens and indenter is mounted on the tipping body (3). On this body is fixed punch (4) and vertical illuminator (9). The punch is made so that it was possible to mount indenter (5). It is possible to screw a lens (10) in the vertical illuminator corresponding to the requirements of the relevant measuring methodology. These parts are surrounded by an interrupted retainer that serves to hold the test item (6) on the test table (7). The retainer also protects the indenter and lens from damage. The tipping body is pivotally mounted in a pressure housing (2) which is guided in the guide housing (1). Tilting is provided by the motor that is coupled to the tipping body with the swingarm (8). When the measurement is started, the motor turns the tipping body from the position of the lens to the position with the indenter. Then begins the indentation of indentor into the material according to the relevant methodology. Upon completion, the engine turns the tipping body to the starting position. The imprint is then evaluated using the software.

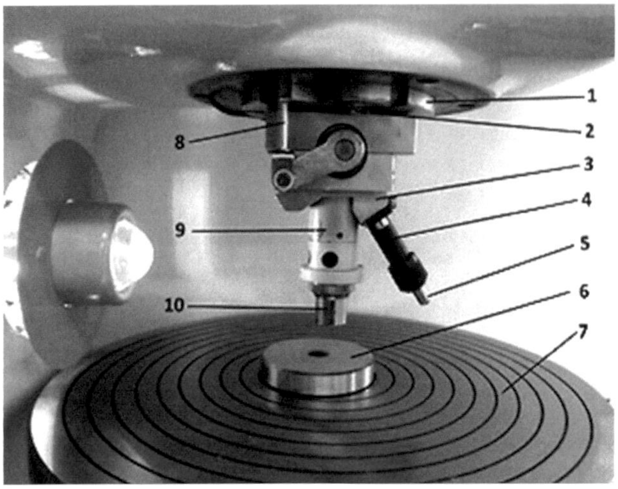

Fig. 2. The original mechanism of the device

In order to perform a hardness measurement, it is necessary to fit in the measuring device a lens that is suitable in the present case and a suitable indenter according to the respective measuring method. This requires manual removal of the retainer, lens, and indenter by service worker. Attachment of the respective lens, indenter and subsequent reassembly of the retainer are followed. Making such exchanges is time-consuming [3].

3 Constructional design of mechanism innovation

Some requirements have been made for the design of the hardness tester mechanism innovation. The innovated device have to make a imprint according to the appropriate method (Vickers, Brinell, Rockwell) and then capture it using a CCD camera (Vickers, Brinell) or using a sensor (Rockwell). From a imprint, evaluate the hardness using a computer program and secure archiving in the computer's memory or external memory. The device should be able to carry all the necessary imprints and lenses and, if necessary, enable them to be swiftly exchanged.

The request to innovate the mechanical parts of the hardness measuring instrument was given by one of the managers of AQUASTYL SLOVAKIA s.r.o. The reason was that customer firms demanded the removal of manual change of indenters and lenses by service workers in changing the hardness measurement method. This exchange is very time consuming and reduces the efficiency of measurement. The constructional design of the innovation is a rotary mechanism that is capable of simultaneously carrying 3 lenses and 3 indenters. Rotation the rotary head and precise positioning of indenters and lenses is ensured through the gear wheel that drives the precise stepper motor.

Figure 3 shows a constructional design of mechanism innovation. Designing the mechanism's innovation was complex because the design of innovation was not the whole device but only a part of it. The innovated mechanism had to be adapted to the original concept of the device, which in some cases caused complications. When designing all the components of the mechanism, optimum materials have been prescribed to meet the requirements of the machine. The mechanical part has been carefully designed to ensure high precision and durability of the device. Professional literature [4,5,9,10] was also used in the constructional design of individual components.

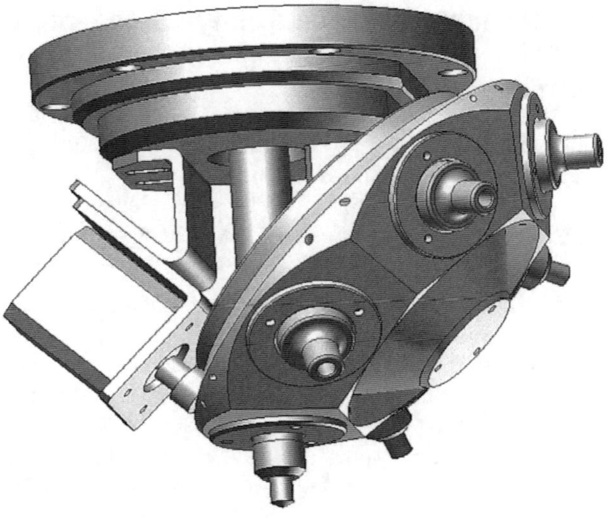

Fig. 3. Model of innovation mechanism

4 Comparison of mechanisms

The original device mechanism is capable of simultaneously carrying only one lens and one punch with the indenter. When changing the lens or the indenter, the operator of the device must first dismantle the device retainer and then replace the indenter and the lens. After replacing the indenter or the lens, it is necessary to screw back the retainer.

The innovated mechanism of the device, unlike the original, is capable of carrying 3 lenses and 3 penetrating bodies simultaneously. When changing the hardness measurement method, the appropriate method is selected in hardness tester software. The mechanism automatically sets the punch with the corresponding indenter into the measuring position. When changing the lens, the appropriate lens is selected in the device software as required, and the mechanism of hardness tester automatically sets the selected lens into the evaluation position of device. Since the rotary head is capable of simultaneously carrying 3 penetrating bodies and 3 lenses, the operator's time is no longer required to remove the retainer and then manually replace the indenter or lens. Therefore, the time required for the replacement of the relevant element is substantially shorter, which ultimately results in an increase in the efficiency of the device during the measurement. The comparison between the mechanisms is shown in Table 1.

Table 1. Comparison between the mechanisms

	Original mechanism	**Inovated mechanism**
Number of lenses	1	3
Number of indenters	1	3
Replace the indenter	manually with the assist of the operator	automatically by selecting the operator
Replace the lens	manually with the assist of the operator	automatically by selecting the operator

5 Tests of functionality and capability of the measurement instrument

Function tests on the HPO-250 KR/AQ instrument were performed in an accredited calibration laboratory. All tests were performed under standard conditions. They followed the initial tests of the innovated device mechanism.

The capability of the device has been verified for all 3 relevant methodologies by which the device is capable of evaluating the test material. According to relevant literature [6, 7,8], for a capability device, the capability index should be $C_{gmk} \geq 1.33$. At least 25 repetitive measurements should be made in the assessment of capability. For each hardness measurement method, 30 measurements were made and the appropriate values were recorded in the tables. From the measured values, the basic statistical characteristics of the process X_a and S_w were calculated for each method. The C_{gm} and C_{gmk} capability indexes were further calculated (Table 2) to determine the hardness tester capability. According to the established criterion, the lower of the calculated values is always taken into account in determining the capability of the device. The results show that the hardness tester can be considered as suitable by all three methods (Rockwell, Brinell, Vickers). The indexes C_{gmkU}, C_{gmkL} and C_{gmk} have the same value for each of the three methods. The same value was issued because the average calculated value is the same as the value of the respective calibration sample.

Table 2. Table of calculated values of the C_{gmk} and C_{gm} indexes

Measurement method	Index C_{gmk}	Index C_{gm}	Capability
Rockwell	1.63	1.63	suitable
Brinell	1.47	1.47	suitable
Vickers	1.97	1.97	suitable

For the Rockwell method, the graph of the variations of the measured deviations was constructed from the measured values (Fig. 4). From the measured deviations was calculated the mean deviation $\mu = 0.01$ (0.02% of the measured value) and the corresponding standard deviation $s = 0.1539$ (0.31% of the measured value).

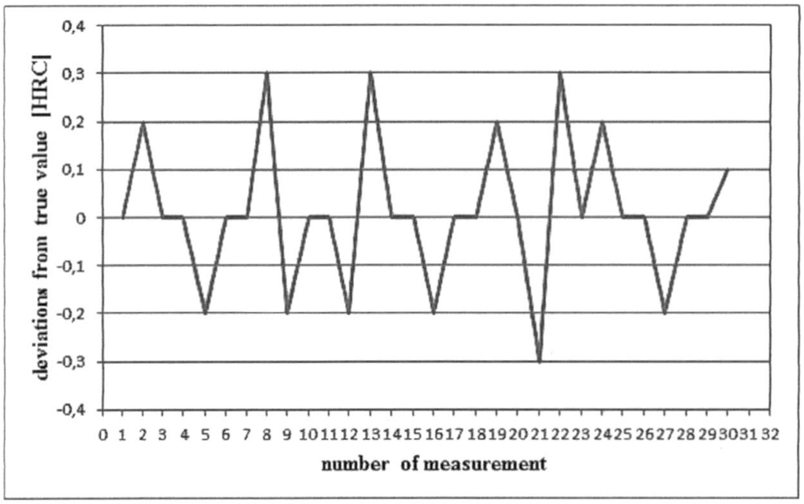

Fig. 4. Graph of measured deviations for the Rockwell method

6 Conclusion

At present, when modern technology puts great demands on the precision, quality and efficiency of the measuring instruments, it is necessary to produce products with the required shapes, dimensions and properties that will meet the highest quality requirements. In production, however, is produced materials whose mechanical properties are not accurate and exhibit certain deviations from the nominal values. The task of measuring instruments is to measure and evaluate deviations that arise during the production of materials and components due to various negative influences.

This article discusses the design of the device mechanism innovation, which is a rotary mechanism. The constructional design was based on the original concept of the device. The proposed mechanism is capable of simultaneously carrying 3 indenters and 3 lenses. The device will allow individual elements to be automatically exchanged using computer software and it will save time for operators. The benefit of the mechanism's innovation is to increase efficiency and save time when measuring, and thus economic benefit. The time required for measurement with exchange the indenter is 65% less than that of the original mechanism. Therefore, it will be possible to measure the hardness over a much larger number of measured samples at the same time. The upgraded device will therefore greatly benefit companies that need fast, accurate, and efficient hardness evaluation.

This work has been supported by the Slovak Research and Development Agency under the contract No. APVV-14-0096 and KEGA project No. 015ŽU-4/2017.

References

1. P. Veles, *Mechanical properties and testing of materials.* Bratislava: ALFA. 408 p. (1989)
2. http://www.aquastyl.sk (cit. 17.05.2018)
3. F. Bárnik, *Innovation mechanical parts of hardness tester HPO250 KR/AQ*. Diploma thesis. University of Zilina, (2015)
4. S. Černoch, *Mechanical engineering manual Volume 1*. Prague: SNTL. 1296 p. DT 621, (1977)
5. S. Černoch, *Mechanical engineering manual Volume 2.* Prague: SNTL. 2586 p. DT 621-7, (1977)
6. M. Dovica, et al., *Metrology in mechanical engineering*. Kosice:EMILENA, 351p (2006)
7. M. Ťavodová, *Assessment of measurement instrument capability*. (2006) http://www.tuzvo.sk/files/FEVT/fakulta_fevt/tavodova.pdf (cit. 11.04.2015)
8. A. Sapietová, V. Dekýš, M. Sapieta, P. Pecháč, *Application of computational and design approaches to improve carrier stability*. Procedia Eng. **96**, 410-418 (2014)
9. B. Leitner, M. Vaško, *Design and Modelling of Tank Wagon Assembly Operations in CAM Environment*. Transport Means 2015 **I & II**, 87-90 (2015)
10. P. Kopas, L. Jakubovičová, M. Vaško, M. Handrik, *Fatigue Resistance of Reinforcing Steel Bars*. Procedia Engineering **136**, 193–197 (2016)

MATEC Web of Conferences **244**, 01023 (2018) https://doi.org/10.1051/matecconf/201824401023
ITEP'18

Static structural analysis of rolling ball bearing

Peter Šulka[1,*], *Alžbeta* Sapietová[1], *Vladimír* Dekýš[1], *Milan* Sapieta [1]

[1]Department of Applied Mechanics, Faculty of mechanical engineering, University of Žilina, Univerzitná 8215/1, 010 26 Žilina, Slovak Republic

Abstract. In this paper, an effort has been put to analyze the rolling ball bearing using finite element analysis the contact pressure level, stress or displacement behavior of rolling ball bearing. The obtained results were then compared with the analytical results obtained through the methodology of Hertz theory in order to demonstrate an almost identical similarity of results and correct setting up of the given simulation in computational software ANSYS Workbench.

Keywords: Static structural analysis, Contact, Hertz theory, ANSYS Workbench.

1 Introduction

Rolling bearings are among the most important components of conveyor belts, working machines or various rotary mechanical applications and, therefore, the trend of increasing demands in terms of their production, accuracy, load-bearing capacity and reliability is increasing.

Their application lies in all branches of industry, from the production of heavy duty machinery in mining and stamping, in all modes of transport or in automotive or aviation, to the application of miniature bearings to high speed dental drills.

First of all, who drew and invented the ball-bearings was Leonardo da Vinci around the year 1500. He designed the ball bearings to be incorporated into his design for the helicopter – it would make rotating the large screw on top much easier by reducing the friction by several orders of magnitude [9]. Philip Vaughan invented the ball bearing in the year 1794, it was 294 years after Leonardo's design. Vaughan's patent described how iron balls could be placed between the wheel and the axle and carriage. The balls let the carriage wheels rotate freely by reducing friction. Later, in August 1869, Parisian bicycle mechanic Jules Suriray received the first French patent for ball bearings. The bearings were then fitted to the winning bicycle ridden by James Moore in the world's first bicycle road race. The modern, self-aligning design of the ball bearing is attributed to Sven Wingquist of the SKF ball-bearing manufacturer in 1907. In the year 2002 INA/FAG established by Dr. Georg Schaeffler and Dr. Wilhelm Schaeffler E1 spherical ball and roller bearing, which is base of current, modern rolling bearings.

* Corresponding author: Peter.Sulka@fstroj.uniza.sk

Reviewers: *Josef* Soukup, *Robert* Grega

Rolling bearings allow for the transfer of forces between the moving and fixed parts of the mechanical system and are designed to minimize friction arising from power and energy transfer.

Due to their multiple use, increasing emphasis is placed on their development and improvement, because in many cases the bearings do not provide the required durability according to the calculation values. There may be many reasons, for example, overload, load lower than required, inadequate lubrication, ineffective sealing, inadequate over fitting, or impact load on the bearing that leads to permanent deformation in the form of imprints in the bearing ring path paths. This very latest case of damage is becoming more and more frequent, for example automobile chassis bearings and it is necessary to examine the problem in order to examine the given load condition using the FEM method and to obtain optimal results for further research and modification of bearings with the intention of eliminating impact damage. Researchers still continue with the intensive development and research in field of rolling bearings with purpose to investigate new theory of bearing aspects, solving significant parameters and values of rolling bearings in order to ensure maximum quality and accurate prediction.

2 Hertz theory

In the elastic area, it is possible to calculate the contact pressure and the resulting deformation at the contact points of the rolling elements and orbits by means of Hertz theory. concerning to the contact of the volume elastic bodies. The general conclusions of this theory are based on three main assumptions.

1. The material of the bodies to be touched must be homogeneous and isotropic.
2. Only normal stress, not shear, can be transmitted in the contact area.
3. The proportional limit of the material must not be exceeded, that is, there can be no plastic deformation [1].

The contact area of the point contact of the load between the two curved bodies in point contact is generally elliptical (see Fig. 1). The elliptical contact surface for two components with identical elastic modulus E and Poisson constant is defined according to Hertz theory with respect to the main axis.

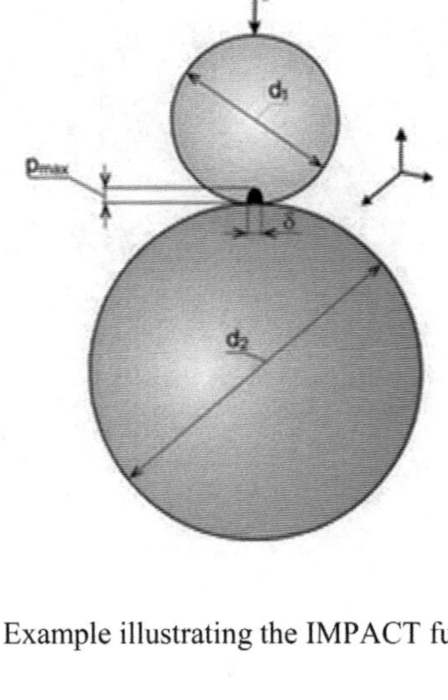

Fig. 1. Example illustrating the IMPACT function

From the Fig. 1 is possible to detect significant parameter δ, which is defined as wide of contact area. In case of point contact, the maximum contact pressure (see Fig. 2) is situated in the middle of contact area with appropriate value defined by following equation [2].

$$p_{max} = 1.5 * \frac{F}{\frac{\pi}{4}\delta^2} \quad (1)$$

p_{max} is the maximum contact pressure [MPa],
F is the external load [N],
δ is the wide o contact area [mm].

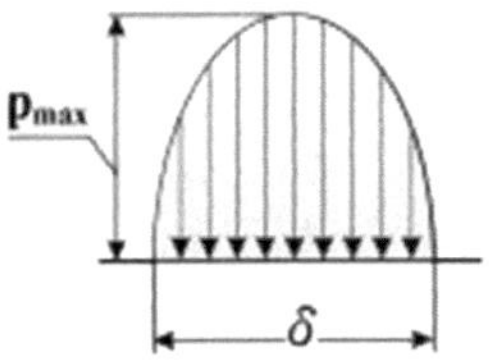

Fig. 2. The elliptical contact pressure

3 Computation of contact pressure based on Hertz theory

The stress calculation was performed using the following relationships dependent on bearing parameters, external load and mathematical coefficients. External load was defined for value 40 000 N according to the empirical formula of curb-impact of specific ball bearings (see Fig. 3).

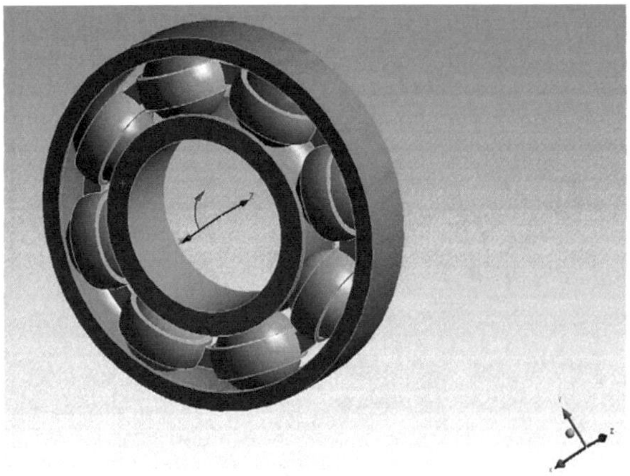

Fig. 3. 3D geometry of analysed rolling ball bearing

$$Fr_{max} = \frac{4.37*F}{Z} = \frac{4.37*40}{8} = 21.85\ kN \quad (2)$$

Fr_{max} is the maximum radial load of the rolling element [N],
F is the size of the external radial load [N],
Z is the number of rolling elements [-],
4.37 is the Striebeck coefficient for single row ball bearings [-] [10].

$$F_n = \frac{Fr_{max}}{\cos(\alpha)} = \frac{21.85}{2*\cos(45)} = 15.45\ kN \tag{3}$$

F_n is the maximum normal load [N],
α is the value of contact angle [°].

$$Q_{max} = F_n = 15.45\ kN \tag{4}$$

Q_{max} is the maximum normal load of rolling element [MPa].

$$\sigma_{max} = 4.37 * \varepsilon_\sigma * \left(\frac{Qmax}{de^2}\right)^{\frac{1}{2}} = 4.37 * 40.4686 * \left(\frac{15450}{6^2}\right)^{\frac{1}{2}} = \mathbf{3663.62MPa} \tag{5}$$

σ_{max} is the maximum contact pressure according to the Hertz theory [MPa],
d_e is the diameter of rolling element [mm],
ε_σ is the empirical coefficient of contact pressure [-] [3].

From these empirical relations, the maximum contact pressure was determined to the value **σ_{max} = 3663.62 MPa**.

3 Computation of contact pressure based on finite element method

The FEM analysis was performed in ANSYS Workbench. The simulation of the investigated bearing was solved as a linear model with given material properties of linear structural steel E = 210,000 MPa, μ = 0.3.

The contact pairs have been customized between the individual roller elements and rings as well as the cage through the contact region function. The contact between the elements was given as friction contact f = 0.05, with a normal stiffness factor of 0.1 and was defined by the Lagrange formula [3,8].

The boundary conditions of the bearing have been specified for each part as follows:

The outer ring and also inner ring displacement was removed in the z-axis direction and allowed in the direction of the x and y axes in the global Cartesian coordinate system. In the cylindrical global coordinate system, the displacement of the outer ring was taken in the y-direction and allowed in the x-axis and z-axis directions. In the cylindrical global coordinate system, the bead and cage displacement was taken in the x direction and allowed in the y and z direction. The coordinate systems are marked in previous figure (see Fig. 3).

External load was set with value 40 000 N as Bearing Load in the negative y-axis direction with a bearing time of one second for simulation of impact load (see Fig. 4).

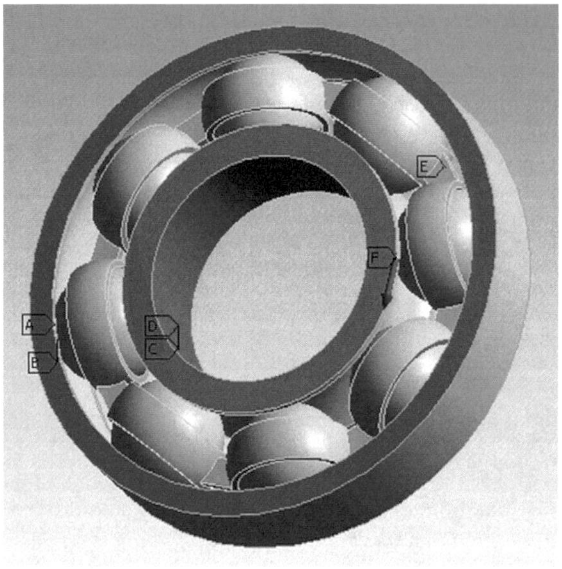

Fig. 4. The boundary conditions of investigated ball bearing

The results of performed FEM analysis with purpose to right investigate and compare analytical and numerical solution of contact characteristics display following figures.

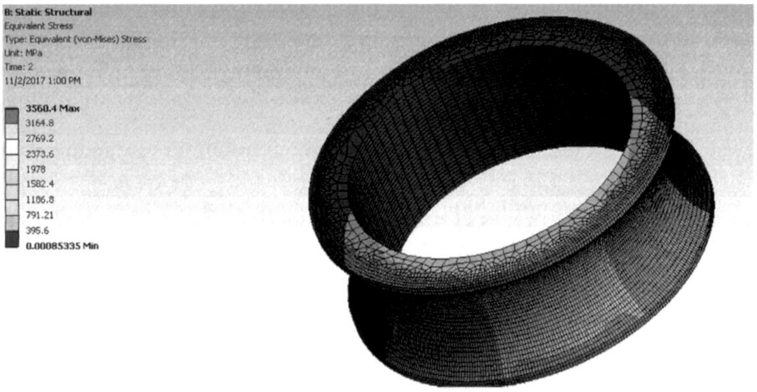

Fig. 5. Equivalent von-Mises Stress

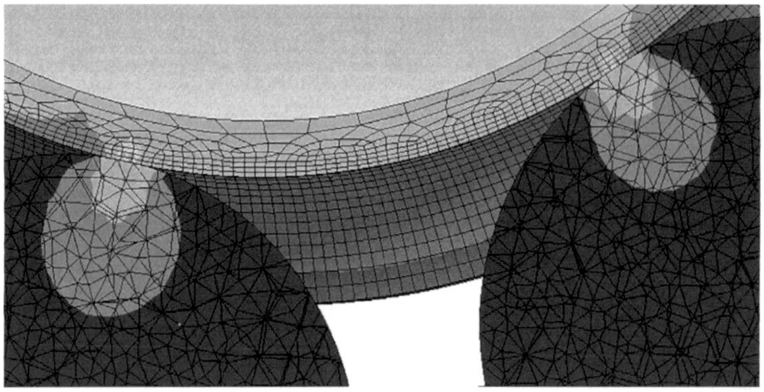

Fig. 6. Equivalent von Mises Stress in the cut

From the previous figures (see Fig. 5, 6) is clearly possible to determine value of Equivalent von Mises stress on different parts of investigated rolling bearings. Accordingly is permited to see the elliptical shape of contact areas adn also under surface stresses. Equivalent von-Mises Stress is often used in design work because it allows any arbitrary three-dimensional stress state to be represented as a single positive stress value. Equivalent stress is part of the maximum equivalent stress failure theory used to predict yielding in a ductile material. Equivalent stress is related to the principal stresses by the following equation [4,6].

$$\sigma_e = \left[\frac{(\sigma_1-\sigma_2)^2+(\sigma_2-\sigma_3)^2+(\sigma_3-\sigma_1)^2}{2}\right]^{1/2} \tag{6}$$

Fig. 7. Elastic deformation of inner ring

In the Fig. 7 is displayed the von Mises or equivalent strain ε_e which is computed by next equation, where ϑ is effective Poisson´s ratio [5,7].

$$\varepsilon_e = \frac{1}{1+\vartheta}\left(\left[(\varepsilon_1-\varepsilon_2)^2+(\varepsilon_2-\varepsilon_3)^2+(\varepsilon_3-\varepsilon_1)^2\right]\right)^{1/2} \tag{7}$$

From the following figures (see Fig. 8, 9), it is possible to detect the value of contact pressure, contact area and it´s shave and as well penetration of rolling elements into orbital paths of bearing rings by the Contact Probe function of Post-processing function.

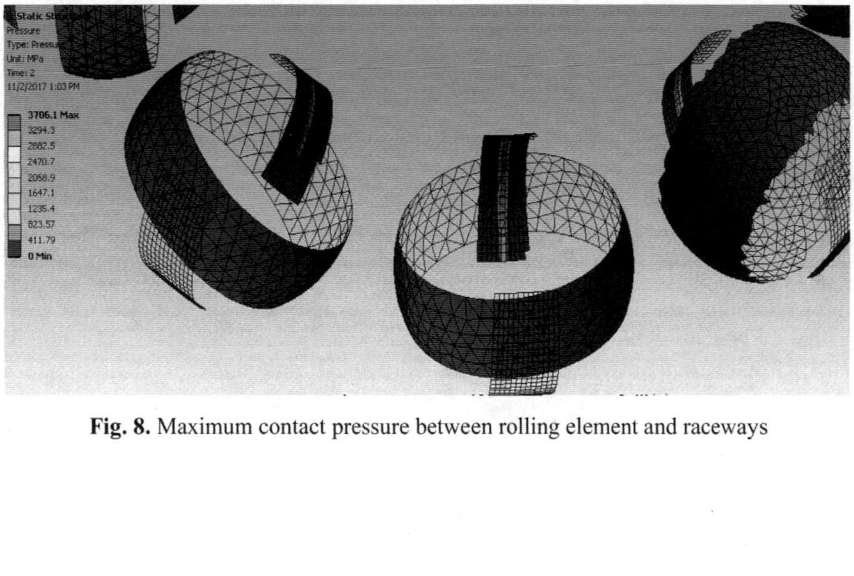

Fig. 8. Maximum contact pressure between rolling element and raceways

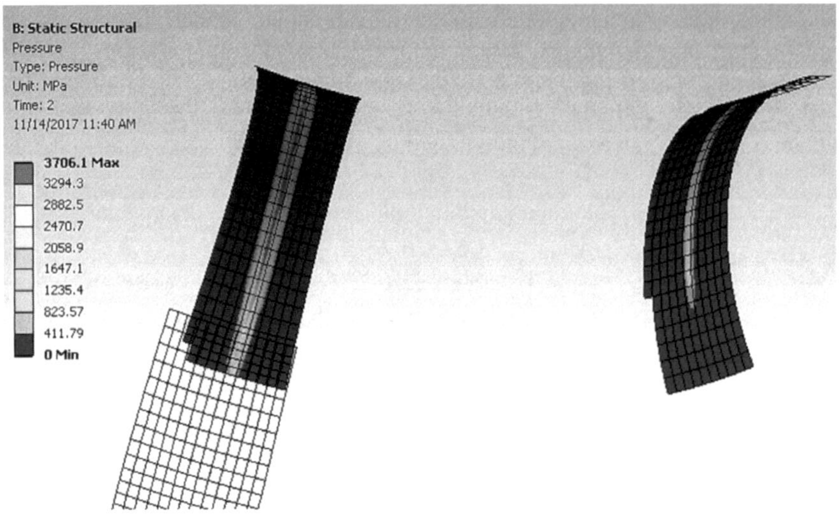

Fig. 9. The detail of contact pressure in contact areas

4 Conclusion

From the simulation and graphical outputs (see Fig. 5- 9), the results of the stress-strain analysis using the FEM software can be clearly seen. Through the analysis it was found that the value of the maximum contact pressure gets the size σ_{FEM} = 3706.1 MPa (see Fig. 9).

By comparing the results of the analytical and FEM (numerical) solutions, it can be detected that the results are almost identical.

$$\sigma_{max} = 3663.62 \text{ MPa} == \sigma_{FEM} = 3706.1 \text{ MPa}$$

The results of the MKP analysis are correct, which is proved by comparison of the analytical and numerical solution.

This paper was supported by VEGA 1/0795/16 and KEGA 017ŽU-4/2017.

References

1. P.K. Gupta, *Computer graphics modeling of rolling bearing dynamics.* Presented at the STLE Annual Meeting, Pittsburg, Pennsylvania (1994)
2. P.R. Anoopath, A.K. Vishwanath, *Hertz contact stress of deep groove ball bearing.* 7th Conference of materials processing and characterizaton, Published by Elsevier ltd., Vol. **5**, No. **1**, Pages 3283-3288 (2017)
3. T. Harris, *Rolling Bearings Analysis.* 4th Edition, John-Wiley & Sons Inc., New York (2001)
4. H. Hertz, *About the contact of elastic bodies.* Collected Works, Vol.**1**, Leipzig, Germany (1960)
5. P. M. Johns, R. Gohar, *Roller bearings under radial and eccentric loads.* Tribology International, Vol. **14**, No. **3**, 131-136 (1981)

6. A. B. Jones, *A general theory for elastically constrained ball and radial bearings under arbitrary load and speed conditions*. ASME Journal of Basic Engineering, 309-320 (1960)
7. V. Dekys, M. Sága, M. Žmindák, A.Hrčeková, *Structural optimization by finite element method.* Acta Mechanica Slovaca, 3-**B2005**, 61-67, (2005)
8. M. Žmindák, R. Melicher, V. Dekýš, *FEM modeling and testing of cyclic plasticity of materials.* Proceedings of the conference on occasion of 55th anniversary of foundation of the Faculty of Mechanical Engineering VSB – Technical University of Ostrava, 311-316 (2005)
9. The ball bearing. Leonardo da Vinci´s Ball Bearing. Accessable on the internet (http://www.leonardodavincisinventions.com/mechanical-inventions/leonardo-da-vinci-ball-bearing/)
10. H. Meerkmamm, Schaeffler Technical Pocket Guide, Schaeffler Technical Documentation (2014)

Analysis of roughness profile on curved surfaces

Mário Drbúl[1,*], *Pavol* Martikáň[1], *Jozef* Bronč-ek[1], *Ivan* Litvaj[2], *Jaroslava* Svobodová[3]

[1]University of Žilina, Faculty of Mechanical Engineering, Univerzitna 1, 010 26 Žilina, Slovakia
[2]University of Žilina, Faculty of Electrical Engineering, Univerzitna 1, 010 26 Žilina, Slovakia
[3]J. E. Purkyně University in Ústí and Labem, Faculty of Mechanical Engineering, Pasteurova 7, 400 96 Ústí and Labem, Czech Republic

Abstract. Production and control of accurate surface bring some issues. There are some technological limits of machining, but we have to know how to control these machined surfaces due to functional and life-time properties, which are affected mainly by roughness surface. Most important part of system of roughness surface measuring is suppression of nominal shape of scanned profile, its filtration according to standard ISO 4288 and evaluation according to the relevant standards. This article described process of filtration of roughness surface profile. If operator of the measuring instrument omits some important aspects at this stage, we obtain incorrect roughness surface profile, which significantly distorts the results of measurement leading to a rough error. The article is aimed at verifying roughness meters and indicates if a certain amount of data would be lost as if this loss affected the measurement result. Objectively, such a loss of data was simulated in the evaluation by considering every 7th scanned point at constant velocity of measurement.

Keywords: roughness surface, filter, wavelength

1 Introduction

The surface of each workpiece consists of different geometric elements, although it appears smooth and shiny at first sight. During the manufacturing process the number of effects occur (e.g. used technology, manufacturing process, human factor, etc.) that hamper the production of components ideal shape. Thus manufactured products show certain allowed deviations of desired geometry. Verification of such deviations is substantial in terms of functionality and interchangeability and is performed in compliance with GPS norms (Geometrical specifications and product verifications). If these norms are not respected, verification results are informative only. [1,2]

The quality of the manufactured surfaces can be assessed by the surface roughness as characteristics of slip, wear resistance, adhesion of protective coatings such as varnishes, corrosion resistance and fatigue resistance, among others, are related to roughness parameters. For accurate measurements of the surfaces, there is need to have an objective method of evaluation and roughness or quality parameter. [3,4]

* Corresponding author: mario.drbul@fstroj.uniza.sk

Reviewers: *Ingrid* Görögvá, *Nataša* Náprstková

One of the important parameters that assess the quality of the products surface is roughness profile. To a certain extent, future durability, reliability, running accuracy, noise level, corrosion resistance, wear and correct functionality of product surfaces depend on this parameter. Relief created on the product surface is bearer of certain information as well as reflection of used technology. During each manufacturing process (machining, material coating) tools leave traces on the surface. The main parameter is the Rz parameter, which is also referred to as the height roughness parameter on the roughness standards for roughness meter verification. Those are spatially arranged and their verification is technologically difficult. Because of the difficulty, the most used method is simplified measuring of roughness (often called as 2D) in one section of measured profile. It is standardized method of surface profile verification. [5].

2 Filtration of primary profile

Before applying appropriate filter we must remove nominal shape of evaluated surface (Fig.1) resp. its tilt, which contains scanned profile (we usually use suitable mathematical method, e.g. method of least squares or polynomial of the n-th degree).

Shape of interlacing element are:

- In 2D: line, circle, arc, ellipse, parabola and others.
- In 3D: plane, cylinder, sphere, ellipsoid, paraboloid and others.

Actual filtering process is following: at first, short-wave filter λ_s is applied on scanned profile, through which are short-wave elements of profile (e.g. micro cracks, pores, scratches, corrosion and others) filtered. The primary profile is created.

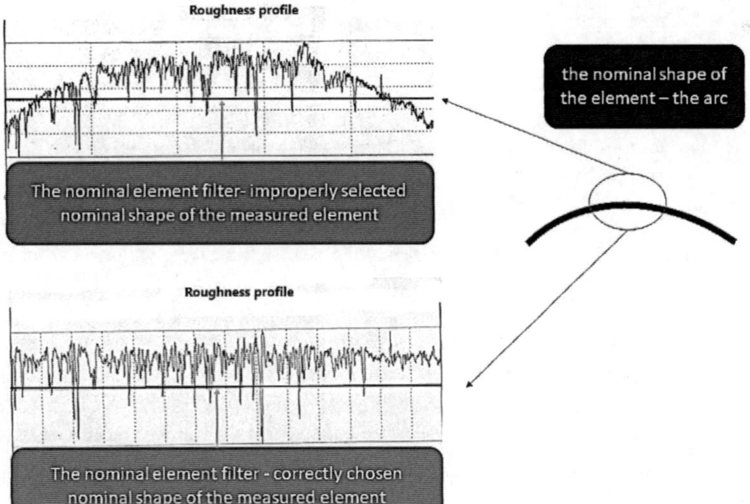

Fig. 1. Removal of the nominal shape of evaluated surface

Transmission characteristics of roughness and waviness profiles are at 50 % level of cut-off wavelength. Main differences between roughness and waviness are illustrated in Fig. 2. Long-wave elements of profile are suppressed after applying of λc filter on primary profile. Roughness profile remains after separation of waviness element.

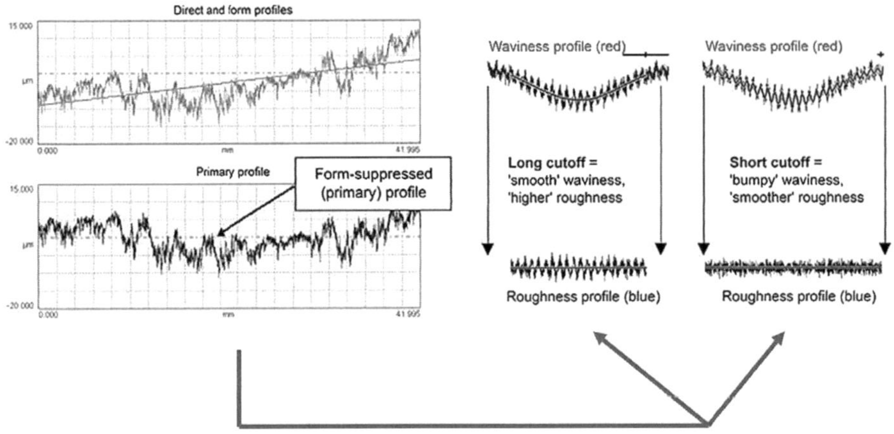

Fig. 2. Illustration of the profile filter transmission characteristics [6]

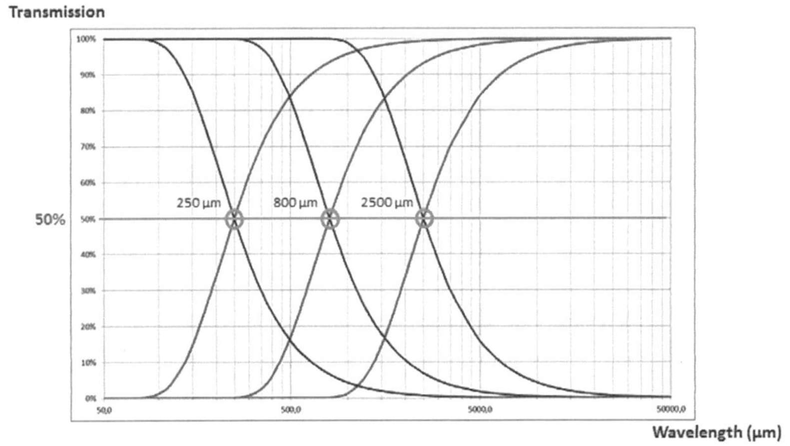

Fig. 3. Transmission characteristics for the roughness filter (blue) and waviness filter (red), for several cut-off values. These curves cross each other at 50 % of transmission. [7]

Phase-corrected filter is filter weighting function and has the equation of Gaussian density function with cut-off wavelength (filter transmission characteristic for long-wave profile elements). Transmission tolerance is not defined for these filters. Instead of tolerances there is stated graphical illustration of deviations of used filter in comparison to Gaussian filter (Fig. 3) in percentual values.

2 Roughness profile filtrations

The most important part of a surface roughness measuring system is filtration of scanned profile and evaluation according to given norm (attention, some roughness measuring machines work with older norms, where calculations of roughness parameters are not in conformity with newer norms. Method further described allows to extract needed parameters calculated with older norms or perform other analysis of given profiles). If a software of surface roughness measuring machine fails at this point, we get wrong roughness profile and thus incorrect informations upon which we perform our decision.

For verification of informations gained by surface roughness measuring machine MahrSurf PS1 (Fig. 4), there was performed measurement. The measuring device is equipped with diamond tip with 2 μm radius. Device was placed in such direction, that it scanned transverse direction of surface relief. As an evaluation sample, we used sample which surface was formed by planning with periodic program.

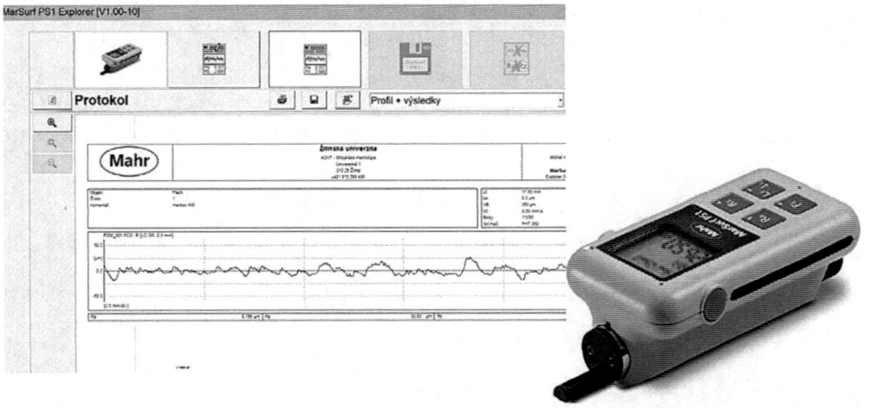

Fig. 4. MahrSurf PS1 [8]

Data significant for verification measuring are saved in text file (ASCI data). These form input data for performing the filtration. Device recorded 11 200 points for measured length. These points form scanned profile in Mahr format. The device MarSurf PS1 measures 2 lengths more, i.e. 11 200 points contain 5 measured segments and 2 segments representing sensor onset and offset (onset and offset lengths depend on device producer).

Value λc was set to 0.8 mm, i.e. length of one segment is lr = 0.8 mm. The overall evaluated length ln (length of 5 measured segments – without onset and offset of sensor) equals 4 mm.

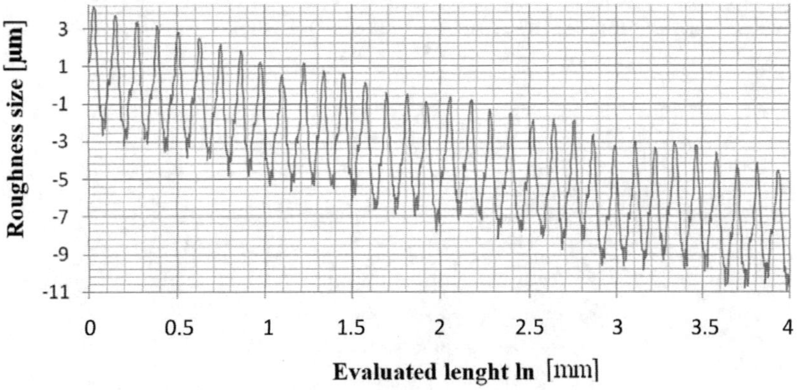

Fig. 5. Scanned profile

Another important step is to calculate the number of points on the base section ln. It is known, that number of scanned points is 11 200 and overall length of scanning is lt = 5.6 mm. The proportion of the total length and the number of scanned points creates sample distance between individual points, which is equal to 0.0005 mm. The number of points in one segment is then defined as quotient of one evaluated segment (0.8 mm) with sample distance (0.0005 mm). This means that in every segment is 1600 points. The value of onset will be contained in 0 – 1600 points range and the value of offset will be contained in 8600 – 11200 points range. By removing onset and offset range, range of 8000 points remains.

Scanned profile in Fig. 5 consists of roughness, waviness and shape profile. The next task will be to obtain, by gradual filtration, separate roughness profile for definition of basic roughness parameters. In illustration of scanned profile can be seen a slope, which is likely to be due to inaccurate machining of sample resp. misalignment of sensor arm and evaluated surface (parallelism is required).

Such profile, however, cannot be evaluated and therefore we must align it. This is one of the first steps in obtaining roughness profile. Possible skew or nominal shape supression is removed through method of least squares. Scanned profile is interlaced with a line (trend line) represented by equation $y = k.x + q$. In Excel, the operation is performed directly in rendered graph. Click Add Trendlines and Show regression equation (Fig. 6).

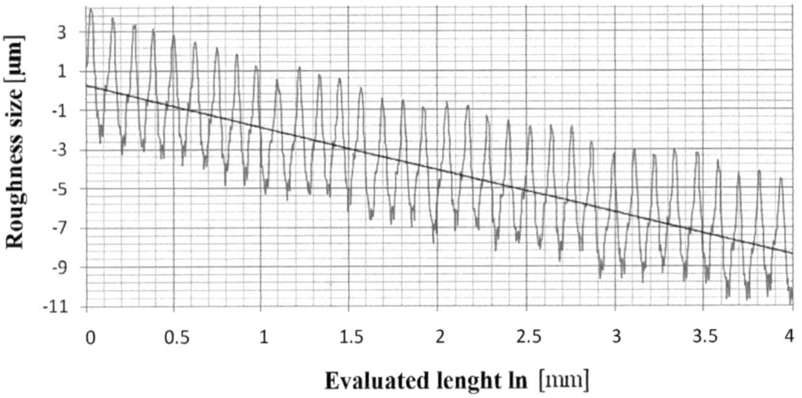

Fig. 6. Scanned profile interlaced with trendline

After application of the method of least squares, regression equation is obtained by which we gain point distances resp. position of every point on Y-axis relative to interlaced line. The distance values of position changes in the Y-axis from scanned profile compile primary profile (Fig. 7).

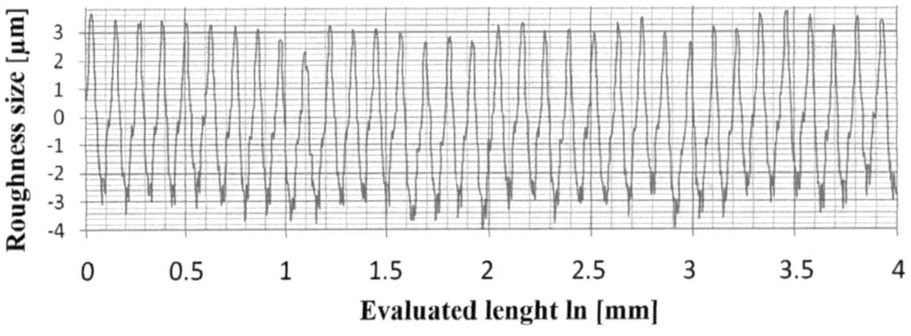

Fig. 7. Primary profile

Now, the primary profile contains only these elements: roughness and waviness. The next task will be to decompile given profile to harmonic elements using Fast Fourier Transformation (FFT) see Fig.8., and create mathematical model of filter (long-wave Gaussian filter to obtain roughness profile). The advantage of the FFT is the ability to decompile any profile to baseband of sine and cosine with final length.

First row of FFT is phase shift. The following rows are corresponding harmonic elements. The advantage of FFT, in addition to decompiling of signal to harmonic sinusoids, is ability to define the order of harmonic sinusoids necessary to determine the individual wavelengths. Value of basic wave length of first harmonic element is 4 mm, with increasing the harmonic element grows the number of periods, leading to proportional decrease of wavelength. In our case, first harmonic element has 4 mm wavelength, second has 2 mm, third has 1.3333mm....and the last one has 0.0078125mm.

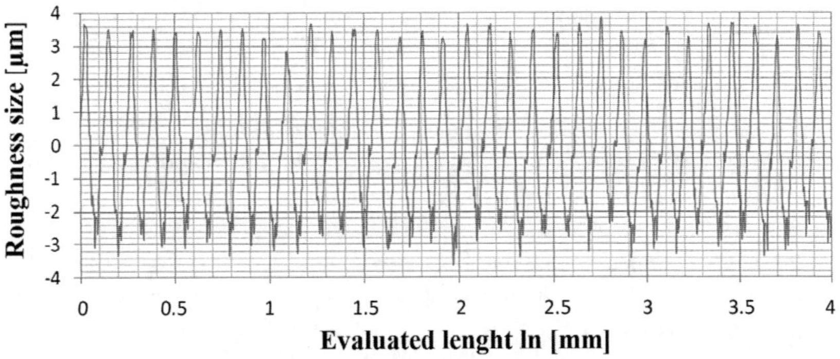

Fig. 8. Roughness of the profile

After defining all harmonic element wavelengths, we can continue with creating of long-wave filter. Using functions of Excel 2007 we have defined long-wave filter as: 1-(EXP((-PI())*(POWER(((0,4697*I1)/K2);2)))). This way we get element of long-wave filter for every wavelength used for removing waviness from primary profile. Afterwards, using function IMPRODUCT (used because of ability to product complex numbers with absolute) allows to merge filter elements with FFT harmonic elements after application of long-wave filter. The result is filtered roughness profile in certain frequency range (Fig.7).

For transition from frequency range to signal area we used a function inverse to FFT and so we have created data for filtered roughness profile (Fig. 8). Further evaluation of results is based on Rt (maximum height of profile) and Rz (average maximum height of profile) values. Table 1. shows very little difference between values obtained by program Excel and values obtained by device MahrSurf PS1.

Table 1. Rt and Rz values

Monitored values of surface roughness	Evaluated by MS EXCEL [µm]	Evaluated by device MahrSurf PS1 [µm]	Difference [µm]
Rt	7.48	7.54	-0.06
Rz	7.28	7.25	0.03

3 Summary

Setting of cut-off parameter and suppression of nominal shape of evaluated surface play important role in filtration of scanned profile. These settings have a decisive influence on the correct specification of surface evaluation and directly affect process of roughness profile filtration and thus its parameters. E.g. cut-off parameter sets the amplitude and phase shift, with which signal passes through evaluated profile. The results of the article can be used as a tool for verifying the accuracy of evaluation of roughness meter results.

The research work reported here was made possible by the grant project No. KEGA 031ŽU-4/2016

References

1. A. Czán, M. Sajgalík, J. Holubjak, K. Kouril, *Studying of cutting zone when finishing titanium alloy by application of multifunction measuring system*. Manufacturing Technology, **13**, 428–431 (2013)
2. R. Cep, A. Janasek, J. Petru, M. Sadilek, P. Mohyla, J. Valicek, M. Harnicarova, A. Czan, *Surface Roughness after Machining and Influence of Feed Rate on Process*. In Key Engineering Materials, **581**, 341-347 (2013)
3. A. Piratelli-Filho, G. Sternadt, R. Arencibia, *Removing deep valleys in roughness measurement of soft and natural materials*. Science and Engineering Journal, **21**, 29-34 (2012)
4. J. Valicek, S. Hloch, M. Gombar, R. Cep, etal. *Non contact optical measurement method application and obtained results interpretation*. Metal 2007 (Conference), 1-5 (2007)
5. D. Stancekova, M. Sajgalik, J. Petru, N. Naprstkova, P. Balaz, *Implementation of coating for failure elimination of dial gauges*. In METAL 2015: 24th international conference on metallurgy and materials,TANGER, 1162-1169, (2015)
6. [online] 2018, [cit. 2018-08-21]. Available at: https://www.renishaw.com/cmmsupport/knowledgebase/en/surface-finish-measurement--22135
7. [online] 2018, [cit. 2018-08-21]. Available at: https://guide.digitalsurf.com/en/guide-filtration-techniques.html
8. [online] 2018, [cit. 2018-08-21]. Available at: http://www.ipfonline.com/products/index/mobile_roughness_measuring_unit_1

The mould for production of plastic spout cap with internal thread by injection moulding

Vladimír Bulej[1,*], *Nadežda* Čuboňová[1], *Manfred* Eberth[1], and *Karol* Vasilko[2]

[1]University of Žilina, Faculty of Mechanical Engineering, Department of Automation and Production Systems, Univerzitna 1, SK-010 26, Zilina, Slovakia
[2]Technical University of Košice, Faculty of Manufacturing Technologies of the Technical University of Kosice with a seat in Prešov, Bayerova 1, SK-080 01, Prešov, Slovakia

Abstract. The main aim of this article is the mould design for production of plastic spout cap with inner thread by injection moulding technology. There are described the basic steps of whole design process with respect to fulfil all technology and functionality requirements. The application of hot runner system as well as hybrid ejection system based on rotary cores and stripper plate can be considered as the main innovative features of the designed mould. The mould flow analysis was carried out for the functionality demonstration.

Keywords: injection moulding technology, hot runner system, rotary cores

1 Introduction

Injection moulding technology grows very fast in recent decades. The possibilities and applications for utilizing injection moulding to produce plastic parts are seemingly infinite and offer manufacturers a cost effective method for producing reliable components [1-3]. Nowadays many components previously made of metallic materials are replaced by plastics thanks to the new technologies and improved material properties [4-8]. That is also the reason why we can meet plastics in many areas as well as in common usage at home [9, 10]. Reasons that injection moulding is so popular include: capability of high production rates, simple to complicated shapes are possible, high tolerances can be maintained, a wide range of materials and extremely cost efficient with low labour rates.

The mould is the key element and it is unique for each product. Its design process is accompanied by a lot of calculations and analyses to identify possible errors already during the preproduction phase and thus affect the final product quality [11]. Time spent during the designing process will save later costs for additional adjustment of the mould. The standard moulds are mostly composed from two parts with a sliding ejection system. Their gating system is based on so-called cold runner system. The result of such mould is set of final parts normally connected together by sprue (the waste). The sprue is typically removed by additional manufacturing process – cutting, what has the negative influence on usability of the melted material, the energy balance as well as on final efficiency of whole process. The

* Corresponding author: vladimir.bulej@fstroj.uniza.sk

Reviewers: *Tomasz* Kmita, *Radek* Knoflíček

designers of injection moulds are then utilizing progressive methods of mould design [12]. One of these methods is also the application of so-called hot runner system, which eliminates the waste of materials and the need for separation of individual parts as well.

2 Design of injection mould for part with inner thread

The injection mould design must be solved according to fulfil customer requirements and demands as well as technological requirements and with respect to available machinery equipment of the company, the data written in product drawing, the additional information about material, quantity of production.

2.1 Product for injection moulding technology

This article describes the design of the injection mould for production of Spout Cap 28/410 for PET bottles (Fig. 1) designed by company YORKER Packaging from low-density polyethylene [13]. Weight of this component is $m = 3.53\ g$ and the volume $V = 3735.15\ mm^3$. It was necessary to solve the right method for part ejection from the mould cavity with respect to the inner thread on a cylindrical surface.

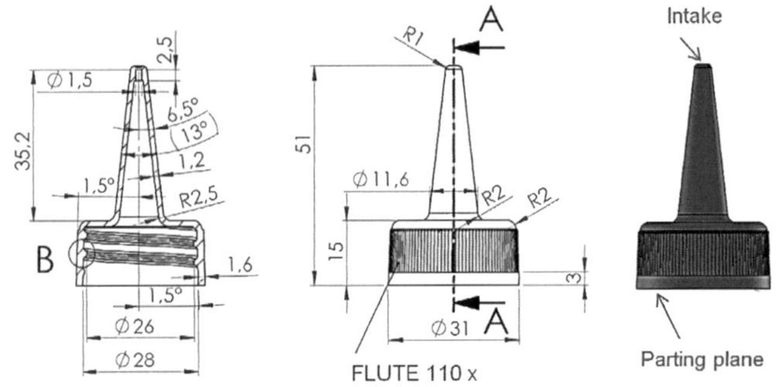

Fig. 1. The final product for injection moulding technology - Spout Cap 28/410 [13]

As it was mentioned above, the final product will be made of low-density polyethylene (list of basic properties in Table 1). Among polyolefin, polyethylene (PE) is one of the most widely used plastics, as the molecular structure of PE is simple and it has good chemical resistance and processing properties.

Table 1. Basic material parameters and properties for product Spout Cap 28/410 [13]

Material	LDPE (Low Density Polyethylene)				
Parameter	**Unit**	**Value**	**Parameter**	**Unit**	**Value**
Density	[$g.cm^{-3}$]	0.91 - 0.94	Shrinkage	[%]	1.5 - 5
Melt Temperature	[K]	450 - 550	Injection Pressure	[MPa]	50 – 150
Mold Temperature	[K]	290 - 345	Injection Speed	Fast speeds	

PE can be classified into ultrahigh-molecular-weight polyethylene (UHMWPE), high-density polyethylene (HDPE), low-density polyethylene (LDPE), and so on. The LDPE we can find in wide variety of products. LDPE is produced by the polymerization of ethylene at high pressure and temperature, the material is semicrystalline-crystalline. The crystallinity level is low because of chain branching. The material is tough but possesses moderate tensile properties and exhibits creep. However, it has good impact and chemical resistance. It is an easy flow material because of long chain branching. Commercial materials have densities in range of 0.91 - 0.94 g.cm^{-3}. LDPE is permeable to gases and vapours [5]. Very close tolerances are not possible with this material and its relatively large coefficient of thermal expansion makes it less suitable for long term applications. Shrinkage is of the order of 0.02 - 0.05 mm.mm^{-1} (1.5 - 5%). Actual shrinkage values are dependent on the moulding conditions. LDPE is resistant to many solvents at room temperatures but aromatic and chlorinated hydrocarbons cause swelling. Like HDPE, it is also susceptible to environmental stress cracking. For this material is recommended the injection pressure between 50 and 150 MPa and fast injection speeds [5,7,11]. Drying before processing is not necessary. All conventional types of runners and gates may be used, so LDPE is well suited also for hot runner systems.

2.2 Machinery

Standard injection moulding machine consists of two main parts - an injection unit and a clamping unit in general. We choose an injection moulding machine *ENGEL victory 60/200 ecodrive* (Fig. 2) with clamping force of 600 kN and injection unit up to 200 MPa. Its basic technical parameters are shown in table 2.

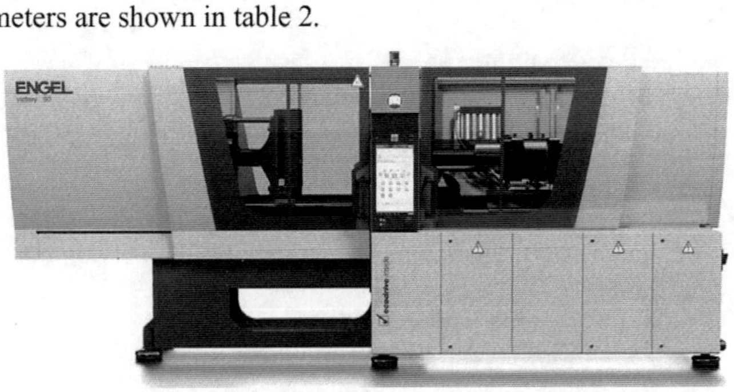

Fig. 2. Injection moulding machine ENGEL victory 60/200 [14]

The comparatively small injection moulding machine is due to a special design feature of the ENGEL victory machine, so-called "tie-bar-less clamping unit". Because there are no tie bars in the way, the mould mounting platens can be fully used up to their very edges. Motors for core-pulls or other functional components can even jut out beyond that. Further benefits of the tie-bar-less technology include fast mould set-up processes and efficient automation solutions because robots can access the mould area directly from the side without having to circumvent any obstacles. The patented Force Divider flexlink ensures high precision and process consistency [14].

Table 2. Basic parameters of injection moulding machine ENGEL victory 60/200 [14]

Injection machine	**ENGEL victory 60/200 ecodrive**				
Parameter	**Unit**	**Value**	**Parameter**	**Unit**	**Value**
Clamping force	[kN]	600	Injection pressure max.	[MPa]	200
Opening stroke max.	[mm]	450	Volume / one stroke	[cm^3]	78
Max. mould weight	[kg]	450	Injection speed max.	[mm.s^{-1}]	113
Mould mounting platens (max. dimensions)	[mm]	670x600	Injection screw diameter	[mm]	25

2.3 Mould design

The design process consists of several steps [13, 15]:

- definition of parting plane - how the products will be placed inside the mould cavity,
- definition of mould multiplicity – the number of product done in one sequence,
- shrinkage calculation (not mentioned in article),
- design of the inlet system, ejector system and tempering system.

2.3.1 Parting plane

The parting or dividing / split plane must be selected at the beginning together with intake position. It defines the surface where are touching both main mould plates while the mould is closed and fixed by clamping unit. In our case we decided to align it with the bottom edge of the cap as is shown in figure 1 – right. In the mould design will be used the above-mentioned hot runner system, and a system of movable cores with ejector stripper board.

2.3.2 Mould multiplicity

Mould multiplicity refers to the number of products which can be produced per one working cycle. For calculation of multiplicity we must take into account two main criteria:

a) Multiplicity according to the clamping force

This method is based on the size of buoyant force caused by the melted material in the parting plane which must be restrained by clamping unit of the injection machine [7].

$$n_{cf} = \frac{0.8 \cdot F_c}{A_{real} \cdot p_{inj}} = \frac{0.8 \cdot 600\,000\ N}{754.7\ mm^2 \cdot 100\ MPa} = \frac{480\,000\ N}{75\,470\ N} = 6.36 \approx 6 \tag{1}$$

Where:
n_{cf} - multiplicity according the clamping force [-],
F_c - clamping force [N],
A_{real} - area of a whole cavity cross-section in parting plane [mm^2] (equation 2),
p_{inj} - injection pressure [MPa],
A_{part} - area of one part cross-section projected into the parting plane [mm^2],
A_{intake} - cross-section area of intake projected into the parting plane [mm^2]; In our case (for hot runner system) is equal to 0,
d_{part} - diameter of the product projected into the parting plane [mm].

$$A_{real} = A_{part} + A_{intake} = A_{part} + 0 = \frac{\pi \cdot d_{part}^2}{4} = \frac{\pi \cdot (31\ mm)^2}{4} = 754.7\ mm^2 \tag{2}$$

b) Multiplicity according to the maximum volume of melted material injected in one cycle

This is based on ratio between the maximal and needed volume of injected material [7].

$$n_V = \frac{V_{max}}{V_{real}} = \frac{78\,000\ mm^3}{3\,735.15\ mm^3} = 20.8 \approx 20 \tag{3}$$

Where:
n_V - volume multiplicity [-],
V_{max} - maximal volume injected within one stroke of piston [mm^3],
V_{real} - real volume of a product (include the part of intake) [mm^3] (equation 4),
V_{part} - volume of one part [mm^3],
V_{intake} - volume of intake per 1 part [mm^3]; for hot runner system is equal to 0,

$$V_{real} = V_{part} + V_{intake} = 3\,735.15\,mm^3 + 0\,mm^3 = 3\,735.15\,mm^3 \quad (4)$$

According to get high quality products we will choose from value of both criteria the lower one (result from equations 1 and 3). We can see that the lower one is the multiplicity according to the clamping force. So our mould will be designed with multiplicity equal to 6 (6 products per one working cycle).

$$n_{FINAL} = \min of\,(n_{cf};\, n_V) = n_{cf} = 6 \quad (5)$$

2.3.3 Intake system

Intake (or gatting) system connects the end of injection nozzle with the mould cavity by one or more channels. The main functional task of intake system is achieving homogenous filling of whole mould cavity. There should be achieved easy removal of final product from the mould as well. A second aim is to minimize the volume of the sprue. Thanks to hot runner system (Fig. 3) we can reduce the sprue to absolute possible minimum.

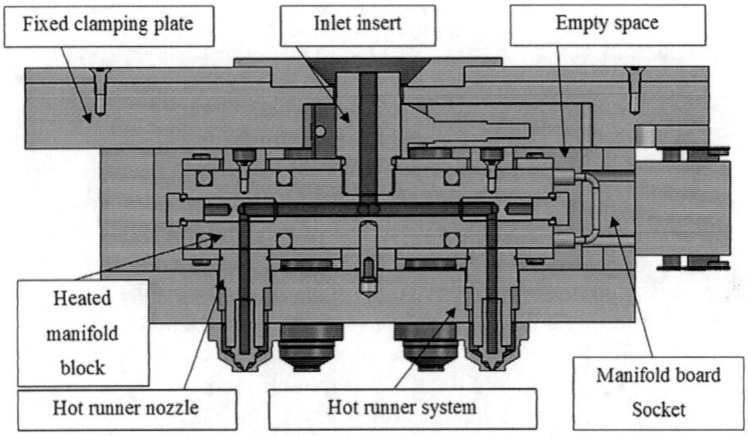

Fig. 3. Cross section and principle of intake system with hot runner nozzles

The principle of the hot runner system is that the melted plastic material remains in the whole intake system in the liquid state. This allows us to use of small cross section for the mouth with minimal or no cold channels what means minimal or no waste of material. Normally the nozzle is directly connected to the mould cavity. In the nozzle centre there is a shut-off valve needle. The connection of nozzle and final product is shown below (Fig. 4).

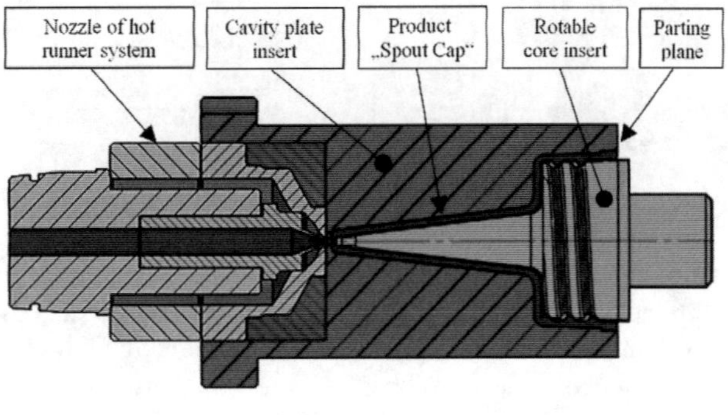

Fig. 4. The cross-section Hot runner system

The manifold with heated distribution channels is used in combination with a heated and insulated nozzle pre-chamber. Their shape is related to the required position of intake channels toward the fitting nozzles – shape called I, H, X, Y and others. For our system we designed hexagonal manifold with six nozzles as is shown in figure 5, right.

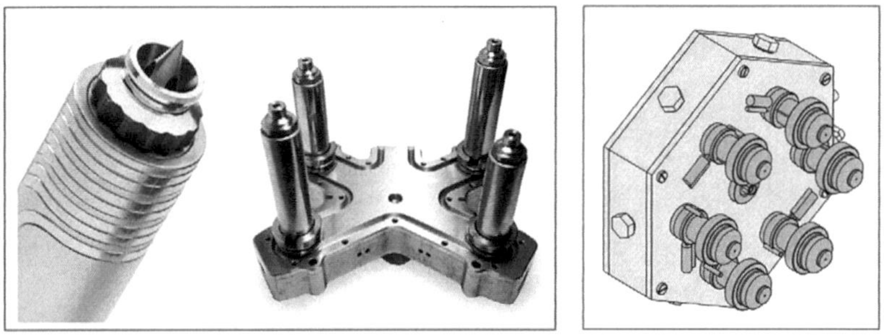

Fig. 5. Hot runner system: an example of hot runner system - a nozzle and nozzles in manifold (left) and our design of hexagonal manifold with 6 nozzles (right)

2.3.4 Ejection system

In general, for removing of final product from the mould cavity, there exist several kinds of ejection systems, e.g. pin ejection, sleeve ejection, stripper plate ejection, blade ejection, rotation core ejection, air ejection, etc. An internal thread form is basically an undercut that does not allow free ejection from the mould. Besides relatively rare method of unscrewing by hand and direct stripping of flexible parts with shallow thread profile there are common two different functional principles of inner threaded parts removing:

- the systems with collapsible threaded core and two-stage ejector - can be used to mould internal thread forms and freely eject small to medium size parts,
- the system with rotating cores driven by gear transmission and stripper plate - this system is based on transformation of linear motion generated by clamping mechanism into the combined rotational and linear motion of core inserts.

Finally we decided for the second alternative, so our mould is designed as hybrid system composed from rotating cores driven by gear transmission and stripper plate. Next step is therefore the design of gear transformation mechanism (see figure 6).

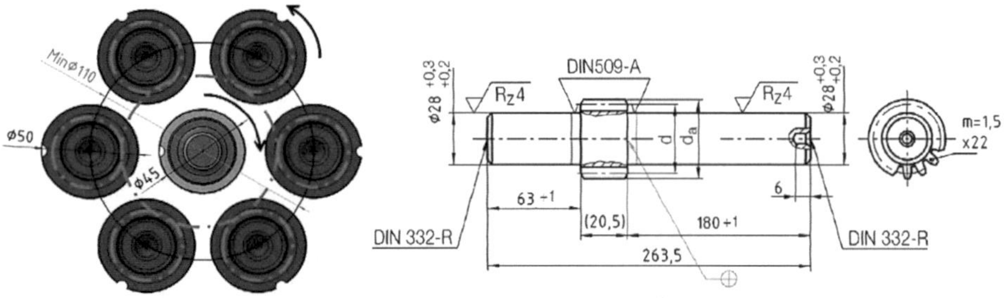

Fig. 6. Transmission driven mechanism for rotating cores motion

The transformation mechanism is composed from the steep lead screw (in our case chosen HASCO Z1500 – [16]) fixed in the axis of static part of mould. The nut combined with pinion gear is supported by bearings inside the moving part of mould. This unit is set to motion by opening of mould. Subsequently the rotational motion is transmitted to the driven gear shafts linked with core inserts arranged in a "planetary" configuration around the central screw according to the mould multiplicity (Fig. 6, left). The calculation of gear transmission system is based on the diameter of cavity plate inserts (Fig. 6, left - 6 circles) as well as the maximum diameter of steep lead screw HASCO Z1500 [16].

$$d_1 = m.z_1 = 1.5 \cdot 22 = 33\ mm \qquad d_2 = m.z_2 = 1.5 \cdot 55 = 82.5\ mm \tag{6}$$

$$a = \frac{d_1 + d_2}{2} = \frac{33 + 82.5}{2} = 57.75\ mm \tag{7}$$

Where: $d_{1,2}$ - pitch diameter of gear 1 or 2 [-],
$z_{1,2}$ - teeth number of gear 1 or 2 [-],
m - module of gears tooth-size,
a - the shortest distance between the gears axes [mm].

As the transmission elements were chosen the gears HASCO Z1550-28x180 and Z1553-36-20. Then the module of both gears is set to m=1.5 and the teeth number is 22 for the first one and 55 for the second one (pitch circle of both gears is calculated in Eq. 6). Our design of whole ejection system as well as cross-section of whole mould is shown in Fig. 7.

The ends of shafts are guided by the threaded inserts which ensure the helical movement of the core inserts. In exactly defined position of the mould during the opening will start to move also the stripper plate (Fig. 7) which help to eject the products Spout cap.

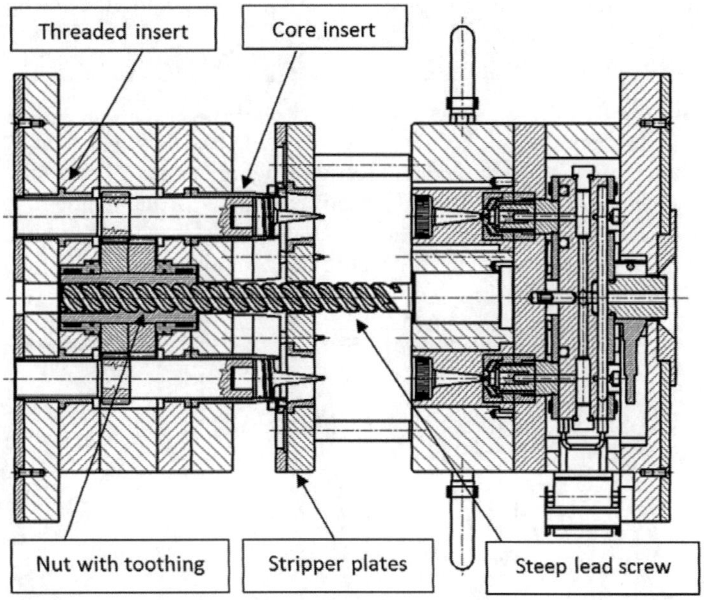

Fig. 7. Final mould design (cross-section view) and ejection system principle

2.3.5 Mould flow simulation results

Nowadays, the simulation play very important role in pre-production phase and can have massive influence on the final costs and efficiency of processes [17-20]. SolidWorks Plastics provides an effective injection moulding simulation tools which can help to identify production limits and predict potential failures and consequently eliminate them even in the early stages of product development [15, 21]. The software is suitable for designers of plastic parts as well as mould tool manufacturers. SolidWorks Plastics is fully integrated with SolidWorks CAD software. Changes are instantly reflected in the 3D model and can easily be passed on to the manufacturability assessment [20- 22]. We decided to simulate injection moulding process only for one part because there are produced six identical parts with small volume and with using of hot runner system in one cycle. Below there are described the results of filling time, injection pressure and cooling time analyses.

Filling time

The results of this analysis indicate for what time the mould cavity is filled in with melted plastic material. Various colours display the flow position of melted plastic at regular intervals (Figure 8, left). Using this analysis, we can understand the influence of the intake mouth location. The intake should be placed regarding that the mould is filled evenly (uniform filling) and in the shortest possible time. Further, it is possible to determine where the cold zones (connections) will occur and where the venting slots should be installed [23, 24]. The analysis revealed that the intake position was chosen correctly and the total filling time is about 0.75 seconds. In our case, the risk of cold zones is quite low and the mould should not be aerated.

Filling pressure

The result of this analysis shows how much pressure will be needed to fill the mould cavity for the over mentioned 0.75 second time period. A maximum pressure of about 35 MPa is in the mouth of the hot intake as expected, while at the furthest point of the mould it is about 5 MPa (Figure 8, middle). Of course this can be modifying by injection ramp controlled by injection control unit. After first real testing of the mould, it could be determined exactly the pressure profile suitable to get the best results.

Cooling time

As a result of this analysis, the colour range was again based on which we can determine the need for the tempering system and its location (Figure 8, right). From this analysis, we identified the places that will be needed to cool with parts of the tempering system. The results of the analysis fit quite well with our assumptions. In the area with the coarsest diameter, it will be necessary to temper or, chill the component to reduce the cooling time from about 12 seconds to lower one.

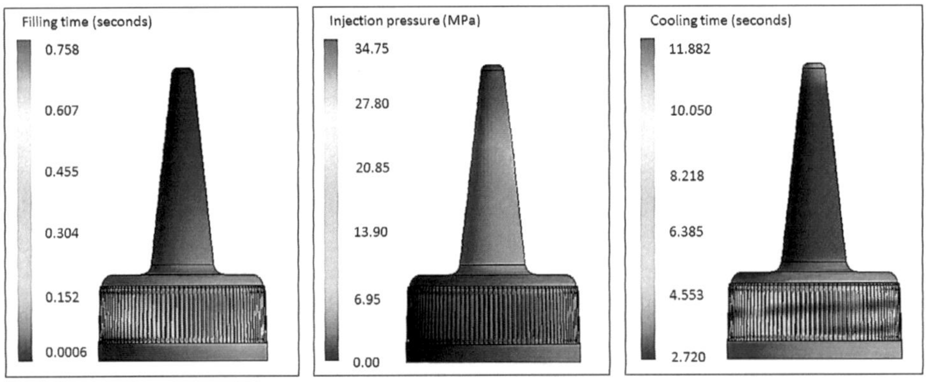

Fig. 8. Simulation results: time of filling (left), injection pressure inside the cavity while filling (middle), time of cooling (right)

The analysis provides much more information about the injection process. However, given the limited scope of this article, only basic parameters have been described. According to the analysis, it can be seen that the filling of the cavity is relatively homogenous, the pressure as well as temperature distribution are within the recommended range. Places where air bubbles will be formed can be resolved by adding venting slots if needed. Cold zones can be removed by appropriate adjustment of the technological parameters (eg. by increasing the melt temperature or injection speed).

3 Conclusions

The main aim of this article was the mould design for selected component Spout cap 28/410 production by injection moulding technology. Regarding to the inner thread on this product it was necessary to solve mainly the gating system and ejection system. The mould design was carry out for injection press *ENGEL victory 60/200 ecodrive*. Thanks to the used hot runner system we were able to reduce the sprue to absolute possible minimum. Regarding to an internal thread was selected the hybrid or combined ejection system composed from stripper plate and rotary cores driven by gear transmission which is suitable for such kind of products. The final design was verified by simulation in mould flow module of CAD software SolidWorks.

References

1. D.E. Dimla, M. Camilotto, et al, *Design and optimization of conformal cooling channels in injection moulding tools*. J. of Mat. Proc. T. **164-165**, p. 1294–1300 (2005)
2. S. Marques, A. Fagali de Souza, M. Jackson, I. Yadroitsau, *Design of conformal cooling for plastic injection moulding by heat transfer simulation*. Polímeros, **25**(6), 564-574 (2015)
3. D. Homar, et al, *Cooling simulation of conformal cooling injection mould insert produced by hybrid manufacturing*. Technological Gazette, **24**(4), p. 981-986 (2017)
4. N. Čuboňová, *Utilization of genetic alghoritms by the tool path programming*. Ann. of DAAAM and Proc. of the Int. DAAAM Symp., 63-64, (2011)
5. J.Wang, D. Chen, S. Wang, et al, *Insert injection molding of low-density polyethylene single-polymer composites reinforced with ultrahighmolecular-weight polyethylene fabric*. J. of Thermopl. Compos. Mat., **31**(8), 1013-1028 (2018)
6. S. Teklehaimanot, *Simulation and Design of a plastic injection Mold*. (University of Applied Sciences ARCADA, Finland [Degree Thesis], 2011)
7. D. Stančeková, R. Kameník, A. Sapietová, J. Pilc, T. Zlámal, J. Cais, *Comparison of properties of steel 100Cr6 and C56E2 in the process of machining*. METAL 2017: 26th Int. Conf. on Met. and Mat. p. 819-825, (2017)
8. J. Korzekwa, et al, *Electromechanical properties of PLZT x/90/10*. Eu. Phys. J: Spec. Top. 154(1), p. 127-130 (2008) DOI: 10.1140/epjst/e2008-00530-5
9. A. Rengevič, M. Fúra, N. Čuboňová, *Analysis of printing parameters for production of components with Easy3DMaker printer*. Adv. in Sci. and Techn.-Res. J. **10**(32), p.1-8, (2016), DOI: 10.12913/22998624/65114
10. A. Rudawska, P. Pawlak, I. Miturska, D. Stančeková, M. Chyra, *Comparative analysis of welded and adhesive joints strength made of acid-resistant stainless steel sheets*. Adv. in Sci. and Techn.-Res. J. **11**(4), p. 97-102 (2017), DOI: 10.12913/22998624/78165
11. M. Bruna, A. Sládek, *Hot tearing evaluation of al - based alloys*. Man. Techn. **16**(2), p. 323-327, 2016
12. D. Stančeková, A. Czán, M. Derbas, *Investigation of defects in forging tools by non-destructive detection method*. Metal 2013: 22nd Int. Conf. on Metal. and Mat. P. 485-490, (2013)
13. M. Skalicky, *Mechanical Design of Mould for Plastic Component Produced by Injection Moulding Technology*. (University of Žilina, Žilina [Master thesis], 2014)
14. Engel Austria GmbH, *ENGEL victory - Machine system with freedom to move*. Available on: https://www.engelglobal.com/en/at/solutions/injection-moulding-machines/victory.html

15. J. Stanček, V. Bulej, *Design of driving system for scissor lifting mechanism*. Acad. J. of Man. Eng. **13**(4) p. 38-43, (2015).
16. Hasco, *Z1500 Product info*. Available on https://www.hasco.com/hasco/en/Product-catalogue/Z/Demoulding/Gear-products/Transmission-elements/Z1500-d1xp1-L-l2/p/Z1500_d1xp1_L_l2.
17. M. Jančušová, *Moulds for processing of plastics (in Slovak)*. (University of Žilina, Žilina, 2010)
18. Z. Tůma, R. Knoflíček, J. Tůma, F. Bradáč, P. Blecha, *The Process Simulation Using by Virtual Reality*. Procedia Eng. **69**, (2014), DOI: 10.1016/j.proeng.2014.03.084
19. Š. Václav, J. Jurko, Š. Lecký, *Allowance treatment static designed couple and repeatable precision in assembly*. MATEC Web of Conferences **137**, Art. No. 04007 (2017) DOI: 10.1051/matecconf/201713704007
20. V. Bulej, et al, *Study of the Workspace of Hybrid Mechanism Trivariant*. App. Mech. a. Mat. **436**, p 366-373 (2013), DOI: 10.4028/www.scientific.net/AMM.436.366
21. J. Lipina, V. Krys, R. Pastor, *Increasing the Load Capacity of Screw Connections in Parts Produced by the Rapid Prototyping Method*. INES 2016 - 20th IEEE Int. Conf. on Int. Eng. Sys. 147-150 (2016)
22. M. Kelemen, T. Kelemenova, I. Virgala, *Rapid Control Prototyping of Embedded Systems Based on Microcontroller*. Modelling of Mech. and Mechat. S. **96**, p. 215-220 (2014), DOI: 10.1016/j.proeng.2014.12.146
23. M. Staněk, D. Maňas, M. Maňas, J. Javorík, *Simulation of injection molding process*. Proc. of the 13th WSEAS Int. Conf. on Aut. Con., Mod. & Sim. p. 231-234 (2015)
24. F. Dupret, A. Couniot, et al, Modelling and simulation of injection molding. Rheology Series, **8**, p. 939-1010 (1999), DOI: 10.1016/S0169-3107(99)80012-4

Contribution to the determination of the thermal emissivity of the composite material using long-wave infrared camera

Ondrej Stalmach[1,*], *Vladimir* Dekys[1], *Frantisek* Barnik[1], *Jan* Simon[1],

[1]University of Zilina, Faculty of Mechanical Engineering, Univerzitná 1, 010 26, Žilina, Slovakia

Abstract. This paper deals with measurement of emissivity of the composite samples. Three composite samples with the different thickness are heated to the temperature sixty-five degrees Celsius. On each sample must be applied a known emissivity coating. Unknown emissivity is calculated using the information obtained during the measurement. The aim of this paper is to show one of the possible approaches to find out the emissivity of the composite material.

Keywords: Thermography, thermogram, emissivity, composite, Onyx

1 Introduction

The world in which we live is characterized by features while, the most notable being the rapid frequency of differential innovation processes and the diversification of needs [1]. All sectors of the economy increase the proportion of the automation and robotics in their industries. More and more emphasis is placed on the complexity and complexity of the machines, while the reducing the number of production personnel. This results in increased requirements for operating parameters.

One of the key activities of the property care program is prevention. It consists of monitoring the status of monitored objects and diagnosing faults in the systems using the infrared thermography. Thermodiagnostics places a major emphasis on the area of temperature measurement by contactless and contacting methods. However, it is mainly aimed at displaying temperature fields, including its quantification. [1]

2 Thermography

Thermography or more accurate infrared thermography is a separate field of study that deals with contactless imaging, measurement, and temperature field analysis on the object surface. The role of thermography is the analysis of infrared energy that is radiated by the object. The infrared camera makes it possible to visualize the infrared radiation of an object that is not visible to the human eye, transforming it into electrical signals and displaying this radiation in the infrared camera. Thermography has proved to be a preventive maintenance tool for

* Corresponding author: Ondrej.Stalmach@fstroj.uniza.sk

Reviewers: *Josef* Soukup, *Radim* Halama

tracking mechanical and electrical equipment, production processes, building objects, it is use in the field of science, research and development, but also in medicine.

2.1 Definitions of the basic parameters

Emissivity is defined as the ratio of the radiation intensity of the real body to the intensity of radiating an absolute black body with the same temperature. The emissivity of the absolute black body is 1, the emissivity of the ideal mirror is 0.

Ideal black body is a theoretical body that absorbs the total incident electromagnetic radiation from the surrounding of all wavelengths to its surface, with the same amount of energy emitting back to the environment.

Reflected apparent temperature is the apparent temperature of other objects that is reflected from the surface of the object to the infrared camera.

Thermogram is a temperature map or image of an object in which the distribution of the radiated energy from the surface of the object being measured is expressed by means of gray tinting or color imaging. [1]

3 Basic principle of the thermography

The basic principle of thermography is the detection of the radiation energy emitted from the measured object [1]. When measuring temperature by the infrared camera is measured the amount of energy that radiates or reflects the measured object to the measuring instrument. In this measurement, the emissivity of the material is important. Emissivity is a dimensionless quantity that determines how much heat the material radiates (emitting) to the environment. [2, 3]

Figure 1 describes the interaction of different types of radiation with the object.

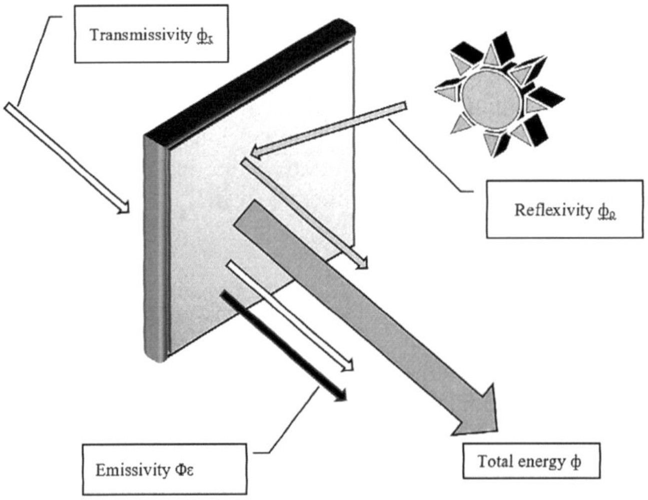

Fig. 1. Interaction of different types of radiation with the body

If an object is irradiated by the flux of Φ, then part of this flow will be absorbed $\Phi\varepsilon$. Part of the flow will reflect $\Phi\rho$ a part of let $\Phi\tau$. The Energy Conservation Act Kirchhoff's law of thermal radiation applies:

$$\Phi = \Phi\varepsilon + \Phi\rho + \Phi\tau = \varepsilon\Phi + \rho\Phi + \tau\Phi \quad (1)$$

It follows that:

$$\varepsilon + \rho + \tau = 1 \tag{2}$$

ε – Emissivity,
ρ – reflexivity,
τ – transmissivity.

For the objects that are not transparent $\tau = 0$, the following applies:

$$\varepsilon + \rho = 1 \tag{3}$$

In general, emissivity, reflectivity, and transmittance are the function of the wavelength, angle and temperature of the object. In our case, we are limited to wavelengths in the range 8-14μm, 10deg from normal to normal specimen area and 60°C.

Modern composite materials are used as a substitute for classical materials. The production of structural elements made on 3D composite printers is relatively simple and inexpensive. It is therefore necessary to know the thermal emissivity of these materials to the thermal analysis. In our case this material is MarkForged's Onyx.

4 Measurement of the reflected apparent temperature

Reflected apparent temperature is the reflected radiation of objects around the measured object. To measure the reflected apparent temperature, the following conditions must be fulfilled:

- The calibrated infrared camera must allow to input the reflected apparent temperature and the emissivity.
- It is necessary to have a reflector that can be created by shrinking and re-aligning the shiny aluminium foil.

The procedure of the method for measuring the reflected apparent temperature is then as follows:

1. Set the emissivity value equal 1 on the infrared camera.
2. Position the camera so that the object is measured in the camera's field of view and that the camera is within the required distance from the measured object.
3. Position the reflector in front of the object being measured so that it is parallel to the surface of the object being measured.
4. Without changing the position, measure the temperature on the reflector using the camera, this temperature is the reflected apparent temperature (Trefl).
5. For the sake of greater precision, repeat this procedure from the 2nd point to the 4th point at least three times and then the average of the measured values.
6. The mean value of the reflected apparent temperature thus obtained is then entered directly into the camera or into the appropriate software. [1]

5 Calculation of the emissivity from an available information

Composite materials play a major role in the industry. Knowledge of emissivity in these materials is essential for quantitative thermography, [5-8]. Three composite samples of different thicknesses are used in the experiment. The material from which the samples are made using a 3D printer is named onyx. The material from which the samples are made using a 3D printer is named onyx. The long-wave infrared camera Flir E50 is used in the experiment.

The first sample has a thickness of 0.4 mm and has 100% fill. The second sample has a thickness of 1 mm and has a 60% fill with a triangular structure. The last one has a thickness of 6 mm and has a 60% fill with a triangular structure.

On the samples must be applied a known emissivity coating. In the experiment are used 3M Scotch super 33+ PVC electrical insulation tape with the known emissivity $\varepsilon = 0.95 \pm 0.05$ and the ThermaSpot with known emissivity $\varepsilon = 0.96 \pm 0.04$. The Thermaspot are applied on the all samples but the 3M Scotch super 33+ PVC electrical insulation tape are applied only on the first and second samples.

The procedure of the method for obtaining information which are important to calculate unknown emissivity of the composite samples as follows:

1. The electric heater is placed in a box made of polystyrene to eliminate the reflections from the environment.
2. The sample is placed on the electric heater and it is heated to the 65 degrees Celsius.
3. The reflector is placed in front of the composite sample so that it is parallel to the surface of the sample.
4. With the infrared camera Flir E50 the 100 snapshots are taken at 30Hz scanning rate.
5. The reflector is removed, and another 100 snapshots are taken.
6. Steps 4. and 5. are repeated until do not have the 500 snapshots with the reflector and 500 without a reflector.
7. The obtained data is then processed in the software MATLAB.
8. Every snapshot can be uploaded to the MATLAB as a 2D matrix with the size 240x180 because the infrared camera Flir E50 has a resolution of 240 * 180 pixels.
9. From the snapshots are selected areas of interest representing the region of the known emissivity and the region with the unknown emissivity.
10. From these data a 3D matrix is created, and then its average value and standard deviation are calculated. These parameters are used to calculate the unknown emissivity value.

The figure 2 illustrates an example where the infrared camera records the radiations from the object on which a known emissivity coating is applied.

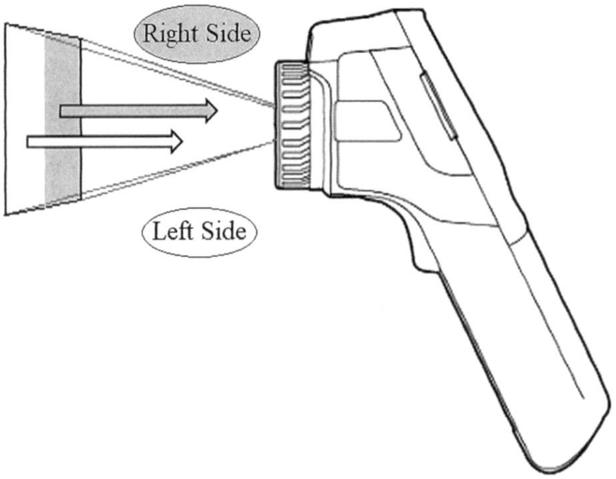

Fig. 2. Object with the regions of known and unknown emissivity.

Emissivity calculation is based on following formulas:

$$T_1 = \varepsilon_1 . T_{obj} + (1 - \varepsilon_1) . T_{refl} \quad (4)$$

$$T_2 = \varepsilon_2 . T_{obj} + (1 - \varepsilon_2) . T_{refl} \quad (5)$$

T_1 - the temperature measured by the infrared camera on the surface of unknown emissivity (Left side).
T_2 - the temperature measured by the infrared camera on the surface of known emissivity (Right side),
ε_1 - the unknown emissivity,
ε_2 - the known emissivity,
T_{refl} - the reflected apparent temperature,
T_{obj} - the temperature of the object.

If the assumption is that the temperature at the surface of the object (T_{obj}) is the same, the unknown emissivity ε_1 is calculated as follows:

$$\varepsilon_1 = \frac{\varepsilon_2(T_1 - T_{refl})}{(T_2 - T_{refl})} \tag{6}$$

The value of the unknown emissivity is not a constant, but it is the value from the interval. To know this interval, it is necessary to calculate the standard deviation of the indirect variable (ε_1) that depends on the multiple of variables ($T_1, T_2, T_{refl}, \varepsilon_2$). The formula is as follows:

$$\delta_{\varepsilon_1} = \sqrt{\left(\frac{\partial\varepsilon_1}{\partial\varepsilon_2}\right)^2 . \delta_{\varepsilon_2}{}^2 + \left(\frac{\partial\varepsilon_1}{\partial T_1}\right)^2 . \delta_{T_1}{}^2 + \left(\frac{\partial\varepsilon_1}{\partial T_{refl}}\right)^2 . \delta_{T_{refl}}{}^2 + \left(\frac{\partial\varepsilon_1}{\partial T_2}\right)^2 . \delta_{T_2}{}^2} \tag{7}$$

$$\delta_{\varepsilon_1} = \sqrt{\begin{array}{c}\left(\frac{(T_1 - T_{refl})}{(T_2 - T_{refl})}\right)^2 . \delta_{\varepsilon_2}{}^2 + \left(\frac{\varepsilon_2}{(T_2 - T_{refl})}\right)^2 . \delta_{T_1}{}^2 + \\ \left(\frac{\varepsilon_2.(T_1 - T_{refl})}{(T_2 - T_{refl})^2} - \frac{\varepsilon_2}{(T_2 - T_{refl})}\right)^2 . \delta_{T_{refl}}{}^2 + \left(\frac{\varepsilon_2.(T_1 - T_{refl})}{(T_2 - T_{refl})^2}\right)^2 . \delta_{T_2}{}^2\end{array}} \tag{8}$$

The results are shown in the Table 1.

Table 1. Results of the unknown emissivity

	ThermaSpot	3M Scotch super 33+
Sample 1	$\varepsilon_1 = 0{,}91 \pm 0{,}04$	$\varepsilon_1 = 0{,}91 \pm 0{,}05$
Sample 2	$\varepsilon_1 = 0{,}93 \pm 0{,}04$	$\varepsilon_1 = 0{,}94 \pm 0{,}05$
Sample 3	$\varepsilon_1 = 0{,}94 \pm 0{,}05$	-

6 Conclusion

The aim of this article is to show how to calculate the unknown emissivity of composite samples by the so-called scientific method and to indicate the values for Onyx material used by Markforged in their 3D printers. Knowledge of emissivity is required for the quantitative assessment of object temperatures.

It was found that the emissivity of the printed samples ranged from 0.91 to 0.94. This interval is explained by the possible variability of print process parameters. Two-layer cover area was used on the sample. If identical print parameters are not followed, the assumption $\tau = 0$ may not be correct because in some cases the printed non-zero-transmitted surface may be printed. This case arises, for example, then the individual filaments have very small slits between them and the two-layer printed surface has very fine mesh properties.

This contribution is a continuation of the computational work on composite materials, infrared camera measurements as well as non-destructive testing of materials performed at the Department of Applied Mechanics of the Faculty of Mechanical Engineering of the University of Zilina (Zilinska univerzita v Ziline), [9-14].

This paper was supported by VEGA 1/0795/16 and KEGA 017ŽU-4/2017 and by the Slovak Research and Development Agency under the contract No. APVV-0736-12.

References

1. V. Peťková, J. Svoboda, *Thermodiagnostics.* ISBN 978-80-8126-132-9 (2016)
2. M. Honner, P. Honnerova, *Survey of emissivity measurement by radiometric methods.* Applied Optics, **54**, 669-683, (2015)
3. M. Honner, P. Litoš, M. Švantner, *Thermography analyses of the hole-drilling residual stress measuring technique.* Infrared Physics & Technology, **45**, 131-142, (2004)
4. P. Honnerova, J. Martan, M. Kucera, Martin, M. Honner, J. Hameury, *New experimental device for high-temperature normal spectral emissivity measurements of coatings,* Measurement Science and Technology, **25**, Art.No. 095501 (2014)
5. P. Pastorek, P. Novak, P. Kopas, M. Mocilan, *Finite element analysis of bond behavior in a steel reinforced concrete structure strengthened carbon fibre reinforced polymer (CFRP) strips.* METALURGIJA, **56**, 405-408, (2017).
6. Z. Murcinkova, P. Novak, V. Kompis, M. Zmindak, *Homogenization of the finite-length fibre composite materials by boundary meshless type method.* Archive Of Applied Mechanics, **88,** 789-804, (2017)
7. P. Novak, M. Zmindak, Z. Pelagic, *High-Pressure Pipelines Repaired by Steel Sleeve and Epoxy Composition.* Applied Mechanics and Materials1, **486**, 181-188, (2014)
8. L. Radziszewski, M. Saga, *Modeling of non-elastic properties of polymeric foams used in sports helmets.* Procedia Engineering. **177**, 314-317 (2017)
9. M. Kekez, L. Radziszewski, A. Sapietova, *Fuel type recognition by classifiers developed with computational intelligence methods using combustion pressure data and the crankshaft angle at which heat release reaches its maximum* Procedia Engineering, **136**, 353-358, (2016)
10. R. Halama, P. Gál, P. Paška, Z. Sedlák, *A new accelerated technique for validation of cyclic plasticity models.* MATEC Web of ConferencesVolume, **157**, Art. No. 05008, (2014)
11. I. Glodova, T. Liptak, J. Bocko, *Usage of finite element method for motion and thermal analysis of a specific object in SolidWorks environment.* Procedia Engineering, **96,** 131-135, (2014)
12. M. Handrik, M. Vasko, P. Kopas, V. Mozer, *The linear and nonlinear stability loss of structures due to thermal load.* Procedia Engineering. **136**, 359-364 (2016)
13. Z. Murcinkova, P. Novak, V. Kompis, M. Zmindak, *Homogenization of the finite-length fibre composite materials by boundary meshless type method.* ARCHIVE OF APPLIED MECHANICS, **88,** 789-804, (2017)
14. P. Novak, M. Zmindak, Z. Pelagic, *High-Pressure Pipelines Repaired by Steel Sleeve and Epoxy Composition.* Applied Mechanics and Materials1, **486**, 181-188, (2014)

Application of artificial intelligence principles in mechanical engineering

Ivan Zajačko[1,*], *Tomáš* Gál[1], *Zuzana* Ságová[1], *Vasyl* Mateichyk[2], *Dariusz* Wiecek[3]

[1]University of Žilina, Faculty of Mechanical Engineering, Department of Automation and Production Systems, Univerzitná 1, 010 26 Žilina, Slovak Republic
[2]National Transport University, Faculty of Automotive and Mechanical Engineering, 01010 Kyiv, 1 Omelianovycha-Pavlenka str., Ukraine
[3]ATH – University of Bielsko Biala, Faculty of Mechanical Engineering and Computer Science, 43-309 Bielsko Biala, Willowa 2, Poland

Abstract. The article deals with methods of Artificial Intelligence and their utilisation in technical diagnostics. Special meaning will be given on methods such as Deep learning. The deep learning method seems to be a very good candidate for defect detection and pattern recognition. The method was applied for technical diagnostic in automotive factory and the problem will be described in the paper.

Keywords: Artificial Intelligence, Deep convolutional neural network, Vision system, Machine learning

1 Introduction

The primary Innovation is today one of the key benchmarks of successful companies. Conventional approaches and methods applied in the production environment are gradually becoming non-sufficient due to the increasing competitive pressure. For this reason, it is necessary to apply innovative methods and approaches to address the entire spectrum of tasks in the production environment. At the same time, market demand for increased flexibility in the behavior of production companies towards the customer requires very close intercomnection of individual levels of company management, and the required increase in profitability for companies providing goods or services requires the most accurate knowledge of customer requirements. [1]

The aforementioned conditions and requirements for companies have recently been addressed through the penetration of IT technologies and automation into the existing conventional technologies and practices of producing companies. At the end of the 20th century and in the first decade of the 21st century, we saw a significant increase in the use of above-mentioned approaches in the production environment, which gradually eliminated the differences between the industry leaders, with these tools even the small companies gradually eliminated competitive advantage of big actors by the gradual partial introduction of these

* Corresponding author: ivan.zajacko@fstroj.uniza.sk

Reviewers: *Jozef* Jurko, *Nicoale* Ungureanu

technologies of by gradual substitution of conventional methods with more progressive ones. [2,3]

The natural development is the improvement of established technologies and methods. In addition to implementing optimizations at different levels, improvements can be made by introducing innovations into individual tasks or complex task groups. An innovative idea linking the conventional automation and IT technology is the introduction of a higher degree of automation. [4] With a higher degree of automation, we understand automation that is not only applied to automate the routine work, but which has the decision-making ability at a defined level. [5] In this area, the highest degree of automation could be achieved by applying artificial intelligence to the production environment. In our context, we can understand the Artificial Intelligence application as follows: Applied Artificial Intelligence in a production environment is a program (or part of it) that can solve a task without introducing rules for solution it into the program. [6,7]

When applying artificial intelligence in its entirety, we would achieve self-acting and self-governing units in the production environment. When achieving an absolute degree of automation through devices with a high degree of artificial intelligence, we could achieve a self-organized production company that would be in full control of the inputs necessary for production, production itself, as well as quality control and expedition. [8,9] In the real world, achieving this goal is just utopian at the moment. [4] Our task is to ensure innovation through partial application of artificial intelligence techniques and achievement of a degree that we can characterize as Advanced Intelligence.

Within the framework of cooperation with the industry, a research task was proposed to fully automate the process of quality control of produced tires by the manufacturer. At present, this task is performed by an operator who has an inspection stand consisting of a rotary mandrel, pneumatic drives, PLC controlling individual drives and lighting. The inspection stand (Fig. 1.) serves only as a device assisting the operator in handling the checked tire (securing it on the rotary thruster and rotating the tire in the selected direction). The entire evaluation process is in the full competence of the operator and is carried out solely on the basis of a quality assessment corresponding to the range of knowledge of the operator of the product. Our goal is to ensure complete automation of the control process. In the initial analysis of the task, we determined that for the elimination of an operator, it will be necessary to automate three different tasks:

1. handling of evaluated tire
2. automatic quality control process
3. automatic detection and removal of any residuals from the molding and vulcanization process of the tire

Fig. 1. Inspection stand used today

1. Manipulation of the evaluated tire will be implemented by extending the inspection stand with a serial kinematic structure robot, supplemented by a corresponding effector optimized for tire manipulation. The robot will ensure that the tire is removed from the production line or from the in-service container, by placing it on a rotary mandrel, and after completing the quality control process, it will ensure that the tire is stored in the corresponding container.
2. An automated quality control process is described in detail the next part of our article
3. Automatic detection and removal of any residuals from the molding and vulcanization process of the tire will be accomplished by a single-purpose handling device with trimming capability. The co-ordination and synchronization of the tire and handling device will be coordinated on the basis of the analysis of the position, shape and size of unwanted residuals, this information will be obtained from the data obtained in the quality control process.

2 Classification method

An automatic quality control process will be created on the principle of similarity with the way this process is currently carried out by an operator. Our task is to perform an analysis of the control process after the assignment of a research task and to transform it in order to completely replace the presence of an operator in this process. Based on the analysis, it was determined that the operator of the equipment provides a visual inspection of the tire during its rotation and, based on the experience and knowledge of the operator, the tested tire is evaluated. [10,11]

Our proposal for transforming this process consists of designing and assembling a camera system with appropriate illumination to get the image of the tested tire with required quality. Due to the fact that the occurrence and type of errors on the produced tires are different and at the same time it is necessary to consider the quality of the evaluation process in regard to deviation of the tire characteristics from the defined values. It was necessary to choose the method of data processing with the required performance and the necessary functions that would be able to replace the evaluation by humans. In our case, we decided to apply machine learning methods - specifically for the application of deep convolutional neural networks (DCNN) with a high probability of successful implementation due to their character and potential performance. [12]

Machine learning - is a subfield of artificial intelligence that develops algorithms and techniques that enable the computer system (machine) to "learn". In this context, we understand the learning as ability to change the internal state of a system in a way that improves the ability to adapt to changes in the surrounding environment or allow the computer system after learning to extend the content of its knowledge base based on the findings obtained [13].

One approach to interpret the knowledge needed for machine learning is to acquire and subsequently analyze the images obtained through the camera. Classical forms of visual data detection are using gradient-based techniques such as gradient-oriented histogram (HoG), scale-invariant feature transform (SIFT), spatial pyramids and, where appropriate, representation of basic functions such as Gabor's filters. However, in recent years, we have more powerful tools for extracting and analyzing application-specific information by extending deep learning systems, and specifically by creating and implementing deep convolutional neural networks.[14]

In our case, we plan to use a deep convolutional neural network to detect and classify the various errors encountered on the surface of the controlled tires. By applying this type of neural network, we assume that there will be a significant increase in the performance of the machine learning system, based on which results the quality control system will be built.[14]

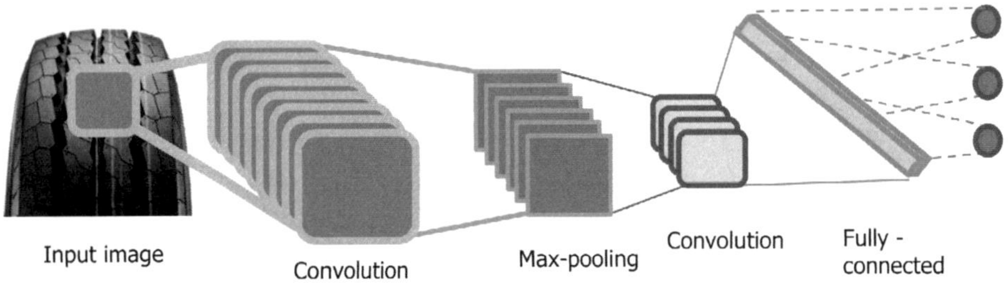

Fig. 2. Convolutional neural network with max-poolig

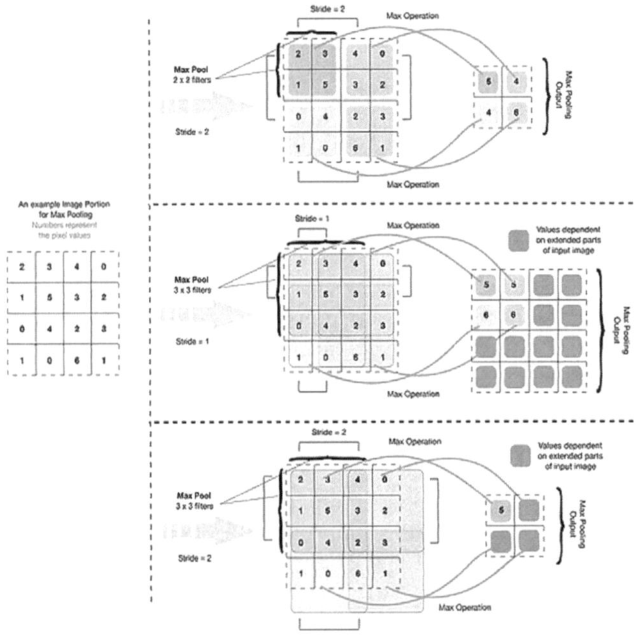

Fig. 3. Max-pooling principles (Sukrit Shankar, University of Cambridge)

The deep convolutional neural network (DCNN) based on the classical convolutional neural network, according to the LeCun consists of three main components:

- the convolutional layer - is connected with the next layer in a similar way to a traditional bipartite multilayer neural network with the key difference that the scales are shared between the connection sets. Each set of weights sharing links is a filter that is linked to the input data. Typically, there are several such filters trained in parallel on each layer. Convolution filters are moved through small local sections created on input image data in the image training process. Each filter acts as an element detector. The result of filter application and convolution across images will create a map of elements (Fig. 2.);
- activation function – Is used to discriminate between image classes by being stored at the output of a convolutional filter and performing a nonlinear transformation of the

data space. Some examples of activation functions are hyperbolic tangent function (Tanh), sigmoid function and rectified linear unit (ReLU);

- max-pooling - is a sample-based discretization process (Fig. 3.). The objective is to down-sample an input representation, reducing its dimensionality and allowing for assumptions to be made about features contained in the sub-regions. A convolutional layer and max pooling layers are successfully created to create the DCNN architecture. Compared to shallow architectures, DCNN has several layers that can represent complex functions with higher efficiency and accuracy of generalization. [13]

3 Implementation

In our solution, we plan to apply training methods that are slightly different from conventionally applied methods. The standard gradient descent method involves optimizing the error across the training set. Because this procedure can be computationally and cost-intensive for a large network, an approximation method called minibatch stochastic gradient descent is often used for DCNN training. [15] A significant difference lies in the fact that instead of calculating the error gradient in the entire training set, each iteration calculates the error gradient for a small portion of the samples. We denote b and n the minibatch size and the total number of training samples. For the number of minibatch gradients, total $T = n / b$ iterations during training. Therefore, the w parameters are obtained by optimizing the approximate expected value of the error function f defined as:

$$E_t[f(w)] = \frac{1}{b}\sum_{i=(t-1)b+1}^{tb} f(w; x_i) \qquad (1)$$

Where $t \in \{1, \ldots T\}$ is iteration index and x_i is i-th training sample.
In every iteration weight will be adjusted by gradient descent rule:

$$w^{(t+1)} = w^{(t)} - \mu \nabla_w E_t\left[f(w^{(t)})\right] \qquad (2)$$

Where μ is learning rate

While the standard gradient descent drops across all the samples in the training set to obtain a single update to value of w for each iteration, stochastic gradient descent uses only one training sample, and the minibatch stochastic gradient descent method uses each sample iteration b (i.e. minibatch size). The stochastic gradient descent method is computationally much cheaper than standard gradient descent. A minibatch gradient descent method may often be as rapid as a stochastic gradient descent if appropriate vectorization is used to calculate derivative terms. [15]

Problems with non-convex object functions have sometimes shown a stochastic gradient value can escape from the local minimum that often forms a trap for a batch gradient. Therefore, stochastic gradient descent can be better deployed in applications such as DCNN training. [14]

Our data for the automatic evaluation of the quality of the produced tires will be generated by sets of images that will be captured by a high-resolution camera system. The inspection stand will be extended by the aforementioned system of cameras. At the same time, the inspection stand will be extended by a system of additional local lighting to achieve the required image quality (Fig. 4.).

After the completion of construction and assembly modification to the inspection stand, it will be necessary to carry out the so-called 'before implementation phase' that allow us to gather the unambiguous definition of acquired images for the process of defining all potentially occurring mistakes arising from the production of tires. At the same time, it will be necessary at this stage to define the ranges of acceptable deviations from the ideal state of the product. All captured and defined examples of images will need to be broken down into classes. A prerequisite for successfully grouping examples into classes is to create classes according to the apparent similarity of its members (geometric, technological, spatial, ...).

After successful completion of the before implementation phase associated with creation of a training set of data, we will perform implementation tests to reveal a suitable variation of the DCNN model. The different test models will be different in the basic parameter combinations (number of element maps, size of applied filters, number of network layers, number of fully interconnected layers). After performing these tests, parameters will be selected to allow the highest accuracy of classification through DCNN.

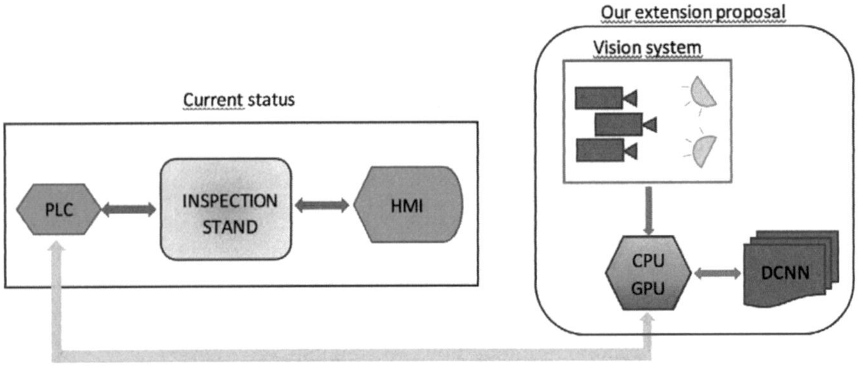

Fig. 4. Our extension proposal of inspection stand

4 Conclusion

We assume that the DCNN application in an automated quality control process by implementing an automated error detection system on inspected tires will solve the most complex set of tasks associated with the development of an automated inspection device excluding manual inspection. Our assumption is based on the fact that the inspection stand will be extended by proposed camera system, that will generate a huge amount of input data (represented by large sets of images), DCNN will be used to effectively extract and recognize image elements, and to automatically detect the presence of errors and deviations in the controlled product. We have explored the use of DCNN to skip complicated procedures for extracting features required in a classical learning approach; instead, we will use unprocessed images as the only input to the classification model, and then optimize the network using the minibatch gradient descent method.

The solution presented in this article is a concept based on the cooperation with the practice We predict the very high demands on part of the practice to given solution. In the next research we will therefore work out other methods of processing the acquired data through the methods of machine learning.

Application of a machine learning principles has the potential to create automatic detection systems that are sufficiently generic to allow us to address presently very difficult to detect failures and errors in evaluating different tasks in production environment.

The outputs of paper was supported by grant VEGA 1/0504/17 "Research a development of methods for multicriterion accuracy diagnostic of CNC machines.

References

1. J. Brethé, D.Lefebvre, *Risk ellipsoids and granularity ration for industrial robot.* International Journal of Factory Automation Robotics and Soft computing, **2**, 93-101 (2007)
2. I. Kuric, V. Bulej, M. Sága, P. Pokorný, *Development of simulation software for mobile robot path planning within multilayer map system based on metric and topological maps.* International journal of advanced robotic systems, **14**, (2017)
3. B. Mičieta, M. Edl, M. Krajčovič, Ľ. Dulina, P. Bubeník, L. Ďurica, V. Biňasová, *Delegate MASs for Coordination and Control of One-Directional AGV Systems: A Proof-of-Concept.* International journal of advanced manufactur. technology, **94**, 415-431 (2018)
4. W. S. Newman, C. E. Birkhimer, R. J. Horning, A. T. Wilkey, *Calibration of a Motoman P8 robot based on laser tracking.* Robotics and Automation, 2000. Proceedings. ICRA'00. IEEE International Conference on. **4**, 3597-3602 (2000)
5. M. Císar, I. Zajačko, I. Kuric, *Diagnostics based on poositioning performance during circular motion.* Inżynier XXI wieku: projektujemy przyszłość. 536-542 (2016)
6. A. Sapietová, M. Sága, I. Kuric, Š. Václav, *Application of optimization algorithms for robot systems designing.* International Journal of Advanced Robotic Systems, **15** (2018)
7. V. Tlach, M. Císar, I. Kuric, Zajačko, *Determination of the Industrial Robot Positioning Performance.* MATEC Web of Conferences. **137** (EDP Sciences, 2017)
8. A. Nubiola, I. A. Bonev, *Absolute calibration of an ABB IRB 1600 robot using a laser tracker.* Robotics and Computer-Integrated Manufacturing, **29**, 236-245 (2013)
9. M. Abderrahim, A. Khamis, S. Garrido, L. Moreno, *Accuracy and Calibration Issues of Industrial Manipulators.* Industrial Robotics-Programming, Simulation and Applications, 131-146 (2007)
10. J. Józwik, D. Ostrowski, P. Jarosz, D. Mika, *Industrial robot repeatability testing with high speed camera Phantom v2511.* Advances in Science and Technology Research Journal, **10**, 86-96 (2016)
11. D. Kumičakova, V. Tlach, M. Cisar. *Testing the performance characteristics of manipulating industrial robots.* Transactions of the VŠB - Technical University of Ostrava: mechanical series, **62**, 39-50 (2016)
12. A. Şirinterlikçi, M. Tiryakioğlu, A. Bird, A. Harris, K. Kweder, *Repeatability and Accuracy of an Industrial Robot: Laboratory Experience for a Design of Experiments Course.* Technology Interface Journal, **9** (2009)
13. Charu C Aggarwal, *Neural Networks and Deep Learning.* Springer, 2018
14. Ragav V., Baoxin L., *Convolutional Neural Networks in Visual Computing.* CRC Press, 2017
15. A. Nubiola, M. Slamani, I. A. Bonev, *A new method for measuring a large set of poses with a single telescoping ballbar.* Precision Engineering, **37**, 451-460 (2013)

Possibilities of applying motif group parameters for description roughness of turned surfaces

Maciej Kowalski[1,*], *Magdalena* Wiśniewska[1], *Paweł* Karolczak[1], *Jozef* Holubjak[2]

[1]Wrocław University of Science and Technology, Faculty of Mechanical Engineering. ul. Łukasiewicza 5, 50-371 Wrocław, Poland
[2]University of Žilina, Faculty of Mechanical Engineering, Univerzitna 1, 010 26, Žilina, Slovakia

Abstract. This article explores the methodology of assessing the applicability of roughness parameters from the motif group to the evaluation of one-way and periodic geometrical surface structures. The results of surface roughness measurements of aluminum samples turned with variable kinematic parameters were presented. Usability of using surface motifs in combination with selected parameters described in ISO standards for assessment of geometrical structures characteristic for longitudinal turning was shown.

Keywords: aluminum alloys, surface geometrical structure, roughness parameters

1 Introduction

In modern manufacturing techniques the main goal is to achieve good, repeatable quality of products. One of the main problems associated with this is the quality of the surface layer of machined object. It combines whole of issues related to methods of surface manufacturing fulfilling specific operation requirements.

Characterizing surface roughness is currently taking a lot of significance in many areas related to the construction, technology and operation of machines and devices. Currently applied methodology of surface roughness analysis was developed around 1970. Since then, over the years, nearly 300 different roughness and waviness parameters were offered to the users [1]. In 90s of the last century, trend for creating new roughness parameters was called "rash of parameters" or "parametric fever"[2]. A large number of defined and possible to use roughness and waviness parameters makes it impossible to give the value to all of them. On the other hand, the complexity of the geometric structure of the surface means that use the one, selected parameter does not give full knowledge about its condition and even more about its operational capabilities [3]. Therefore a few, suitably selected parameters should be used to description of the usable surface state. Because there are no standards or procedures that would inform when and in what cases given parameters should be used, the selection should be representative to given application. For each element to be manufactured, it is necessary to decide which features of the of the geometric structure of the surface are relevant and which roughness parameters correspond to them.

* Corresponding author: maciej.kowalski@pwr.edu.pl

Reviewers: *Mária* Čilliková, *Michal* Hatala

In industry, the *Ra* parameter is most commonly used parameter. It is given very often in connection with the description of surface machining method by which it can be obtained. If production conditions are stabilized this type of evaluation makes it possible to maintain constant, within defined limits, surface roughness characteristics. It is due to quite strong statistical associations between parameters describing different profile features for surfaces machined in similar conditions. In case of the parts operating under specific conditions and determining the functional characteristics of the entire product e.g. the surface of the exhaust fins of internal combustion engines, two or more roughness parameters are used to assess the surface [4].

Another important aspect increasing the possibilities of describing surface geometrical conditions are differences in the standards regarding surface roughness used in individual countries. Universal globalization however means that many standards have been unified and the most universal are being pursued. Differences, however, remain. An example here can be the so - called French methods for roughness calculating. They were created for the French automotive industry at the beginning of 80s of the last century. In Poland they are called *motif method.*

1.1 Motifs method

This method relies on division primary profile into so – called motifs. According to the norm motif is concluded between the highest points of two locals not necessarily neighboring peaks of the profile (fig.1) Motif is characterized by:

- its length *AR* or *AW* measured parallel to direction of primary profile ,
- its two depths H_j i H_{j+1} measured perpendicularly to direction of primary profile,
- its characteristic depth *T*, which is the lower depth of this two ($T = \min(H_j, H_{j+1})$.

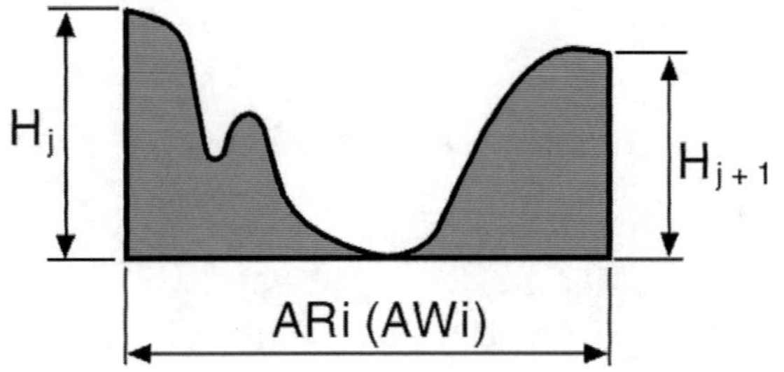

Fig. 1. Roughness motif

Fragments received this way are next joined according to specific rules among which following conditions are distinguished: envelope, length, magnification and equality [5,6].

Envelope condition – concerns peaks which are higher than neighboring ones. Joint peak of two neighboring motifs must be smaller or equal from one of adjacent peak.

Length condition – concerns limitation of length of one motif to the limit value A (conventional boundary between roughness and waviness) or limit value B (conventional boundary between waviness and shape deviations). Sum of the length of two adjacent motifs $AR_i + AR_{i+1}$ have to be smaller or equal to A for roughness and $AW_i + AW_{i+1} < B$ for waviness.

Magnification condition – concerns suppression of the smallest peaks however the greatest possible motifs are sought - after. This condition prevents combining of two motifs when its characteristic depth *T* is smaller than depth of one of the two initial motifs. The effect of this is suppression of the smaller peak lying between large ones. Motifs can be combined only if the characteristic depth T of the created motif is greater or equal to both depths characteristic for component motifs.

Equality condition – prevents combination of motifs with similar depths that occur for example with periodic profiles. Prevents suppression of motifs which depth is similar to depth of adjacent one. The combination of the motifs in this calculation method should be used until all possible further combinations are exhausted. Motifs can be combined only when one of the depths is smaller or equal to 60% of the characteristic depth T of the tested connected motif.

Knowing the conditions for combining motifs it is possible to proceed of calculating parameters from motif group. In this work four of them were designated (fig. 2):

Pt – sum of the height of the highest profile peak Zp and depth of the lowest valley Zv profile inside evaluation length ln

R – arithmetic mean of the depth value H_j of roughness motifs inside evaluation length which value is calculated from the formula 1:

$$R = \frac{1}{m}\sum_{j=1}^{m} H_j \tag{1}$$

Rx – the largest depth H_j from all the roughness motifs inside the evaluation length.

AR – the arithmetic mean of the length value AR_i of roughness motifs inside evaluation length

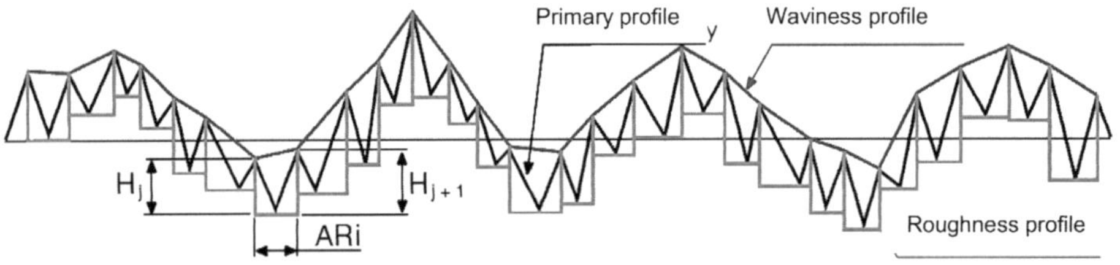

Fig. 2. Graphic description of parameters *R*, *Rx*, *AR*

The undoubted advantages of the motif method are: no need to apply filtration, a small number of parameters and mapping of each local peak on the elevation at omitting those which are in profile valleys and which are not participating in contact of two surfaces cooperating with each other.

2 Research methodology

A shaft of aluminum alloy 7075 was tested. As a machining method turning was chosen. Research was carried out on a CNC TUR 560 MN numerical lathe. For tests turning inserts TCGX16T304-Al H10 were used with uncoated cemented carbide and they were mounted in a holder marked STGCR 2020K16. The following cutting parameters were used: v_c = 105; 210 ; 420 ; 530 and 660 m.min^{-1}; f = 0.08; 0.19; 0.37 mm.rev^{-1}. Depth of cut was a_p =0.5 mm.

Roughness measurements were conducted using Form Talysurf 120L tester (fig. 3) applying following settings: Gaussian digital filter, λc = 0.8 mm, λs = 0.0025 mm, bandwidth: 300:1. During calculating motif parameters following settings were used: shape: straight line LS, filter R & W, A = 0.25 mm, B = 1.4 mm, λs = 0.0025 mm. For analysis of usefulness of motif method for description of examined surfaces, five amplitude parameters, apart from already described parameters from motif group, were chosen: *Ra, Rz, Rp, Rv, Rq,* two statistical: *Rku, Rsk* and spatial: *RSm, RS,* as well as hybrid: *RΔa i RΔq*.

Fig. 3. Measuring stand –Form Talysurf-120L tester, Taylor Hobson Limited company

Two indicators were used for assess suitability of individual roughness parameters. First of them is the correlation coefficient ρ, which defines the level of linear relation between two variables. It is a quotient of covariance and the product of standard deviations of this variables what was shown on equation 2:

$$\rho_{XY} = \frac{cov(X,Y)}{\sigma_X \sigma_Y} \tag{2}$$

It is considered that when the correlation coefficient is equal 0 linear correlation does not exist but a curvilinear correlation may occur. The closer values to the number 1 or -1 will achieve correlation coefficient – the greater linear correlation will be. Above 0.9 and below -0.9 it is considered that very strong linear correlation occurs.

However, one should be aware of the fact of frequent over - interpretation connected with drawing conclusions only on the basis of the value of the correlation coefficient. Therefore to assess the suitability of use of roughness parameters from individual groups second indicator which defines significance level of change of characteristic profile feature was proposed [7,8]. It was marked with a symbol *t*. It is expressed in the percent and described with equation:

$$t = \frac{R_{max} - R_{min}}{R_{śr}} * 100\% \tag{3}$$

where:

- R_{max} – maximum value of tested roughness parameter,
- R_{min} – minimum value of tested roughness parameter,
- $R_{śr}$ – average value of tested roughness parameter,

3 Tests results

Making qualification of suitability of a given roughness parameter for description of roughness changes, the closeness of the correlation coefficient ρ to one was assumed and whether the level of significance of the change of tested parameter *t* is greater than 30%. Tables 1-4 presents surface roughness tests results along with calculated values of suitability indicators of the calculated parameters.

Table 1. Influence of the feed *f* on values of selected amplitude roughness parameters 2D

Machining parameters		Roughness parameters							
		AMPLITUDE					**MOTIF**		
v_c [m.min^{-1}]	f [mm.rev^{-1}]	*Ra*	*Rz*	*Rp*	*Rv*	*Rq*	*Pt*	*R*	*Rx*
105	0.08	0.53	2.989	1.569	1.42	0.64	3.863	1.95	3.21
	0.19	1.183	5.896	3.298	2.598	1.407	6.82	4.796	6.276
	0.37	4.946	19.235	12.414	6.820	5.667	21.976	9.436	21.076
210	0.08	0.551	3.353	1.862	1.490	0.674	4.141	2.236	3.846
	0.19	1.167	5.688	3.412	2.275	1.376	7.134	5.254	6.839
	0.37	4.981	18.966	12.243	6.723	5.694	21.972	9.851	21.583
420	0.08	0.521	3.399	2.030	1.368	0.643	5.473	2.169	4.41
	0.19	0.559	3.564	2.112	1.451	0.713	6.897	2.113	5.813
	0.37	4.917	19.256	12.599	6.657	5.639	21.345	9.773	21.043
530	0.08	0.545	3.09	1.747	1.352	0.661	4.113	2.213	3.796
	0.19	1.194	5.937	3.514	2.423	1.431	6.966	5.22	6.54
	0.37	4.936	19.479	12.72	6.759	5.664	21.94	9.663	21.643
660	0.08	0.610	3.396	1.846	1.550	0.741	4.283	2.17	3.836
	0.19	1.198	5.986	3.593	2.392	1.439	7.233	5.23	6.796
	0.37	4.906	19.461	12.809	6.651	5.637	22.143	9.076	21.463
ρ		0.96	0.96	0.96	0.97	0.96	0.97	0.98	0.97
t [%]		201%	172%	184%	153%	197%	158%	138%	166%

Analyzing presented tests results and roughness measurements it can be stated that according to theory along with increasing of feed all basic amplitude roughness parameters, both classic ones and those of group of motifs also increased. Level of correlation coefficient in each case was greater than 0.96 and indicator *t* oscillated between 138–201%, which clearly indicates good description of the relevance of changes in the profile by these parameters.

Statistical parameters such as skewness *Rsk* and kurtosis *Rku*, showed insufficiently high linear correlation with feed. In their case indicator *t* assumed values greater than limit of 30%, however they were smaller than in case of the amplitude parameters.

Along with increasing feed, change of spatial parameters *RSm* and *RS* was also observed. This indicates a change in the distance between successive profile peaks as well as about increasing width of profile valleys. Linear correlation coefficient assumed for them values greater than 0,95 whereas indicator *t* higher than 154%.

Table 2. Influence of the feed *f* on values of selected spatial, statistical and hybrid roughness parameters 2D

Machining parameters		Roughness parameters						
		STATISTICAL		SPATIAL		MOTIF	HYBRID	
v_c [m.min^{-1}]	f [mm.rev^{-1}]	*Rku*	*Rsk*	*RSm*	*RS*	*AR*	*RΔa*	*RΔq*
105	0.08	2.313	0.075	60.686	16.93	85.133	5.83	7.33
	0.19	2.232	0.290	180.23	36.983	176.133	5.426	7.106
	0.37	2.07	0.655	364.516	192.89	171.533	7.116	8.81
210	0.08	2.766	0.205	67.726	17.5	88.8	6.053	7.69
	0.19	2.360	0.429	187.21	36.03	183	5.526	7.35
	0.37	2.033	0.6407	373.146	200.44	179.3	7.153	8.86
420	0.08	3.467	0.563	71.733	17.91	90.9	5.756	7.533
	0.19	9.734	0.993	79.78	18.206	87.8	5.943	7.676
	0.37	2.085	0.663	373.496	209.873	178	7.243	9.01
530	0.08	2.504	0.346	58.103	16.603	85.57	6.366	7.99
	0.19	2.398	0.415	179.62	36.656	183.47	5.69	7.686
	0.37	2.09	0.661	365.643	292.283	175.83	7.203	9.003
660	0.08	2.384	0.174	64.95	17.413	86.36	6.213	7.823
	0.19	2.439	0.469	177.073	39.133	182.45	5.586	7.663
	0.37	2.117	0.6811	372.566	245.6	163.07	7.173	9.003
ρ		-0.83	0.81	0.99	0.95	0.77	0.78	0.87
t [%]		45%	102%	154%	226%	67%	25%	18%

Remaining parameters including the only spatial parameter from the motif group and hybrid parameters *RΔa* and *RΔq* due to low level of calculated significance indicators were considered to be slightly dependent on the feed value. By analyzing tests results configured by constant feed and variable cutting speed (Table 3 and 4) it can be seen that no parameter has shown sufficient correlation with cutting speed v_c. For none of the parameters correlation coefficient above 0,9 was not obtained. The highest value was obtained for *RΔq* parameter and it was at 0.86. For all parameters apart hybrid, values of indicator *t* higher than 30% were obtained however when linear correlation coefficients were too low. It can be concluded that influence of the cutting speed on surfaces features cannot be described by these parameters.

Table 3. Influence of the cutting speed v_c on the values of selected amplitude roughness parameters 2D

Roughness parameters		Roughness parameters							
		AMPLITUDE					MOTIF		
v_c [m.min^{-1}]	f [mm.rev^{-1}]	*Ra*	*Rz*	*Rp*	*Rv*	*Rq*	*Pt*	*R*	*Rx*
0.08	105	0.53	2.989	1.569	1.42	0.642	3.863	1.95	3.21
	210	0.551	3.353	1.862	1.494	0.674	4.14	2.23	3.84
	420	0.521	3.399	2.030	1.368	0.643	5.47	2.16	4.41
	530	0.545	3.099	1.747	1.352	0.661	4.113	2.213	3.796
	660	0.610	3.396	1.846	1.552	0.741	4.283	2.17	3.836

0,19	105	1.183	5.89	3.298	2.598	1.407	6.82	4.796	6.276
	210	1.167	5.688	3.412	2.276	1.376	7.13	5.25	6.83
	420	0.559	3.564	2.112	1.459	0.713	6.89	2.116	5.813
	530	1.194	5.937	3.514	2.421	1.431	6.966	5.2	6.5
	660	1.198	5.986	3.593	2.397	1.439	7.233	5.23	6.796
0,37	105	4.946	19.235	12.414	6.829	5.667	21.976	9.436	21.076
	210	4.981	18.966	12.243	6.723	5.694	21.97	9.85	21.58
	420	4.917	19.256	12.599	6.654	5.639	21.34	9.77	21.043
	530	4.936	19.479	12.72	6.751	5.664	21.94	9.663	21.643
	660	4.906	19.461	12.809	6.659	5.637	22.143	9.076	21.463
ρ		-0.05	0.4	0.48	-0.23	-0.02	0.29	0.04	035
t [%]		81%	65%	85%	73%	75%	54%	107%	56%

Additionally correlation between classical parameters and those from motif group was checked. Such dependence or independence of parameters is very important and plays (or may play) decisive role in selection of identification method and describing geometrical surfaces features. In a situation when two parameters are strongly dependent from each other they can be treated as a complement of main indicator used for description a given feature of parameter. As the main indicator ideal for describing given feature such parameter should be chosen that is absolutely independent of others because only in such case its values can be determined in unequivocally and objective way.

Table 5 shows linear correlation coefficient between classical amplitude parameters, hybrid ones and from motif group. Table 6 shows correlation between spatial parameters and parameter AR from motif group.

Analyzing data given in tables it can be stated that close dependence and correlation is between parameters *Pt, Rx* and *R* from the motif group of parameters and amplitude parameters at the lack of close connection with statistical and hybrid parameters.

Table 4. Influence of cutting speed v_c on selected values of spatial, statistical and hybrid roughness parameters 2D

Cutting speed		Roughness parameters						
		STATISTICAL		**SPATIAL**		**MOTIF**	**HYBRID**	
v_c [m.min^{-1}]	f [mm.rev^{-1}]	*Rku*	*Rsk*	*RSm*	*RS*	*AR*	*RΔa*	*RΔq*
0.08	105	2.313	0.075	60.686	16.93	85.133	5.83	7.33
	210	2.766	0.205	67.726	17.5	88.8	6.053	7.69
	420	3.467	0.563	71.733	17.91	90.9	5.756	7.533
	530	2.504	0.346	58.103	16.603	85.57	6.366	7.99
	660	2.384	0.174	64.95	17.413	86.3	6.213	7.823
0.19	105	2.232	0.29	180.23	36.983	176.133	5.426	7.106
	210	2.36	0.429	187.21	36.03	183	5.526	7.353
	420	9.734	0.993	79.78	18.206	87.8	5.943	7.676
	530	2.398	0.415	179.62	36.656	183.4	5.69	7.686
	660	2.439	0.469	177.073	39.133	182.4	5.586	7.663

0.37	105	2.07	0.655	364.516	192.89	171.533	7.116	8.81
	210	2.033	0.64	373.146	200.44	179.3	7.153	8.86
	420	2.085	0.663	373.496	209.873	178	7.243	9.01
	530	2.09	0.661	365.643	292.28	175.83	7.203	9.003
	660	2.117	0.681	372.566	245.6	163.07	7.173	9.003
ρ		0.32	0.49	0.05	0.27	-0.18	0.56	0.86
t [%]		251%	332%	95%	201%	93%	25%	23%

Table 5. Matrix of correlation coefficients for amplitude, hybrid and motif roughness parameters

	Ra	*Rz*	*Rp*	*Rv*	*Rq*	*Rku*	*Rsk*	*RΔa*	*RΔq*
Ra	1								
Rz	0.9995	1							
Rp	0.9991	0.9998	1						
Rv	0.9987	0.9990	0.9978	1					
Rq	1.0000	0.9996	0.9993	0.9989	1				
Rku	-0.3367	-0.3307	-0.3223	-0.3475	-0.3343	1			
Rsk	0.5267	0.5373	0.5462	0.5182	0.5296	0.5309	1		
RΔa	0.8932	0.8854	0.8900	0.8746	0.8920	-0.2324	0.4603	1	
RΔq	0.9269	0.9237	0.9288	0.9119	0.9266	-0.2399	0.5424	0.9775	1
Pt	0.9957	0.9964	0.9970	0.9937	0.9960	-0.2584	0.5990	0.8864	0.9250
R	0.9657	0.9687	0.9657	0.9736	0.9664	-0.3971	0.5010	0.7653	0.8367
Rx	0.9974	0.9979	0.9984	0.9956	0.9977	-0.2780	0.5830	0.8884	0.9275

Some observed relations are obvious. For example relation *Pt*-*Rz* results from their definition – *Pt* refers to the primary profile, and *Rz* refers to roughness profile. In 2D system their dependence is defined as $Pt = 1.27Rz$. Equally high correlation occurs between parameters *Rx* and *Rz*. Their dependence was calculated as $Rx = 1.17Rz$.

The only spatial parameter from motif group showed independence from classical parameters. Correlation between *AR*, spatial and hybrid parameters is less than 0.78.

Table 6. Matrix of correlation coefficients for spatial roughness parameters and from motif group

	RSm	*RS*	*RΔa*	*RΔq*
RSm	1			
RS	0.937385076	1		
RΔa	0.774039748	0.891210473	1	
RΔq	0.841180807	0.925931889	0.977547659	1
AR	0.8500291	0.570359323	0.248876757	0.384685697

4 Summary

- For evaluation of usefulness of individual roughness parameters for description geometric changes of surface layer and geometric state of the surface not only correlation coefficient ρ should be taken into account but also t indicator determining the level of significance of change in characteristic profile feature. Applying of two indicators will allow for more complete assessment of parameters usefulness.
- Use of the parameters from the group of motifs for description of the surface with oriented periodic character is possible and may complement description of this type of surfaces carried out with using parameters described in ISO standards.
- For amplitude parameters best describing surface characterized by features resulting from variable feed during machining include: *Ra, Rz, Rp, Rv, Rq* and motif parameters: *Pt, Rx, R.*
- Spatial parameters that describes in best way surface characterized by features resulting from variable feed may be *RSm and RS. The* only *motif* parameter from that group: *AR* despite the value of indicator t equal 67%, due to low correlation does not describe all changes of surface geometrical features.
- Due to the close dependence (regardless of turning parameters) of *Pt* and *Rz* parameters as well as *Rx* and *Rz* it seems unreasonable using all three of them simultaneously.
- Hybrid parameters: *RΔa* and *RΔq* and statistical: *Rku* and *Rsk* do not show sufficiently strong relations with feed and cutting speed, therefore it is not advisable to describe the state of geometric structure of surface by themselves.

The article was made under support grant project KEGA 022ŽU-4/2017 Implementation of on-line education in the area of precise technologies with an impact on educational process to increase skills and flexibility of students of engineering fields of study

References

1. M. Wieczorowski, A. Cellary, J. Chajda, *Przewodnik po pomiarach nierówności powierzchni czyli o chropowatości i nie tylko*. Zakład Poligraficzno - Wydawniczy M - Druk. Poznań (2003)
2. K.J. Stout, L. Blunt,: *A contribution to the debate on surface classifications – random, systematic, unstructured, structured and engineering*. International Journal of Machine Tools & Manufacture, **41**, 2039-2044, (2001)
3. K.E. Oczoś, V. Liubimov, *Struktura geometryczna powierzchni. Podstawy klasyfikacji z atlasem charakterystycznych powierzchni kształtowanych*. Oficyna Wydawnicza Politechniki Rzeszowskiej, Rzeszów (2003)
4. P. Pawlus, *Struktura geometryczna powierzchni cylindrów podczas eksploatacji silnika spalinowego*. Oficyna Wydawnicza Politechniki Rzeszowskiej, Rzeszów (1999)
5. S. Lou, X. Jiang, P. J. Scott, *Correlating motif analysis and morphological filters for surface texture analysis*. Measurement **46** 993-1001, (2013)
6. W. Grzesik, J.Rech, K. Żak, *High-Precision Finishing Hard Steel Surfaces Using Cutting, Abrasive and Burnishing Operations*. Procedia Manufacturing Volume **1**, 619-627 (2015)
7. A. Rudawska, M. Reszka, T. Warda, I. Miturska, J. Szabelski, A. Skoczylas, D. Stancekova, *Milling as a method of surface pre-treatment of steel for adhesive bonding*. J. Adhes. Sci. Technol., **23**, 2619-2636 (2016)
8. M. Kowalski: *Metodyka wyboru parametrów chropowatości do opisu topografii powierzchni*. Politechnika Wrocławska, Instytut Technologii Maszyn i Automatyzacji, Wrocław (2005)

Research on water jet cutting of composites based on epoxy/microparticles from coconut shell

Miroslav Müller[1,*],, *Petr* Valášek[1], *Miloslav* Linda[2], *Viktor* Kolář[1]

[1]Czech University of Life Sciences Prague, Faculty of Engineering, Department of Material Science and Manufacturing Technology, Kamýcká 129, Praha 6 – Suchdol, Czech Republic
[2]Czech University of Life Sciences Prague, Faculty of Engineering, Department of Electrical Engineering and Automation, Kamýcká 129, Praha 6 – Suchdol, Czech Republic,

Abstract. The paper deals with a machining of polymeric composite materials with a reinforcement based on a waste from a coconut processing, i.e. a coconut shell by means of an unconventional technology, i.e. a water jet (WJ). These composite systems are desirable in the material engineering. Not only the research of their subsequent mechanical properties is important, but also searching for effective ways of machining and dividing to required shape and size. The composite material was tested in a form of boards made by means of a vacuum infusion. The aim of the research is to study an influence of the abrasive water jet and water jet at its impact on a surface of the machined polymeric composite material. The research was proved on CNC cutting machine AWJ CT 0806. Cutting conditions, i.e. various traverse speeds 50, 100, 250, 500, 750 and 1000 mm/min, an influence of cutting by the water jet with abrasive particles and without them on an integrity of the tested material and the surface topography were evaluated within the research. A surface analysis is also evaluated by means of a scanning electron microscopy (SEM) depending on a traverse speed and a mass flow of the abrasive grains.

Keywords: abrasive water jet, traverse speed, kerf width, SEM, water jet

1 Introduction

Composite materials with natural phases /a filler have been developed in a present in the material engineering. The natural filler is used in a form of microparticle filler or plant / animal fibres.

A substitution of synthetic fillers/reinforcements with natural ones is important in the material research. Biocomposites are also made with the use of various other ingredients, e.g. egg shell [1, 2, 3], seeds of Phoenix [4], Ensete ventricosum [5], oil palm fibres [6, 7], sisal fibres [8, 9], flax [10, 11] and other narural fibres (cellulose fibres, rice – husk, jute etc.) [12, 13, 14]. Many research studies deal with mechanical properties of composite materials. However, it is necessary to focus also on other operations in an engineering production, i.e. material dividing and subsequent machining [15]. The material dividing is a common attribute of production companies [16]. Classically applied methods of machining are not

* Corresponding author: muller@tf.czu.cz
Reviewers: *Anna* Mičietová, *Sergej* Hloch

always technically and economically the most suitable for machining of new materials [17]. Various variants of cutting by means of a water jet technology belong among prospective methods of the composite materials dividing.

It does not come to a heat affecting of machined materials, to an evaporation and to a degradation of the material. A temperature is a problematic factor mainly at machining of polymers and polymeric composites. The liquid-jet is the working tool for the machining. It is completed with an ingredient of abrasive substances at the abrasive water jet technology [16]. This technology consists in a principle of generating high water pressure which goes through a nozzle of a small diameter so the water gains very high kinetic energy similarly as at the water jet [17-19]. Abrasive particles several fold increasing its effect are added into the high-speed water stream [16]. An interaction between a matrix and a reinforcement can be failed at acting of high-speed liquid with abrasive grains with the surface of the workpiece at the composite materials cutting [19-20]. The AWJ technology proved to cut in the effective way also the reinforcing glass fabric, corundum and glass-bead particles without their delamination from the matrix. The delamination and significant removal of the material show themselves in the first stage of the cutting. Subsequent cut is regular [21]. However, the results proved the delamination of layers at various traverse speeds.

The aim of the research is to study an influence of the abrasive water jet and water jet at its impact on a surface of the machined polymeric composite material with the coconut shell microparticle filler.

2 Methodology

The filler from a woody shell of the coconut which is of brown colour and it is covered with peeling bay bundles was used for the research. The bay stays on the coconut as a remainder of a green packaging in which the coconut was originally placed during its growth and harvesting.

Waste from a food industry, i.e. the coconut shells were grinded and milled. Subsequently the microparticles of grinded coconut shells were fractionalised on vibrating sieves with a mesh size 200 μm. Microparticles were dried before the application at the temperature 105 °C for the time 24 h.

A vacuum infusion was used for the production of the composite material. A two-component epoxy resin with a low viscosity suitable for a laminating technology including the vacuum infusion was used as the matrix. The resin was hardened with cycloaliphatic polyamine, the main component isophorone di-amine.

The water jet with a dash of abrasive grains (a garnet MESH 80) was used for the research. The research was performed on the device CNC cutting machine AWJ CT 0806 (fig. 1). The speed of the cut was 50, 100, 250, 500, 750 and 1000 mm/min. The diameter of the nozzle was 0.8 mm. The working pressure was 380 MPa. The water jet with the abrasive grains affected perpendicularly to the cutting edge of the sample. The distance of the nozzle above the cut material was 3 mm.

Cuts of the composite material by water jet WJ technology and abrasive water jet AWJ technology at which the mass flow of the abrasive grains was 90.62 ± 1.19 g/min were tested within the research. The testing cut was made on the composite board of dimensions 90 x 130 mm. A length of the cut was 60 mm.

A width of a kerf inlet (fig. 1 B) and outlet (fig. 1 C) of the abrasive water jet was evaluated within the research. It considerably differed. A parameter of the kerf width was evaluated by means of a stereoscopic microscope Arsenal SZP 11-T ZOOM which is equipped with a digital camera and an evaluating software Quick Industrial Photo 2.3. The evaluated kerf was measured in 30 places 10 mm from the beginning and the end of the cut. A thickness of the cut composite board was 4.5 ± 0.1 mm. A topography of the machined surface by means of a surface analysis by means of a scanning electron microscopy (SEM) depending on the traverse speed and AWJ or WJ technology was evaluated within the research. The measurement was performed on the scanning electron microscope TESCAN MIRA 3 GMX. SE detector were used for the measurement, the accelerating voltage was 5 kV. The samples were dusted with gold by means of the equipment Quorum Q150R ES - Sputtering Deposition Rate using Gold. Measured values were processed by means of statistical analysis methods by use of Anova F-test (p-values).

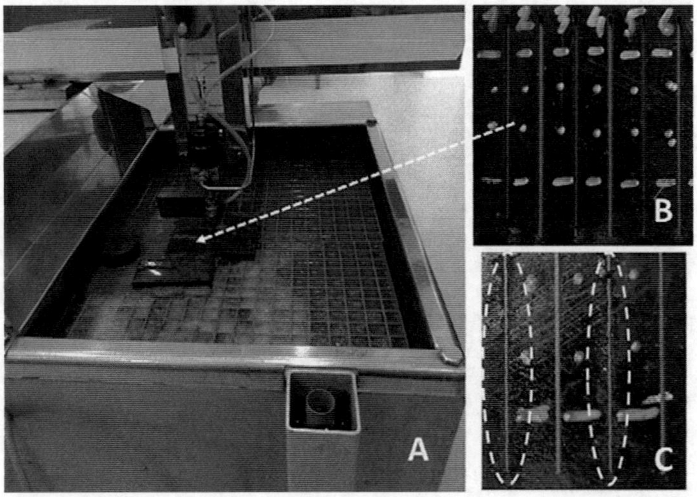

Fig. 1. Machining of composites based on epoxy / microparticles from coconut shell by means of water jet technology: A: CNC AWJ CT 0806, B: cut in tested composite sample – inlet of abrasive water jet, C: cut in tested composite sample – outlet of abrasive water jet

3 Results

Experiments results proved a quiddity of a combination of the traverse speed and the use of abrasive particles at dividing the biological reinforcement based composite materials. The composite reinforcement is evident from fig. 2. Fig. 2 A, B represent the microparticle filler on the basis of the woody coconut shells. The filler is of a specific shape which is visible in fig. 2 A, B. The bay fibre from the coconut shell is evident from fig. 2 C.

Results of a kerf width at cutting of the composite material by means of the water jet technology WJ are obvious from fig. 3. The kerf width inlet did not significantly differ depending on the traverse speed in the interval from 50 to 1000 m/min. The kerf width difference was maximum of 11.15 %. The kerf width outlet was of high influence depending on the traverse speed. The kerf width influence was 5.31 to 791.19 %. A significant destruction of the outlet side (fig. 1 C) occurred at higher speeds at cutting by means of the water jet without the abrasive grains. It was the speeds from 500 mm/min.

It is obvious from the statistical testing of the kerf width inlet and outlet that they are statistically non-homogeneous groups (p = 0.0000), i.e. there is the difference between the kerf width inlet and outlet in the significance level 0.05.

Results of the kerf width at cutting of the composite material by means of AWJ technology are evident from fig. 4. The kerf width inlet did not significantly differ depending on the traverse speed in the interval from 50 to 1000 mm/min. The difference of the kerf width was maximum 15.31 %. More significant influence of the kerf width was outlet depending on the traverse speed. The difference in the kerf width was up 34.69 %. The difference was increasing with increasing traverse speed.

It is obvious from the statistical comparison of the kerf width inlet and outlet that they are statistically non-homogeneous groups (p = 0.0000), i.e. there is the difference in the kerf width inlet and outlet in the significance level 0.05. The influence of the traverse speed at AWJ as well as WJ technology was statistically proved (p = 0.0000).

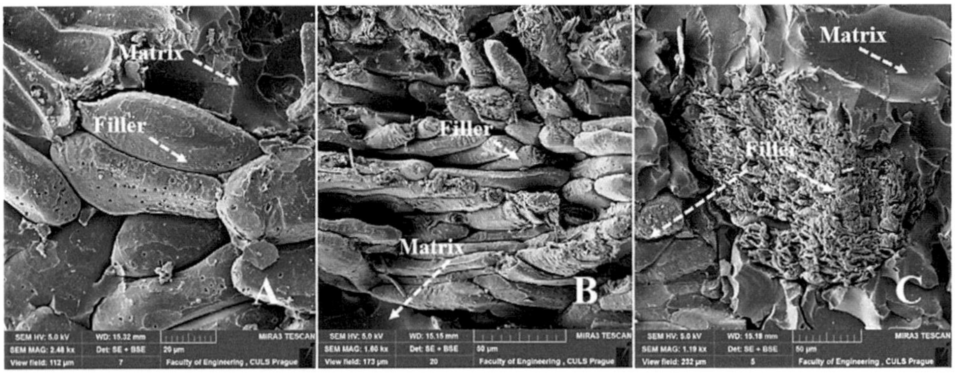

Fig. 2. SEM images of composite material with filler based on microparticles from coconut shell: A: microparticles of coconut woody shell – longitudinal orientation (MAG 2.48 kx), B: microparticles of coconut woody shell – transverse orientation (MAG 1.60 kx), C: bay fibre from coconut shell (MAG 1.19 kx)

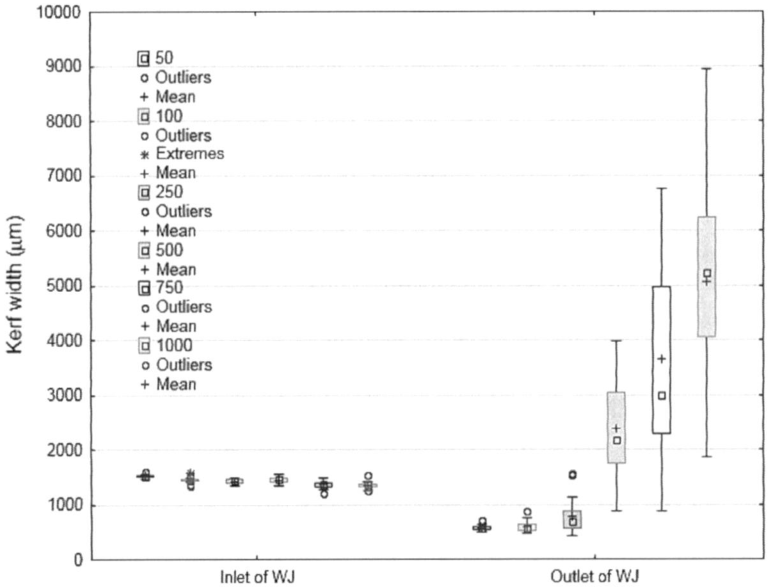

Fig. 3. Kerf width of cut by WJ technology of polymeric composite material with coconut shell microparticle based filler

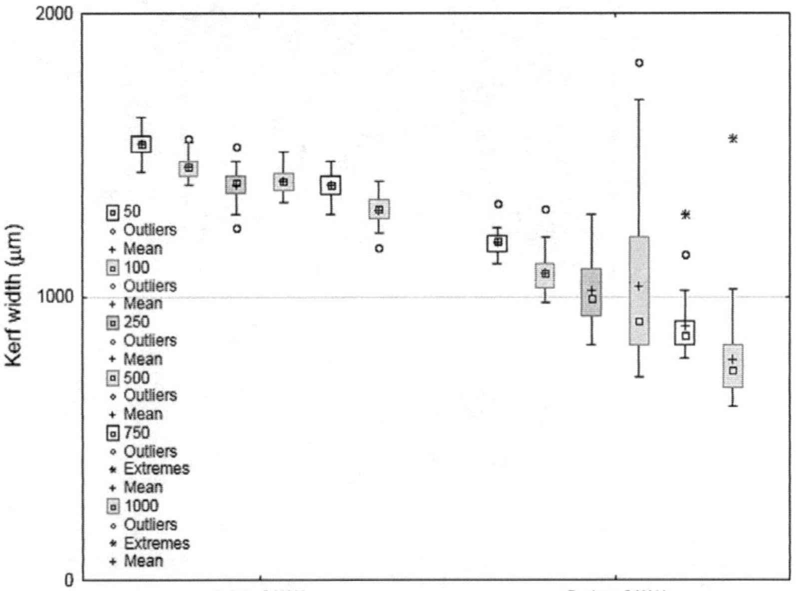

Fig. 4. Kerf width of cut by AWJ technology of polymeric composite material with coconut shell microparticle based filler

The cutting zone showed huge differences between WJ and AWJ as it is visible from fig. 5. A regular cut by the use of AWJ technology is evident from fig. 5 A. An irregular cutting zone is obvious from fig. 6 A and B. Outlet there is obvious an area of tearing in the end of the cut. This negative state occurred at higher speeds, i.e. 500, 750 and 1000 mm/min. The water jet had not already sufficient kinetic energy for the cut and it came to a deformation leading to a material fracture. Therefore the results in fig. 3 are of higher scattering, namely up of 791.19 %.

A dual cutting zone is obvious from fig. 7 A, i.e. inlet is relatively smooth and outlet there are visible significant grooves. It is usual cutting zone [22]. A significant grooving of the surface (fig. 7 B) occurs in the bottom part of the cut from which a principle of AWJ cutting is visible, i.e. after the abrasive water jet impact into the composite material its penetration and water flow into the crack occur which causes its spreading. The effect of WJ and AWJ on the filler of the composite materials in the form of the coconut shell microparticles is evident from fig. 7 C, D and fig. 8 A, B. Fig. 7 C, D and E represent the cut by the abrasive water jet and it is evident that it comes to cutting of the filler. Fig. 8 B represents the cut by water jet without abrasive grains and it is evident that it comes to tearing partial segments of the filler by acting the kinetic energy to the cut. The difference between AWJ and WJ technology is obvious from fig. 7 D and fig. 8 B.

A delamination often occurs as a result of the impact with high speed of the water jet [16, 20]. The delamination is an important factor which can change a mechanical behaviour of composite systems during their technological processing which can be e.g. water jet cutting. Experiment results showed that the delamination could be caused also by an absence of abrasive grains at cutting by water jet technology.

SEM analysis of the surface and the evaluation of the kerf width certified that the abrasive water jet was much better than pure water jet. The research showed that quality cuts of polymeric composite materials could be prepared by AWJ technology [23].

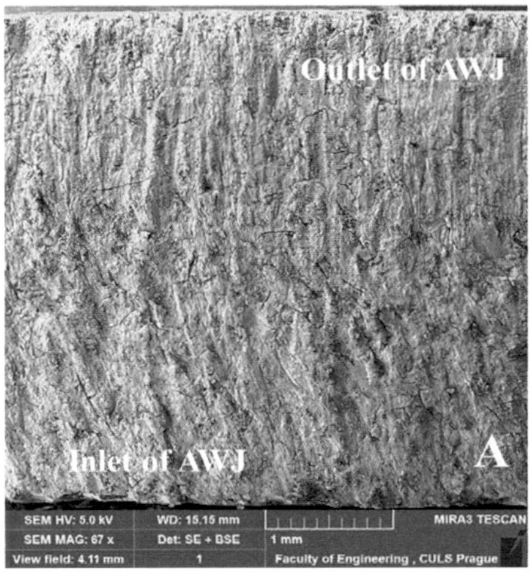

Fig. 5. SEM images of cut of composite material with coconut shell microparticles based filler A: cut by abrasive water jet (AWJ) technology, speed 750 mm/min (MAG 67 x)

Fig. 6. SEM images of cut of composite material with coconut shell microparticles based filler A: cut by water jet (WJ) technology without abrasive grains, speed 750 mm/min – right side of cut (MAG 75 x), B: cut by water jet (WJ) technology without abrasive grains, speed 750 mm/min – left side of cut (MAG 75 x)

The kerf width inlet was higher than outlet. These conclusions were proved at cutting composite materials when increased removal of the material in the initial phase of the cut was manifested itself. Except for variants where it did not come to the cut but to the destruction of the material due to the kinetic energy acting [24].

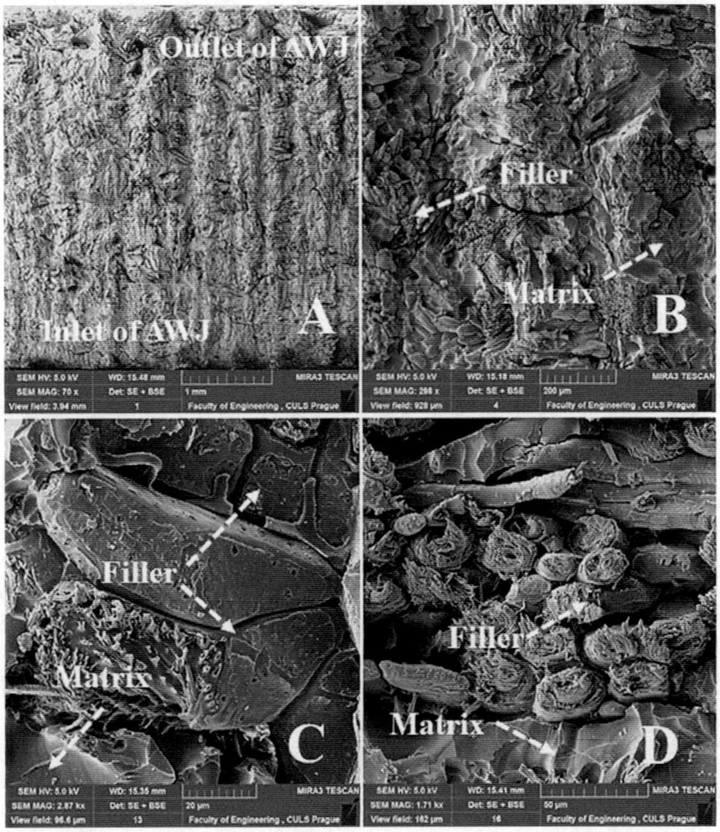

Fig. 7. SEM images of cut of composite material with coconut shell microparticle based filler, speed 100 mm/min: A: cut by abrasive water jet technology (AWJ) (MAG 70 x), B: detailed view on cut by abrasive water jet technology (AWJ) (MAG 298 x), C: cut of woody coconut shell microparticle by AWJ technology– transverse orientation (MAG 2.87 kx), D: cut of woody coconut shell microparticle by AWJ technology – longitudinal orientation (MAG 1.71 kx)

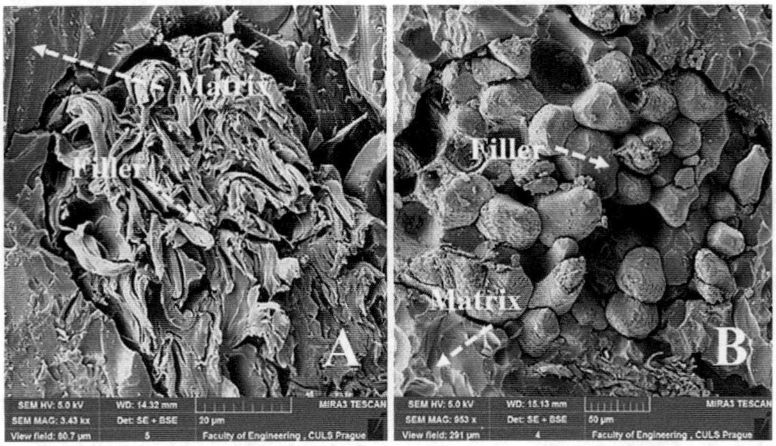

Fig. 8. SEM images of cut of composite material with coconut shell microparticle based filler, speed 100 mm/min: A: cut of bay fibre from coconut shell by AWJ technology (MAG 3.43 kx), B: cut of woody coconut shell microparticle by WJ technology – longitudinal orientation (MAG 953 x)

4 Conclusions

The research dealing with the polymeric composite material with the filler based on coconut shell microparticles machined by means of unconventional technology, i.e. the abrasive water jet (AWJ) and the water jet (WJ) proved a quiddity of the combination of the traverse speed and the mass flow of the abrasive grains to the efficiency of the cut and connected cut surface. The significant destruction of the material occurred under unsuitable cutting conditions and at the absence of the abrasive grains in the water jet inlet, i.e. on the bottom side of the cut.

Tested composite boards serve as a design element in the furniture industry due to their properties and use of the natural filler potential, i.e. an upper functional board which can substitute synthetical lamino.

This paper has been made with the assistance of the grant IGA TF CZU (Research on mechanical properties of polymeric composite materials cut by water jet, No. 31140/1312/313112).

5 References

1. M. Zieleniewska, M.K. Leszczynski, L. Szczepkowski, A. Bryskiewicz, M. Krzyzowska, K. Bien, J. Ryszkowska, *Development and applicational evaluation of the rigid polyurethane foam composites with egg shell waste.* Polymer Degradation and Stability, **vol. 132**, 78-86 (2016)
2. D.A. Oliveira, P. Benelli, E.R. Amante, *A literature review on adding value to solid residues: egg shells.* Journal of Cleaner Production, vol. **46**, 42-47 (2013)
3. J.M. Badanoa, C. Betti, I. Rintoulb, J. Vich-Berlanga, E. Cagnola, G. Torres, C. Vera, J. Yori, M. Quiroga, *New composite materials as support for selective hydrogenation; egg-shell catalysts.* Applied Catalysis A: General 390, 166-174 (2010)
4. A. Ruggiero, P. Valášek, M. Müller, *Exploitation of waste date seeds Phoenix dactylifera in form of polymeric particle biocomposite: Investigation on adhesion, cohesion and wear.* Composites: Part B, vol. **104**, 9-16 (2016)
5. Č. Mizera, D. Herák, P. Hrabě, M. Müller, A. Kabutey, *Mechanical Behavior of Ensete ventricosum Fiber Under Tension Loading.* Journal of Natural Fibers, vol. **14,** 287-296 (2017)
6. G. Koronis, A. Silva, M. Fontul, *Green composites. A review of adequate materials for automotive applications.* Composites: Part B, vol. **44**, 120-127 (2013)
7. C.A. Kakou, F.Z. Arrakhiz, A. Trokourey, R. Bouhfid, A. Qaiss, D. Rodrigue, *Influence of coupling agent content on the properties of high density polyethylene composites reinforced with oil palm fibers.* Materials and Design, vol. **63**, 641-649 (2014)
8. X. Lu, M. Qiu Zhang, M. Zhi Rong, G. Shi, G. Cheng Yang, *Melt processable composites of sisal.* Composites Science and Technology, vol. **63**, 177-186 (2003)
9. X. Zhao, R. Li, S-L. Bai, *Mechanical properties of sisal fiber reinforced high density polyethylene composites: Effect of fiber content, interfacial compatibilization, and manufacturing process.* Composites: Part A, vol. **65**, 169-174 (2014)
10. A.S. Doumbia, M. Castro, D. Jouannet, A. Kervoëlen, T. Falher, L. Cauret, A. Bourmaud *Flax/polypropylene composites for lightened structures: Multiscale analysis of process and fibre parameters.* Materials and Design, vol. **87**, 331-341 (2015)
11. A.A. Nasir Abdul, A.I. Azmi, A.N.M. Khalil *Measurement and optimisation of residual tensile strength and delamination damage of drilled flax fibre reinforced composites.* Measurement, vol**. 75**, 298-307 (2015)

12. M.P. Ho, H. Wang, J.H. Lee, C.K. Ho, K.T. Lau, J. Leng, D. Hui *Critical factors on manufacturing processes of natural fibre composites.* Composites Part B: Engineering, vol. **43**, 3549-3562 (2012)
13. H-Y. Cheung, M-P. Ho., K-T. Lau, F. Cardona, D. Hui. *Natural fibre-reinforced composites for bioengineering and environmental engineering applications.* Composites: Part B, vol. **40,** 655-663 (2009)
14. K.A. Iyer, J.M Torkelson *Green composites of polypropylene and eggshell: Effective biofiller size reduction and dispersion by single-step processing with solid-state shear pulverization.* Composites Science and Technology, vol. **102**, 152-160 (2014)
15. M. Müller, P. Valášek *Research on aluminium alloy alcu4mg surface machined by abrasive water jet.* Manufacturing technology, vol. **17**, 925-930 (2017)
16. M. Müller, R. D´Amato, A. Rudawska *Machining of polymeric composites by means of abrasive water-jet technology.* In 16th international scientific conference: engineering for rural development 24.05.2017, Latvia univ agr, fac engn, Jelgava, Latvia. Jelgava, Latvia: Latvia univ agriculture, Faculty engineering, inst mechanics, 5 j cakstes blvd, Jelgava, lv-3001, Latvia, 121-127 (2017)
17. M. Kušnerová, J. Foldyna, L. Sitek, J. Valicek, S. Hloch, M. Harnicarova, M. Kadnar *Innovative approach to advanced modulated waterjet technology.* Technicki Vjesnik-Technical Gazette, vol**. 19 (3)**, 475-480 (2012)
18. S. Hloch, J. Hlavacek, K. Vasilko, J. Carach, I. Samardzic, D. Kozak, I. Hlavaty, J.J. Scucka, Klich, D. Klichava *Abrasive waterjet (AWJ) titanium tangential turning evaluation.* Metalurgija, vol. **53(4)**, 537-540 (2014)
19. J. Wang, D.M. Guo *A predictive depth of penetration model for abrasive waterjet cutting of polymer matrix composites.* Journal of Materials Processing Technology, vol. **121**, 390-394 (2002)
20. D.K. Shanmugam, T. Nguyen, J.A. Wang *Study of delamination on graphite/epoxy composites in abrasive waterjet machining.* Composites: Part A, vol. **39**, 923-929 (2008)
21. D.K. Shanmugam, F.L. Chen, E. Siores, M. Brandt *Comparative study of jetting machining technologies over laser machining technology for cutting composite materials.* Composite Structures, vol. **57**, 289-296 (2002)
22. M. Hashish *A model study of metal cutting with abrasive water jet.* In: ASME Journal of Engineering Materials and Technology, vol. **106**, 88-100 (1984)
23. E. Lemma, F.L. Chen, E. Siores, J. Wang *Study of cutting fiber reinforced composites by using abrasive waterjet with cutting head oscillation.* Composite Structure, vol. **57**, 297-303 (2002)
24. J. Wang *Abrasive waterjet machining of polymer matrix composites-cutting performance, erosive process and predictive models.* Int J Adv Manuf Technol., vol. **15**, 757-768 (1999)

Non homogeneity of surface integrity after hard milling of high tempered bearing steel

Mária Čilliková[1], *Miroslav* Neslušan[1,*], *Anna* Mičietová[1], *Juraj* Uríček[1]

[1]Faculty of Mechanical Engineering, University of Zilina, 010 26 Žilina, Slovakia

Abstract. This paper deals with surface integrity after hard milling of high tempered bearing steel 100Cr6. Surface integrity is expressed in terms of shape deviation as well as Barkhausen noise emission. Furthermore, components of cutting forces are also measured as a function of tool wear and their non homogenity within the contact between the milling cutter and workpiece is discussed. The results of measurements indicate that shape as well as structure remarkable non homogeneity should expected after hard milling due to missing full contact between milling cutter and workpiece. This non homogeneity can be monitored via Barkhausen noise technique. On the other hand the competitive grinding cycles produced more homogenous surface and surface state is a function of infeed speed.

Keywords: hard milling, grinding, surface integrity, Barkhausen Noise

1 Introduction

Cyclic magnetization in a ferromagnetic material produces magnetic pulsation as a result of irreversible and discontinuous Bloch Walls (BW) motion. Discontinuous BW motion is due to interference of BW with microstructure features such as precipitates, dislocation cells, grain boundaries and other lattice defects. This phenomenon is named Barkhausen noise (BN). BN technique has found high industrial relevance mainly for monitoring surfaces loaded near their physical limits. BN features are usually correlated with residual stresses, microhardness or structure transformations. This technique is mostly adopted for monitoring of surfaces after grinding cycles due to strong correlation among thermal over tempering, associated surface burn and the corresponding BN features [1, 2, 3, 4]. Effect of thermal softening after grinding due to over tempering decreases dislocation density (and the corresponding surface hardness), transforms carbides shape and their morphology as well as produces tensile stresses [2, 4]. All the above mentioned aspects contribute to the higher magnitude of BN.

Nowadays, hard machining (mainly turning and milling) can substitute grinding cycles. Development in machine tools as well as in process technology raised industrial relevance of hard machining. However, hard turning or milling cycles can suffer from formation of white layers (WL) induced at low flank wear VB or unexpected catastrophic tool failures. Machined surface after hard machining is mainly a function of VB and cutting speed. Tools of high VB

* Corresponding author: miroslav.neslusan@fstroj.uniza.sk
Reviewers: *Ildiko* Maňková, *Nataša* Náprstková

produce relative thick white layer (WL) as well as the corresponding heat affected zone (HAZ) [5, 6]. On the other hand, grinding cycles can suffer from thermal over tempering. Ground surface can sometimes exhibit thermally softened layers (HAZ). However, thickness of HAZ and WL after hard milling is about 1 order lower than that induced by grinding. It is also worth to mention that the ratio between WL and HAZ thickness after hard machining is much higher as opposed to grinding [5, 7, 8]. Hard milling surfaces were already investigated and reported [5, 7, 8, 9]. However significant aspect associated with non homogenity due to missing full contact between milling cutter and sample has not been discussed yet. For this reason this study is mainly focused on evaluation of surface integrity after hard milling with inserts of variable VB and compares shape deviations as well as dynamic BN signals with ground samples.

2 Experiments and methods

Experiments were conducted on samples made of bearing steel 100Cr6. Ten pieces of dimension 100x70x5 mm were prepared for long term test. Bearing steel was heat treated on hardness 45 HRC (austenitizing temperature 830°C – oil 62°C, followed by tempering for 2 hours temperature 470°C). Cutting process was monitored as a long term test where such aspects as flank wear VB, structure alterations and corresponding surface integrity expressed in magneto elastic responses (BN) of the hard milled surface were investigated. Cutting and other conditions: milling machine - FA4 AV, dry cutting, cutting tool made of cemented carbides R300-1240E-PM, R300-050Q22 - 12M 262489 of diameter Ø 50mm with 2 inserts of variable flank wear VB (in the range 0.05 to 0.8 mm), $a_p = 0.25$ mm, $v_f = 112$ mm.min^{-1}, n = 500 min^{-1}, dry cutting. Grinding was carried out under the following conditions: grinding machine BPH 20, A9860J9V, single crystal diamond dresser, $a_p = 0.005$ to 0.04 mm (10 passes), $v_f = 8$ m.min^{-1}, $v_c = 25$ m.s^{-1}.

BN measurement was performed by the use of RollScan 300 and software package ViewScan in the frequency range of 70 to 200 kHz (magnetizing frequency 125 Hz, magnetizing voltage 10 V). Each surface was measured in tangential direction. BN values indicated in the paper represent the effective (rms) value of BN signal. BN measurements were carried out in the dynamic mode. This means that during BN measurement components move (speed on the motion 5 mm.s^{-1}). Cutting forces were measured by the use of three component piezoelectric dynamometer KISTLER. The signal was sampled by frequency 10 kHz. Brief sketch of components of cutting force can be found in Fig. 1. Surface profile after machining was measured on MP 60 measuring device.

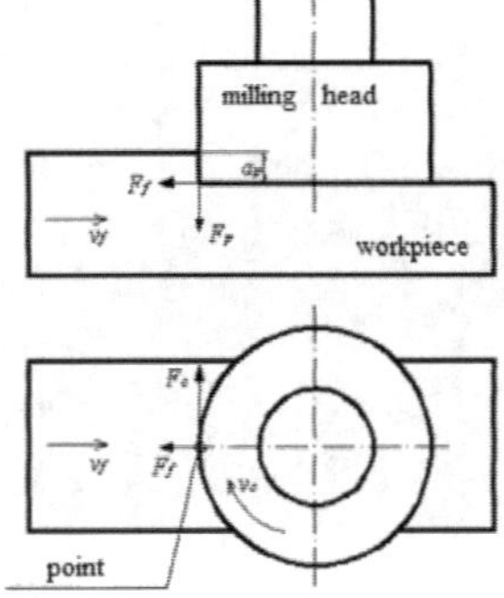

Fig. 1. Brief sketch of measurement of cutting forces during milling

3 Results of experiments

Fig. 2 and Fig. 3 illustrate the typical surface which can be obtained after hard milling. The main problem is associated with the contact between milling cutter and the sample. As Fig. 2 and Fig. 3 shows that at certain position cutting inserts start to miss the full contact with the sample and run out. This position (in Fig. 2 indicated by grey color) corresponds to the cutter diameter (in this case 50 mm). Due to limited toughness of the milling system milling head tends to be tilted against the machined surface in a certain small angle which changes the condition in the contact between milling cutter (cutting edge) and sample and results into the different appearance of machined surface in this zone. Fig. 3 also indicates that this zone can be easily found by visual inspection. Such character of machined surface can be found for all flank wears and the different sample hardness. Fig. 4 and also details illustrated in Fig. 5 and Fig. 6 show that this effect can be also found in the records of cutting forces components (only passive components indicated in this study, but all other components exhibit the similar behaviour). Fig. 4 and Fig. 5 show that during full contact F_p component oscillates in the range from 200 to 700 N (see also Fig. 6a). As soon as the full contact is missing the run out of the insert causes fall of F_p to zero (see also Fig. 6b). In the further process F_p tend to decrease due to shortening path of the contact between the cutting insert and sample (see also Fig. 6c,d).

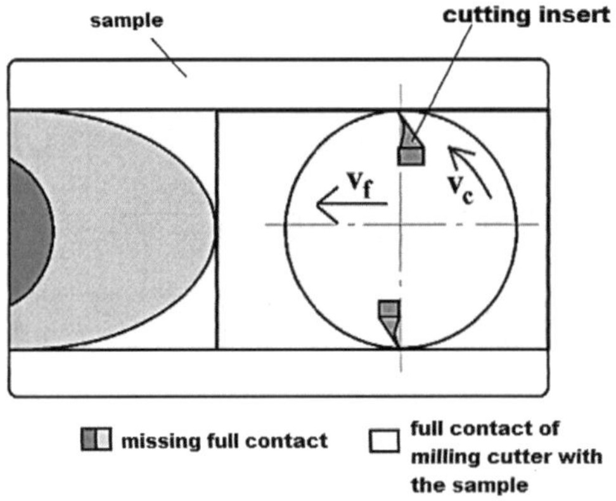

Fig. 2. Non homogenous contact between milling cutter and sample during milling

Fig. 3. Sample photo after hard milling

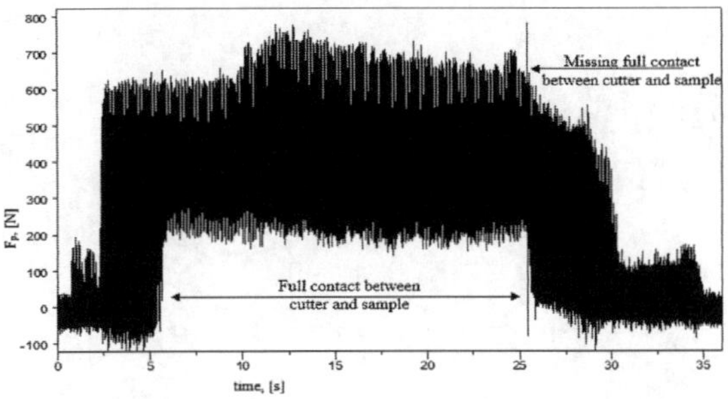

Fig. 4. Record of passive component of cutting force

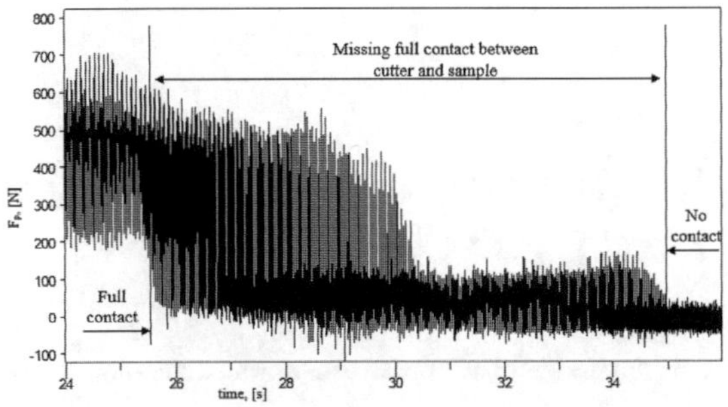

Fig. 5. Detail of passive component F_p

Further measurements of the hard milled surface produced under the variable flank wear VB indicate remarkable shape as well as structure non homogeneity of the machined surface. Fig. 7 shows record of the surface profile in which can be clearly recognized zone of full contact between the milling cutter and the sample as well as strong decrease on the surface height. This point is kept constant and corresponds to the milling cutter diameter. Quite high height of the measured profile is due to high passive forces which are typical for hard machining operation and the corresponding strong influence of passive force on precision of produced components.

Milling operations are usually employed for roughing cycles followed by consecutive final grinding. One might argue that such surface would be finished and surface non homogeneity will be removed by grinding since grinding cycles produced much better surface state from the point of view of shape deviations (see Fig. 8). However, quite high height of the profile increases allowances for the consecutive grinding cycles which in turn means longer grinding times and the corresponding costs for technological operations.

Fig. 9 shows that non homogeneity of machined surface is not only a matter of surface height (of generally surface precision expressed in any term) but also non homogeneity of microstructure and stress state should be expected since BN emission is sensitive to the alterations of microstructure expressed in all terms as well as residual stress state. Being so, it can be found that surface integrity expressed in state of microstructure as well as stress state after hard milling also exhibits remarkable non homogeneity. lace the figure as close as possible after the point where it is first referenced in the text. If there is a large number of figures and tables it might be necessary to place some before their text citation.

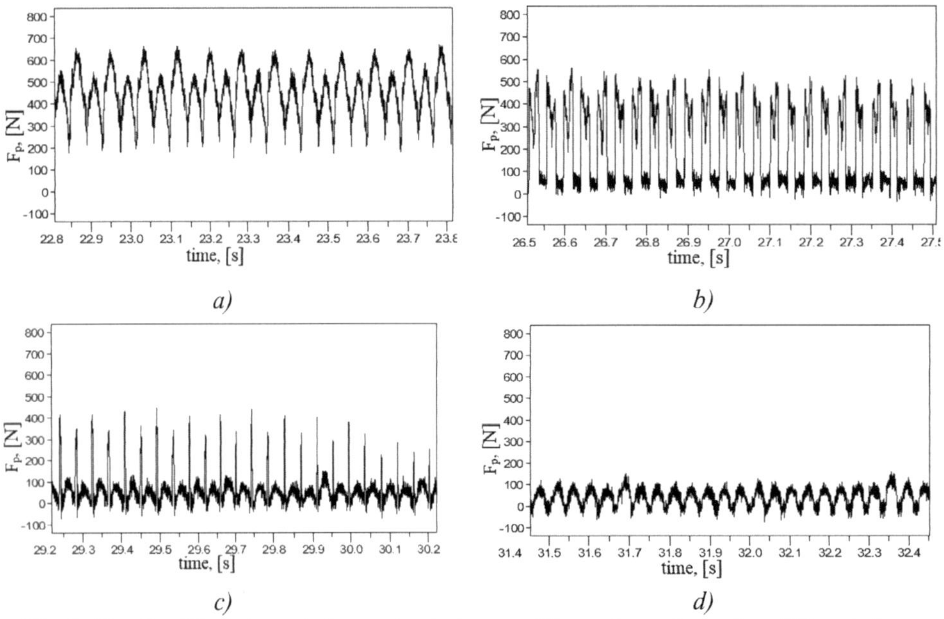

Fig. 6. Details of F_p during missing contact between milling cutter and sample – different phases

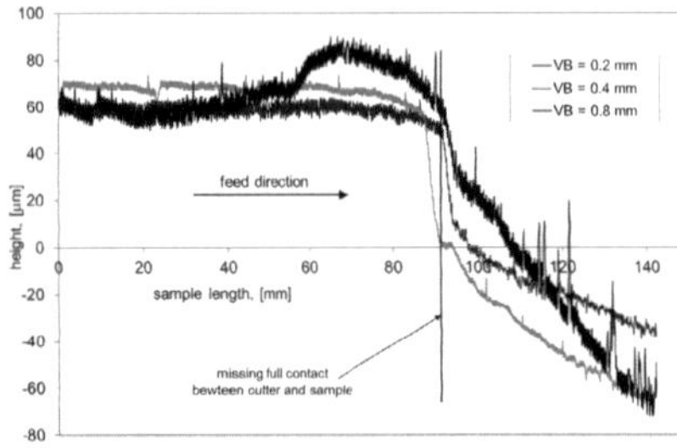

Fig. 7. Surface profile after hard milling

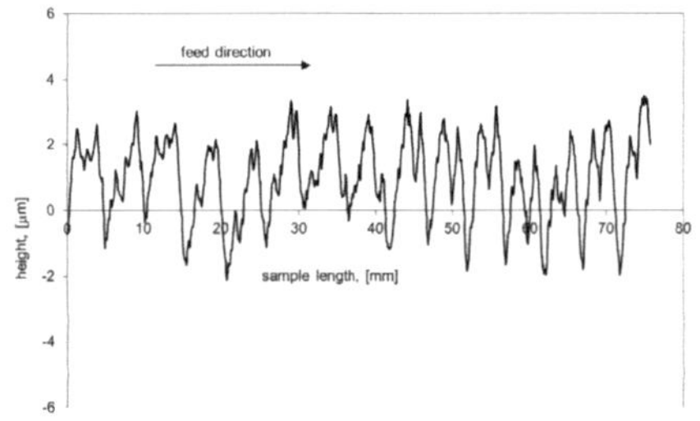

Fig. 8. Surface profile after grinding, a_p = 0.01 mm

Quite similar BN values (taking into consideration 1 sample) can be found for full contact between milling cutter and the sample. As soon as full contact between milling cutter and the sample is interrupted the progressive increase or decrease of BN values occurs. It is worth to mention that progressive change of BN values in the zone of missing full contact between milling cutter and the sample correspond to the progressive change of the sample height as that illustrated in Fig. 9. On the other hand, evolution of BN in the zone of missing contact is not the same for all VB. As can be seen in Fig. 9, BN decreases in the feed direction in this zone for VB = 0.05 mm whereas more or less visible increase can be found for more developed VB. The main reason can be found in the different thermal load versus VB. As is was reported, machined surface produced by insert of VB = 0.05 mm does not contain WL (considering hardness 45 HRC). As soon as VB attains 0.4 mm thin WL can be found on the machined surface. It is assumed that the differences in BN evolution in the zone of missing contact should be associated just with this aspect.

Fig. 10 illustrates the BN emission and so the corresponding surface state expressed in terms of microstructure and stress state is more homogenous for grinding cycles (when compared to hard milling cycles). BN values are kept low about 200 mV for a_p = 0.01 and 0.02 mm. On the other hand, increased cutting depth in grinding results into surface overheating and remarkable surface burn associated with the intensive surface oxidation appearing as coloured surface, see Fig. 11. Surface overheating and the corresponding surface re-hardening results into initiation of WL. As it is well known WL after grinding decreases BN emission and contributes to the lower BN [3] for high infeed rates as that depicted in Fig. 10. Higher infeed rates during grinding contribute to visible waviness of BN signal due to grinding process instability.

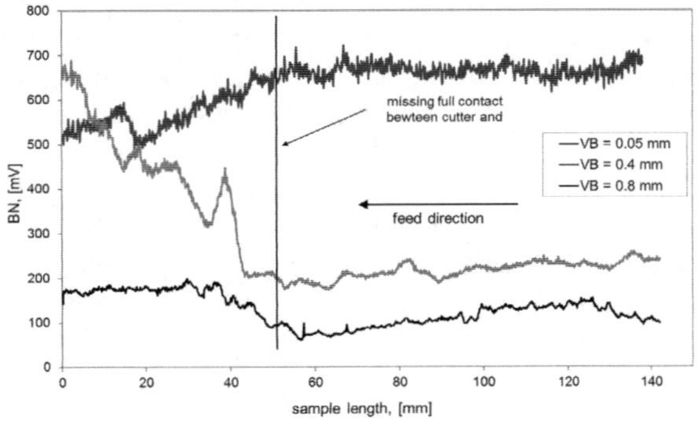

Fig. 9. Dynamic BN record after hard milling

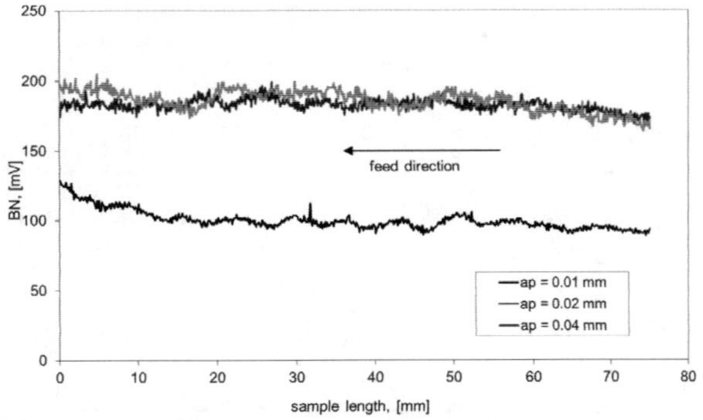

Fig. 10. Dynamic BN record after grinding

It is also worth to mention that character of WL initiated by hard milling and grinding is different. Thin WL after hard milling increases BN due to preferential orientation of matrix and corresponding BWs alignment into the direction of measurement [9]. As soon as thickness of WL after hard milling increases (along with more developed VB) BN decreases due to increasing volume of retained austenite as a non ferromagnetic feature strongly hindering BWs motion.

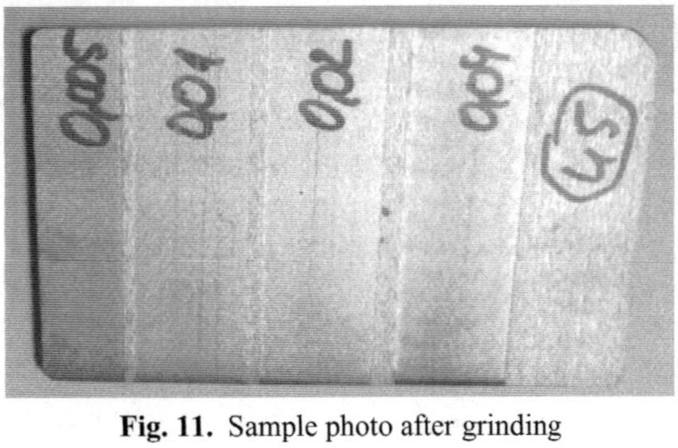

Fig. 11. Sample photo after grinding

4 Conclusions

This study reports about certain disadvantages of hard milling process. Strong non homogeneity of hard milled surface can complicate the consecutive grinding cycles not only from the point of view of the allowances but also initiates complications when non destructive monitoring of the surface after grinding is carried out via BN technique. Industrial experience indicates that the high BN emission after hard milling or turning (when hard turning or milling is performed as a roughing cycle before grinding) can result in certain difficulties when surface after finishing grinding is monitored via BN technique. As it was mentioned, the high BN emission after grinding is associated with surface burn due to over-tempering. Typical thickness of the layer ground off by finishing grinding is about 0.2 ÷ 0.3 mm. However, in production large bearings (or large components) some areas affected by previous hard milling or turning are not fully removed by grinding and retain below the ground surface. Being so, it could be difficult to distinguish between the high values originated from hard milling or over-tempering during grinding. As a consequence of this some parts after grinding could be referred as thermally damaged despite correctly performed finishing grinding.

This study was carried out under the financial support of VEGA agency project n. 1/0121/17 and KEGA project n.008ŽU-4/2018.

References

1. B. Karpucheewski, *Introduction to micro magnetic techniques*, ICBN 01, Hanover. (1998).
2. M. Mičúch, M. et all., *Micro magnetic study of cutting conditions and grinding wheel wear influence on surface integrity*, Manufacturing Technology, Vol. **14**, 66-71 (2014)
3. V. Moorthy, et all.. *Evaluation of heat treatment and deformation induced changes in material properties in gear steels using magnetic Barkhausen noise analysis*, ICBN 03, Tampere, (2001)
4. M. Neslušan, J. Čížek, K. Kolařík, P. Minárik, M. Čilliková, O. Melikhova, *Monitoring of grinding burn via Barkhausen noise emission in case-hardened steel in large-bearing production*, J. Mater. Proc. Technol., Vol. **240**, 104-117. (2017)
5. Y.B. Guo, J. Sahni, *A comparative study of hard turned and cylindrical ground white layers*, Int. J. of Mach. Tool & Manuf., Vol. **44**, 135-145. (2004)
6. D. Brandt, D., *Randzonenbeeinflussung beim Hartdrehen*, PhD. Dissertation, Universität Hanover, (1995)
7. M. Neslušan, M. Rosipal, V. Ochodek, *Analysis of some aspects of surface integrity after grinding and hard turning trough Barkhausen noise*, ICBN 09, Hejnice. (2011).
8. M. Neslušan et. all., *Application of Barkhausen noise for analysis of surface integrity after hard turning*, Manufacturing technology, Vol. **12**, 60-65 (2012)
9. M. Neslušan, *Magnetic anisotropy in hard turned surfaces*, Acta Physica Polonica A, Vol.**6**, 188 – 189. (2014)

Influence of Ca, Sb and heat treatment on AlSi9CuMnNi alloy in frame of their properties from view of machining

Nataša Náprstková[1,*], *Pavel* Kraus[1], *Jan* Sviantek[1], *Tuong* Nguyen Van[2]

[1]Jan Evangelista Purkyne University in Usti nad Labem, Faculty of Mechanical Engineering, Pasteurova 7, 400 96 Usti nad Labem, Czech Republic
[2]Faculty of Mechanical Engineering, Nha Trang University, Nha Trang, Vietnam

Abstract. Treatment of metal alloys is one of the ways to affect the quality and properties of the material At the Faculty of Mechanical Engineering of the Jan Evangelista Purkyně University in Ústí nad Labem, one part of the research consists of searching for the influence of various modifying and inoculating elements on selected aluminium alloys. One of the alloys undergoing the present research is the hardening aluminium alloy AlSi9CuMnNi. The article describes one from these experiments, than AlSi9CuMnNi alloy was modified with various amounts of calcium (0.05, 0.1 and 0.15 wt. % Ca) and 0.2 wt. % Sb. The alloy without modification and with 0.2 wt. % Sb was heat-treated, too. In the experiment were made three castings for each type of alloy. Experimental samples were evaluated form view of final hardness, were machined, too and there were analysed tool wear of using cutting inserts and final ship from machining (turning). Presented experiments are part of the extensive research undertaken at FME at JEPU in Ústí nad Labem.

Keywords: modification, alloy, machining, chip

1 Introduction

For various reasons, the alloy are adjusting (alloyed, inoculated and modified), which of course also applies to aluminum alloys. These reasons are eg. to improve the physical and technological properties. At the Faculty of Mechanical Engineering in this context implement various experiments just with aluminum alloys. One of them is research the effect of different amounts of selected elements to certain aluminum alloys and using of heat treatment, if it is possible, in particular of Al-Si alloys. In this case investigated elements were Ca and Sb and investigated alloy AlSi9CuMnNi. [1-3] These elements improve the morphology of silicon in the Al-Si alloy. Ca also improves the castability of aluminum alloys, but also it has negative effects, e.g. deterioration of the homogeneity of the resulting structure and is therefore considered as the harmful element. Ca and Sb in the role of modifiers can be considered relatively less common in the present area, Na and Sr are more commonly used. [4,5]

* Corresponding author: natasa.naprstkova@ujep.cz

Reviewers: *Karol* Vasilko, *Miroslav* Neslušan

Some types of aluminium alloys can be heat-treated. Heat treatment of aluminium alloys can be in the sense for CSN 42 0056 defined as the process by which a product or product parts in solid state is subjected to one or more cycles of annealing to achieve the desired structure or substructure and properties. [1, 6-9]

Modification of aluminum alloys is process by which the melt is deliberately modified by various elements in order to influence the mechanism of eutectic solidification. Modification changes the morphology and size of crystals of silicon (in the case of Silumin), causing a significant increase in mechanical properties compared to unmodified alloys. Large size silicon of crystals is lower strength properties of Al-Si al-loys. Strength and plastic properties of the modified alloys are therefore higher in comparison to unmodified alloys. Modification is meaningful only for aluminum alloys with a silicon content of more than 5%. Changing the structure may also affect the machinability of the modified alloy. [1-3, 10]

The article describes the one part of the experiments that are carried out at FME JEPU in Usti nad Labem in presented area based on the requirements of companies and deals with the analysis of tool wear after experimental alloys machining (turning) and their final chips.

2 Experiment

For the experiment were made castings from alloy AlSi9CuMnNi for each type of alloy (0.05, 0.1 and 0.15 wt. % Ca and 0.2 wt. % Sb and alloy without modification and with 0.2 wt. % Sb with heat-treated, too). AlSi9CuMnNi belongs to the Al - Si - Cu group, which can be heat treated. In these alloys, the mechanical properties after the heat treatment vary greatly depending on the copper content. These alloys also have good foundry and mechanical properties, but the main disadvantage is low corrosion resistance. They are used for thin-walled castings in engine manufacturing, such as blocks, crankcase, air-cooled cylinder heads or carburettor parts. These alloys are heat-treatable (hardenable) and, in this state, they can achieve maximum strength properties. [1-3, 7, 11]

The alloy AlSi9CuMnNi was made at FME from individual components. There was not applied master alloy. Seven casting groups were made, which two were unalloyed (one was heat-threated), three were alloyed by varied amounts of calcium (0.05, 0.1 and 0.15 wt. %), and two groups were alloyed by 0.02 wt % Sb (one was heat-threated).

As a kind of heat treatment was carried out solution annealing leading to an increase in material hardness and uniformity of the chemical composition. In the first part, the hardening of the material supercritical cooling, and in the second part of the artificial aging. The entire course of the heat treatment process took approximately 16 hours. [1, 6-9]

Tab. 1 Characteristics of cutting plate (insert) DCMT 070202 E-UR

Dimensions of insert [mm]					f [mm·rev^{-1}]		a_p [mm]	
l	d	d_l	s	r_z	f_{min}	f_{max}	a_{pmin}	a_{pmax}
7,8	6,350	2,8	2,38	0,2	0,05	0,12	0,2	1,0

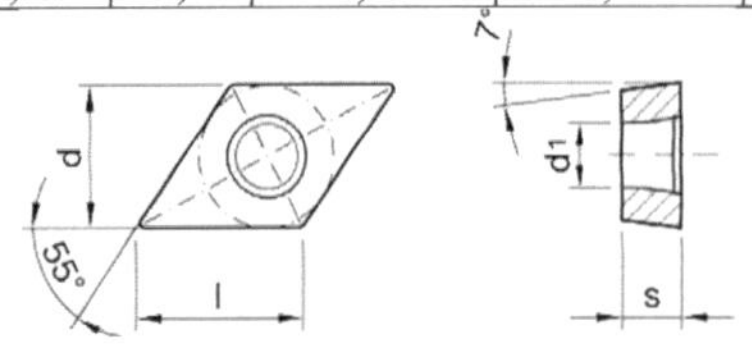

Experimental samples were machined on a lathe Emco Mat - 14 S (at the FME available). Set cutting conditions were based primarily from the type of machine and tool (tab. 1) and based on the material to be machined and used machine and tool was set depth of cut a_p=1 mm and feed per revolution f = 0.12 mm. On the base of possibilities and calculations the cutting speed for actual machining v_c was adapted to used lathe for resulting value $v_c = 200\ \mathrm{m \cdot min^{-1}}$ (maximum possible cutting conditions). [10, 12]

The wear of cutting inserts after machining was measured in the back and were examined the following parameters:

- wear in the peak VBc,
- back wear VB,
- maximum back wear VBmax.

Each area for measuring of tool wear shown in Fig. 1 (only for ilustration). Evaluation was carried out according to the wear characte-ristics of the standard ISO 3685. Evaluation carried out according to the wear characteristics of the standard ISO 3685. On Fig 2 is example of chip lenght measuring. For measurement of tool wear and chip length was used the software Quick PHOTO CAMERA 3.2. [9, 13-15]

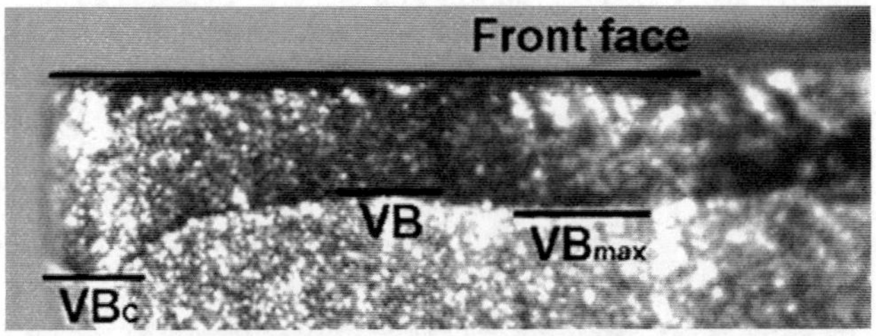

Fig. 1. Principle of measurement of wear values of cutting plate (insert)

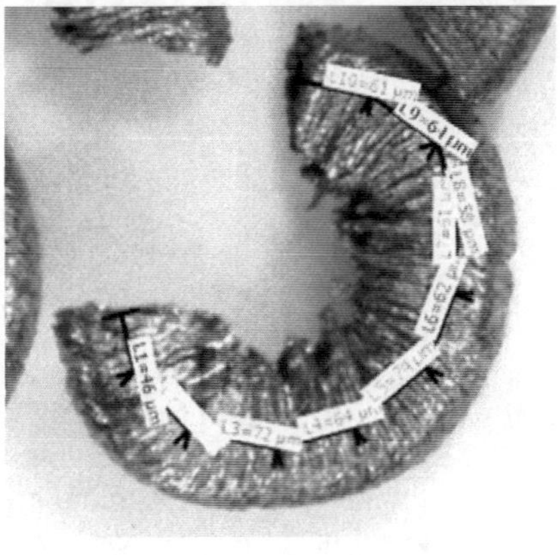

Fig. 2. Principle of chip leght measurement

The Brinell hardness of samples was also measured according to standard EN ISO 6506-1 on the hardness tester Ernst AT 250X. The hardness measurement was performed on turned samples by injection a ball with 2.5 mm diameter at a load 612.916N for 10 seconds. [15, 16]

3 Results and discussion

Other step was measuring of cutting inserts tool wear after cutting of individual alloys. Generally speaking, that the each all plates wear was very small, by visual observation difficult to see. The main reason was that the machined samples were from aluminium alloy, and the time during which the plate was in the work was about 25 minutes. Even though it was possible to identify and measure the tool wear using a microscope in order to identify differences between the different types of alloys. Generally true that on the plates formed up edge (built-up), see Fig. 3.

Fig. 3. Up edge (built-up) on the tool after machining (without modification and HT)

Graph on the Fig. 4. shows the tool wear VB for each group of samples. From this is can be seen that the greatest tool wear at this parameter is for the unmodified alloy. For this type of alloy after heat-threated is tool wear smaller and it is smaller, than for alloy modified by Ca. Between unmodified and Ca modified alloys is for this parameter very small different in favor modification of Ca (the worst is for 0.15 wt. % Ca). Better results for this parameter show Sb modification and heat treatment.

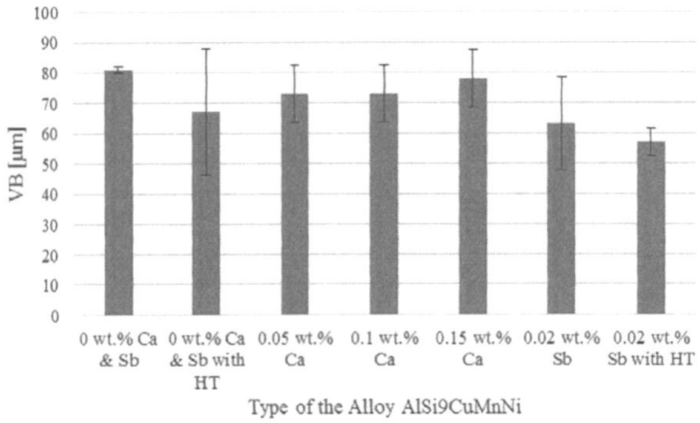

Fig. 4 Tool wear parameter VB (after 25 min of machining)

Graph on the Fig. 5 shows the tool wear VBc for each group of samples. From this is can be seen that the greatest tool wear at this parameter is for the unmodified alloy, like for parameter VB. For this type of alloy after heat-threated is tool wear smaller and it is smaller, than for alloy modified by Ca, like for VB too. Between unmodified and Ca modified alloys is for this parameter very small different in favor modification of Ca, too. Better results for this parameter show Sb modification and heat treatment. For all types of alloy the same course as for parameter VB could be observed.

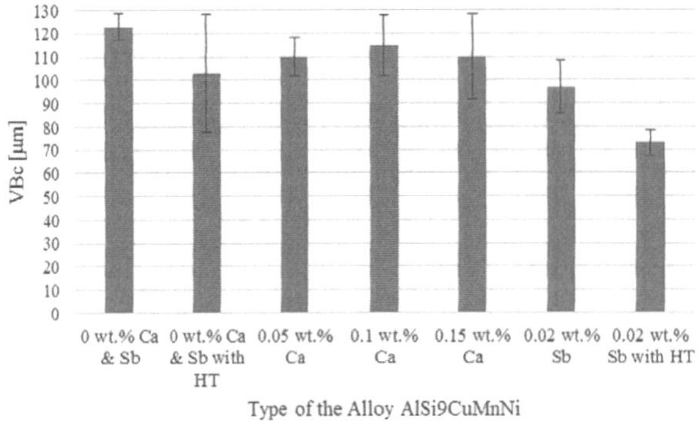

Fig. 5 Tool wear parameter VBc (after 25 min of machining)

Graph on the Fig. 6. shows the tool wear VBmax for each group of samples. The greatest tool wear at this parameter is for the unmodified alloy, too, like for VB and VBc. In this case the tool wear is the smallest for 0.1 wt. % Ca and next for 0.5 wt. % Sb with heat-treatment. There the difference is very small.

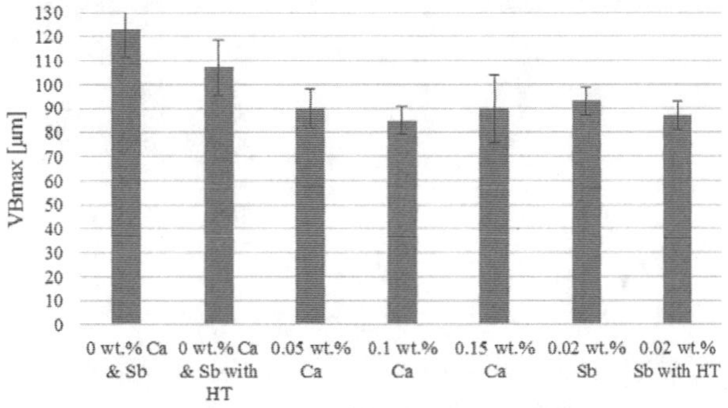

Fig. 6 Tool wear parameter VBmax after 25 min of machining)

From the point of view of wear, where the VBmax criterion is more often used than the VBc criterion for the wear evaluation of tools, it is possible to add that, in terms of wear, better results show (although there is a relatively small difference) a modification of 0.1 wt. % Ca. Practically the best results for all adjustments due to the evaluated wear criteria of the tool show modification 0.2.wt. % Sb with using heat treatment.

Another evaluation was evaluation in terms of length and shape of the obtained chips. The chips collected after machining of experimental castings had spiral trace for every type of cast alloy. For modified types of alloy chips were rather long shape in comparison with unmodified alloy. The heat-treatment has positive effect on the chip length. Wherein the effect of calcium on the chip shape was observed primarily a larger quantity of chips more twisted into a spiral when this amount is increased with higher calcium content. The Fig. 7. shows a graph comparing the average chip size for each type of the experimental alloy. It is evident that the heat-treated alloys had the shortest chip. The shorter chips also had modification 0.2 wt. % Sb. The calcium modification had a negative effect on the chips length. The longest chip was for 0.1 wt. % Ca modification.

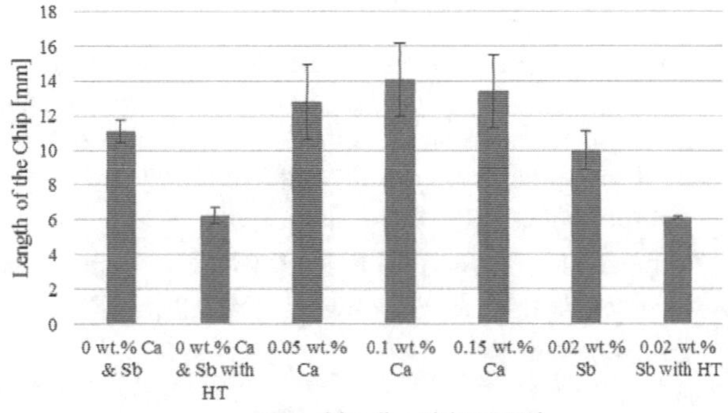

Fig. 7 Comparison of chip sizes for each for individual alloys

Next was the hardness measurement of each alloy type. The measured values are summarized in the graph in Fig. 8. The highest values had heat treated alloys. The modification of Ca had the opposite effect (for each amount of Ca in alloy).

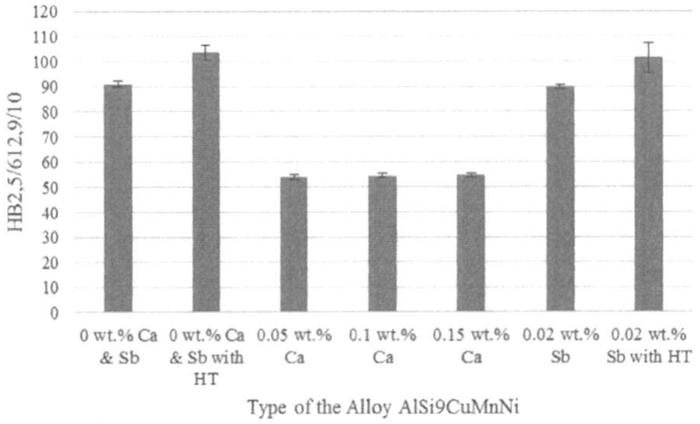

Fig. 8 Comparison of hardness for individual alloys

Further, when comparing the hardness for each type of alloy, it appears that the unmodified alloy with heat-treatment has the highest hardness. The second is modified alloy by 0.2 wt. % Sb with heat-treatment. The Sb modification itself also had a positive effect on hardness. For modification of calcium to AlSi9CuMnNi alloy is possible to that the overall hardness was reduced.

4 Conclusion

In the context of the presented experiment the alloy AlSi9CuMnNi was made at FME using pure metals and were prepared samples with defined amount of Ca and Sb and some types of presented alloy were heat-treated. Castings were analyzed in terms of their influence to machining (tool wear, chip length).

Subsequently, the alloys were machined and the tool wear of cutting inserts and the shape and length of the chip were evaluated. The hardness of the experimental alloys was also measured.

The tool wear measurement, when the VB, VBc and VBmax parameters were measured, showed that the largest tool wear was reported during machining of the unmodified alloy, which was likely to be caused by large unmodified Si boards. Modification Ca showed a little positive effect on insert tool wear, however, the difference between individual alloys with varying amounts of Ca was not significant. Somewhat better results are Sb modifications and heat treatment.

The chips collected after machining were in terms of their shape tended to be longer with a trace in a spiral shape, with the effect of calcium at chip shape were observed mainly higher incidence spiral chip with increasing calcium content. The impact of Ca on the chip length had a negative impact because the shortest chip had an unmodified alloy. Somewhat better results are Sb modifications and heat treatment, like for tool wear.

The Ca modification had a negative impact on alloy hardness. since the unmodified alloy had a higher hardness. As expected, heat treatment had a positive effect. Good effect had modification by Sb, too.

The presented results are part of a larger research, which is realized at FME JEPU in Ústí nad Labem.

Authors are grateful for the support of grant No. CZ.1.05/4.1.00/11.0260 EDIMARE and SGS FME JEPU in Ústí nad Labem for 2018.

References

1. S. Michna, I. Lukac, P. Louda, V. Ocenasek, H. Schneider, J. Drapala, R. Koreny, A. Miškufova, et al. *Aluminium Materials and Technologies from A to Z.* Adin, Prešov (2007)
2. M. Michalski, F. Romankiewicz. *AlSi21CuNi silumin modification with phosphor and strontium micro additions E3S.* Web of Conferences, 19, Article number 03026 (2017)
3. J. Formanek,, L. Kucerova, M. Tochylin. *The Hazards of Defects and Poor Quality of Aluminium Materials.* Proceedings of 24th Internatioal Conference on Metallurgy and Materials Metal, 1552-1557 (2015)
4. I. A. Petrov, D. V. Berezhnoi, A. P. Ryakhovskii, V. S. Moiseev. Effect of modification on the solidification of Al–Si aluminum alloys. Russian Metallurgy (Metally), Springer Nature 2017(**3**), 184-187 (2017)
5. N. Naprstkova, J. Cais, M. Ingaldi. *Modification of AlSi9CuMnNi Alloy by Antimony and Heat Treatment and their Influence on Tool Wear after Turning.* Manufacturing Technology, FME JEPU 16(**1**), 209-214 (2016)
6. N. Naprstkova, J. Cais, J. Sviantek, H. Cierna. *Change of selected resulting properties of AlSi9CuMnNi alloy by heat treatment and antimony.* Proceedings of 15th international scientific conference Engineering for Rural Development, Latvia, 1133-1138 (2016)
7. V. Weiss, E. Střihavková. *Influence of the homogenization annealing on microstructure and mechanical properties of AlZn5,5Mg2,5Cu1,5 alloy.* Manufacturing Technology, 12(**12**), 297-302 (2012)
8. P. Wieroński, J. Pezda, L. Ponikwia. Effect of Heat Treatment on Machining Properties of the AlSi9Cu3(Fe) Alloy. Archives of Foundry Engineering, Walter de Gruyter 16(**3**), 137-140 (2016)
9. N. Naprstkova, D. Kalincova. *Influence of Additional Chemical Components on Machining Properties of Selected Aluminium - Silicon Alloy.* Proceedings of 14th international scientific conference Engineering for Rural Development, Latvia, 766-771, (2015)
10. N. Naprstkova, S. Kusmierczak, J. Cais. *Modification of AlSi7Mg0.3 alloy by strontium.* Manufacturing Technology, FME UJEP, 13(**3**), 373-380 (2013)
11. J. Petru, J. Schiffner, T. Zlamal, M. Sadilek, D. Stancekova. *Investigations of Cutting Tool Wear While Machining Inconel 718.* Manufacturing Technology, 15(3), 396-403 (2015)
12. N. Naprstkova, S. Kusmierczak, J. Cais. *Modification of AlSi7Mg0.3 alloy by strontium.* Manufacturing Technology, FME UJEP, 13(**3**), 373-380 (2013)
13. Zetek, M., Bakša, T., Schorník, V., Zetková, I. *Monitoring and Evaluating Cutting Tool Wear using a IFM G4 Microscope.* Manufacturing Technology, FME UJEP, 16(**5**), 1211-1216 (2016)
14. J. Petru, T. Zlamal, R. Cep, D. Stancekova, M. Pagac, O. Vorel. *Mechanisms of cutting insert wear and their influence on cutting ability of the tool during machining of special alloys.* Proceedings METMG 2015, Canada, 36-40, (2015)
15. S. Kusmierczak. *Usage of Technical Equipment in Teaching Technical Subjects.* Proceedings of 14th international scientific conference Engineering for Rural Development, Latvia, 748-752 (2015)
16. I. Novakova, J. Stverak, J. Moravec. *Hardness of the High Pressure Die Castings from Alloy AlSi9Cu3 in dependence on the Subsequent Processing Technology.* Manufacturing Technology. FME JEPU, 15(**4**), 647-652 (2015)

Optimization of machining processes preparation with usage of Strategy Manager

Tomáš Dodok[1*], *Nadežda* Čuboňová[1], *Dorota* Wiecek[2]

[1]University of Žilina, Faculty of Mechanical Engineering, Department of Automation and Production systems, Univerzitná 1, 010 26 Žilina, Slovak Republic
[2]University of Bielsko-Biala, Faculty of Mechanical Engineering and Computer Science, Department of Industrial Engineering, Willowa 2, 43-309 Bielsko-Biala, Poland

Abstract. At present, the optimization has an important role in machining processes preparation of NC programs. The article deals with the possibilities of automate creation of NC programs by using the NC strategies in CAD/CAM system Edgecam 2017 R2. The main emphasis is placed on the suitability of use, the limitation of the creation and application of the NC strategies. The experiments deal with the comparison of effectiveness of the NC strategies applied to the components with different geometric features. The experimental components are divided into two groups. First group is represented by components with 2D features. Second group is based on components with 3D features. The motivation for this type of research is to increasing efficiency of the machining processes creation by automation tools in CAD/CAM systems.

Keywords: Strategy Manager, Optimization, Edgecam 2017, NC strategies.

1 Introduction

CAD/CAM systems allow to shorten time of machining preparation including technological documentation,testing and manufacturing optimalization on the CNC machine. The main advantages of deploying CAD/CAM systems include the ability to create NC programs for the machining of complex components that include simple featres as well as components that contain complex features or surfaces.

The necessary condition for increasing efficiency is to optimize the machining process. Optimization is a complicated multi-stage process that uses multiple methods [1-2]. From the point of view of the machining process, it is possible to optimize the choice of workpiece, machines, tools, operations, technological conditions, tool paths and so on [3-4].

In addition to optimizing the machnining process, there are also possibilyties to optimize processes within pre-production stages. One of these options is to optimize the NC program creation process so that repetetive tasks performed by the programmer can be removed when new machining processes are made. Therefore, some CAD/CAM systems, such as PLX system NX 10, CAD/CAM system Edgecam 2016 and Mastercam 2017, make it possible to create NC strategies that can be used to optimize the preparation of NC programs.

* Corresponding author: tomas.dodok@fstroj.uniza.sk

Reviewers: *Stanisla* Legutko, *Peter* Pokorný

2 The NC strategies

NC strategy is defined as algorithm with conditions that define features on which the NC strategy will be used. The NC strategy contains also machining cycles [5-6]. Those machining cycles can be used for roughing and finishing operations face milling, thread making and so on. The NC strategy also contains conditions for choice of tool base e.g. for machining holes [7-8]. The NC strategies can be divided into three groups, based on the way how they are used:

a) ***Manual NC strategy*** - The NC strategy is applied manually (it does not have the necessary conditions for its automatic application). For the proper functioning of the NC strategy, input parameters are required from the user.
b) ***Semiautomatic NC strategy*** - The NC strategy is applied automatically to the feature for which it was determined, but requires input data from user.
c) ***Automatic NC strategy*** - the NC strategy works independently, based on entry conditions. It does not require user input and can be applied automatically to the required features.

The process of creation of NC strategies consist of several steps in Strategy Manager (Figure 1):

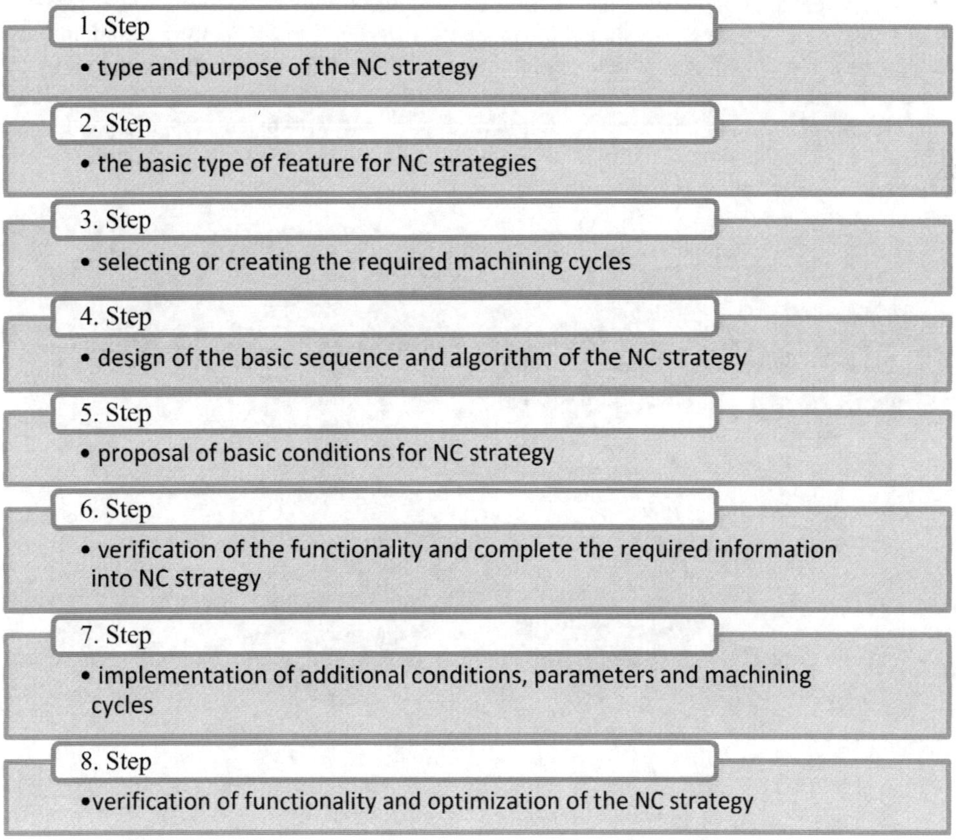

Fig. 1. The methodology of NC strategy creation in CAD / CAM system Edgecam 2017 R2

The first step, in process of creating the NC strategy, is to design the primary purpose of the NC strategy. The purpose of the NC strategy can be to machining a specific type of feature e.g. drilling holes. In the second step, the user determines for what type of feature the NC strategy will be designed for [9]. The user can design the NC strategy for one particular type of feature or for several different types of features. In the third step and fourth, the user must define the necessary machining cycles and create the basic algorithm of the NC strategy. The fifth step is to define the basic conditions. The first conditions will be focused on the choice of feature or group of features for which the NC strategy will be applied. In this step is also necessary to determine the conditions for selecting the appropriate type of tools, dimensions of selected tools and cutting parameters. In the sixth step, it is necessary to verify the basic functionality of algorithm and conditions. In the seventh step, additional conditions and parameters can be implemented [9]. These conditions and parameters allow specifying additional parameters e.g. other subtypes of feature [10-12]. In the last eighth phase, the user performs a check on the functionality of the NC strategy.

3 Application

Several types of components were selected to verify the functionality of the experimental NC strategies and their impact on the machining process creation (Figure 2, Figure 4 and Figure 6). Two parameters were selected to compare the effectiveness of the NC strategies. The first parameter to be compared is the time necessary to prepare machining process with and without implementation of NC strategies (Figure 3, Figure 5, Figure 7). The second parameter is machine time (Figure 8). Designed components, marked A to F (Figure 2), contain only 2D features. The dimensions of these componets are determined by the maximum workpiece that can be machined on the EMCO MILL 105, dimensions of workpiece are 100x50x30 mm. For experimental components K to N (Figure 6), workpiece was designed with dimensions of 500x400x110 mm. For these types of components, it is possible to use experimental NC strategies without modification.

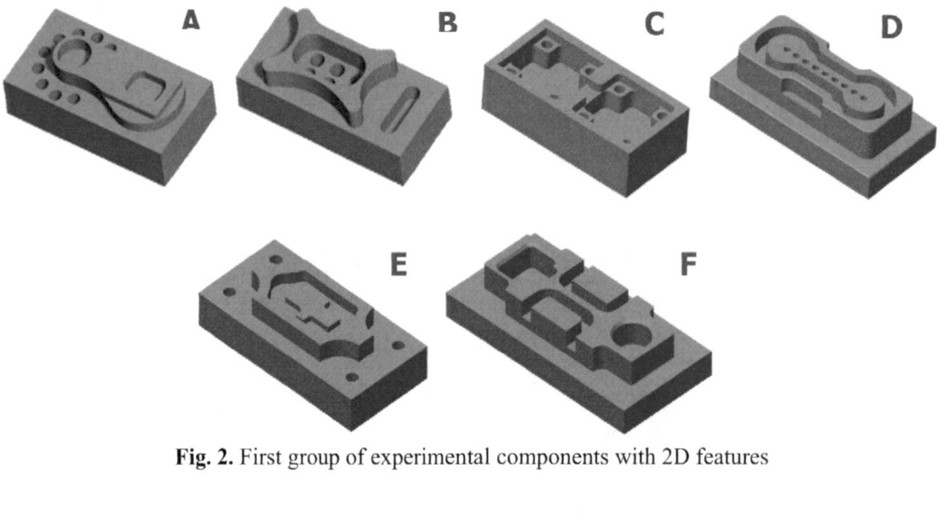

Fig. 2. First group of experimental components with 2D features

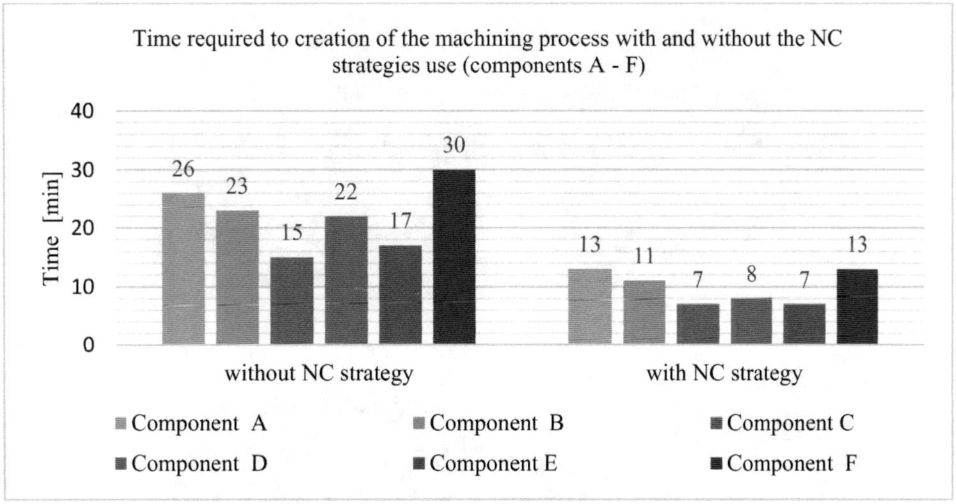

Fig. 3. Time required to create complete machining sequence with and without the NC strategies use for components A - F

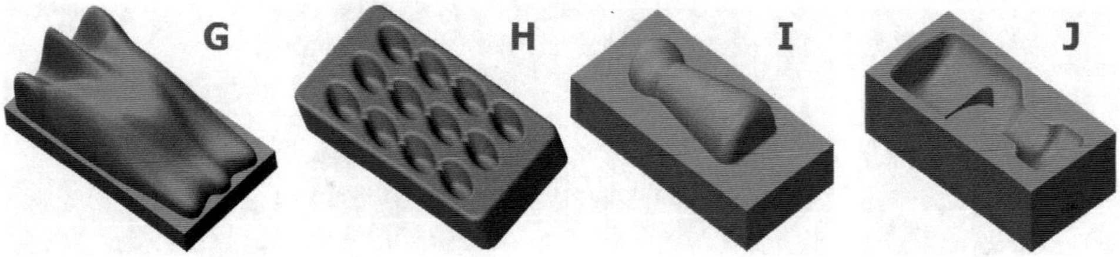

Fig. 4. Second group of experimental components with 3D features and surfaces

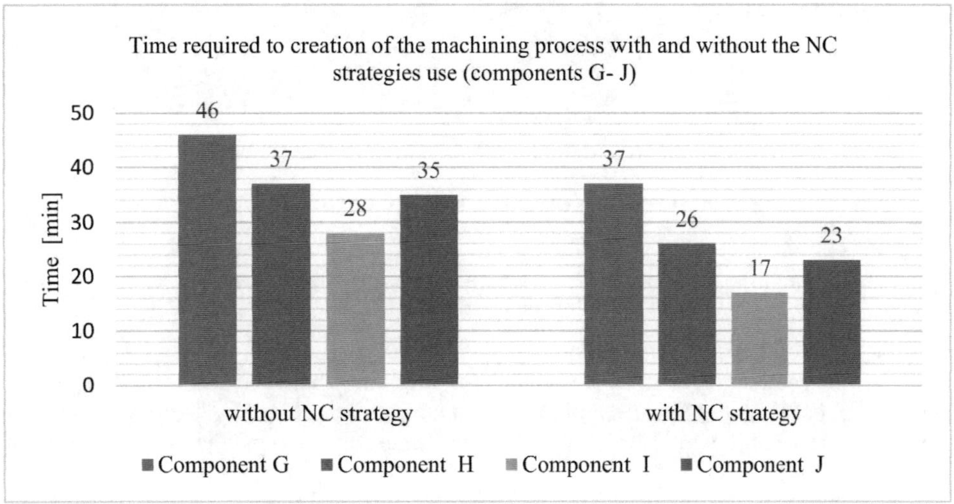

Fig. 5. Time required to create complete machining sequence with and without the NC strategies use for components G - J

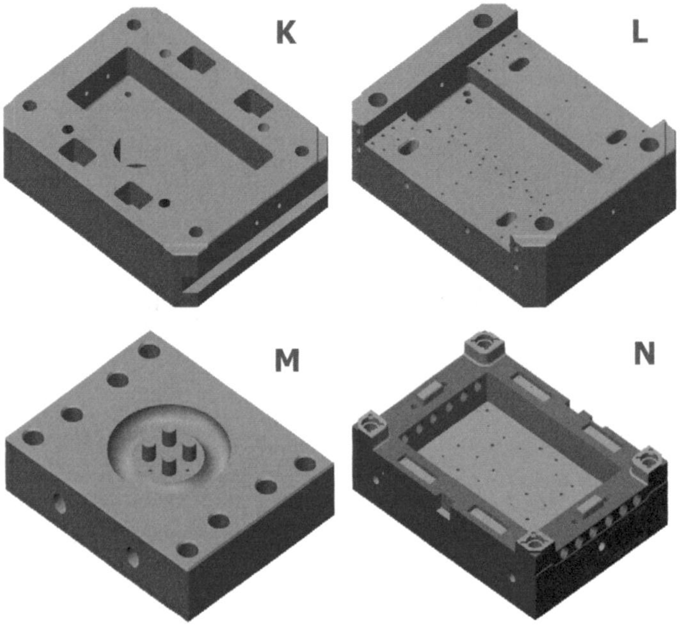

Fig. 6. Third group of experimental components with complex features

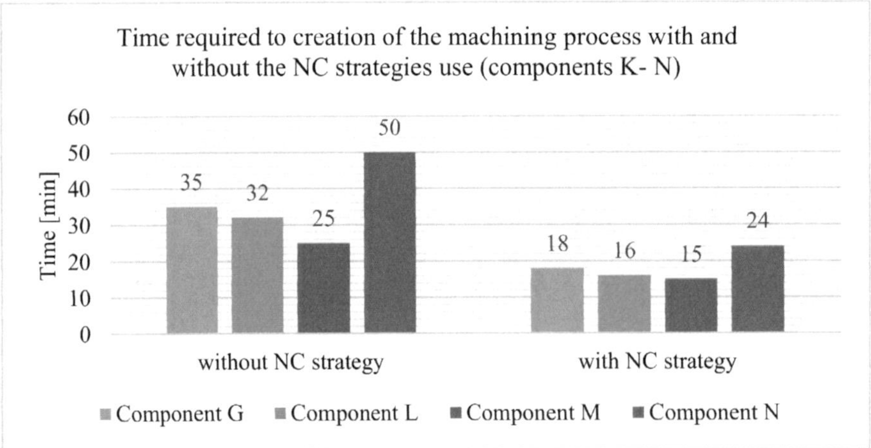

Fig. 7. Time required to create complete machining sequence with and without the NC strategies use for components K - N

The results of the experiments showed that the differences between the machining process created without the NC strategies and the machining process created by using NC strategies were minimal according to comparison of machine time generated in the CAD/CAM system Edgecam 2017 R2.

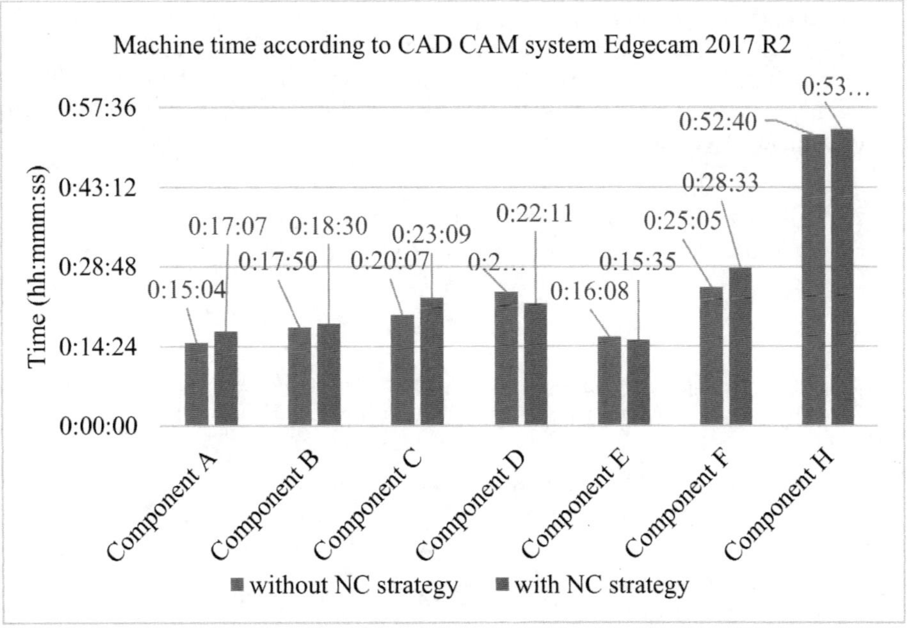

Fig. 8. Machine time according CAD/CAM system Egdecam 2017 R2

4 Conclusion

After getting acquainted with possibilities of preparation of the NC strategies in the CAD/CAM system Edgecam 2017 R2, it can be formulate the following findings:

1. The Strategy Manager constitutes a comprehensive solution for creating of NC strategies that can be used by less experienced users as well as advanced users of the CAD/CAM system Edgecam 2017 R2.
2. The process of creating an NC strategy requires basic knowledge of work with the Strategy Manager module and knowledge of the creation of the algorithms and conditions that will be used in NC strategies.
3. Before creating the NC strategy, it is necessary to perform an analysis of machined parts to determine the appropriate components or features for which the NC strategies will be created. The analysis result can be used to determine suitable technological parameters of machining cycles and conditions for NC strategies.

Base on the achieved results by the solving of this task, the following conclusion can be formulated:

a) ***Recognition errors of 3D features*** - CAD/CAM system Edgecam 2017 R2 is not capable of correctly recognizing 3D features in certain cases. These are mainly features that are composed of several sub-features or multiple 3D surfaces. In this case, the user has several options to create the 3D feature manually. But these user-defined-features don't have same properties as automatically recognised features. What causes further problems addition problem for creation of the NC strategies and conditions.

b) ***Missing shapes for slots, fillets and chamfers.*** The most significant problem is recognition of fillet or chamfer. CAD/CAM system Edgecam 2017 R2 recognizes these features as 2D Step or 3D Steps. The problem is caused by the absence of rules for recognition these geometric features. Same problem can be notice when recognizing 2D Slots.

c) ***2D features.*** 2D features represent the most appropriate group for which NC strategies can be created. They contain enough properties to determine the conditions for choosing tools and machining cycles.

References

1. N. Čuboňová, M. Císar, *Design of camera mount and its application for monitoring machining process.* Advances in Science and Technology Research Journal, **9**(26), 2015, 34-40.
2. B. Babic, N. Nesic, Z. Miljkovic, *A review of automated feature recongnition with rule-based pattern recognition.* Computers in Industry, **59**(4), 321–337, 2008.
3. G. Krolczyk, S. Legutko, A. Stoic, *Influence of the cutting parameters and conditions onto surface hardness of Duplex Stainless Steel.* Technicky Vjesnik – Technical Gazette, **20**(6), 1077-1080, 2013.
4. P. Twardowski, S. Legutko, G. M. Krolczyk, *Investigation of wear and tool life of coated carbide and cubic boron nitride cutting tools in high speed milling*. Advances in Mechanical Engineering, **7**(6), article number: 1687814015690216, 2015.
5. V. Bulej, J. Uríček, V. Poppeová, R. Zahoranský, M. Rupiková, *Study of the workspace of hybrid mechanism Trivariant.* Applied Mechanics and Materials. **436** 366-373, 2013.
6. T. Dodok, N. Čuboňová, I. Kuric, *Workshop Programming as a Part of Technological Preparation of Production.* Advances in Science and Technology Research Journal, 11(1), 2017, 111-116.
7. D. Kumicakova, A. Rengevic, M. Cisar, V. Tlach, *Utilisation of kinetic sensors for the design of a human-robot collaborative workcell.* Advances in Science and Technology Research Journal, **11** 4), 270-278, 2017.
8. I. Kuric, V. Bulej, M. Sága, P. Pokorný, *Development of simulation software for mobile robot path planning within multilayer map system based on metric and topological maps*. International Journal of Advanced Robotic Systems. **10**(6), pp. 14, 2017.
9. N. Náprstková, *Students connecting to production problems resolutions in CAD/CAM area.* 9th International Scientific Conference - Engineering for Rural Development 2010, Jelgava, Litvia, 310-314, 2010.
10. A. Rudawska, M. Reszka, T. Warda, I. Miturska, J. Szabelski, D. Stančeková, A. Skoczylas, *Milling as a method of surface treatment in adhesive bonding.* Journal of Adhesion Science and Technology, **30**(23), 2619-2636, 2016.
11. J. Stanček, V. Bulej, *Design of driving system for scissor lifting mechanism*. Academic Journal of Manufacturing Engineering. **13**(4), 38-43, 2015.
12. M. Drbúl, D. Stančeková, O. Babík, J. Holubjak, I. Görögová, D. Varga, *Simulation possibilities of 3D measuring in progressive control of production*. Manufacturing Technology, **16**(1), 53-58, 2016.

The effect of chip former on power loading of re-cessing tool

Jiří Kratochvíl [1,*], *Marek* Sadílek [1], *Jozef* Mrazik [2]

[1]VSB-TUO, Fakulty of Mechanikal Engieniring 17. listopadu 15/2172, 708 33 Ostrava-Poruba, CZ
[2]University of Žilina, Faculty of Mechanical Engineering, Univerzitna 1, 010 26, Žilina, SK

Abstract. The article handles the influence of a chip former on cutting force and its components. This is allowed by using three grooving inserts from ISCAR company, which differ only with the chip former. These inserts have the same base material, coating and they are designed for the same proportion of recessed slot. The conditions of experiment were used in intersection of recommended cutting conditions of the 3 inserts. During the experiment were observed the components of cutting force at changing feed and cutting speed. The results were compiled in tables and graphs were created for each of the used inserts. The particular force loads of the inserts were compared in the end and conclusions thereof were made.

Keywords: Recessing, Forces at cutting, Chip formers

1 Introduction

The cutting production technologies are still being used in factories, one of them is turning. Turning operation serves for production of rotation parts, where one of the operation sectors is the recessing. This article concludes the comparison of recessing inserts from ISCAR CR s.r.o. company. These inserts have the same base material, coating and they are designed for the same proportion of recessed slot. The comparison was done based on force loading at the defined recess.

2 Experiment conditions

For introducing the experiment it is necessary to get familiar with entry conditions, which can influence the experiment. For this experiment were used the following conditions:

- Machine - DMG MoriSeiki NLX 2500/700.
- Dynamometr Kistler 9129AA.
- Material at experiment 1.6722 - 16NiCrMo16-5.
- Tool - Turning recession cutter MAHR 25 with adapter HGPAD 4R-T12 and insert type DGN 4003J, DGN 4003C a GRIP 4004Y.

[*] Corresponding author: Jiri.kratochvil@vsb.cz

Reviewers: *Peter* Demeč, *Anna* Rudawska

2.1 Material

For this experiment material 1.6722 - 16NiCrMo16-5 was chosen. Rounded bar with diameter 150 mm and length 250 mm was used.

It concerns alloy cast construction steel with significantly high yield strength and tensile strength. Among further characteristics of the material belong sufficient moulding capacity and tenacity [1]. Numeric expression of mechanical features is resumed in Table 1.

Table 1. Mechanical properties [1]

Tensile strength Rm (MPa)	Yield strength Re (MPa)	Elongation A (%)
1450	1150	12

Chemical composition of the steel can be seen in Table 2.

Table 2. Chemical composition (Weight %) [1]

C (%)	Si (%)	Mn (%)	P (%)	S (%)	Cr (%)	Ni (%)	Mo (%)
0.12-0.18	≤ 0.50	≤ 0.50	≤ 0.030	≤ 0.025	1.00-1.40	4.00-4.50	0.15-0.45

Typical application of alloy structural steels is in the field of ship, vehicle, airplane, guided missile, weapons, railway, bridges, pressure vessel, machine tools, mechanical components with a bigger sectional size and so on.

2.2 Tool

The turning recessing tool MAHR 25 (Fig. 1.a) with adapter HGPAD 4R-T12 (Fig. 1.b) was chosen for the experiment.

a) b)

Fig. 1. a) Turning recessing tool MAHR 25 [2], b) Adapter HGPAD 4R-T12 [2]

The adapter is provided with inserts. During the experiment were used 3 types of inserts, the inserts type DGN 4003J, DGN 4003C and GRIP 4004Y (Fig. 2). Particular dimensions and recommended cutting conditions were summarized in Table 3. All compared inserts had the same base material IC 808 and the same coating TiAlN/AlTiN+TiN applied by PVD method [2-4].

Table 3. Dimensions and recommended cutting conditions of the inserts [2]

Fig. 2	Typ	CW	RE	INSL	f_{zmin}	f_{zmax}	v_{cmin}	v_{cmax}
		mm	mm	mm	mm	mm	m.min^{-1}	m.min^{-1}
b)	**DGN 4003J**	4	0.3	18.9	0.05	0.18	80	180
c)	**DGN 4003C**	4	0.3	18.8	0.1	0.3	80	180
d)	**GRIP 4004Y**	4	0.4	19	0.09	0.24	80	180

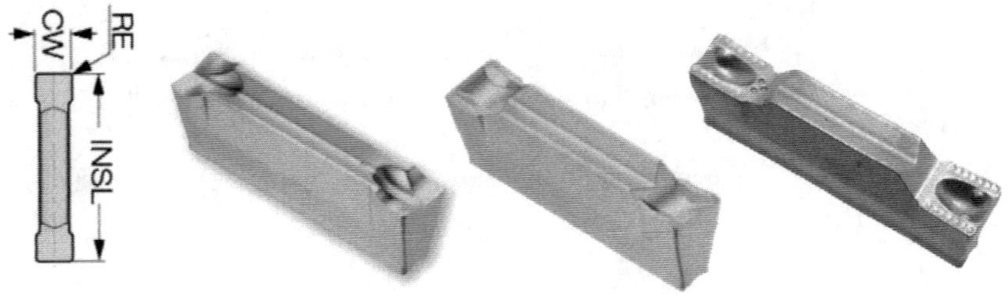

a) b) c) d)

Fig. 2. a) Scheme with dimensions, Picture of insert type: b) DGN 4003J, c) DGN 4003C, d) GRIP 4004Y [2]

2.3 Selection of cutting conditions

Within parameter selection there were three cutting speeds chosen, minimum v_{c1} = 80 m.min^{-1}, maximum v_{c3} = 180 m.min^{-1} and medium value v_{c2} = 130 m.min^{-1} of the speed. At feed step selection three different feeds were chosen, which approximately come out of intersection of feeds recommended by producer. From this intersection was again selected minimum f_1 = 0.1 mm, maximum f_3 = 0.2 mm and medium value f_2 = 0.15 mm of the feed.

3 Course of experiment

During the experiment all three values of feed were used for each value of the selected cutting speed.

For each combination of cutting conditions there were 3 recessing of width 4 mm and depth 2 mm turned, always with the new cutting edge for avoidance of results interference by wearing off the tool. Between particular recessing were gaps 2 mm.

When turning the recessing the components of cutting force were measured with usage of dynamometer Kistler 9129AA [5-7]. The minimum, maximum and medium values of individual cutting force components were evaluated at each recessing, with help of DynoWare program provided with dynamometer (the range of evaluated data was determined by two vertical lines, which is visible on Fig. 3. For further processing the average value of particular cutting force components at each recessing was figured. As there were three recessing turned for each combination of cutting conditions, it was necessary to eventually calculate the average of values from the three recessing. The passive component of cutting force proved as minimal during the measurement and therefore it was excluded for further processing, see Fig. 3.

In Fig. 3 it is possible to see the progress of force components for the set of three recessing, where green colour marks the cutting component of force F_c, blue colour means feed component of force F_f and black is passive component of force F_p. By means of the formula (1) was then calculated the final cutting force F.

$$F = \sqrt{F_c^2 + F_f^2} \tag{1}$$

F - final cutting force (N),
F_c - cutting component of force (N),
F_f - feed step component of force (N).

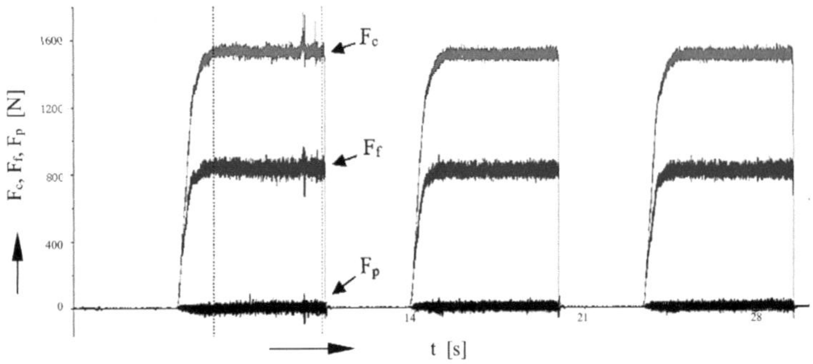

Fig. 3. Example of measured data record (DGN 4003J, v_c = 130 m.min^{-1}, f = 0.15 mm)

4 Evaluation of experiment

In this chapter can be found the factual measured and calculated values matching with the particular cutting conditions for single inserts.

4.1 Insert DGN 4003J

This insert is the first choice for soft material, parting the tubes, small diameters and thin-walled parts. It is generally applied on low carbon, alloy and austenitic stainless steel by using low-to-medium feeds [2].

The values measured for insert DGN 4003J are compiled in the Table 4. From the calculated average values was subsequently created a diagram of dependency of particular components and total cutting force on the used cutting conditions (Fig. 4).

Table 4. Measured and calculated values of individual cutting force components for insert DGN 4003J

v_c	f	F_{f1}	F_{f2}	F_{f3}	F_{c1}	F_{c2}	F_{c3}	F_f	F_c	F
m.min^{-1}	mm	N	N	N	N	N	N	N	N	N
80	**0.1**	858.3	895.7	890.7	1292	1322	1316	882	1310	**1579**
80	**0.15**	1015	1007	1018	1683	1662	1667	1013	1671	**1954**
80	**0.2**	1093	1092	1097	1845	1850	1852	1094	1849	**2148**
130	**0.1**	742	734	731	1148	1141	1139	736	1143	**1359**
130	**0.15**	835	820	818	1533	1511	1508	824	1517	**1727**
130	**0.2**	905	903	900	1881	1871	1866	903	1873	**2079**
180	**0.1**	632	629	627	1068	1064	1060	629	1064	**1236**
180	**0.15**	758	758	755	1445	1445	1441	757	1444	**1630**
180	**0.2**	851	838	837	1815	1795	1795	842	1802	**1989**

In the diagram (Fig. 4) is apparent a huge increase of force components with the increasing feed, especially the cutting component of cutting force. With the increasing cutting speed however both followed cutting force components slightly drop down.

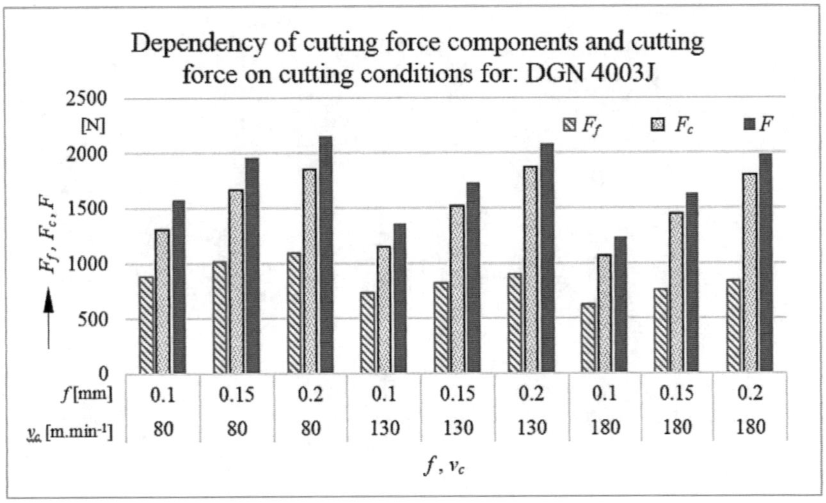

Fig. 4. Dependency of cutting force components and cutting force on cutting conditions for: DGN 4003J

4.2 Insert DGN 4003C

This insert is widely used for grooving and parting bars, hard materials and tough applications. A positive rake and single cavity with a negative land and shoulders make the cutting edge extremely resistant. The insert is designed for medium-to-high feed.

The measured and calculated values of individual cutting force components for insert DGN 4003C are displayed in Table 5 and in graphical view in Fig. 5.

Table 5. Measured and calculated values of individual cutting force components for insert DGN 4003C

v_c	f	F_{f1}	F_{f2}	F_{f3}	F_{c1}	F_{c2}	F_{c3}	F_f	F_c	F
m.min^{-1}	mm	N	N	N	N	N	N	N	N	N
80	**0.1**	859	851	848	1243	1234	1231	853	1236	**1502**
80	**0.15**	970	955	954	1597	1582	1576	960	1585	**1853**
80	**0.2**	1071	1065	1063	1886	1889	1892	1066	1889	**2169**
130	**0.1**	807	809	808	1154	1150	1147	808	1150	**1406**
130	**0.15**	874	869	868	1478	1470	1469	870	1472	**1710**
130	**0.2**	999	994	990	1817	1809	1804	994	1810	**2065**
180	**0.1**	762	775	771	1085	1086	1081	769	1084	**1329**
180	**0.15**	909	910	926	1439	1434	1441	915	1438	**1704**
180	**0.2**	992	1012	1007	1762	1767	1762	1004	1764	**2029**

For Graphical view in Fig. 5. were used the average values of particular cutting force components and cutting force calculated (see formula 1) out of data measured during cutting with insert DGN 4003C.

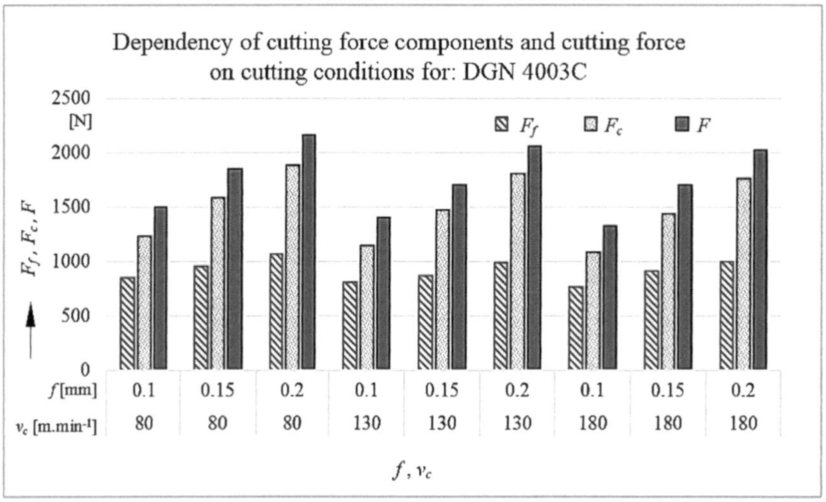

Fig. 5. Dependency of cutting force components and cutting force on cutting conditions for: DGN 4003C

From the graphical view we can observe the significant growth of cutting component of cutting force with the increasing feed and moderate growth of cutting force feed component. On the other hand, the higher cutting speed leads to a slight drop down in cutting component of the cutting force and also to decrease of the total cutting force. The influence of higher cutting speed on feed component is rather small (minor decrease of values).

4.3 Insert GRIP 4004Y

Thanks to its curving construction are the inserts GRIP 4004Y the only two-sided inserts on market, which can groove in bigger depth than is their own length [2].

The results of measurements and calculated values for the insert can be found in Table 6 and graphically displayed in Fig. 6.

Table 6. Measured and calculated values of individual cutting force components for insert GRIP 4004Y

v_c	f	F_{f1}	F_{f2}	F_{f3}	F_{c1}	F_{c2}	F_{c3}	F_f	F_c	F
m.min^{-1}	mm	N	N	N	N	N	N	N	N	N
80	**0.1**	695	694	694	1181	1176	1174	694	1177	**1367**
80	**0.15**	854	855	856	1567	1562	1561	855	1563	**1782**
80	**0.2**	1013	1001	996	1881	1889	1890	1003	1887	**2137**
130	**0.1**	667	668	672	1099	1097	1095	669	1097	**1285**
130	**0.15**	796	805	812	1473	1464	1466	804	1468	**1674**
130	**0.2**	922	933	934	1831	1827	1822	930	1827	**2050**
180	**0.1**	668	689	706	1055	1060	1066	688	1060	**1264**
180	**0.15**	846	885	893	1442	1448	1447	875	1446	**1690**
180	**0.2**	922	931	937	1786	1789	1795	930	1790	**2017**

For graphical view in Fig. 6. were used the average values of single cutting force components and cutting force (see Formula 1) calculated out of data measured in cutting with insert GRIP 4004Y.

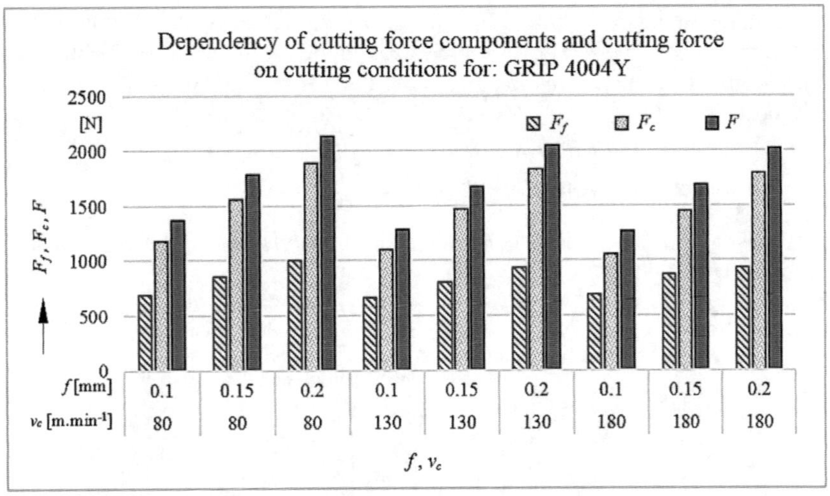

Fig. 6. Dependency of cutting force components and total cutting force on cutting conditions for: GRIP 4004Y

Similar to the previous insert the main portion of effect on the force components increase takes the feed. Raising the cutting speed has slightly decreasing effect on cutting force components and the total cutting force itself.

5 Comparison of force load of inserts chip formers

With regards to the previously proved higher effect of feed step on the proportion of cutting force component the values of cutting force were lined ukrap in order to let the feed dependencies show up.

5.1 Comparison of cutting component of force

Thanks to the graphical view of the calculated data in Fig. 7 it is evident that the insert type DGN 4003J shows higher force load at most of the tests. For lower cutting speeds and lower feed the insert type GRIP 4004Y has smaller force load. However the differences of cutting component between particular inserts are minor.

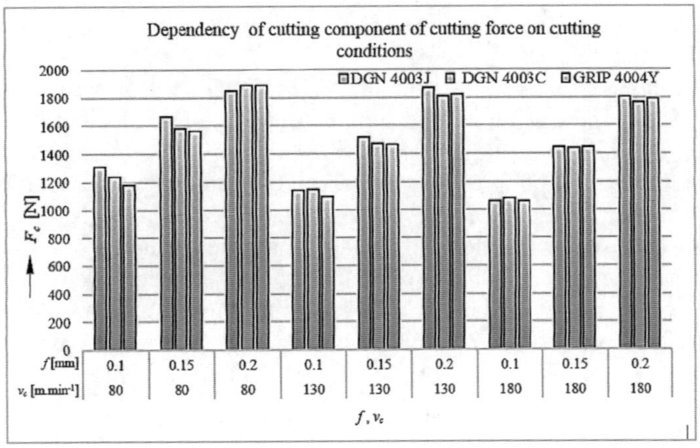

Fig. 7. Dependency of cutting component of force on cutting conditions

At closer look on the insert type DGN 4003J, see Fig. 2.b is visible sharper cutting edge compared to the other two insert types, which are equipped with facetas on the main and side edge. Despite of its geometry this insert has the highest cutting component of the cutting force in almost all cases.

5.2 Comparison of feed component of force

To enable the comparison of feed component of cutting force for the individual applied inserts was created the Fig. 8.

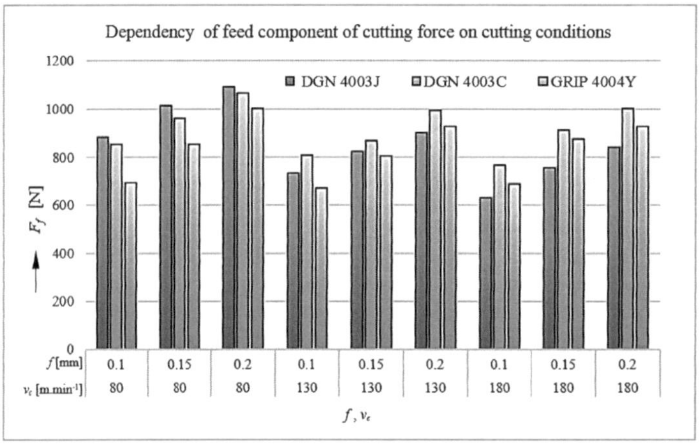

Fig. 8. Dependency of feed component of cutting force on cutting conditions

The biggest differences between individual inserts and their chip formers are visible on Fig. 8. For feed component of cutting force and for higher cutting speed the highest force load shows at insert type DGN 4003C. The lowest force load at the highest cutting force has in this context the insert DGN 4003J. On the other hand the insert GRIP 4004Y has the lowest force load in lower cutting speeds.

5.3 Comparison of total force

If the final cutting force is calculated from the recorded values according to formula (1), the result will be dependency of this force on cutting conditions (Fig. 9). At cutting speed v_c = 80 m.min^{-1} were recognized the biggest differences between the followed inserts. The differences were however lower with higher feed. The lowest force load was measured on insert type GRIP 4004Y.

Also with cutting speed v_c = 130 m.min^{-1}.the lowest force load was detected on insert type GRIP 4004Y in the whole range of applied feeds. The force load of individual inserts was however lower than with cutting speed v_c = 80 m.min^{-1}.

The lowest values of cutting force were reached at cutting speed v_c = 180 m.min^{-1}.. At this cutting speed the lowest force load was achieved with insert DGN 4003J. Though the differences in force load among the followed inserts are the smallest.

It comes out of the graph that for lower cutting speeds and with regards to the final force load is more feasible the insert GRIP 4004Y, while for higher cutting speed it is the insert DGN 4003J. However the difference in force load with insert type GRIP 4004Y is not significant.

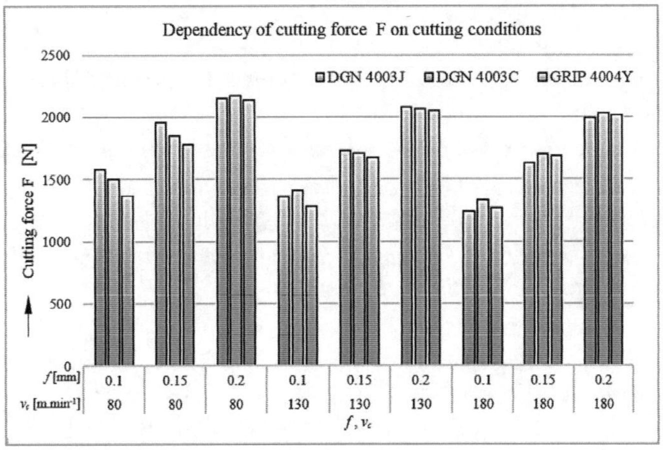

Fig.9. Dependency of force of cutting F on cutting conditions

The drop down at higher cutting speeds can be explained by the chip forming processes, which was well summarized by T. Skopeček [8] including the examples confirming this fact. To get closer understanding this is possible only for plastic materials, in which thermal softening processes apply. The increase in cutting speed leads to higher temperature at the point of cut and to bigger material softening compared to lower speeds. Thanks to this material softening it is not necessary to use such high shearing force to cut off the chip. For confirmation and evaluation of the theory about thermal softening process it would be suitable to install pyrometer or thermal imager e.g. according to the drawing in the article [9,10].

6 Conclusions

For comparison of inserts chip formers there were confronted three inserts at differing cutting conditions formed by intersection of producer recommended cutting conditions. From the diagrams created with measured and calculated values comes out the finding that feed influences the force load of insert more than cutting speed does. For lower cutting speeds is more suitable the insert type GRIP 4004Y, which with regards to the minor force differences at higher speeds appear to be the most universal one and therefore also the chip former on this insert is the best constructed for material 1.6722.

For practical use we can make the conclusion that with the lowest force load it should get to the lowest abrasive wear-out and thus to the highest durability of the machined tool. Research in this field is interesting both to producers of the tools and the companies, which are using this technology and this material, which means mostly companies operating in automotive and aerospace industry.

Article has been done in connection with projects Education system for personal resource of development and research in field of modern trend of surface engineering - surface integrity, reg. no. CZ.1.07/2.3.00/20.0037 financed by Structural Founds of Europe Union and from the means of state budget of the Czech Republic and by project Students Grant Competition SP2018/150 and SP2018/136 financed by the Ministry of Education, Youth and Sports and Faculty of Mechanical Engineering VŠB-TUO.

References

1. *Technical specifications 1.6722 /16NiCrMo16-5* [online]. http://www.ccsteels.com/Structural_steel/2593.html
2. Iscar.com. Iscareletroniccatalog [online]. https://www.iscar.com/eCatalog
3. J. Novotný, J. Cais, N. Náprstková, *Analysis of aluminium alloys AlSi7Mg0.3 and AlMg3 by means of X-Ray Diffraction*, In Manufacturing Technology, Vol. **14**, 392-397 (2014)
4. S. Kuśmierczak, *Methods of evaluation degraded parts.* 14th International Scientific Conference Engineering for Rural Development 790-794 (2015)
5. A. Gorog, D. Stančeková, M. Kapustova, I. Gorogova, I., Miturska, *Influence of selected cutting conditions on cutting forces.* In. Advances in Science and Technology Research Journal, **11**, 179 – 185 (2017)
6. D. Stančeková, M. Šajgalík, J. Petru, N. Náprstková, P. Balaz, *Implementation of coating for failure elimination of dial gauges.* In: METAL 2015. (2015)
7. D. Kalincová, M. Ťavodová, H. Čierna, P.Beňo, *Analysis the causes of distortion castings after heat treatment* , Acta Metallurgica Slovaca, **23**, 182-192 (2017)
8. T. Skopeček, *Vliv vybraných parametrů procesu na řezné síly při HSC.* In: MM PRŮMYSLOVÉ SPETRUM https://www.mmspektrum.com/clanek/vliv-vybranych-parametru-procesu-na-rezne-sily-pri-hsc.html
9. N. Čuboňová, M. Císar, *Design of Camera Mount and its Application for Monitoring Machining Process.* In: Advances In Science And Technology-Research Journal, **9**, 26 (2015)
10. J. Jerzy, I. Kuric, S. Grozav, V. Ceclan, *Calibration of 5 axis CNC machine tool with 3D quickSET measurement system.* Academic Journal of Manufacturing Engineering Vol. **12**, 20-25, (2014)
11. M. Sapieta, A. Sapietová, V. Dekýš, *Comparison of the thermoelastic phenomenon expressions, in stainless steels during cyclic loading.* Metalurgija. **56,** 203-206 (2017)

Impact study of single stage and multi stage abrasive machining on static strength of lap adhesive joints of mild steel

Anna Rudawska[1], *Izabela* Miturska[1,*]

[1]Department of Production Engineering, Faculty of Mechanical Engineering, Lublin University of Technology, address: ul. Nadbystrzycka 36, 20-618 Lublin

Abstract. The article presents the results of research on the influence of one of the technological factors, which is preparation of C45 steel surface before bonding. Seven methods of surface preparation performed by mechanical abrasive grinding tools of different gradations: P40, P120, P320 and P500 in various combinations one-stage or multi stage and degreasing were analysed. The obtained results were subjected to a statistical one-way analysis of variance, which proved that the strength of the adhesive joints depends on the appropriate choice of tools used for abrasive processing, while preparing the surface of the joints before bonding.

Keywords: abrasive machining, surface preparation, adhesive joints, static strength

1 Introduction

Cohering is one of the methods of bonding that often uses a different material than the one the bonded elements are made of. Adhesives have the capacity to produce the force of adhesion to the bonded elements. Adhesion process involves putting a thin layer of an adhesive between the surfaces that are bonded by adhesive and cohesive forces, i.e. the force of internal cohesion of materials. Moreover, adhesives can also protect from corrosion and serve as leak stoppers [1, 2].

The usage of adhesives on an industrial scale dates back to the 20th century, when modern types of glue, such as synthetic phenolic and epoxy adhesive were invented. It made it possible to use the adhesion process for bonding elements made of both wood and metal. The first sectors of industry where structural adhesives were used on an industrial scale were aviation, automotive and rail industries. A need for such type of bonding method stems from using new types of materials in different constructions, such as polymeric composites, for bonding of which adhesion in the most effective and optimal method [2]. However, adhesion cannot be used everywhere. There are lots of sectors where, due to technological and constructional reasons, this method of bonding cannot be used. It results from some limitations of adhesive bonds, as well as their disadvantageous features, such as fast ageing

* Corresponding author: i.miturska@pollub.pl

Reviewers: *František* Holešovský, *Dana* Stančeková

or instability of various structural and mechanic properties in specific exploitation conditions [2].

The adhesion process involves numerous technological and constructional factors that impact the adhesive joints' strength. They are, among others: the method of the bonded materials' surface preparation, type of adhesive and method of its application on the bonded surfaces, conditions of the adhesive-bonded joint's hardening depending on, among others, adhesive type (temperature, time and pressure) and seasoning conditions. The most vital constructional factors are shape and dimensions of a joint (depending on, among others, geometric features of the bonded materials, as well as the type of an adhesive joint load.

Changes within aforementioned factors may have different impact on the specific joints' properties. Problems describing impact of these factors on the adhesive joints have been described in many works and articles [3, 4]. However, due to specification of the joints that are the subject of the present article, it is necessary to carry out a study related with impact of these factors on particular cases and applications. A change within the aforementioned factors in case of a particular joint may impact its properties in a quite different way, e.g. another material, including its strength properties.

The present article is focused on the examination of impact of the C45 low alloy steel preparation to the process of bonding, which constitutes one of the technological factors. A method of the steel sheet surface preparation used in the study was one- and multi-stage mechanical working with use of coated abrasives with abrasive grits of different gradation. Also, measurements of the surface roughness after the considered types of the surface treatment were carried out. The analysis included seven methods of surface preparation by means of mechanical treatment with use of coated abrasives with abrasive grits of different gradation: P40, P120, P320 and P500 in various combinations, as well as degreasing.

1.1 Factors impacting the adhesive joints' strength

When making adhesive joints, it is necessary to get familiar with numerous problems related with adhesive technology, as well as to adhere to guidelines concerning a particular type of joint. One of the most vital matters is knowledge of time and properties of the bonded components as they directly influence the choice of a proper method of surface preparation and of type of adhesive that should be used. The adhesive joint construction's geometry and dimensions have direct influence on the adhesive application method, choice of technological instrumentation to be used, as well as on the value of pressure put during the curing process. The assumed exploitation conditions connected with a type of load on the adhesive joint and type of environment in which the joint will be exploited also influence the choice of a particular type of joint and type of adhesive.

1.2 Preparation of surface for the adhesive process

Surface preparation operations are one of the first technological stages of the bonding process. They are of great importance due to the necessity of establishing proper adhesive properties of the surfaces that have influence on their adhesive capacity, in turn, leads to creating a durable adhesive joint. Surface treatment before bonding may be divided into three groups: initial surface preparation, basic surface treatment and additional surface treatment. A division of methods in each of the groups is presented in Figure 1.

The choice of method of surface preparation for the bonding process depends on numerous factors. The most important one, though, is a type of the material. When it comes to the low alloy steel, there are some recommendations related with the surface preparation process [5-7]. Due to the C45 steel characteristics that make mechanical working easier, the recommended methods are: grinding, abrasive machining, vapour blasting, sandblasting, shot blasting, superfinishing, polishing. Abrasive machining with use of coated abrasives of different graininess, as well as degreasing were selected as the surface preparation methods for the study needs.

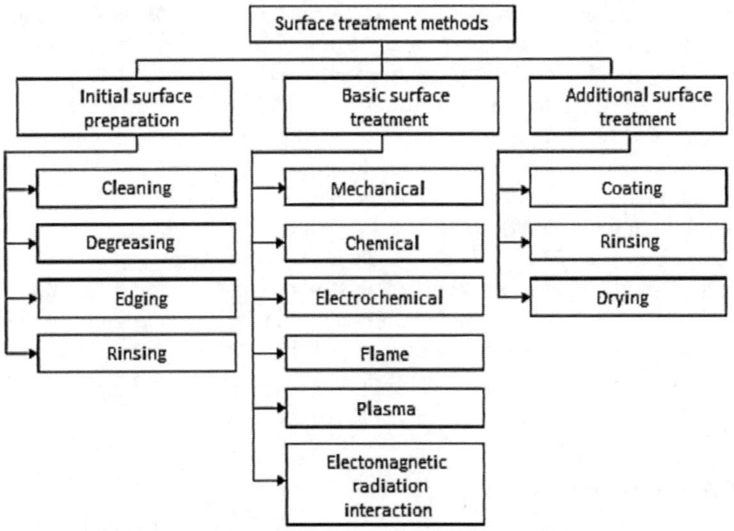

Fig. 1. Types of surface treatment used in the bonding technology [4]

Proper preparation of the material surface is a vital condition that has to be fulfilled in order to obtain adhesive joints with the assumed properties. After the surface is prepared properly, all joining mechanisms should start working - both physical, including the mechanical, as well as chemical ones. This very first stage of the bonding process should ensure obtaining as strong adhesive joints as possible. The aim of surface preparation is also to remove all kinds of impurities from the joined elements' surface, as they weaken adhesive joints. Another reason for this is to change the surface geometric structure (by increasing the surface roughness properly), thanks to which the real surface moistured by adhesive is increased and penetration of adhesive into the joined surfaces' structure is better (Fig. 2).

Fig. 2. Scheme of an adhesive joint of surface area of two solids joint with adhesive [4]

The increase in the given material's surface roughness results in a higher number of cavities with irregular shapes the adhesive may penetrate. Adhesive penetration depends on its adhesiveness, temperature, pressure, penetration time, and the aforementioned cavities' section and depth.

1.3 Degreasing and mechanical working

Mechanical working is classified as one of the basic methods of surface treatment of numerous materials aimed at bonding [7]. Mechanical working makes it possible to increase the surface area, which results in possible increase in the surface energy and change in the range of the intermolecular forces saturation [8]. Surface geometry is an inseparable aspect of mechanical working. It is realised by the profile and roughness measuring. Problems related with surface roughness during creating adhesive joints and with the mechanical adhesion theory have been referred to in the references [9-11]. One of the most popular methods of mechanical working is usage of the coated abrasives, which are widely used in treatment of materials aimed at joining due to, among others, a possibility of application in various conditions. Another important factor is the ease of obtaining non-directional structure of treatment marks, which is beneficial and desired especially in structural joining.

Degreasing is often used as a method of surface preparation before the joining process, including bonding, as it removes all kinds of dust and adhesive impurities that are often residues from the previous working, including machining, or anti-corrosion protection and lubricants [12, 13]. Degreasing may be conducted in different ways using specific degreasing agents. Producers of adhesive often offer ready-to-use products of certain characteristics and aimed at specific use.

2 Experimental studies

Experimental studies included creating adhesive joints of C45 steel sheets, whose surface was subjected to different methods of surface treatment. Surface structure was assessed by measuring the roughness parameters and making the surface analysers. The force breaking joints was being measured. Then the strength of the examined joints was determined.

2.1 Characteristics of adhesive joints and the joined material

Adhesive joints made of C45 low alloy steel sheets were used in the study. Chemical composition and some of its physical properties are presented in Table 1.

The joined samples were of the following dimensions: 2 mm thick, 100 mm long and 20 mm wide.

Table 1. Chemical composition and physical properties of C45 steel [14]

C45 Steel				
Chemical composition, %		Physical properties		
		Designation	Unit	Value
C	0.44	Rm	MPa	638
Mn	0.55			
Si	0.21			
P	0.01	Re	MPa	369
S	0.02			
Cr	0.16			
Ni	0.24	Hardness	HB	255
Cu	0.07			

The subjects of the study were lap adhesive joints subjected to shearing, whose scheme is presented in Figure 3.

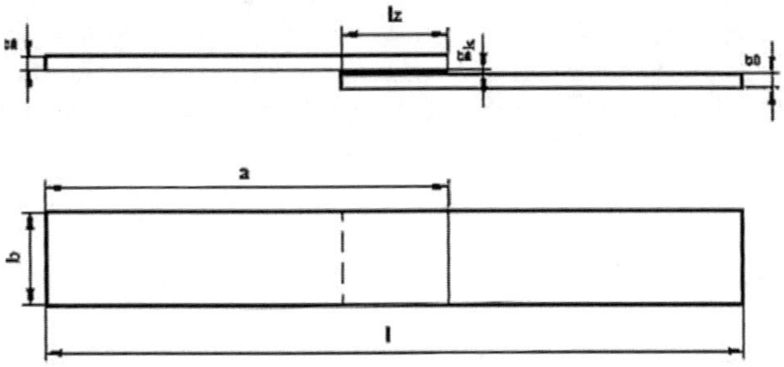

Fig. 3. Scheme of the created adhesive joints of C45 steel where:
a = 100 mm, b = 20 mm, g = 2 mm, lz = 15 mm, gk = 0.1 mm

An assumed lap length was lz = 15 mm, and an adhesive-layer was about 0.1 mm long. Bonding field and the adhesive-layer thickness were verified after the joints were made.

Adhesive joints were made with use of the adhesive composition Epidian 57/PAC/1:1, whose preparation included mixing of the Epidian 57 epoxy resin and PAC curing agent in 1:1 weight proportion with use of a turbine bi-plate mixing arm during 3 minutes. The adhesive composition was prepared directly before the bonding process. It was applied manually on one of the surfaces to be joined with use of a polymeric toothed tray. The composition's thickness was the same on the whole joined surface. The expiry date, by then of which adhesive does not change its properties was observed during the bonding process. The bonding and curing processes were performed in stable conditions at the temperature of 25° ± 1°C and 25% ± 1% relative humidity. The curing time was 7 days and the pressure was 19.6 MPa. Then the strength tests were performed. Thanks to this it was possible to determine the values of failure forces, based on that, to determine the joints' strength. A trial of the lap adhesive joints' cutting was performed on the testing machine Zwick/Roell Z150, according to the norm DIN EN 1465.

2.2 Type of mechanical working

Surface preparation was performed by means of mechanical working with use of coated abrasives with abrasive grits of different gradation and, then, degreasing. Types of the surface treatment operations adopted in the study are presented in Table 2.

Table 2. Chemical composition and physical properties of C45 steel [14]

No.	Variant of working	Type of working
1.	P40	Mechanical working with a coated abrasive of P40 gradation
2.	P120	Mechanical working with a coated abrasive of P120 gradation
3.	P320	Mechanical working with a coated abrasive of P320 gradation
4.	P500	Mechanical working with a coated abrasive of P500 gradation
5.	P40/P120	Mechanical working with a coated abrasive of P40 and P120 gradation
6.	P40/P120/P320	Mechanical working with a coated abrasive of P40, P120 and P320 gradation
7.	P40/P120/P320/P500	Mechanical working with a coated abrasive of P40, P1230, P320 and P500 gradation

Before bonding all the samples' surfaces after mechanical working were degreased with a degreasing agent Loctite 7063. The processes of mechanical working and degreasing are described in the references [5, 15-17]. There were 16 samples in total - 4 of each surface preparation variant served to the surface roughness examination. The remaining samples were used to make adhesive joints. There were 42 adhesive joints made in total.

3 Test results

3.1 Surface roughness

The surface roughness was examined on 4 selected samples of each surface preparation variant. Selected roughness parameters were measured: Ra, Rz, Rmax, Rq and RSm according to the standard PN-EN ISO 4287:1999/A1:2010.
According to this norm, selected parameters of the roughness profile are defined as:
• Ra - arithmetical mean of the profile's ordinates: arithmetical mean of the absolute ordinates' value inside the roughness sample length,
• Rz - roughness height according to 10 profile points: mean distance from the 5 peak vertices to the 5 lowest valley marks observed on the roughness sample length, measured from the reference line parallel to the average line,
• Rmax - maximum depth of a valley: it is the most important roughness depth of single pits on the whole roughness sampling length,
• Rq - average profile square deviation from the average line along the roughness sampling length,
• RSm - width-frequency parameter - average width of the profile elements: average value of the width of the elements inside the roughness sampling length.
Characteristics of the selected parameters of the roughness profile are presented in Table 3.

Results of the selected roughness parameters of the examined samples after different surface preparation methods are presented in Table 4. Surface roughness measurements were aimed at determining some parameters of the geometrical structure of the bonded samples' surface. Table 5 presents exemplary profilographs of C45 steels sheets after the surface treatment methods analysed above.

Table 4. Roughness parameters of C45 steel after different surface preparation methods

No.	Surface preparation variant	Surface roughness parameters [μm]				
		Ra	Rz	Rmax	Rq	RSm
1.	P40	1.907	13.05	18.01	2.55	125.50
2.	P120	1.444	8.39	10.19	1.66	97.25
3.	P320	1.404	7.50	11.10	1.80	111.25
4.	P500	1.267	6.80	8.57	1.60	103
5.	P40/P120	1.728	11.80	16.21	2.47	134.5
6.	P40/P120/P320	1.700	10.61	14.43	2.07	121.8
7.	P40/P120/P320/P500	1.467	9.66	12.96	1.92	101

Based on the surface roughness measurements' results it may be observed that the highest roughness height parameters were obtained for the samples prepared by the variant: mechanical working P40 and mechanical working P40/P120. Considerably lower surface roughness parameters were obtained for the mechanical working P500.

Table 3. Characteristics of the selected parameters of the roughness profile [18]

	Definition	**Measurement scheme**	**Calculations**
Ra	Average profile deviation from the average line measured along the roughness sampling length or measuring length	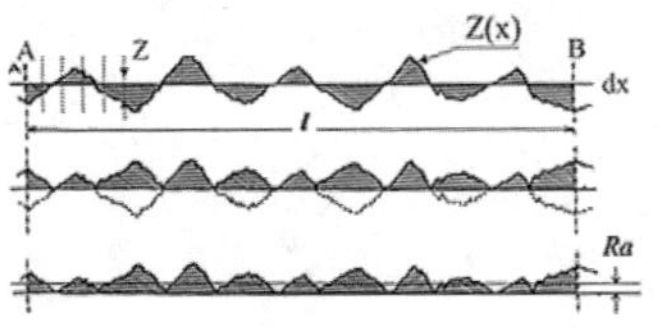	$Ra = \frac{1}{l}\int_A^B \lvert Z(x)\rvert dx$ $\approx \frac{1}{n}\sum_{i=1}^{n} \lvert Z_i\rvert$
Rz	Roughness height from the average line along the roughness sampling length or measuring length	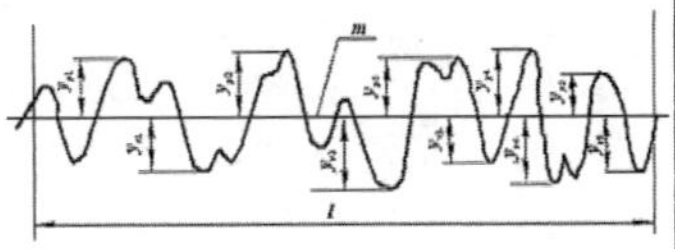	$Rz = \frac{\sum_{i=1}^{5}\lvert y_{pi}\rvert + \sum_{i=1}^{5}\lvert y_{vi}\rvert}{5}$
Rma x	Maximum roughness height	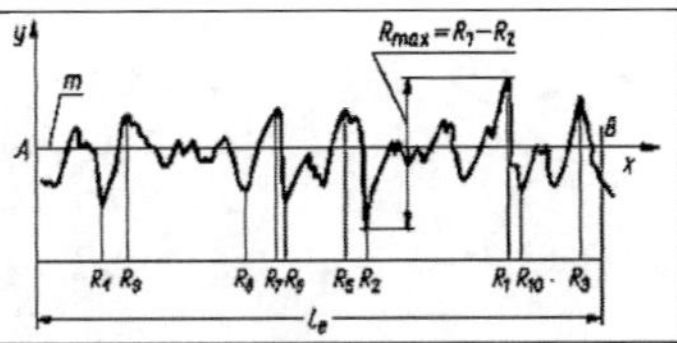	$Rmax = R_1 - R_2$
Rq	Average profile square deviation from the average line along the roughness sampling length or the measuring length	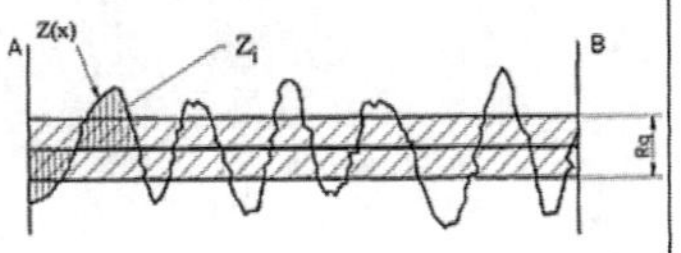	$Rq = \sqrt{\frac{1}{l}\int_0^l \lvert Z^2(x)\rvert dx}$ $\approx \sqrt{\frac{1}{n}\sum_{i=1}^{n} Z_i^2}$
RSm	Average value from mean spacing of irregularities RSmi occurring in the range of the roughness sampling length I	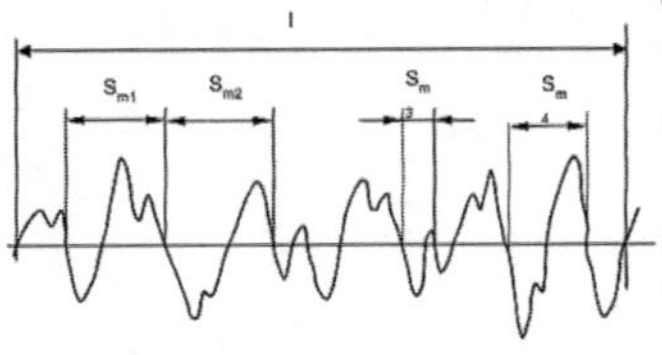	$RSm = \frac{1}{n}\sum_{i=1}^{n} Rsmi$

Based on that it may be observed that single-stage mechanical working lets obtain more repeatable surface roughness profile, thanks to which adhesive penetration while making adhesive joints will be better. As a result, it may be assumed that adhesive joints of elements of such roughness profiles will present better strength results.

Table 5. Exemplary surface analysers of the examined steel sheets after different methods of surface preparation

Surface preparation variant	**Exemplary surface analysers, Lc = 8mm**
P40	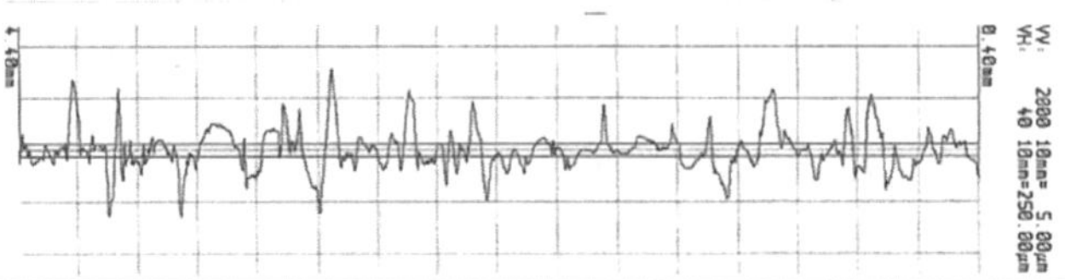
P120	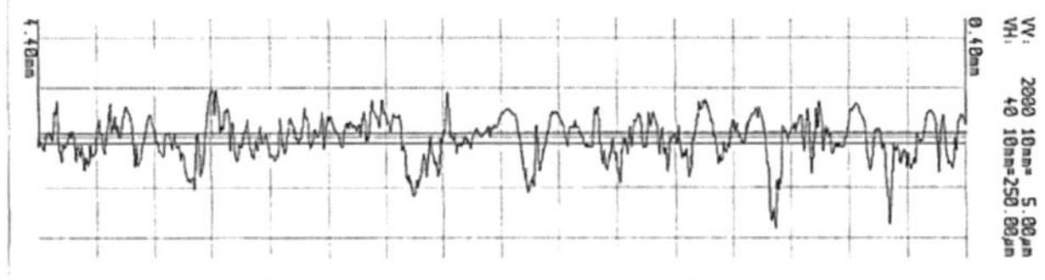
P320	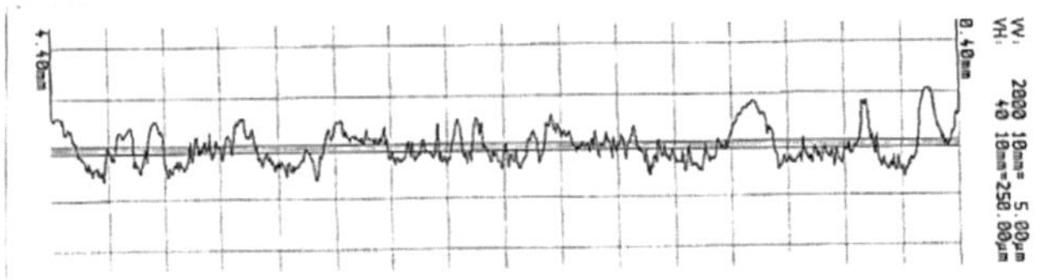
P500	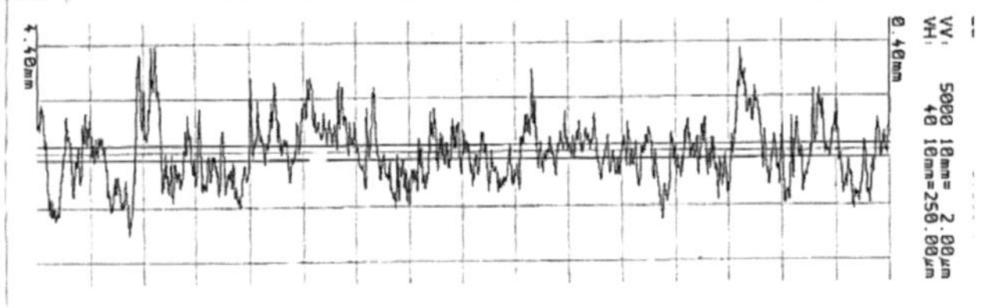
P40/P120	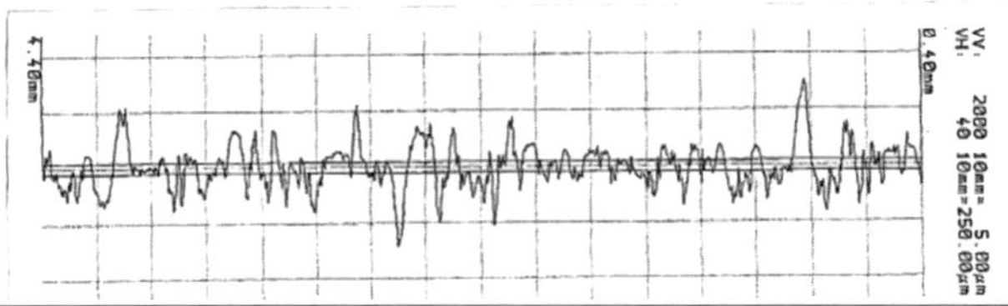
P40/P120/P320	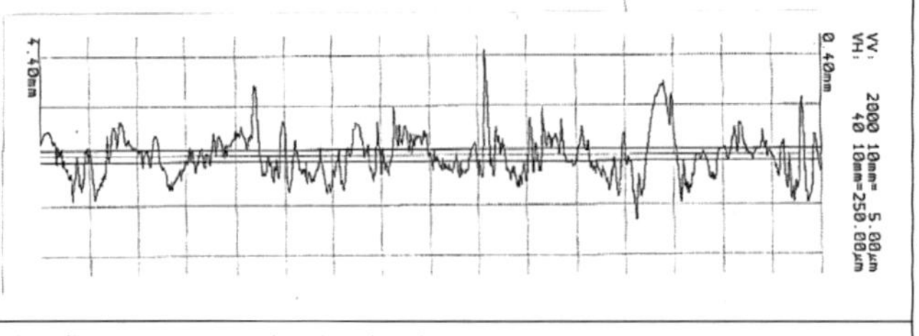
P40/P120/P320/P500	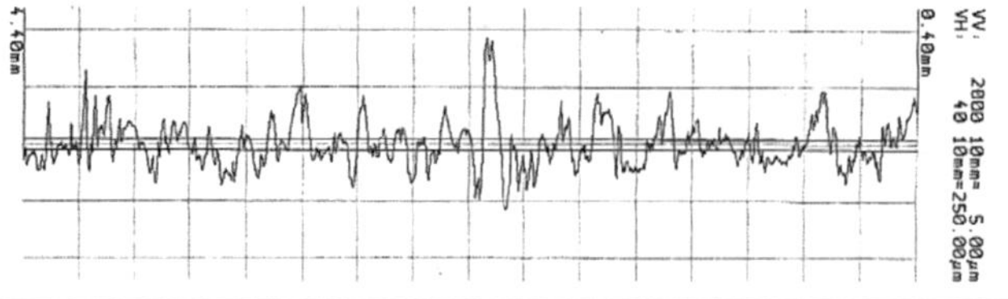

3.2 Strength test

Results of the strength tests conducted in the strength test are presented in Figure 4 and Table 6.

Based on the test results it may be observed that single-stage mechanical working has beneficial impact on the adhesive joints' strength of C45 steel in comparison with the joints in case of which the surfaces of the joined samples were subjected to multi-stage mechanical working. However, attention should be paid to substantial standard deviation of particular variants of surface preparation. The biggest discrepancies occur in case of multi-stage mechanical working, especially for mechanical working with use of coated abrasives P40 and P120.

When comparing the obtained values with the surface roughness parameters, it may be stated that they cannot serve as an indicator when it comes to the surface preparation method in all cases.

It turns out that in spite of quite low height parameters in case of the mechanical working P500, the strength of adhesive joints was the highest.

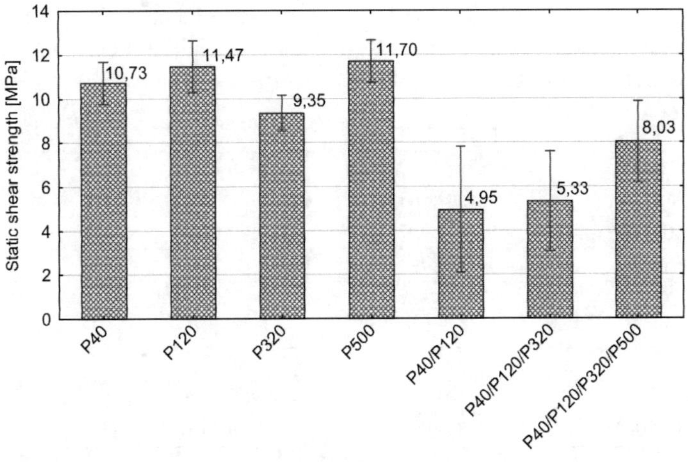

Fig. 4. Strength of the lap adhesive joint made of C45 steel after different surface preparation methods

Table 6. Average failure force, standard deviation and strength of adhesive joints of steel sheets made of C45 steel

No.	Surface preparation variant	Failure force, N		Shear strength, MPa	
		Mean	Standard deviation	Mean	Standard deviation
1.	P40	854.03	369	10.73	1.0
2.	P120	810.90	455	11.47	1.2
3.	P320	559.78	224	9.35	0.8
4.	P500	1323.60	140	11.70	1.0
5.	P40/P120	490.32	817	4.95	2.9
6.	P40/P120/P320	272.13	220	5.33	2.3
7.	P40/P120/P320/P500	324.35	177	8.03	1.9

Geometric structure of surface, as well as geometric dimensions of micro irregularities has to be taken into consideration too as they may be crucial due to possible penetration by adhesive. When analysing surface analysers, it may be observed that the structure after mechanical working with use of an abrasive of P500 gradation is different than the one after other variants of working. Such structure may ensure better mechanical adhesion in comparison to the abrasives of different gradation.

Also, it may be observed that in terms of the strength of adhesive joint of the joined materials, beneficial surface structure was obtained for mechanical working with an abrasive of P120 gradation. However, in order to confirm the existence of similarities or discrepancies between the obtained strength of the adhesive joints, it is necessary to conduct a statistical analysis of the obtained results.

3.3 Statistical analysis of the obtained results

Strength of adhesive joints is a vital criterion of such joints' assessment. Shear strength is defined as a deviant strain of an object caused by tangential stresses. After strength tests, an analytical analysis at a confidence level $\alpha = 0.05$ was obtained. In the beginning, the analytical analysis was conducted in order to check the normality of distribution with use of a Shapiro-Wilk normality test for small samples. Its results are presented in Table 7.

Table 7. Normality test - adhesive joints' strength

Surface preparation variant	Shapiro-Wilk statistics W	Probability level p	Normality of distribution
P40	0.927	0.555	Yes
P120	0.993	0.996	Yes
P320	0.928	0.565	Yes
P500	0.857	0.180	Yes
P40/P120	0.918	0.489	Yes
P40/P120/P320	0.970	0.894	Yes
P40/P120/P320/P500	0.908	0.423	Yes

Compliance analysis of the empirical distribution with the normal distribution examined by means of the Shapiro-Wilk test did not reject the hypothesis of the normality of distribution of the adhesive joints' strength for the examined methods of surface preparation ( ). Another stage of study was to check the variance homogeneity with use of the Levene's test, whose results are presented in Table 8.

Table 8. Variance homogeneity Levene's test

	SS Effect	df Effect	MS Effect	SS Error	df Error	MS Error	F	p
Shear strength	9.16	6	1.53	38.66	35	1.105	1.382	0.249

The Levene's test does not reject the variance equality hypothesis. Taking this into consideration, the next step was to conduct the ANOVA statistics in order to check the impact of the surface preparation of the joined elements on the adhesive joints' strength. The Tukey's and HSD test was conducted and the interaction graph was drawn in order to interpret and compare the results precisely.

Table 9. Tukey's Honestly Significant Difference test at α = 0.05

Surface preparation variant	**{1}**	**{2}**	**{3}**	**{4}**	**{5}**	**{6}**	**{7}**
P40 {1}		0.989	0.801	0.956	0.001	0.001	0.123
P120 {2}	0.989		0.357	1.000	0.001	0.001	0.022
P320 {3}	0.801	0.357		0.243	0.002	0.005	0.835
P500 {4}	0.956	1.000	0.243		0.001	0.001	0.012
P40/P120 {5}	0.001	0.001	0.002	0.001		1.000	0.052
P40/P120/P320 {6}	0.001	0.001	0.005	0.001	1.000		0.123
P40/P120/P320/P500 {7}	0.123	0.022	0.835	0.012	0.052	0.123	

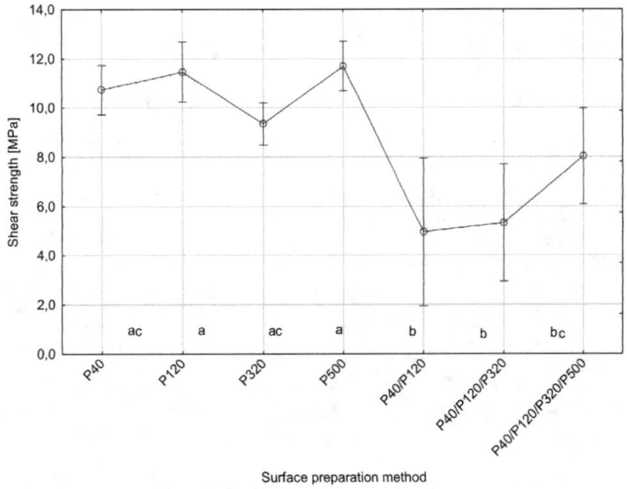

Fig. 5. Graph of interactions of average values of the adhesive joints' strength with designated homogenous groups

Table 10. The post-hoc Tukey's test designating homogenous groups

No.	**Surface preparation**	**Average shear strength**	**a**	**b**	**c**
1	P40	10.73	****		****
2	P120	11.47	****		
3	P320	9.35	****		****
4	P500	11.70	****		
5	P40/P120	4.95		****	
6	P40/P120/P320	5.33		****	
7	P40/P120/P320/P500	8.03		****	****

Based on the obtained results, it may be observed that ANOVA showed significant interaction between the surface preparation method and the adhesive joints' strength. Also, strength of the adhesive joints of samples whose surface had been prepared by means of a single-stage mechanical working differs significantly from strength of the adhesive joints of samples whose surface had been prepared by means of multi-stage mechanical working. Only in case of mechanical working designated as P40/P120/P320/P500, a similar result with the working P40 and P320 was observed (they are in one homogenous group).

4 Conclusions

The study was related with the impact of one-stage or multi stage abrasive machining on the static strength of the lap adhesive joints of C45 steel. Surface treatment method was a variable factor in the study. The other factors like: type of adhesive and other conditions of the joints making were invariable during the study. A method of the steel sheet surface preparation used in the study was one- and multi-stage mechanical working with use of coated abrasives with abrasive grits of different gradation. Also, measurements of the surface roughness after the considered types of the preparatory treatment were carried out.

Based on the research, it was observed that adhesive joints of the C45 steel sheets, whose surface was subjected to mechanical working with a P500 abrasive (11.70 MPa) were the strongest. Statistically insignificant strength was also obtained for adhesive joints of the element subjected to mechanical working with a P120 abrasive (11.47 MPa). The lowest strength was obtained in case of the joined elements whose surface was prepared in a few stages. What is more, this method is more time-consuming. In addition, when analysing the statistical analysis' results, it was observed that the greatest discrepancies occurred in case of treatment with use of several abrasives. The highest repeatability of results was obtained in case of working for which the strength was the highest.

To sum up, abrasives of higher gradation used during the surface preparation have beneficial impact on the adhesive joints of C45 steel sheets. Test results presented above were obtained for the presented conditions of the adhesive joints' making. In the situations when the adhesive joint must carry a specified weight of a known value, it is possible to use the adhesive technology in joining sheets made of low-carbon steel. It will also let avoid difficulties connected with other welding methods.

References

1. Rudawska, *Selected Issues in Constituting Homogeneous and Hybrid Adhesive Joints.* (Lublin University of Technology, Lublin, 2013)
2. Borsellino, et al., *Comparisons of processing and strength properties of two adhesive systems for composite joints.* Int. J. Adhes. Adhes. **27.6**, 446-457 (2007)
3. Dmitruk, P. Mayer, J. Pach., *Pull-off strength of thermoplastic fiber-reinforced composite coatings.* J. Adhes. Sci. Technol. **32.9**, 997-1006 (2018)
4. A. Baldan, *Adhesively-bonded joints and repairs in metallic alloys, polymers and composite materials: adhesives, adhesion theories and surface pretreatment.* J. Mater. Sci. **39.1**, 1-49 (2004)
5. L.F.M. da Silva, R.J.C. Carbas, G.W. Critchlow, M.A.V. Figueiredo, K. Brown, *Effect of material, geometry, surface treatment and environment on the shear strength of single lap joints.* Int. J. Adhes. Adhes. **29.6**, 621-632 (2009)
6. A. Rudawska, *Selected aspects of the effect of mechanical treatment on surface roughness and adhesive joint strength of steel sheets.* Int. J. Adhes. Adhes. **50**, 235-243 (2014)
7. L.H. Lee, ed. *Adhesive bonding.* (Springer Science & Business Media, 2013)
8. A.J. Kinloch, *Adhesion and adhesives: science and technology.* (Springer Science & Business Media, 2012)

9. G.W. Critchlow, K.A. Yendall, D. Barani, A. Quinn, F. Adrews, *Strategies for the replacement of chromic acid anodizing for the structural bonding of aluminium alloy*. Int. J. Adhes. Adhes. **26** (2006)
10. M. Rucka, J. Lachowicz, E. Wojtczak., *Lamb wave propagation in a single lap adhesive joint*. Shell Structures: Theory and Applications Volume 4: Proceedings of the 11th International Conference Shell Structures: Theory and Applications, (2017)
11. M. Shahid, S. A. Hashim. *Effect of surface roughness on the strength of cleavage joints*. Int. J. Adhes. Adhes. **22.3**, 235-244 (2002)
12. A. Rudawska, M. Reszka, T. Warda, I. Miturska, J. Szabelski, A. Skoczylas, D. Stancekova, *Milling as a method of surface pre-treatment of steel for adhesive bonding*. J. Adhes. Sci. Technol., **23**, 2619-2636 (2016)
13. J.M. Arenas, J.N. Julián, A. Cristina, *Optimum adhesive thickness in structural adhesives joints using statistical techniques based on Weibull distribution*. Int. J. Adhes. Adhes. **30.3**, 160-165 (2010)
14. U.A. Khashaba, et al., *Machinability analysis in drilling woven GFR/epoxy composites: Part I–Effect of machining parameters*. Composites Part A. **41.3,** 391-400 (2010)
15. W. Zielecki, et al., *Surface topography effect on strength of lap adhesive joints after mechanical pre-treatment*. ACME **13.2**, 175-185 (2013)
16. G. Rotella, M. Alfano, T. Schiefer, I. Jansen, *Evaluation of mechanical and laser surface pre-treatments on the strength of adhesive bonded steel joints for automotive industry*, . J. Adhes. Sci. Technol., **30**, 747–758 (2016)
17. P. Davies, et al., *Influence of adhesive bond line thickness on joint strength*. Int. J. Adhes. Adhes. **29.7**, 724-736 (2009)
18. M.W. Sayers, *Guidelines for conducting and calibrating road roughness measurements*. (The World Bank Washington, D.C., U.S.A 1986)

Use of high-speed water flows for accelerated mechanical modelling of erosive wear of concrete surfaces

Libor Sitek[1,*], *Petr* Hlaváček[1] and *Lenka* Bodnárová[2]

[1]Institute of Geonics of the CAS, Studentská 1768, 708 00 Ostrava, Czech Republic
[2]Brno University of Technology, Faculty of Civil Engineering, Veveří 331/95, 602 00 Brno, Czech Republic

Abstract. The article deals with an unconventional way of simulating the erosion of concrete surfaces by means of high-speed water flows. The proposed method allows the use of short-term experiments to predict the future failure of concrete elements in specific concrete mixtures exposed in practice to the action of fast-flowing fluids, whether as regards sewers, overfalls of water reservoirs, inlet channels of retention reservoirs, etc. The article shows results of experiments on disintegration of the surface layer of model concrete test specimens of various physical-mechanical properties that were exposed to various types of high-speed water flows. It has been demonstrated that the application of high-speed water flows significantly accelerates the degradation process of the surface layers. This simulation of the real situation can considerably shorten the time needed for the development and testing of new erosion-resistant construction elements.

Keywords: high-speed water flow, cement composites, erosive wear, abrasion, abrasive material, short-time erosion test

1 Introduction

Concrete surfaces exposed to the long-term contact with fast-flowing polluted water (more than 5 $m.s^{-1}$) are affected by gradual erosion and abrasive wear caused by the presence of small particles in the water flow. In addition, at high flow rates (more than 15 $m.s^{-1}$), cavitation occurs as another destruction mechanism. Due to these effects, surfaces of concrete structures are rapidly degraded by initiation of small caverns. Caverns quickly expand and depth failure with consequent concrete breakage occurs. This particular complex type of degradation was found in sewers, water supply channels in water treatment plants, water-power plants, especially in chutes in dam overfalls during floods, but also in overfalls of weirs, dam crests or inlet channels and stilling basins of retention reservoirs.

Development of resistant concrete mixtures and determination of suitable surface treatments to make construction elements resistant to the flowing liquids has been the subject of research for decades. The disadvantage is that the resistance prediction of these elements

* Corresponding author: libor.sitek@ugn.cas.cz

Reviewers: *Anna* Mičietová, *Miroslav* Müller

is complicated and too unreliable, as the degradation of real structures often occurs in a period of several years. The impossibility of verifying the resistance of construction components to erosive and abrasive wear of the flowing fluids in real time complicates the development of new resistant mixtures. At present, requirements for concrete quality are primarily solved by the demand for the minimum concrete strength class, the maximum water/cement ratio and the aggregate type. As there is no precise methodology for testing the influence of the flowing fluids on the surface of concrete structures, it is impossible to test the proposed formulae in laboratories and use them in practice.

Many years of experience with the disintegration of cement composites using the high-speed water jets encouraged us to use this unconventional technology for an accelerated mechanical simulation of the degradation process of concrete surfaces. Thus, the main goal of the research is the effort to find a suitable method providing sufficient information about the resistance of the concrete against the violation by the flowing liquids in a short time. The method based on the action of high-speed water jet on surface structures of concrete should authentically simulate the real situation.

2 High-speed water jets

High-speed water jet technology is able to disintegrate even the hardest materials due to high energy transmitted to an extremely small area (see e.g. [1]). Using water jets, there is no mechanical tool-material interaction in the process of disintegration. The maximum temperature of the jet does not exceed 70°C and zones in the area of the cut are thus unaffected by the heat. No thermal interference of the material to be cut as well as universality of the jet application are the greatest benefits of the abrasive water jets. The erosion capability of the jets is widely used in modern industry. The water jet technology achieved significant progress in the last decades in applications such as cutting of wide range of materials, surface cleaning, removal of surface layers and repair of concrete structures. Nowadays, a number of commercial high pressure systems are available on the market, some of them generating pressures up to 700 MPa, other delivering up to hundreds of litres of water per minute. Water jet cutting and/or cleaning equipment, except the pump, is lightweight and the whole cutting or cleaning process can be easily automated [2].

The water jet technology is also very advantageous for the removal of deteriorated concrete layers from buildings and structures [3]. The first serious concept for the use of water jets for concrete removal and cleaning (i.e. hydrodemolition) was probably that of [4]. The first commercial hydrodemolition unit was finally developed and introduced 10 years later [2]. High-speed water jet technology has been used for the removal of damaged concrete in civil engineering since the 80's of 20th century [5]. The water jets are able to remove the damaged layer selectively without introducing any additional cracks into the construction (in contrast to traditional technologies like jackhammering, grit blasting, milling, etc.). Moreover, adhesion strength of coatings applied on surfaces prepared by water jets comply with values specified in relevant standards concerning the concrete surface treatment prior to repair [6]. Several articles deal with the interaction of water jets with concretes (e.g. [2], [7-9]) and steel-fibre reinforced concretes [10-11], while other describe the simulation of erosion of concretes in hydraulic structures in different ways (e.g. [12-14]). In [15], acoustic emission sensing technique is proposed for on-line monitoring of hydro-abrasive erosion. Wear particles originated from erosion of concrete after the application of high velocity water flow are then studied in the work [16].

Based on previous experience, the following types of jets were selected for testing the erosion damage that simulated the effects of fast-flowing liquids on concrete surfaces:

2.1 Plain water jet

The jet (Fig. 1a) is generated by the change of the pressure energy of water to the kinetic energy by the acceleration of the water flow within the nozzle. The maximum pressures currently used for the generation of the plain water jets are up to 700 MPa. This jet type is suitable for cutting of softer materials (rubber, paper, wood, plastic, etc.), cleaning of materials from paints and coatings and removal of surface layers.

When a steady continuous, inviscid and incompressible liquid jet impinges on a flat rigid surface, the maximum pressure (according to Bernoulli´s equation) at the point of impact equals the stagnation pressure p_s, given by:

$$p_s = \frac{1}{2}\rho V_0^2 , \qquad (1)$$

where p_s is the stagnation pressure, ρ the density of the liquid (water) and V_0 the velocity of the impinging jet [17]. Despite the high velocity of the impinging jet (typically 700 m.s^{-1}) and thus the high stagnation pressure, the force effects of the jet on the impact area are relatively small due to the limited area of the water flow impact. For example, at the pressure of 250 MPa and diameter of 0.33 mm, the final measured force on the impact area was only around 40 N [18].

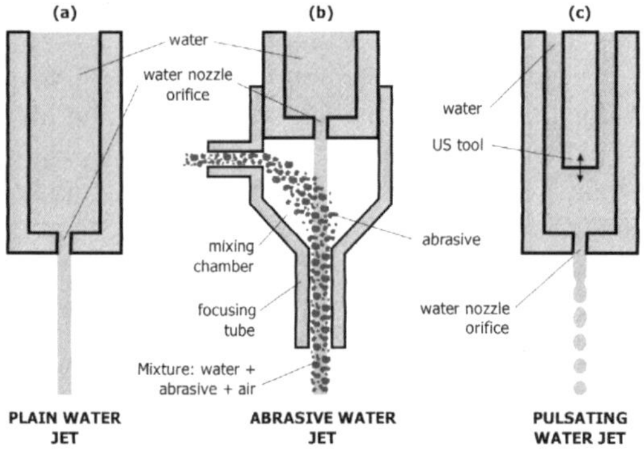

Fig. 1. Types of water jet used in experiments

2.2 Abrasive water jet

Cutting performance of the plain water jet can be significantly increased by adding abrasive particles into the jet. They are added into the jet downstream the water nozzle in a mixing chamber and consequently focused and accelerated in a focusing tube (Fig. 1b). Pressures up to 420 MPa are commonly used to generate this type of the jet. Almandine garnet or olivine concentrates with grain size of about 0.2 mm are usually used as abrasive materials [19]. With a single water jet based system, several machining operations can be performed on a wide range of materials with various geometries. The abrasive water jet can cut, drill, turn, mill, etc. difficult-to-machine materials such as composites, structural ceramics, high-strength alloys, glass, rocks, etc.

Despite the significantly higher cutting performance compared to the plain water jet, the impact forces of the abrasive water jet remain similar. The impact force of the abrasive water jet was measured to be about 30 N at the water pressure of 300 MPa, with the nozzle diameter of 0.33 mm and mass flow rate of 300 g·min^{-1} of the Australian garnet with grain size of 80 mesh (i.e. mean grain size of 0.18 mm according to [19]). The water medium only accelerates and transports abrasive particles to the area of interaction with the material. Mixing of the abrasive with the water jet by the flow through the abrasive head and acceleration in the focusing tube leads to certain losses in energy. The resulting force is thus slightly lower compared to the forces measured in the plain water jet. The velocity of individual abrasive particles in the high-speed water jet at the outlet from the focusing tube primarily depends on the grain size, water pressure in front of the nozzle, concentration of abrasive grains in the jet and the stand-off distance from the nozzle exit [20]. E.g. in the paper [21], the average particle velocity measured in the jet with the water nozzle of 0.25 mm at the water pressures of 80-140 MPa was 117-161 m.s^{-1}.

2.3 Pulsating water jet

One of possible methods how to significantly improve the performance of the plain water jet without the addition of abrasives is the method using a pulsating water jet (Fig. 1c). The impact pressure on a target generated by a bunch of water is considerably higher than the stagnation pressure of the corresponding continuous plain water jet [22, 23].

If a drop or a slug of liquid strikes the same target, the initial impact pressure will be much higher. The so-called waterhammer pressure developed by the impact of the jet with a flat head on a rigid plane surface can be derived by a one-dimensional (Rankine-Hugoniot) analysis, as:

$$p_{\mathrm{i}} = \rho V_0 C_0 \,, \tag{2}$$

where C_0 is the speed of sound (sonic velocity) in a liquid (water) and V_0 is the velocity of the jet. The equation (2) represents the simplest approximation to the maximum impact pressure rise developed in a liquid-solid collision. After very short time (the duration of the impact pressure is affected by the size and shape of the drop), the pressure drops to the stagnation pressure p_{s}. In the pulsating jet, the impact pressure is amplified (see [24])

$$z = \frac{p_{\mathrm{i}}}{p_{\mathrm{s}}} = \frac{2C_0}{V_0} = \frac{2954}{V_0}, \tag{3}$$

where p_{i} is the impact pressure, p_{s} is the stagnation pressure. Velocities of the continuous jets generally do not exceed 700 m.s^{-1}. Thus, the impact pressure of the pulsating jet can be at least 4 times higher at the same velocity. Therefore, significant improvement in cutting performance is expected. Moreover, additional effects such as fatigue failure of the target material due to the cyclic loading contribute to more effective performance of the pulsating water jets.

An increase in the impact force is given, for example, in [17]. The force of the water jet generated at the pressure of 20 MPa, the diameter of 1.98 mm, and the stand-off distance of 40 mm was increased from about 120 N to 400 N.

3 Experiments

Suitability of the above mentioned types of water jets for an accelerated mechanical simulation of the surface layer degradation of selected cement composites by the combined wear was evaluated on the basis of experimental tests. The jets used, including their parameters and the range of values set during the experiments, are shown in Table 1.

Because of large-scale experimental work, all experiments were divided into two stages. During the first stage, a suitable configuration of working parameters of the high-speed water jet to simulate the erosion wear of concretes by the flowing water was determined. In the second stage, the tests were focused on comparing two different types of concretes and their resistance to fast-flowing water. Particular settings of parameters of the water jet and cutting devices used during individual stages are listed in the text.

Table 1. Tested jets and the range of parameters used in experiments

Water jet type	Water nozzle diameter [mm]	Focusing tube diameter and length [mm]	Water jet traverse speed [mm/min]	Water pressure [MPa]	Abrasive mass flow rate [g/min]	Jet incl. angle [°]	Stand-off distance [mm]
Continuous plain water jet	0.33 1.75	-	0-40	30-300	-	0-15	22.5-1000
Abrasive water jet	0.33	0.78/76	0-40	100-300	0-300	0-15	22.5-1000
Pulsating water jet	1.75	-	0-1	30-70	-	2-8	100

3.1 Determination of suitable configuration

Several options to remove surface layers of concrete specimens using the continuous high-speed water jets without and with addition of abrasives were tested. The first series of tests was performed from a relatively short distance. Other tests were carried out from long distances (Fig. 2).

Fig. 2. Short (left) and long (right) stand-off distances of jet action

Concrete cubes of original strength of 40 MPa placed in the NaCl solution for 6 months were selected as test specimens (chlorides can be commonly found in drinking water, besides, they can be of natural origin from rock subsoil. Their source is often waste water, leachates and sheet-washes from winter spreading of roads, etc). The water jet was generated by a standard JETS 7.5/60c high-pressure pump made by PTV Ltd. with a PTV diamond cutting head used for the abrasive water jet generation or, where appropriate, with a SLICE cutting head (water nozzle diameter of 0.33 mm, focusing tube diameter of 0.78 mm, focusing tube length of 76 mm). In the case of the abrasive water jet, Australian garnet GMA concentrate with the grain size of mesh 80 (0.18 mm) was used.

By the water jet action from a short distance, various water jet inclinations to the impact area, jet position to the edge of the specimen and distance of the outlet of the focusing tube from the specimen (stand-off distance) were tested. Abrasive mass flow rate, water pressure in front of the nozzle and traverse speed of the cutting head over the specimen or, in the case of the stationary water jet, duration of the water jet action were also changed.

By the water jet action from long distances, only the abrasive water jet was tested. The position of the jet impact on the specimen, stand-off distance, water jet inclination angle and water pressure in front of the nozzle were changed. Abrasive mass flow rate of 200 g·min^{-1} was constant and the tests were performed with a stationary water jet (i.e. without jet-specimen motion). The abrasive water jet impact on specimens always took 1 minute.

3.2 Comparison of two different types of concretes

At this stage, several options to remove surface layers of two different concrete mixtures using the continuous and pulsating water jets without abrasives and continuous water jets with abrasives were tested. Concrete tiles of various composition and properties (see Tables 2 and 3) - commonly used concrete of low strength (hereinafter referred to as *concrete I*) and progressive solid concrete (hereinafter referred to as *concrete II*) - served as test specimens. For reasons of clarity, the stage was further divided into a series of tests of concrete surfaces treatment by the abrasive water jet and a series of tests using the plain water jet (without the addition of abrasives).

Table 2. Composition of concrete mixtures

Composition (per 1 m^3 of fresh concrete)	Concrete I quantity [kg]	Concrete II quantity [kg]
Cement CEM I 42.5 R	-	385
Cement CEM II/B-S 32.5	310	-
Aggregate 0-4 mm river sand	1135	1110
Aggregate 4-8 mm coarse aggregate rounded gravel	645	640
Water	138	148

Table 3. Properties of concrete mixtures after 28 days of curing

	Concrete I	Concrete II
Density [kg·m^3]	2240	2250
Compressive strength [MPa]	21	52
Flexural strength [MPa]	2.2	6.1

3.2.1 Treatment of concrete surfaces using the abrasive water jet

The tests were carried out using the abrasive water jet from a relatively short distance. The water jet was generated by a standard JETS 7.5/60c high-pressure pump on the basis of a pressure multiplier made by PTV s.r.o. with a SLICE cutting head used for the abrasive water jet generation (water nozzle diameter of 0.33 mm, focusing tube diameter of 0.78 mm, focusing tube length of 76 mm). The Australian garnet GMA with the grain size of mesh 80 (0.18 mm) was used as abrasive material.

3.2.2 Treatment of concrete surfaces using the plain water jet

The series of tests were carried out using the water jet without abrasives from a long distance (100 mm). The water jet was generated by a Hammelmann HDP 253 high-pressure plunger pump. The cutting head was positioned using the ABB IRB 6640-180//2.55 Master robot.

The water nozzle diameter of 1.75 mm was the same in all tests. Several tests were performed using the so-called pulsating water jet with a pulsation frequency of 20 kHz. According to our experience, this type of water jet is able to better simulate the abrasive wear of concrete surfaces.

First, grooves were performed with the jet slightly inclined to the testing surface. The jet moved only vertically from the bottom to the top at the speed of 1 mm·min^{-1}. Subsequently, the surfaces were created using the cutting scheme in Fig. 3. The speeds used for the jet motion are presented in the figure in individual trajectories.

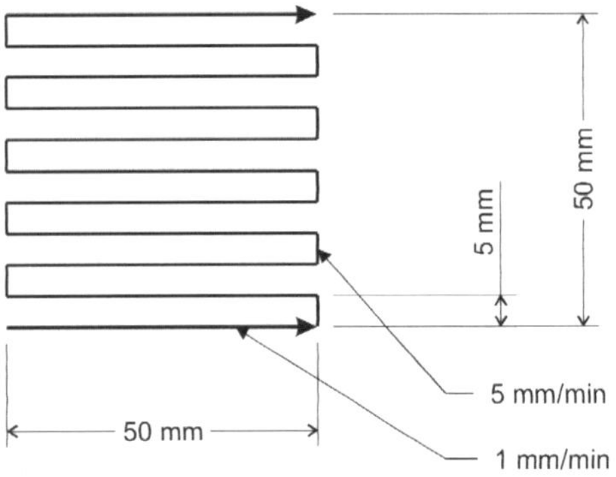

Fig. 3. Scheme of areal treatment of the surface

4 Results and discussion

In the following part, selected images of the treated surfaces (Figs. 4 - 10) including description and setting of parameters of the water jet technology for individual grooves/surfaces are shown.

The results show that abrasive water jets generated at high pressure and acting on the concrete surfaces from short distances are too aggressive and essentially cut the surface layers (Fig. 4) without washing out and taking away the structural elements of the surface layers, as is the case for the real erosion mechanism caused by rapidly flowing clean or polluted water. Unlike the abrasive jet, the plain water jet is partially reflected by the impact on the coarse aggregate after the hardened cement paste was removed. The jet bypasses the aggregate, thus changing its direction. However, erosion abilities of the plain water jet are several times less efficient compared to abrasive water jet (Fig. 5, surface 1), which is confirmed in [2].

When the abrasive water jet acts from large distances (up to 1 m), a considerable amount of the water jet energy is lost and, above all, the flow is expanded and impacts on a larger surface area. This method better reflects the reality as regards the modelling of the effects of the impinging water (Fig. 6, left). However, when the abrasive water jet impacts on the surface at a small angle, the subsurface structural elements are rather cut-off than eroded (Fig. 6, right). For the water jet without abrasives, the selected distance (500-1000 mm) is too long and the water jet is unable to leave any trace on the surface of the studied specimens within a specified period (1 min).

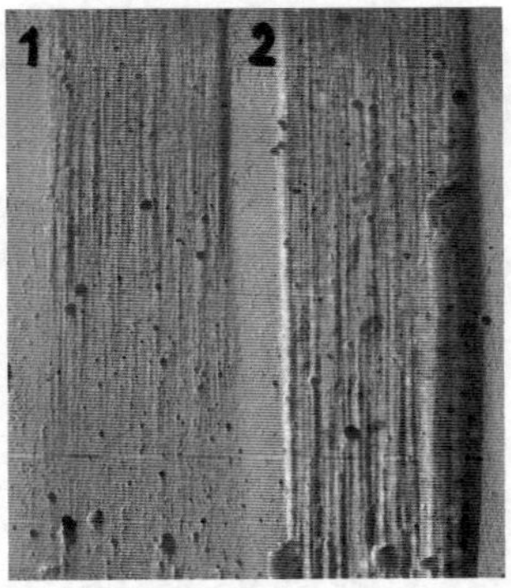

Fig. 4. Influence of water pressure and abrasive mass flow rate on the appearance of the treated surface (direction of jet action – from top to bottom, jet traverse speed of 20 mm·min^{-1}, jet inclination angle of 2° relative to the surface, stand-off distance of 22.5 mm; ***surface 1*** – water pressure of 100 MPa, abrasive flow rate of 100 g·min^{-1}, ***surface 2*** - water pressure of 300 MPa, abrasive flow rate of 300 g·min^{-1})

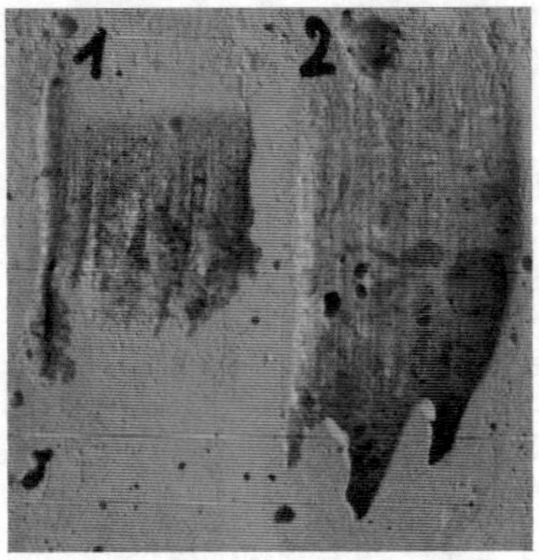

Fig. 5. Effect of abrasive presence on the appearance of the treated surface (direction of jet action – from top to bottom, jet traverse speed of 40 mm·min^{-1}, jet inclination angle of 10° relative to the surface, stand-off distance of 22.5 mm; ***surface 1*** – plain water jet, water pressure of 300 MPa, ***surface 2*** – abrasive water jet, water pressure of 200 MPa, abrasive flow rate of 200 g·min^{-1})

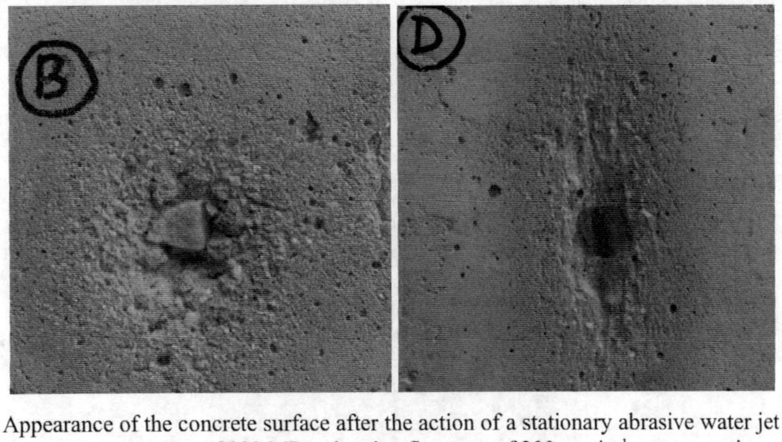

Fig. 6. Appearance of the concrete surface after the action of a stationary abrasive water jet from a long distance (water pressure of 300 MPa, abrasive flow rate of 200 g·min^{-1}, exposure time of the jet acting on surface - 1 min; ***left*** - jet inclination angle - perpendicular to the surface, stand-off distance of 1000 mm, ***right*** - jet inclination angle of 15° relative to the surface, stand-off distance of 500 mm)

Comparison of two different types of concretes shows that less solid concrete is less resistant to the high-speed water jet impact - whether it is the water jet without or with addition of abrasives. Although the abrasive water jet easily penetrated in the surface layers of concretes, the aggregate in the concrete structure was cut (Fig. 7) due to the high speed and energy of abrasive particles, as stated in [25]. Again, this method does not correspond to the resulting degradation of real concrete structures by erosion and abrasion. In this regard, tests using the water jets without abrasives - whether continuous or pulsating – better reflect the reality. The pulsating water jet is more efficient than the continuous water jet (compare left and right surfaces in Fig. 8) [26]. The fine-tuning of the required disintegration and especially the elimination of cutting into the material depend on the water jet parameters setting.

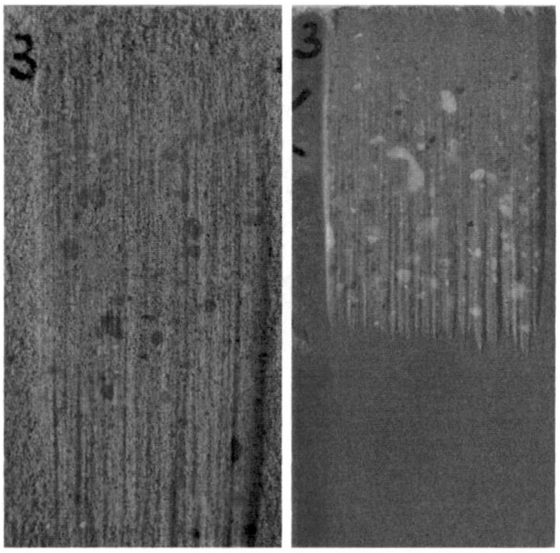

Fig. 7. Influence of concrete type on the appearance of the treated surface (direction of jet action – from top to bottom, water pressure of 110 MPa, water nozzle diameter of 0.33 mm, focusing tube diameter of 0.78 mm, jet traverse speed of 5 mm·min^{-1}, jet inclination angle of 2° relative to the surface, stand-off distance of 12 mm, abrasive flow rate of 50 g·min^{-1}; ***left*** – concrete I, ***right*** – concrete II)

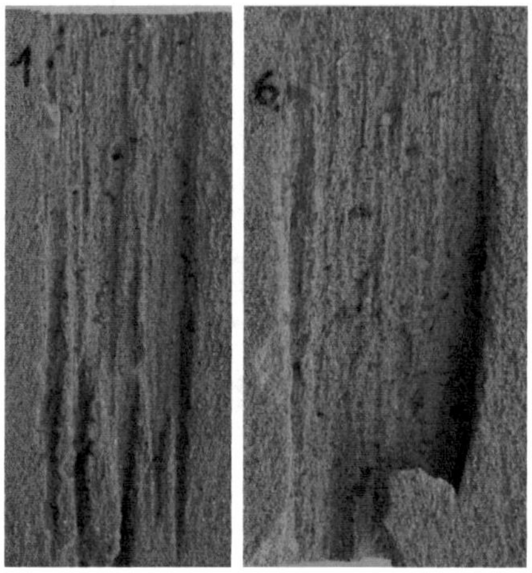

Fig. 8. Influence of pulsations in jet on the appearance of the treated surface (direction of jet action – from top to bottom, water pressure of 70 MPa, water nozzle diameter of 1.75 mm, jet traverse speed of 1 mm·min^{-1}, jet inclination angle of 8° relative to the surface, stand-off distance of 100 mm; jet motion according to scheme in Fig. 3; ***left surface*** – continuous jet, ***right surface*** – pulsating jet)

When a solid concrete was disintegrated by the water jets without abrasives with appropriately set parameters, the water jets did not penetrate into the structure of concrete surface layers, if the concrete surface was smooth. However, on the rough surface (Fig. 9), the material was disintegrated and removed due to the cavitation caused by the water flowing over the surface of the studied specimen [27].

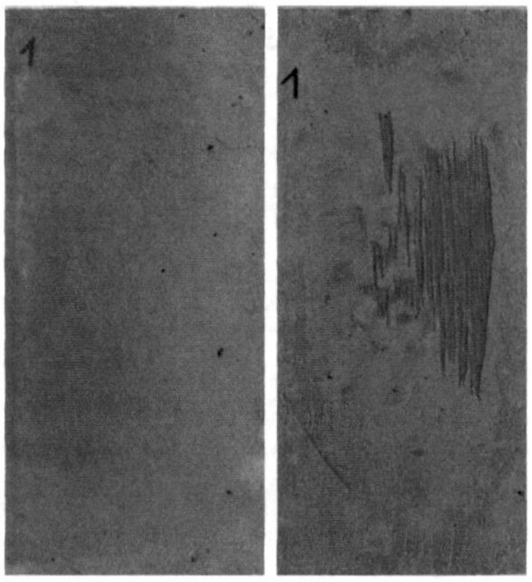

Fig. 9. Influence of surface roughness on the appearance of the treated surface (direction of jet action – from top to bottom, plain water jet, water pressure of 30 MPa, water nozzle diameter of 1.75 mm, jet traverse speed of 1 mm·min^{-1}, jet inclination angle of 8° relative to the surface, stand-off distance of 100 mm, jet motion according to scheme in Fig. 3; ***left*** – smooth surface, ***right*** – rough surface).

 https://doi.org/10.1051/matecconf/201824402007

Regading suitable water jet motion, the impact of a narrow, non-moving jet causes a deep groove on the studied surface. The water jet is thus not appropriate for the simulation of the real degradation process. It is more appropriate if the water jet moves along the surface perpendicular to the direction of the jet impact (e.g. Fig. 10). The composite motion of the water jet along the surface as seen in Fig. 3 is considered as the most suitable

Another possible option is the use of the so-called flat water jets, which spread the energy over a larger area due to a special nozzle [28]. However, this method was not tested during the experiments described. Due to the limited scope of the article, there is also no quantification of the obtained data describing the surface properties of concrete, statistical processing, etc. These data will be the subject of further publications by the authors of the article.

5 Conclusion

On the basis of an experimental program focused on finding an appropriate configuration of the high-speed water flow for an accelerated simulation of erosion failure of concrete surfaces exposed in practice to fast-flowing liquids, the following was found.

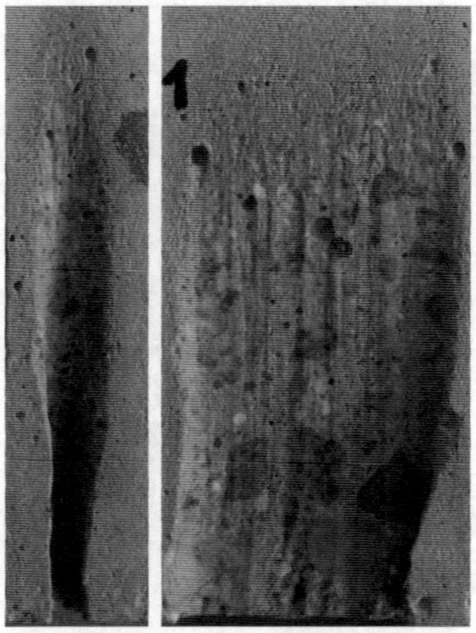

Fig. 10. Influence of jet motion along the concrete surface (direction of jet action – from top to bottom, abrasive water jet, water pressure of 300 MPa, water nozzle diameter of 0.33 mm, focusing tube diameter of 0.78 mm, focusing tube length of 76 mm, jet inclination angle of 10° relative to the surface, stand-off distance of 22.5 mm, abrasive flow rate of 300 g·min^{-1}; ***left groove*** – stationary jet, ***surface 1*** – moving jet, traverse speed of 40 mm·min^{-1})

The high-speed water jet technology is a suitable tool for simulation of erosion and abrasive wear of concretes. The abrasive water jet impacting on the surface from short distances is inappropriate for the simulation of real degradation processes, as the water jet cuts the structural elements of concretes, the aggregate included. However, this type of the jet can be used to simulate the impacting flow, but only if acting from long distances. The plain water jet impacting on the surface at a small angle is, on the contrary, a suitable simulation tool. However, it is desirable to move the plain water jet along the studied surface according to the appropriate scheme as the jet acts at the point of impact only. More efficient than the continuous plain water jet is the pulsating water jet, which shortens the simulation times. Simulating water jet parameters should be particularly set based on the appearance and the state of the real-degraded surfaces so that the result reflects the reality. Generally, rough surfaces are disintegrated more easily and quickly compared to the smooth surfaces due to the turbulence and cavitation caused by the flows of fast-flowing liquids along the surfaces.

This work was supported by grant of the Czech Science Foundation (18-25035S) and by project of the Institute of Clean Technologies for Mining and the Utilization of Raw Materials for Energy Use – Sustainability program, reg. no. LO1406 financed by the Ministry of Education, Youth and Sports of the Czech Republic.

References

1. D.A. Summers, *Waterjetting Technology.* E & FN Spon, London, (1995)
2. A.W. Momber, *Hydrodemoliton of Concrete Surfaces and Reinforced Concrete Structures*. Elsevier, Oxford, (2005)
3. A.W. Momber (ed.), *Water Jet Application in Construction Engineering*. A.A.Balkema, Rotterdam, (1998)
4. L.H. McCurrich, R.D. Browne, *Application of water jet cutting technology to cement grouts and concrete.* In 1st Int. Symp. Jet Cutting Technol. Eds. T. E. Brock, A. Richardson. BHRA, Cranfield. G7/69-G7/91 (1972)
5. R. Medeot, *History, Theory, and Practice of Hydrodemolition.* In 5th American Water Jet Conference, Toronto, Canada. U.S: Water Jet Technology Association. (1989)
6. J. Silfwerbrand, *Improving concrete bond in repaired bridge decks.* Concrete Int., **12**(9), 61-66 (1990)
7. A. Momber, R. Kovacevic, *Fundamental investigations on concrete wear by high-velocity water-flow.* Wear **177**(1), 55-62 (1994)
8. A.W. Momber, R. Kovacevic, J. Ye, *The fracture of concrete due to erosive wear by high-velocity water-flow.* Tribology transactions **38**(3), 686-692 (1995)
9. A.W. Momber, *A probabilistic model for the erosion of cement-based composites due to very high-speed hydro-abrasive flow.* Wear **368,** 39-44 (2016)
10. X.G. Hu, A.W. Momber, Y.G. Yin, *Hydro-abrasive erosion of steel-fibre reinforced hydraulic concrete.* Wear **253**, 848–854 (2002)
11. X.G. Hu, A.W. Momber, Y. Yin, *Erosive wear of hydraulic concrete with low steel fiber content.* Journal of hydraulic engineering **132**, 1331-1340 (2006).
12. ACI Committee 210. *Erosion of Concrete in Hydraulic structure* (ACI 210R-03). American Concrete Institute, Farmington Hill, MI (2003)
13. E.K. Horszczaruk, *Hydro-Abrasive Erosion of High Performance Fibre-Reinforced Concrete.*Wear **267**(1–4), 110-115 (2009)
14. Y.W. Liu, S.W. Cho, T.H. Hsu, *Impact abrasion of hydraulic structures concrete,* Journal of marine science and technology **20**(3), 253- 258 (2012)
15. A.W. Momber, R.S. Mohan, R. Kovacevic, *On-line analysis of hydro-abrasive erosion of pre-cracked materials by acoustic emission.* Theoretical and Applied Fracture Mechanics **31**, 1-17 (1999)
16. A.W. Momber, R. Kovacevic, *Accelerated high speed water erosion test for concrete wear debris analysis.* Tribology Transactions **39**(4), 943-949 (1996)
17. J. Foldyna, P. Jekl, L. Sitek, *Possibilities of utilization of modulated jets in rock cutting.* Proc. of the 1st Int. Conf. Mining Techniques, Filipowicz, Feliks (eds.), AGH, 85–96 (2001)
18. L. Sitek, *Analysis of the process of rock disintegration by high-speed water jets based on laboratory measurements.* Ph.D. Thesis, VŠB – Technical University Ostrava (2000)

19. P. Martinec, J. Foldyna, L. Sitek, J. Ščučka, J. Vašek, *Abrasives for AWJ cutting*. INCO –COPERNICUS No. IC 15-CT98-0821. Institute of Geonics (2002)
20. M. Zeleňák, J. Foldyna, M. Linde, F. Pude, T. Rentsch, J. Fernolendt, H.U. Poort, *Measurement and analysis of abrasive particles velocities in AWSJ*. Procedia Engineering, **149**, 77-86 (2016)
21. M. Zeleňák, J. Ščučka, K. Souček, *Velocity measurements on continuous micro water jet using fluorescent PMMA seeding particles*. Proceedings of the 5th Int. Conf. Water Jet 2017, Klichová, Sitek (eds.), Institute of Geonics, 221-231 (2017)
22. M.M. Vijay, *Power of pulsed liquid jets*. In: Geomechanics 93 Z. Rakowski, Editor. A. A. Balkema Publishers, Rotterdam, Netherlands, 265-274 (1994)
23. M.M. Vijay, J. Foldyna, J. Remisz. *Ultrasonic Modulation of High-Speed Water Jets*. In: Geomechanics 93 Z. Rakowski, Editor. A. A. Balkema Publishers, Rotterdam, Netherlands. 327-332 (1994)
24. J. Foldyna, L. Sitek, P. Jekl, D. Nováková, *Measurement of force effects of modulated jet*. Proceedings of the 39th Int. Conf. Experimental Stress Analysis, Jírová, Jiroušek, Kult (eds.), CTU in Prague, 63–68 (2001)
25. A.W. Momber, *Effects of erodent flow energy and local exposure time on the erosion of cement-based composites at high-speed hydro-abrasive flow*. Wear **378-379**, 145-154 (2017)
26. J. Foldyna, L. Sitek, P. Martinec, J. Ščučka, P. Jekl, M. Młynarczuk, *Rock cutting by pulsing water jets*. In: Eurock 2005 - Impact of Human Activity on the Geological Environment. P. Konečný Editor, A. A. Balkema Publishers, Leiden, Netherlands, 129-134 (2005)
27. A.W. Momber, *Short-time cavitation erosion of concrete*, Wear 241, 47-52 (2000)
28. L. Sitek, L. Bodnárová, J. Válek, M. Zeleňák, J. Klich, J. Foldyna, M. Novotný, *Effects of water jet on heat-affected concretes*. Procedia Engineering **57,** 1036-1044 (2013)

Comparative study of flank wear, tool life and type of chips for two similar turning inserts from different producers

Karel Šramhauser[1,*], *Nataša* Náprstková[1], *Pavel* Kraus[1], *Martin* Marek[1]

[1]J. E. Purkyně Univerzity in Ústí nad Labem, Faculty of Mechanical Engineering, Pasteurova 3334/7, 400 96 Ústí nad Labem,

Abstract. The article describes the analysis of the durability and lifetime of selected interchangeable cutting inserts used for turning based on the wear measurement of the cutting tool, flank wear namely. The experiment was carried out on one type of material, such as stainless steel 1.4404 according to DIN EN 10088-3. The measurements were made on two turning inserts of the same type from two different manufacturers - Sandvik coromant and Pramet tools. Evaluation of turning inserts wear was performed at four different cutting speeds. The value that was observed was the flank wear VB of the tool until the critical value was reached. Other different analysis have been performed, such as the length and shape of the chips analysis according to ISO 3685. Based on the results of the measurements, it has been found that the Pramet insert, which has a thinner layer of coating, has reached a longer life and allows heat to flow out of the cutting area. It can be considered that thinner coating can positively contribute to better machining of stainless steels, which could be helpful for machining companies. Cutting inserts wear research is realized at the Faculty of Mechanical Engineering of Jan Evangelista Purkyně University in Ústí nad Labem.

Keywords: tool life, removable cutting inserts, tool wear, chip's evaluation, turning

1 Introduction

Currently, tools with removable inserts are used quite ordinary. This type of tool has many advantages - it eliminates the problem of joining the soldering insert with the tool holder, including the effect of tension after soldering. It also saves the cost of tool sharpening, saves time to remove the shank and to adjust the tool. In combination with CNC machines, these benefits are crucial. [1, 2]

During the machining process, the tool is under influence of mechanical stress and a large amount of heat on the face and flank of the tool. The cutting insert, due to a combination of mechanical, thermal and chemical factors, gradually loses its cutting properties, [3, 4]. Wear on the cutting edge is changing tool geometry. Furthermore, the roughness of the contact points changes and the tool structure changes. [5] Throughout the machining process, the physical and chemical types of wear cause plastic deformation or brittle fracture which acts

* Corresponding author: karel.sramhauser@ujep.cz

Reviewers: *Stanislaw* Legutko, *Augustín* Görög

suddenly and can cause an immediate machining operation. This result, for example, as the plastic deformation of the tool or nose breaking. Wearing mechanisms also depend on temperature. [6, 7]

The reason for performing tool wear research is to record the actual course of wear, to evaluate the resulting wear and the ability to meet the cutting process requirements. All this in real time and machining environment. The type of selected turning inserts is similar for comparison purposes as well as cutting conditions. Tested materials are different, giving us a number of results that can be used by machining companies for the future choice of inserts purchase with regard to the most commonly used material or materials. [8, 9]

2 Materials and methods

Research is carried out on several types of materials, such as steels for quenching and tempering or normalised steels, etc. The test described in this article is part of the research. Test was performed on steel class 1.4404 according to DIN EN 10088-3. This is corrosion-resistant steel in an industrial environment, particularly resistant to pitting corrosion in the presence of chlorides, and especially resistant to sulfuric acid and phosphoric acid. Steel 1.4404 is weldable without the risk of intergranular corrosion in the area of thermal influence. It is suitable for cold forming and has good workability. Polishing is difficult to achieve mirror surface. Steel 1.4404 is used for welded constructions in aggressive industrial environments, for the manufacture of pressure vessels and components in the chemical, pharmaceutical and textile industries. This material can be used for food contact and dishes where high purity of the product is required. [10, 11, 12]

Table 1. Chemical composition of steel 1.4404 according to DIN EN 10088-3

Element symbol	Percentage content [wt %]
C	≤ 0.03
Mn	≤ 2.0
Si	≤ 1.0
Cr	$16.5 \div 18.5$
Ni	$11 \div 14$
Mo	$2 \div 2.5$
P	≤ 0.045
S	≤ 0.03

In this part of experiment, measurements were made on two rhomboidal-shaped removable turning inserts, namely on CNMG 120408E-M T9325 by Pramet (Fig. 2.) and CNMG 120408-PM 4325 by Sandvik (Fig. 3.). For measurement was used Olympus SZX10 stereomicroscope with measuring software in parallel.

Fig. 1. Removable turning insert Pramet CNMG 120408E-M T9325

Fig. 2. Removable turning insert Sandvik CNMG 120408-PM 4325

As a part of research was also to performed EDS analysis for both turning inserts. EDS analysis of these plates (Figures 1. and 2.) was performed on the Tescan Vega 3 scanning electron microscope with the Bruker X-flash nano EDS analyzer. Based on the analysis carried out it was possible to conclude that Pramet CNMG 120408E-M T9325 (Fig. 3) is formed with a substrate and a double coat. The substrate consists of tungsten carbide, titanium carbide and cobalt binder. The top coating with a thickness of 7 µm consists of Al_2O_3 with a measurable presence of titanium and carbon. The lower coat with a thickness of 6 µm is formed by titanium carbide. The presence of oxygen, aluminum and tungsten can also be observed probably by diffusing Al and O from the topcoat and W from the substrate. Sandvik CNMG 120408-PM 4325 (Fig. 4) is also formed with the substrate and double coat. The substrate is composed of tungsten carbide with a cobalt binder. Top coated layer with thickness 6 µm, is composed of Al_2O_3. Lower coating with thickness 12 µm is from the TiCN. [13, 14, 15]

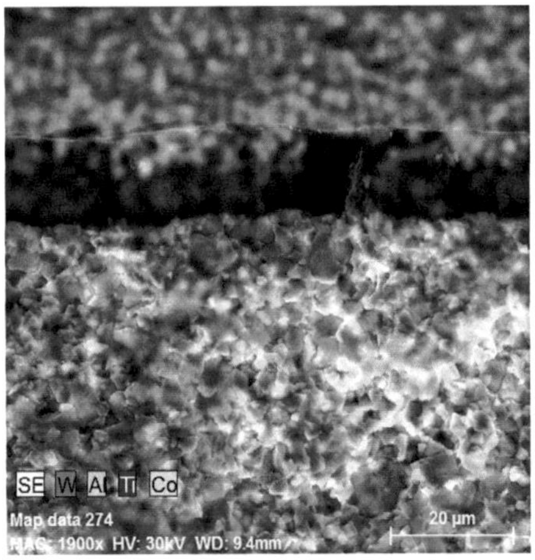

Fig. 3. Cross section of CNMG 120408E T9315

Fig. 4. Cross section of CNMG 432-PM 4325

The experimental material was turned on the three-axis CNC lathe DOOSAN Lynx 220L with the control system FANUC. For all operations was used tool holder PCLNR 2020 K12. The depth of cut a_p was set at 1.5 mm because the system had an instability at depth of 2 mm offset. Feed rate f was set to 0.3 mm·ot^{-1}. The inserts were tested at cutting speeds v_c = 80 m·min^{-1}, 100 m·min^{-1}, 120 m·min^{-1} a 140 m·min^{-1}. The critical wear limit of VB_{MAX} was determined on 0.3 mm.

2.1 Chip shape and length evaluation

Along with the measurement of VB_{MAX} flank wear, the shape and the length of chips was continuously analyzed using the cut-off method. The reason to look at the chip shape is the fact that only some types of chip are ideal for machining because of the smoothness of the cutting process, and these changes of the chip during turning indicates a change in the geometry of the cutting tool. [14, 15]

Chip length measurement by the cut-off method was performed on the Olympus SZX10 stereomicroscope (Fig. 5.). From each sample taken after the cutting cycles, 10 chips were selected that were most present in the sample set (type of shape). Chips were evaluated by ISO 3685. [16]

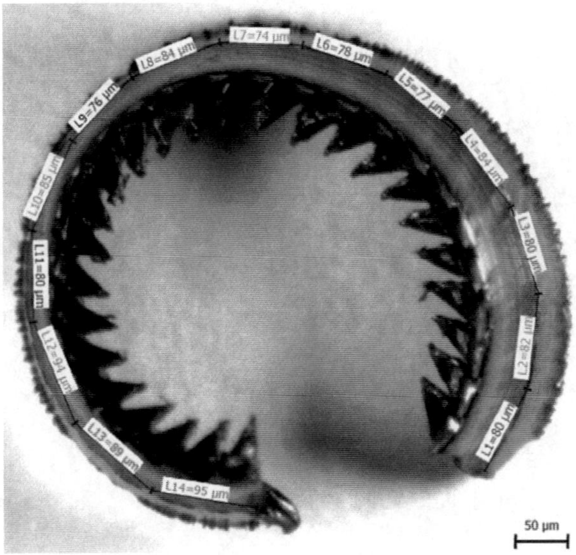

Fig. 5. Chip measurement by the cut-off method

3 The results and discussion

During the measurement the VB_{MAX} wear on the tool's flank was photographed based on visual evaluation. The result of photographic recording is a series of frames with the current wear value, which is used as the input data to create a real-time wear curve to depend on time. During machining stainless steel, the build-up-edge was created, as with titanium alloys. Because of this, it was necessary to distinguish optically between accumulated machined material on the tool and a really worn surface. The VB wear was measured from the initial level (due to plastic deformation) to the visible worn portion of the tool coating.

Fig. 6 shows reached maximal wear limit at Pramet CNMG 120408E-M T9325. The picture shows a flank wear of value 312 µm. There is also built-up-edge (BUE), probably due to the low cutting speed, and at the end of the flank wear area, you can see the notch wear, which is a common wear type when machining stainless steels.

Fig. 7 shows reached maximal wear limit at Sandvik CNMG 120408-PM 4325. The flank wear value is 307 μm. There is built-up edge and flank wear as similar as at Fig. 6. Unlike the Pramet insert, Sandvik CNMG 120408-PM 4325 has a built-up edge closely to tool tip.

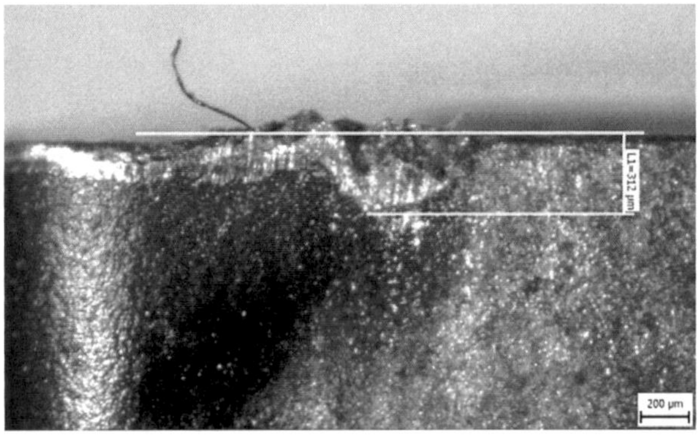

Fig. 6. VB_{MAX} wear of CNMG 120408E-M T9325, (v_c = 100 m·min^{-1}, T = 23 min)

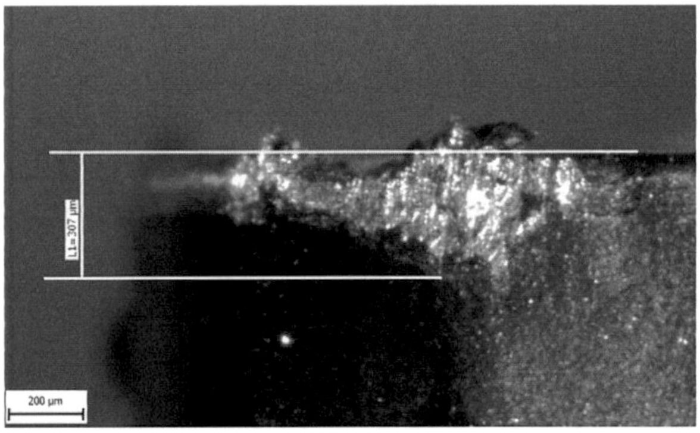

Fig. 7. VB_{MAX} wear of CNMG 120408-PM 4325, (v_c = 100 m·min^{-1}, T = 15 min)

According to the measured values of wear of the Pramet CNMG 120408E-M T9325 (Fig. 7.), it is apparent that the lowest life of the cutting edge was manifested at the cutting speed v_c = 140 m·min^{-1}. Shortly thereafter, the trend of wear was followed at cutting speeds v_c = 120 m·min^{-1} a further v_c = 100 m·min^{-1}. From the course of wear dependence at cutting speed v_c = 80 m·min^{-1} there is a repetitive built-up-edge on the tool, probably due to a lower cutting temperature at the tip of the tool nose.

The wear process data of the Sandvik CNMG 120408-PM 4325 was evaluated in the same way. The lowest lifetime of the cutting edge was at the cutting speed v_c = 140 m·min^{-1}. Additionally, the wear rate at cutting speeds v_c = 100 m·min^{-1} and v_c = 120 m·min^{-1} was followed. The course of wear at cutting speed v_c = 80 m·min^{-1} is similar to that of Pramet CNMG 120408E-M T9325. Again, a recurrent increase in the flank wear of the tool is evident.

Fig. 8 shows the comparison of the lengths of the time T [min] until the VB_{MAX} [μm] is reached, depending on the cutting speeds v_c [m·min^{-1}]. The comparison shows that the same or better results on tested steel class 1.4404 according to DIN EN 10088-3 have been achieved by Pramet CNMG 120408E-M T9325.

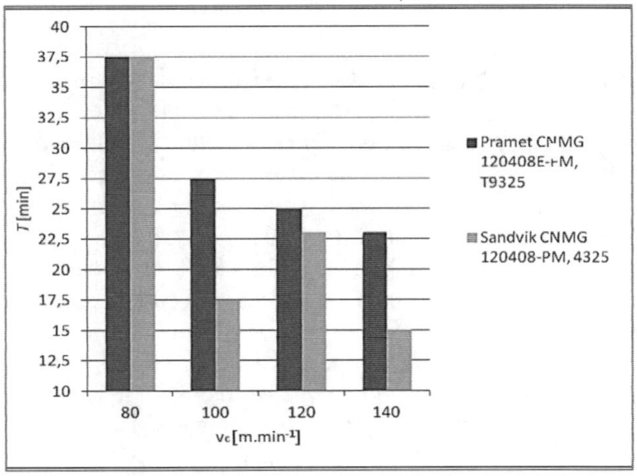

Fig. 8. Comparison of turning inserts wear time depending on cutting speeds

The coated layer on the Sandvik CNMG 120408-PM 4325 is about 5 μm thicker. Machining of stainless steels is difficult in creating built-up edge, making notch wear and also increasing wear due to high temperatures.

When we did machining basic, quality or miscellaneous steels of earlier parts of research, a thicker coated layer contributed to longer tool life. Based on the experiment data, it can be said that during machining stainless steel, the coating layer behave like a thermal insulator and thus mislead the heat from the cutting area. This results in higher wear due to higher temperatures.

In this part of experiment, three types of chips were produced according to ISO 3685 during machining. In majority, arch loose chips and arch connected chips, both of which were approximately the same. Elemental chips were restricted in two cases, at Sandvik CNMG 120408-PM 4325 at a cutting speed of 100 $m \cdot min^{-1}$ and at Pramet CNMG 120408E-M T9325 at a cutting speed 120 $m \cdot min^{-1}$, however, only in a small amount. The size of the analyzed chips based on the cut-off method was very similar to most of the experiment, approximately 6 mm, except for CNMG 120408E-FM VBD turning, T9325 with a cutting speed v_c =140 $m \cdot min^{-1}$, with arched chips appeared to be about twice as large as all other measurements (12 mm). Tab. 2. shows evaluation of the shapes and lengths of the chips on average for the individual cutting speeds.

Table 2. Chip evaluation on cutting speeds

Cutting speed v_c **[m·min⁻¹]**	**Cutting insert type**	
	Pramet CNMG 120408E-FM T9325	Sandvik CNMG 120408-PM 4325
	Chip type / Chip lenght *l* [mm]	
80	arch loose / 6 ÷ 16	arch loose, arch connected / 4 ÷ 7
100	arch loose, arch connected / 4.3 ÷ 6.8	arch loose, arch connected, elemental/ 6 ÷ 8
120	elemental arch loose, / 4.3 ÷ 6.8	arch loose, arch connected / 6 ÷ 8
140	arch loose, arch connected / 4.3 ÷ 6.8	arch loose / 6 ÷ 8

4 Conclusion

The paper describes the method and process of measuring the wear of the selected removable insert for turning. The analysis was performed under the given cutting conditions and parameters. Analyzed cutting inserts were Pramet CNMG 120408E-M T9325 and Sandvik CNMG 120408-PM 4325, tested material was steel class 1.4404 according to DIN EN 10088-3. Based on the entry tests, the wear time dependence for a given inserts was examined at different cutting speeds, at $v_c = 80, 100, 120$ a $140\ \mathrm{m \cdot min^{-1}}$. Better results, or longer lifetime, were achieved by the Pramet CNMG 120408E-M T9325. In addition to the cutting speed $v_c = 140\ \mathrm{m \cdot min^{-1}}$, where the Pramet and Sandvik inserts values would be virtually the same. The reason for the better results of Pramet insert was that the weaker coated layer allowed the heat to drain from the edge to reduce the heat load of the tool. During the measurement, were also evaluated chips by the cut-off method. According to ISO 3685 were most common arch loose, arch connected or elemental chips included. Based on the results of the measurements, it can be said that a thinner coating may contribute positively to better machining of stainless steels.

Authors are grateful for the support of grant No. CZ.1.05/4.1.00/11.0260 EDIMARE and SGS FME JEPU in Ústí nad Labem for 2018.

References

1. K. Kocman, *Technologické procesy obrábění.* (CERM, Brno, 2011)
2. J. Beddoes, M. J. Bibby, *Principles of Metal Manufacturing Processes.* (Antony Rowe, Eastbourne, 2003)
3. J. Madl, V. Shubert, *Experimentální metody a optimalizace v teorii obrábění.* (ČVUT, Praha, 1985)
4. J. Jurko, *Opotrebenie rezných nástrojov.* (Prešov, 2005)
5. G. M. Krolczyk, P. Nieslony, S. Legutko, *Determination of tool life and research wear during Duplex Stainless Steel turning.* Archives of Civil and Mechanical Engineering, **15**(2), 347-354, (2015)
6. J. Brychta, R. Cep, J. Novakova, L. Petrkovska, *Technologie II – 2. díl.* (VŠB – TU Ostrava, 2008)
7. J. Majerik, I. Barenyi, *Wear and Tool Life Investigation of Carbide Inserts while Hard Machining of Armox 500 Steel.* Manufacturing Technology, **18**(2), 273-278, (2018)
8. G. Krolczyk, S. Legutko, P. Raos, *Cutting edge wear examination during turning of Duplex Stainless Steel.* Tehnicki Vjesnik – Technical Gazette, **20**(3), 413-418, (2013)
9. G. Krolczyk, M. Gajek, S. Legutko, *Predicting the tool life in the dry machining of Duplex Stainless Steel.* Eksploatacja i Niezawodnosc – Maintenance and Reliability, **15**(1), 62-65, (2013)
10. J. Leinveber, P. Vavra, *Strojírenské tabulky.* (Úvaly, 2011)
11. N. Naprstkova, J. Cais, P. Kraus, M. Tavodova, *Using of electron microscopy for tool wear evaluation.*, (2017)
12. J. Madl, M. Vrabec, J. Kafka, R. Dvorak, *Technologie obrábění I.díl.* (ČVUT Praha, 1999)
13. A. Michalski, A. Sokolowska, S. Legutko, *The useful properties of TiNx-Ti coatings deposited onto drills at 500 K using the Reactive Pulse Plasma method.* Thin Solid Films, **129**(3-4), 249-254, (1985)

14. P. Twardowski, S. Legutko, G. M. Krolczyk, et al., *Investigation of wear and tool life of coated carbide and cubic boron nitride cutting tools in high speed milling*. Advances in Mechanical Engineering, **7**(6) (2015)
15. G. Krolczyk, M. Gajek, S. Legutko, *Effect of the cutting parameters impact on tool life in Duplex Stainless Steel turning process*. Tehnicki Vjesnik – Technical Gazette, **20**(4), 587-592 (2013)
16. ISO 3685. *Tool-life testing with single-point turning tools.* (1993)

The evaluation of the surface quality after laser cutting

Tomáš Zlámal[1,*], *Šárka* Malotová[1], *Jana* Petrů[1], *Zbigniew Brytan*[2], and *Václav* Musil[1]

[1]VŠB - Technical University of Ostrava, Faculty of Mechanical Engineering, 17. listopadu 15/2172, Ostrava - Poruba, 708 33, Czech Republic
[2]Instytut Materiałów Inżynierskich i Biomedycznych, Konarskiego 18a, 44-100 Gliwice, Poland

Abstract. The contribution deals with the evaluation of the surface quality after laser cutting and determination of the influence individual input laser parameters at the whole process. There were selected samples for experiment made from S235JR material with the thickness 10 mm and 15 mm. There was set up values of process parameters, such as laser power, focal spot position, cutting speed and shield gas pressure. The final surface quality was measured and evaluated on the basis of the selected surface roughness parameters and material ratio. The surface quality was evaluated from point of view of defects emerging after thermal splitting according to standard ČSN EN ISO 17 658.

Keywords: Laser Cutting; Cutting Parameters; Surface Roughness, Abbott-Firestone Curve.

1 Introduction

Currently, there is the development of non-conventional methods of machining including of the laser cutting technology which has width use in most key industry. The laser technology brings many benefits including of high accuracy, low running costs, quiet operation, zero wear of cutting tool, sharp edges, relatively small heat-affected area and especially width range of cut materials, metal or non-metallic material in comparison with conventional methods of chipping machining. In the past, there were realized several researches focused on laser cutting whether for steel, non-ferrous material, heavy machinable or plastic materials and their influence on the surface quality [1, 2, 3].

The cut quality is influenced especially by input parameters, choice of cut material and its thickness, structure or homogeneity. The laser cutting process is very complicated, therefore setting of the parameters such as laser power, shield gas pressure, cutting speed and focal spot position can significantly influence final cut quality. The right combination of these parameters can achieve reduction or even total prevention amount of defects emerging after thermal splitting. The defects, such as slides, melting of upper edge, presence of iron scales, burrs etc. have significantly impact on the evaluate surface from the point of view of surface roughness and material ratio which is characterized by Abbott-Firestone curve [4].

* Corresponding author: tomas.zlamal@vsb.cz

Reviewers: *Anna* Rudawska, *Michal* Hatala

It is important to focus on the final surface roughness and values of core roughness Rk, reduced peak height Rpk and reduced valley depth Rvk. The results of measuring help to predict surface in area of development of cracks, amount of protrusions and recesses which have the influence on primary wear, final life-time of surface cut sample and enable preset of input parameters which ensure higher cut quality [5].

2 The factors influencing laser cutting

As mentioned above, it is necessary to set the process parameteres with regards on selected material and its thickness in order to ensure sufficient cut quality. Incorrect setting of parameters can cause reduction of cut quality and presence of burrs, heat, etc. The evaluation of cut quality after laser cutting occurs in technical publications [6, 7, 8]. In these publication the authors mention the parameters, such as cutting speed, focal spot position, shield gas pressure or laser power and their influence on the surface quality. It is necessary to set sufficient laser power with regard to the material thickness which ensure its total separation for laser cutting. It is important that the value of the laser power was constant during the cutting time and allow to get the best cut quality combined with other set input cutting parameters. The low value of laser power causes significantly worsening of surface quality in the bottom part of the cut (tab. 1). In case of incorrect set (high value) of cutting speed may occur incomplete separation of material or by contrast it may cause worsening of surface roughness and pulling out material on the bottom edge. Just as laser power, the cutting speed has to be adapted to the type and thickness of material. Incorrect selected cutting speed can cause high value of surface roughness and origin burrs, it leads to a poor quality cut [9, 10]. In table 1, there are show of decreased of the cut quality caused unsuitable set cutting parameters.

Table 1 Cutting quality with different settings [11]

Cut quality	Low value	Optimal	High value
Laser power [W]		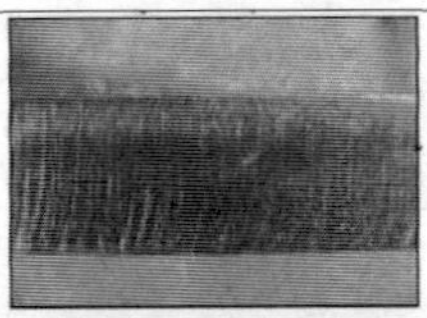	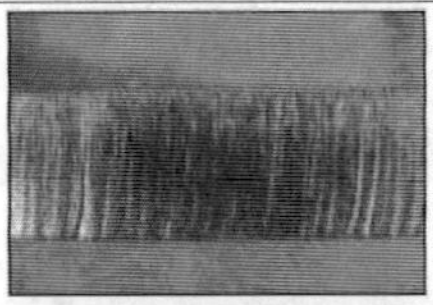
Cutting speed [mm·min^{-1}]	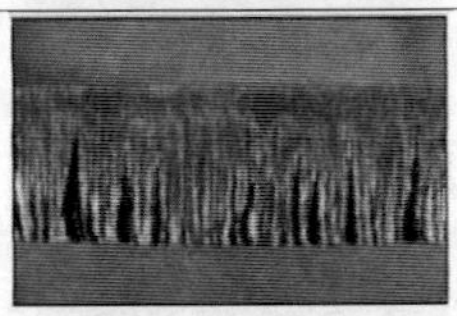		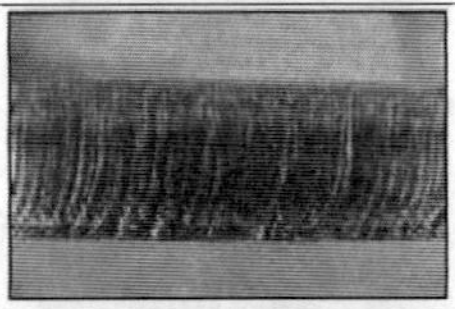
Focal spot position [mm]	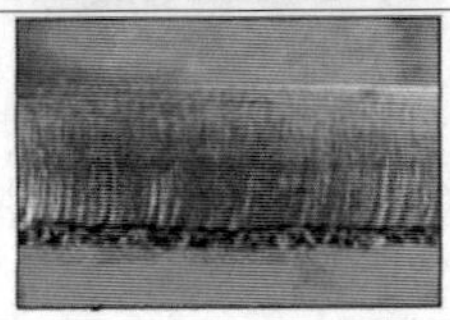	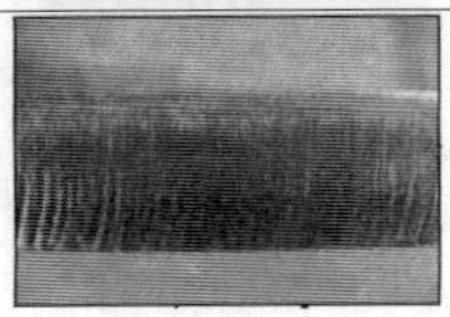	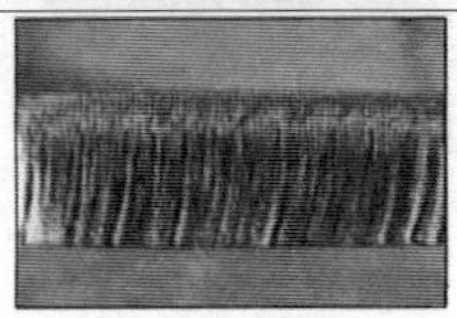

The next tracked parameters are focal spot position and shield gas pressure. If the focal spot position is set too near to the material, there can be created burrs on the bottom edge of the cut. In the opposite case, larger position causes worsened surface roughness and cut quality. The focal spot position is connected with its nozzle standoff height which is determine with regards on the material thickness and type of cutting. The shield gas pressure has an influence on the cut quality, surface roughness and origin of burrs. The shield gas pressure is necessary adapt to the material thickness. The function of shield gas pressure is to remove melting material from contact area of the laser beam together with machining material and ensure protection of lens during cutting. Increasing of the shield gas pressure allows increasing of cutting speed, and it has an impact on the worsened surface roughness [12]. In technical literature [13], there are shown practical experience dealing with evaluation of the laser quality and from the point of view of defects newly-emerge directly in place of the cut.

3 Design of the experiment

The experimental part was focused on determination of the cut quality after laser cutting. The evaluation deals with selected surface roughness parameters and with the defects emerging during laser cutting. There were used samples made from soft carbon constructional steel ČSN 11 375 (S235JR). The dimensions of the samples were 60 x 30 mm with thickness 10 mm and 15 mm. The cuting speed, the shield gas pressure and the focal spor position were changed according to proposal the experimental activity and achieved of the cut quality. The other parameters had the constant value at the whole time of the cutting, see table 2.

Table 2. Designed parameters for laser cutting

Laser power [W]	3 200			
Nozzle standoff height[mm]	0.9			
Thickness of the sample 10 mm				
Cutting speed [mm·min^{-1}]	1800	1700	2000	1900
Shield gas pressure [kPa]	50	50	70	60
Focal spot position [mm]	1.5	1.6	1.7	1.8
Thickness of the sample 15 mm				
Cutting speed [mm·min^{-1}]	1200	900	800	1400
Shield gas pressure [kPa]	60	70	80	60
Focal spot position [mm]	1.6	1.7	1.8	1.4

The significant criterion for the evaluation of the cut quality is the surface roughness. There was used optical microscope Alicona InfiniteFocus G5 for its evaluation. The surface measurement was realized out of area of visible defects emerging during laser cutting, such as heating of the cut edge and presence of the burrs. The evaluation of the influence of cut quality was realized in place where visibly came about the biggest influencing of the surface by laser cutting. There were surface roughness parameters Ra, Rq, Rz and the core roughness Rk, reduced peak height Rpk and reduced valley depth Rvk which were determined from material ratio (Abbott-Firestone curve) which were evaluated (fig. 1).

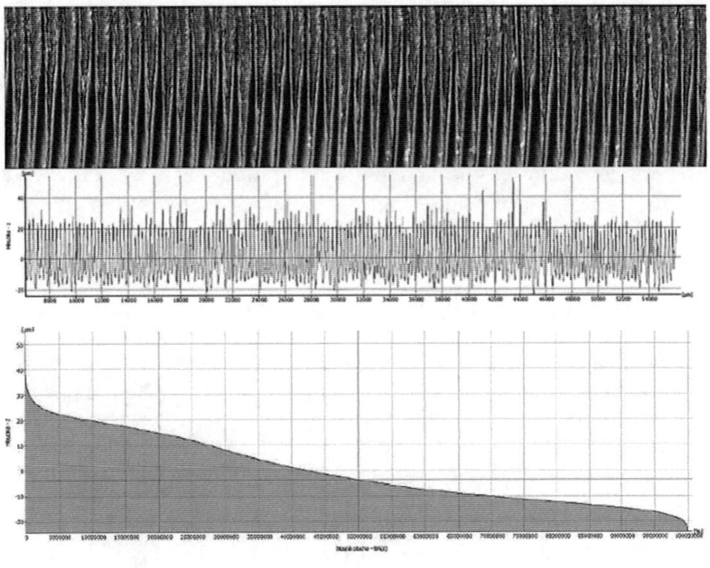

Fig. 1. Example of measurement of the surface quality

4 The evaluation of cut quality after laser cutting

Cutting of the samples was realized under pre-determined cutting parameters, the thickness of sample substantially influences cutting process and the final cut quality. There has been proposed lower cutting speed for thick samples to avoid incomplete splitting of the material. The surfaces of cut were composed of the slides (tab. 3) – the scratches in to the surface forming its roughness. The slides were caused by higher cutting speed and occurred more likely on the bottom cut of the edge. The grooves are almost parallel to the direction of the laser beam, and their occurrence is particularly evident in sample no. 3, where the highest cutting speed was used. With increasing speed is created larger slide against direction of the cutting. The selected value of the laser power was sufficient for cutting of the both set of the samples.

The top edge of individual cuts were covered with overhangs which were caused by overflowing material. The cut edge overhang was accompanied by the adherent slag intersecting the cutting area was in the sample no. 7 (larger thickness of the sample). The cause of her origin was high value of the shield gas pressure combined with lower cutting speed. It caused excessive worsening of the cut quality (tab. 3). The surfaces on samples of 15 mm were created by strong slides again on the bottom sides of the cut which changed into grouped gouges in specific places, especially for samples no. 6 and 7 as a result of set very low cutting speed.

The measured results of the surface roughness of the both sets of samples confirmed that the thickness of cut material has the influence on surface roughness of the cut (tab. 3). For the sample set with smaller thickness (10 mm), there were measured lower values of the surface roughness. The best results were evaluated for sample no. 2, there was set the lowest value of cutting speed and the shield gas pressure. For the same sample, there was also measured the lowest value of the core roughness Rk, which it corresponds to the smallest area comprising of the protrusions and the recesses. This confirm that the surface after laser cutting is the least prone to the wear and clinging of impurities on the surface. With increasing values of the input parameters, there came about significant worsening of the surface quality the other cutting surfaces. The samples of the thickness 15 mm, there were used much lower cutting speed in comparison with previous set of the samples with regard to total separation of the material. Despite lower input parameters, especially the cutting speed, there wasn´t achieved better cut quality. The combination of the larger thickness and smaller cutting speed more than 50 % caused worsened surface roughness of the all cuts. This can observed on the samples no. 6 and no. 7, where the values core roughness Rk acquired very high value confirming surface poor-quality created by strong unevenness. In table, there is shown the values another parameters, such as Rpk - reduced peak height, peak height sticking out above the core roughness, Rvk - reduced valley depth is the depth under the core roughness.

Table 3. The results of the samples with thickness 10 mm and 15 mm

Sample		Ra [μm]	Rq [μm]	Rz [μm]	Rk [μm]	Rpk [μm]	Rvk [μm]
Thickness 10 mm	1	11.57±0.56	13.71±0.22	60.65±0.13	36.23	11.08	4.95
	2	11.53±0.94	13.27±0.29	64.26±0.15	26.60	17.75	3.43
	3	13.04±0.77	15.27±0.90	69.86±0.33	46.38	8.02	8.43
	4	13.30±0.08	15.60±0.14	67.90±0.29	39.92	14.94	3.21
Thickness 15 mm	5	17.16±0.80	20.37±0.41	93.28±0.32	61.29	10.30	11.70
	6	17.37±0.07	21.22±0.54	103.51±0.71	59.72	13.89	17.12
	7	20.47±0.15	25.38±0.65	126.38±0.11	67.47	11.54	29.90
	8	16.11±0.09	19.30±0.83	94.49±0.26	54.34	14.08	9.96

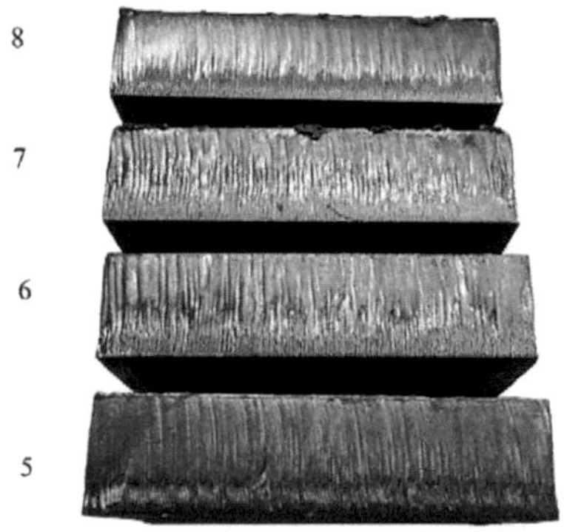

5 Conclusion

The experimental activity was focused on the evaluation of the cut quality using laser technology. For realization of the experiment there was designed testing samples with thickness of 10 and 15 mm which were made from material S235JR and were selected parameters of the laser cutting. The cut quality was evaluated on the basis of the measuring selected surface roughness parameters and selected parameters defined material ratio using optical non-contact microscope Alicona Infinite Focus G5. The cut quality of the surfaces was evaluated on the basis of standard ČSN EN ISO 17 658, which determinates defects emerging after thermal splitting on the material surface.

Not only the choice of material, its thickness but selected parameters of the cutting process have an influence on the surface quality of material after laser cutting. For the experiment, there were set constant laser power and nozzle standoff height apart cut material. The selected parameters, especially cutting speed, were selected with regards to thickness of the material, so that was ensured total splitting of the material. Depending on select values of the cutting speed was achieved better results of the surface roughness for set of samples with thickness 10 mm. The best surface roughness was measured for the sample with the lowest cutting speed and the shield gas pressure. This surface achieved required quality and improved resistance against the wear of surface during process. Despite substantially lower set input parameters of laser, the surface for samples with thickness 15 mm showed strongly worsened roughness and surface quality. The combination of the more than twice the lower cutting speed with shield gas pressure caused poor quality cuts. For these samples, there was measured also high value of the core roughness Rk, which confirmed that the surface is prone to the wear and it created by strong unevenness.

From the point of view of evaluation of surface quality according to standard ČSN EN ISO 17 568, the surfaces were covered with slides, which were caused by higher cutting speed. For the samples with thickness 10 mm, there were the slides mostly parallel with direction of the laser beam. It caused by constant and sufficiently large laser power, where didn´t come about delay of the cutting tracks. For the both sets of the testing samples, there were cut edge covered with overhangs caused by overflowing basic material. As a result of of the higher shield gas pressure and the lower cutting speed was created the adherent slag for the samples with thickness 15 mm. The other way around the samples which were cut with the low cutting speed had worsened surface roughness and strong slides. These slides changed gradually into grouped gouges in specific places.

Generally, it can be confirmed that surface quality created either laser technology or other thermal separation is significantly influenced by material thickness and set process parameters. Not always, the setting their high values is achieved request surface quality. Just suitable combination of cut material and input parameters can achieve significantly lower surface roughness and better cut quality. If there will achieve to find of suitable combination for selected cut material, then the surface won´t need to finish using by chipping machining and it will have an influence decrease producing time and costs at the total production.

Article has been done in connection with projects Education system for personal resource of development and research in field of modern trend of surface engineering - surface integrity, reg. no. CZ.1.07/2.3.00/20.0037 financed by Structural Founds of Europe Union and from the means of state budget of the Czech Republic and by project Students Grant Competition SP2018/150 and SP2018/136 financed by the Ministry of Education, Youth and Sports and Faculty of Mechanical Engineering VŠB-TUO.

References

1. W. Pakiela, T. Tanski, Z. Brytan, K. Labisz, *The influence of laser alloying on the structure and mechanical properties of AlMg5Si2Mn surface layers.* Appl. Phys. A Mat. S&P, **122** (2016)
2. E. Librera, G. Riva, H. Safarzadeh, B. Previtali, *On the use of Areal Roughness Parameters to Assess Surface Quality in Laser Cutting of Stainless Steel with CO2 and Fiber Sources*. Procedia CIRP, **33**, 532-537 (2015)
3. K. Jarosz, P. Loschner, P. Nieslony, *Effect of Cutting Speed on Surface Quality and Heat-affected Zone in Laser Cutting of 316L Stainless Steel.* Proc. Eng., **149**, 155-162 (2016)
4. J. Petru, T. Zlamal, R. Cep, K. Monkova, P. Monka, *Influence of Cutting Parameters on Heat Affected Zone after Laser Cutting.* Tehnicki Vjestnik – Tech. Gaz., **20**, 225-230 (2013)

5. W. Pakiela, T. Tanski, M. Pawlyta, K. Pakiela, Z. Brytan, M. Sroka, *The structure and mechanical properties of AlMg5Si2Mn alloy after surface alloying by the use of fiber laser.* App. Phys. A., **124** (2018)
6. D. Narovec, *Analysis of the surface after laser cutting*: Diploma Thesis, 67 (2018)
7. M. Hatala, J. Zajac, R. Cep, I. Orlovsky. *Research of the technological parameters importance for plasma arc thermal cutting.* In: 2nd ICMAE, 110-116 (2011)
8. M. Tonkovic, *Analysis of the material for laser cutting*: Dip. thesis, 66 (2014)
9. V. Musil, M. Sadilek, R. Cep, S. Malotova, *The influence of cutting parameters on surface quality and nitrogen consumption during laser cutting.* MM SJ. **2018**, 2363-2366 (2018)
10. Trumpf Werkyeugmaschinen GmbH + Co. KG. *Laser machining: CO2 laser - technical information*, **2**, 195 (2016)
11. J. Stastny, *Connection of mistakes cut with cutting characteristic at laser cutting.* Dip. Thesis, 66 (2012)
12. M. Grepl, *Study of the Affected Areas of Superalloys Materials after Laser Beam Machining*. Dis. thesis, 178 (2015)
13. Trumpf Werkyeugmaschinen GmbH + Co. KG. *Laser processing: CO2 laser – tech. info.* **2**, 156 (2007)

Studies of thin-walled parts deformation by gripping force during turning process on an example of bearing ring

Adam Patalas[1,*], *Michał* Regus[1], *Katarzyna* Peta[1]

[1]Poznań University of Technology, Faculty of Mechanical Engineering and Management, Piotrowo 3, 60-965 Poznań, adam.patalas@put.poznan.pl; michal.j.regus@doctorate.put.poznan.pl

Abstract. In this paper thin-walled part deformation during finishing turning process caused by gripping force of hydraulic lathe chuck was investigated. Bearing ring was taken as an example of thin-walled part undergo finishing turning operation. Finite Element Method (FEM) was used to define the deformation of examined part. The aim of presented research was to compare the deformation of bearing ring caused by gripping force of hydraulic 3-jaw chuck and 6-jaw chuck for different values of total gripping force. The data obtained from conducted simulations allowed to evaluate the influence of gripping force on machining part deformation which is directly related with its geometrical accuracy.

Keywords: FEM, thin-walled parts, gripping force, machining part deformation

1 Introduction

Machining is one of the most common methods for producing parts for different branches of industry. Nevertheless, there is a group of parts, which production is associated with high technological difficulties. Thin-walled parts, which are highly susceptible for deformation, can be counted in among these parts. The following factors leads to deformation of thin-walled parts during machining: gripping force, cutting force and heat generated during process [1, 2]. In this paper bearing ring deformation, during finishing turning of inner diameter, caused by gripping force of hydraulic chuck was investigated. In this purpose Finite Element Method was used. The drawing of analysed bearing ring is presented in fig. 1.

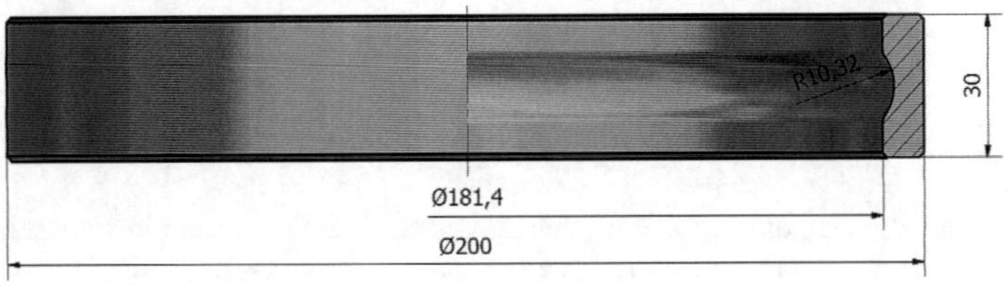

Fig. 1. Simplified drawing of thin-walled bearing ring

* Corresponding author: adam.patalas@put.poznan.pl
Reviewers: *Jana* Petrů,, *Alžbeta* Spietová

A couple of research regarding deformation of thin-walled parts which undergo machining process have been already conducted. In papers [2- 5] behaviour of thin-walled workpiece during milling operation was investigated. In researches [1, 5, 6] deformation of rings under turning process were analysed. R. Thakore et al. describes in [7] potential problems during grinding operation of bearing rings.

In this paper numerical simulations of bearing ring deformation induced by gripping force were conducted for 3-jaw chuck and 6-jaw chuck (Fig. 2).

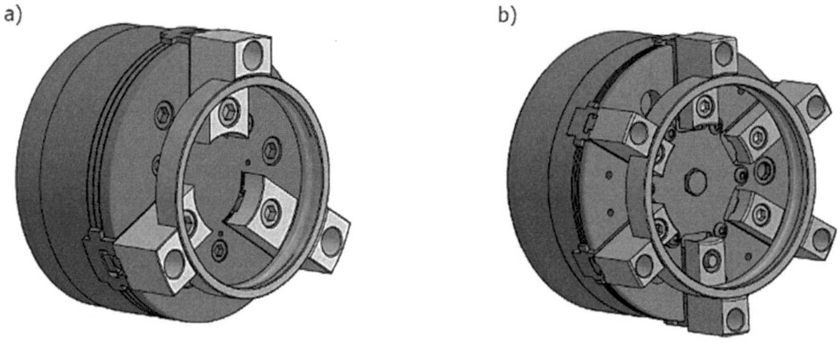

Fig. 2. Clamping strategy; a) 3-jaw chuck SCHUNK ROTA NCD 255 b) 6-jaw chuck ROTA NCR 250

Most of modern CNC lathes are equipped with hydraulic cylinder which control the opening of lathe chuck. Depends on the application, hollow and partially-hollow cylinders are in use. The principle of operation for clamping mechanism on CNC lathe is presented in fig. 3.

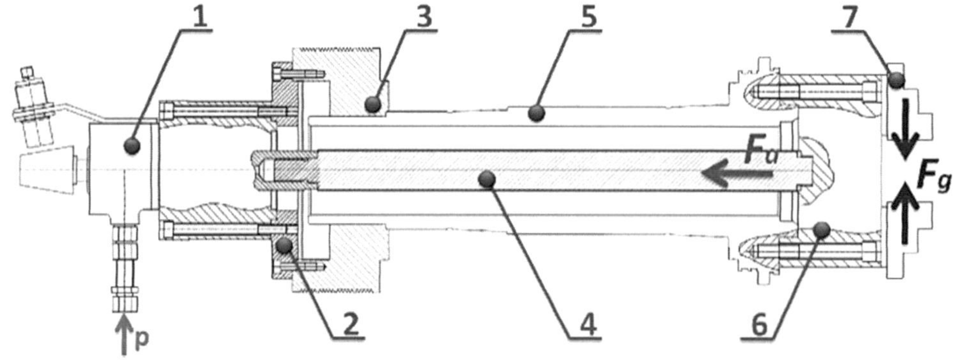

Fig. 3. Schematic drawing of clamping mechanism on CNC lathe; 1) Rotating hydraulic clamping cylinder, 2) Adapter of clamping cylinder, 3) Pulley of main drive, 4) Draw bar, 5) Spindle, 6) Chuck, 7) Jaws; Fa- actuating force of clamping cylinder, Fg - chuck gripping force.

For further analysis partially-clamping cylinder SMW SIN-S 100 was assumed. The operating range of clamping cylinder were limited to 7÷40 bar. Active piston area of selected cylinder is 66 cm^2 [8]. The theoretical characteristic of cylinder SIN-S100 is shown in fig. 4.

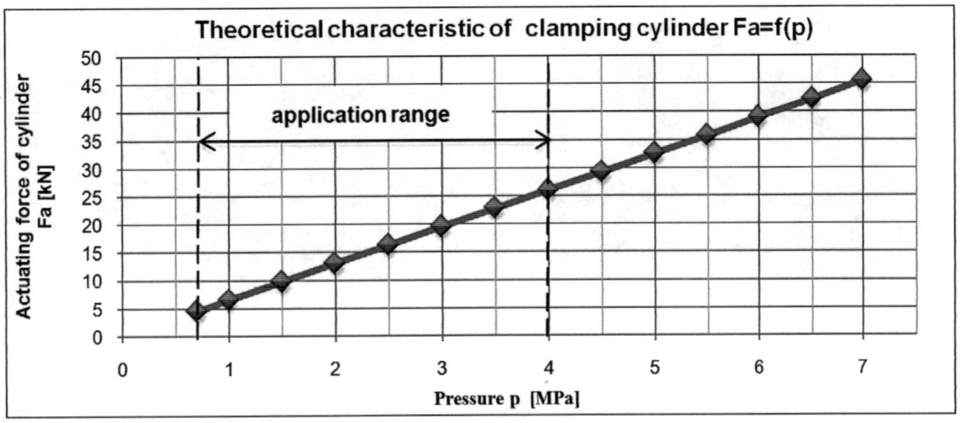

Fig. 4. Theoretical characteristic of hydraulic clamping cylinder SMW SIN-S 100 [8]

Machining process of thin-walled parts require possibly low gripping force to minimize the deformation and keep geometrical tolerances of produced parts [6]. Number of chuck's jaws plays significant role in thin-walled part deformation. The simulations of bearing ring deformation were conducted for 3-jaw chuck (ROTA NCD 255) and 6-jaw chuck (ROTA NCR 250) [8]. The characteristics of chuck total gripping force in function of pressure [Fg = f(p)] are presented in fig. 5.

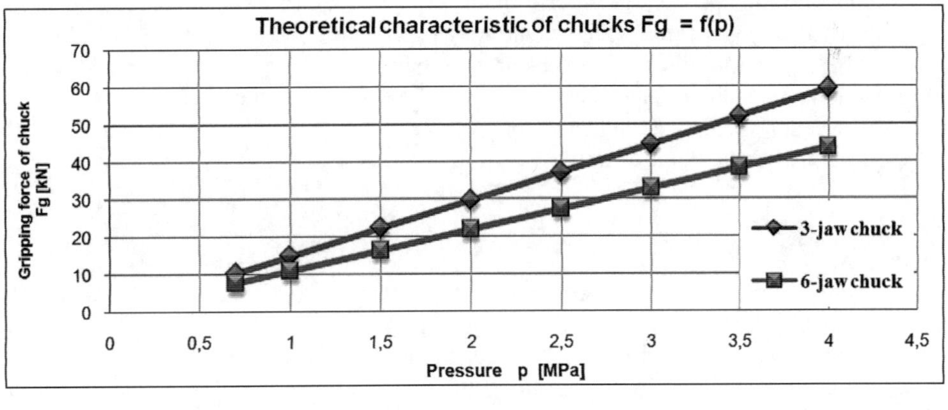

Fig. 5. Theoretical characteristics of chucks in combination with clamping cylinder SIN-S 100 [9]

2 Methodology of research

To evaluate the deformation of bearing ring under finishing turning conditions, Finite Element method (FEM) was applied. ANSYS software was used for simulation. Direct solver with force convergence control was taken. Augmented Lagrange method was applied for contact problem formulation. Due to the contact stress occurrence between the jaws and workpiece, static structural simulation is considered as non-linear. As simulation aims to estimate deformation of machined part, refined mesh was applied only for analysed bearing ring (Fig. 6).

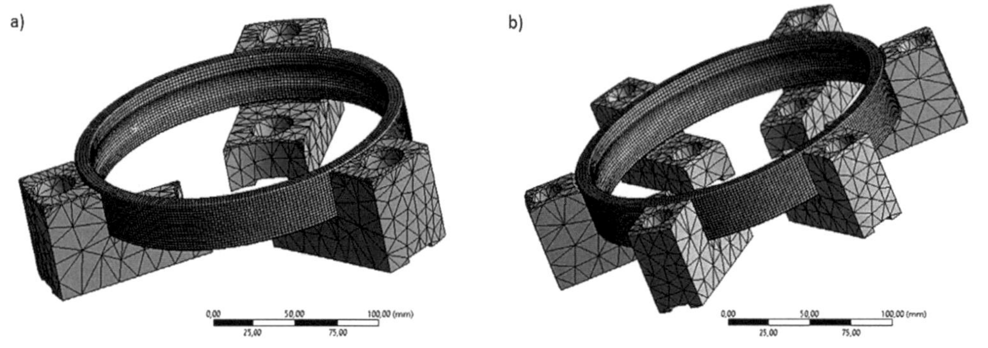

Fig. 6. Meshing problem a) 3-jaw chuck, b) 6-jaw chuck.

As a consequence of the above, for jaws tetrahedron finite element was assumed, while for workpiece hex-dominant type was chosen.

Further assumptions were taken for the simulation:

- linear model of material of bearing ring (alloy steel 100Cr6),
- linear characteristic of hydraulic clamping cylinder,
- linear characteristic of clamping chucks,
- equal clamping force at every chuck's jaw,
- inertial force working on jaws wasn't taken into consideration due to low spindle speed during turning operation at internal diameter of bearing ring (assuming cutting speed $v_c = 240$ m.min^{-1}, spindle speed would be $n = 421$ rpm),
- cutting force influence was ignored during the simulation, due to low value under finishing turning condition, as well as the fact, it works steadily at the circumference of ring,
- thermal deformation during machining process was ignored in simulation, since it affects evenly the shape of ring.

3 Results

Base on the assumptions, presented above, simulations were conducted, to evaluate the internal strains and deformation of the bearing ring. Results of FEM analysis for the lowest gripping force are presented in figure 7. As seen in figure 7 application of 6-jaw chuck allows to reduce significantly the part deformation, and as a consequence, obtain higher roundness accuracy.

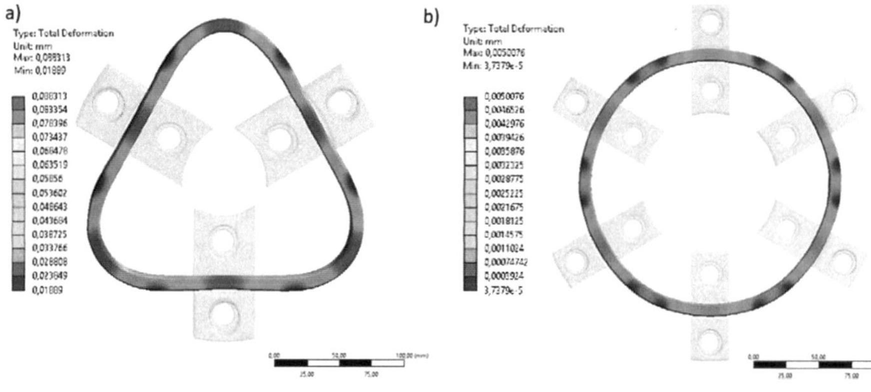

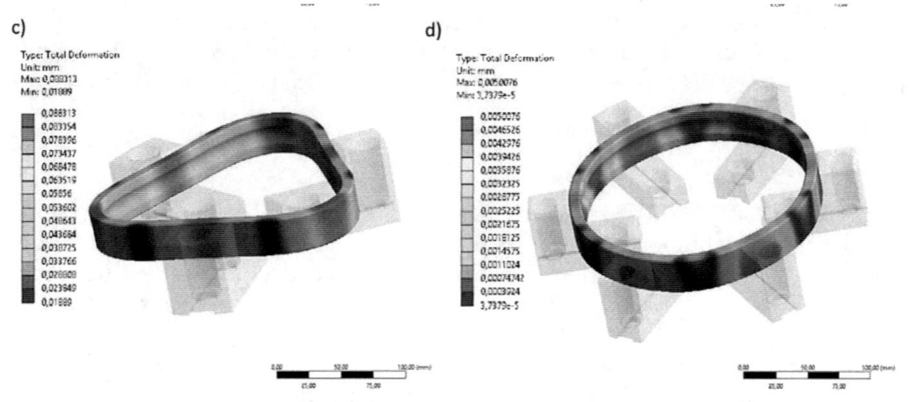

Fig. 7. Results of FEM analysis for total gripping force of lathe chuck Fg = 10 kN a),c) clamping in 3-jaw chuck with visible triangular roundness error of ring profile, b),d) clamping in 6-jaw chuck with less deformation and profile error

Deformations of jaws were omitted, as this problem, however critical for overall fixture performance, was specified as non-essential in this very case. Deformations on diagrams for 3-jaw fixture are up scaled 20 times, while deformations on diagrams for 6-jaw fixture are up scaled 200 times, for better clarity of visualization and in order to emphasis characteristics. Complete results of conducted simulations for different value of gripping force are presented on below diagrams (Figure 8). As it was mentioned in previous part of this paper, the linear model of the material has been assumed. Consequently, the obtained characteristics of stress and deformation are linear. It needs to be highlighted that those results are reliable within Hooke's law application range. It's marked on the graphs below.

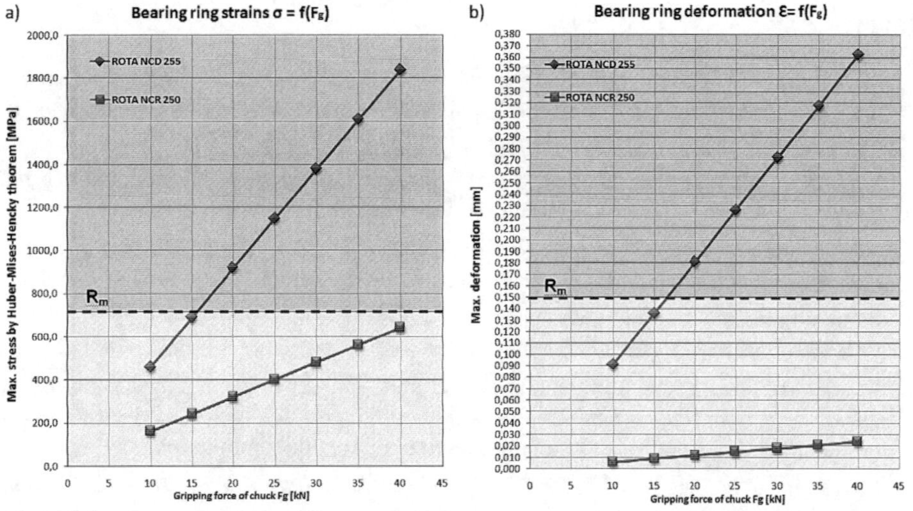

Fig. 8. Results of FEM analysis a) stress of bearing ring as a function of gripping force $\sigma=f(Fg)$, b) deformation of bearing ring as a function of gripping force $\varepsilon=f(Fg)$; area under dashed line represents the field of Hooke's law application

Above this range, yield limit for selected steel is exceeded, therefore linear model of material cannot be considered anymore. To evaluate plastic deformation of bearing ring non-linear model would have to be implemented. However, it's worth to mention, that such simplification is reasonable in this particular case. One of the major assumption of conducted analysis, is that workpiece can undergo only elastic deformation. Exceeding the yield limit results in destruction of bearing ring, what obviously is not permissible in production process.

4 Conclusions

In this paper thin-walled part deformation under finishing turning condition was investigated. As a research object, bearing ring made of alloy steel was taken. To evaluate the influence of gripping force on part deformation FEM analysis was conducted. Base on the obtained results, it can be stated that application of 6-jaw chuck allows to reduce significantly the stress and deformation of thin-walled parts. According to the Figure 8, if we apply the gripping force greater than 15 kN on 3-jaw chuck, yield limit would be exceeded. Consequently, bearing ring would be damaged. Even the lowest gripping force applied to the 3-jaw chuck (10 kN), results in bearing ring deformation of 0.09 mm - which is not acceptable. With the same gripping force applied by 6-jaw chuck we can reduce the maximal deformation of bearing ring to 5 μm. As a result, geometrical tolerances can be fulfilled.
Base on presented mathematical model it's possible to estimate the thin-walled part deformation under gripping force of lathe chuck. Consequently, right value of hydraulic pressure can be adjusted to reach suitable gripping force, that would provide required geometrical accuracy during production process. Furthermore, presented research can be helpful for selection of appropriate CNC lathe equipment, like hydraulic clamping cylinder and power chuck

Acknowledgement
This research was supported by Poznań University of Technology as part of statutory research on Faculty of Mechanical Engineering and Management.

References

1. M. Estremsa, M. Arizmendia, A. J. Zabaletaa, A. Gil, *Numerical Method to Calculate the Deformation of Thin Rings during Turning Operation and its Influence on the Roundness Tolerance*. Procedia Engineering, **132,** 872 – 879 (2015)
2. J. Loehea, M. F. Zaeha, O. Roescha, *In-Process Deformation Measurement of thin-walled Workpieces*. Procedia CIRP, **1**, 546 – 551 (2012)
3. J. Shia, J.Gaoa, Q.Songa, Z.Liua, Y.Wana, *Dynamic deformation of thin-walled plate with variable thickness under moving milling force*. Procedia CIRP, **58**, 311 – 316 (2017)
4. Z. Cai, J. Geng, C. Zhang, Y. Du, *Systematic solving of machining deformation and process optimization for complex thin-walled parts*. Procedia CIRP, **56**, 167 – 172 (2016)
5. W. Bai, X. Zhu, *Finite element simulation and analysis of part deformation induced during milling of thin-walled aerospace monolithic structure parts*. IEEE, Converence on Intelligent Computing and Intelligent Systems; **1**, 440 – 444 (2010)
6. M. Estremsa, M. Arizmendia, W. E. Cumbicusa, A. Lópeza, *Measurement of clamping forces in a 3 jaw chuck through an instrumented Aluminium ring*. Procedia Engineering, **132**, 456 – 463 (2015)
7. R. Thakore, T. Parsana, R. Dave, A review: *Potential Failures in Grinding Process of Bearing Rings and its Solution*. IJERT, **3**, 666 – 670 (2014)
8. www.smwautoblok.com (Access: 30.01.2018)
9. www.schunk.com (Access: 30.01.2018)

Simulation of spark ignition engine performance working on biogas hydrogen mixture

Donatas Kriaučiūnas[1,2], *Saugirdas* Pukalskas[1,*], *Alfredas* Rimkus[1,2]

[1]Vilnius Gediminas Technical University, Department of Automobile Engineering, 03224 J. Basanavičiaus g. 28, Vilnius, Lithuania
[2]Vilnius College of Technologies and Design, Technical Faculty, 01100, Olandų g. 16, Vilnius, Lithuania

Abstract. Numerical simulations of *Nissan Qashqai HR16DE* engine with increased compression ratio from 10,7:1 to 13,5:1 was carried out using AVL BOOST software. Modelled engine work cycles while engine works with biogas (BG) and hydrogen (H_2) mixtures. For biogas used mixture of 35 % carbon dioxide (CO_2) and 65 % methane (CH_4). Three mixtures of biogas with added 5 %, 10 % and 15 % H_2 was made. The simulation of engine work cycles was performed at fully opened throttle and changing engine crankshaft rotation speeds: n_{e1} = 1500, n_{e2} = 3000, n_{e3} = 4500, n_{e4} = 6000 rpm. Simulation results demonstrated what adding hydrogen to biogas increase in-cylinder temperature and nitrogen oxides (NO_x) concentration because of higher mixtures lower heating values (LHV) and better combustion process. Other emissions of carbon monoxide (CO) and hydrocarbons (HC) decreased while adding hydrogen due to the fact that hydrogen is carbon-free fuel.

Keywords: spark ignition engine, biogas, hydrogen, performance

1 Introduction

Huge challenges for developed countries is to reduce usage of fossil fuels and greenhouse gas (GHG) emissions. The biggest constituent of greenhouse gas and largest contribution to global warming is carbon dioxide [1], [2]. After 2012 in Lithuania largest carbon dioxide equivalent source is transport sector Fig.1.

One way to reduce greenhouse gas emission and fossil fuel usage in this sector is to use small, high efficiency engines like 2018 FORD ECOBOOST 999 cm^3 petrol engine with turbocharger which generates 100 HP and emits just 97 g/km of CO_2 or VAG Group 999 cm^3 TFSI engine with 97 HP and 97 g/km of CO_2 [4]. The more efficient engine the less fuel they use and less greenhouse gas that have a harmful effect on the environment they produce. Other way is to increase usage of alternative fuels.

* Corresponding author: saugirdas.pukalskas@vgtu.lt
Reviewers: *Stanislav* Beroun, *Pavol* Kukuča

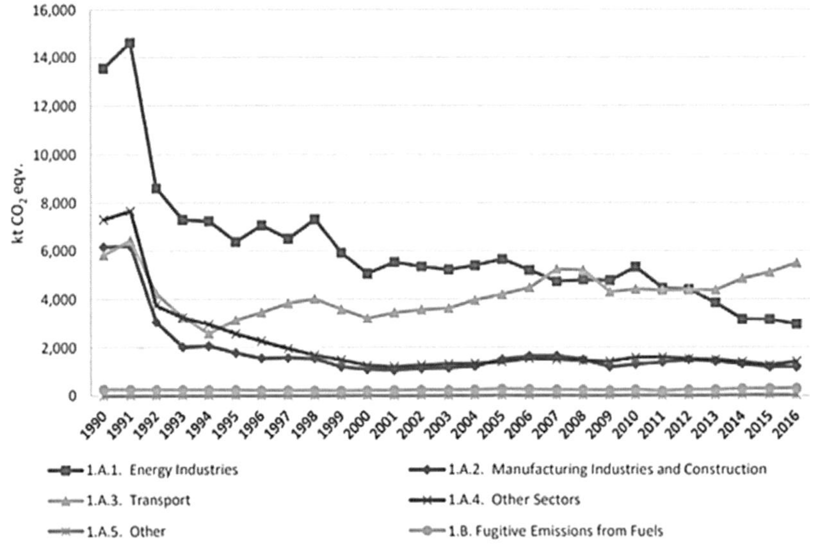

Fig. 1. Total GHG emissions from the different subsectors within the Energy Sector (CRF 1) [3]

In Lithuanian transport sector is one of the largest energy consumers Fig. 2. In 2016 used 215 thousand tons of petrol which includes just 9.9 thousand tons of bioethanol. A similar situation is with diesel fuel, transport sector in 2016 used 1398 thousand tons of diesel fuel which includes just 54.9 thousand tons of biodiesel [5]. Transport sector use just about 4 percentages mainly first generation alternative fuel.

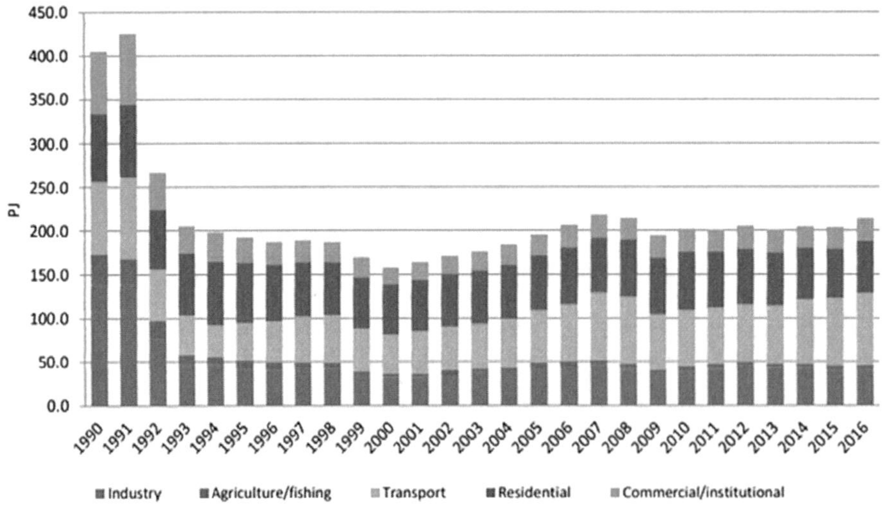

Fig. 2. Final energy consumption in Lithuania [3]

First way works in case of passenger vehicles and they also could use alternative fuel or just a part of it. Freight or public transport must have high engine torque and small engines aren't capable of producing it. So in heavy transport best way of reducing usage of fossil fuels and greenhouse gas emissions should be increased usage of second third or fourth generation alternative fuel. One of possibility is to use biogas. There is a large chose of raw material from which biogas could be produced. Also depending on the raw material it could be second third or fourth generation alternative fuel.

Biogas is mixture of different gases made out of anaerobically digested organic matter from waste materials. Biogas consist mainly from combustible methane and inert CO_2, if CO_2 concentration increase fuel lower heating value decrease [6]. Depending on raw material and parameters of anaerobic digestion process, CO_2 and CH_4 concentration in biogas can variate. Typically, methane concentration is from 55 – 88 %, CO_2 concentration is from 20 to 45 % and hydrogen sulphide (H_2S) concentration is from 1 to 2 %. Also in biogas could be found traces of carbon monoxide (CO), hydrogen and oxygen (O_2) [7]. Properties comparison of biogas with low concentration of methane and other gaseous fuel used in transport sector shown in Table 1.

Table 1. Comparison of hydrogen, LPG, natural gas and biogas properties [8], [9]

Property	Biogas	Methane	Natural gas	Hydrogen (H_2)
Composition (% vol)	CH_4 – 57 % CO_2 – 41 % CO – 0,18 % H_2 – 0,18 %	CH_4 – 100 %	CH_4 – 85 % C_2H_6 – 7 % C_3H_8 – 2 % N_2 – 1 % CO_2 – 5 %	H_2 – 100 %
Lower heating value (LHV) at 1 atm and 15°C (MJ/kg)	17.0	50.0	50.0	120.0
Density at 1 atm and 15°C (kg/m^3)	1.20	0.717	0.79	0.0838
Flame speed (cm/s)	25	36.5	34	270-320
Stachiometric A/F (kg of air/ kg of fuel)	5.8	17.2	17.3	34.3
Flammability limits (vol% in air)				
Leaner	7.5	5.3	5.0	4.0
Richer	14.0	14.0	15.0	75.0
Octane number				
RON	130	125	120	130
MON	-	-	120	-
Autoignition temperature (°C)	650	595	540	858

Raw material for making biogas could be any organic material. But it would be best if raw materials wont includes sugarcane, food crops, animal fats anything what suitable for food and will produce first generation alternative fuel. For next generation biogas production commonly used raw materials includes industrial and commercial waste, municipal solid waste, by-products from farms or water treatment plants and many others. List of raw materials suitable for production of next generation biogas is shown in Fig. 3.

Biogas could be used in internal combustion engines, because large part of biogas composition is methane 55 – 88 % which could be burn in internal combustion engine. Using biogas in compression ignition engine is possible, but because of 15 – 25 % higher auto ignition temperature and low ignitability, comparing with diesel, engine can't operate on 100 % of biogas. Therefore compression ignition engines must have pilot injection of diesel fuel to ignite biogas. There is no such problem then biogas is used in spark ignition engine. The same spark plugs which ignites petrol – air mixture can ignite mixture with biogas. Also biogas has about 20 – 30 % higher octane number than petrol fuel and greater resistance to knock. But in biogas there is a part of non-combustible gas – CO_2 which decrease lower heating value of the fuel. As an example then CO_2 concentration increase from 13 % to 49 % lower heating value decrease 41 % also maximum in cylinder pressure decreases and brake specific fuel consumption increase [6]. For better engine parameters biogas could be cleaned from CO_2 and burned just CH_4, but it's not economical, so there is a need to find measures to improve ecological and energetic parameters of spark ignition engine using produced and untreated biogas.

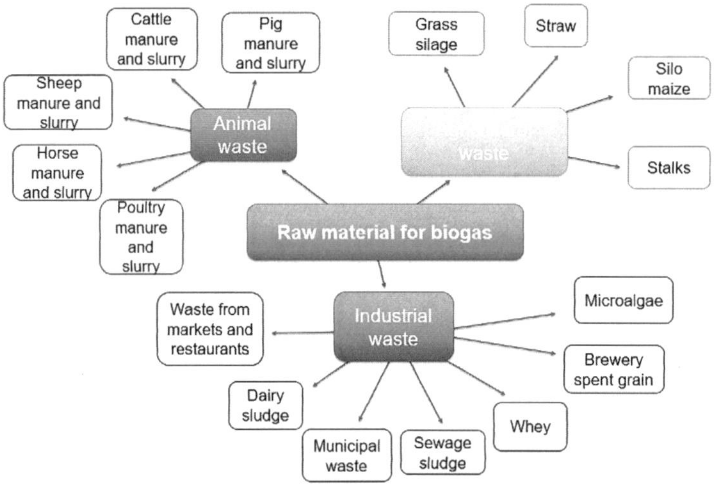

Fig. 3. Raw materials for biogas production

Researchers tried to use different measures to increase engine effectiveness and performance. Porpatham et al. tried to increase oxygen concentration in the intake air 21 %, 22 % and 23 %. Higher oxygen concentration in the intake air improves mixture burning speed and increases power output. The maximum brake thermal efficiently with 22 % of oxygen is 27 % and with 23 % – 28 %, and occurs at an equivalence ratio of 0.95. While with the same equivalence ratio and engine working with biogas and air with 21 % of oxygen brake thermal efficiently was 26.2 %. The main cause of this is the increased combustion rate [10]. Other researchers tried to use biogas on engines with increased compression ratio. Kwon et al. conducted research on a small one-cylinder spark ignition engine and increased its compression ratio by 15 %. Increased compression ratio leads to 21.8 % higher engine power and 7.8 % higher thermal efficiency when engine working on CH_4. While adding CO_2, the engine performance degrades. When engine working with 40 % CO_2 and 60 % CH_4 engine generate same power as it was generating before modification and working with CH_4 [11]. There are also other means used to increase engine effectiveness while using biogas. One of possibility is to use dual fuel biogas – hydrogen blends to improve combustion properties [12]. Adding hydrogen increase fuel blend lower heating value, flame speed and reduce GHG emission, because of hydrogen burning emissions [13]. Leung and Wierzba investigated burning characteristics of biogas – hydrogen mixture. Observed that 10 % of hydrogen in biogas has significantly improved flame stability and gave better results than further increasing part of hydrogen in biogas [14].

2 Materials and methods

For numerical simulation of spark ignition engine AVL BOOST model was made. Fig. 4 presents created AVL BOOST model. The basis of the model was *Nissan Qashqai HR16DE* engine, with displacement of 1598 cm^3, multiport fuel injection system and maximum engine power of 84 kW at 6000 rpm.

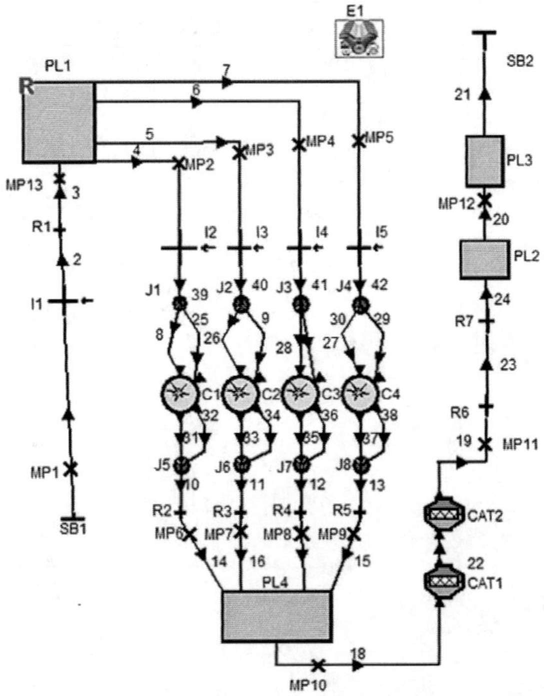

Fig. 4. *Nissan Qashqai HR16DE* engine AVL BOOST model: SB – system boundaries; MP – measuring points; J – junctions; I – port fuel injectors; PL – plenum; R – restriction; CAT – catalytic converter; E1 – engine parameters

Compression ratio was increased from 10.7:1 to 13.5:1, for spark ignition engine (SIE), fuelled with biogas for better power output and thermal efficiency while using biogas [15]. Rest engine specification shown in Tab.2.

Table 2. Engine specification

Items	**Specification**	**Items**	**Specification**
Engine Capacity, cm^3	1598	Stroke, mm	83.6
Rated engine power, kW	84	Compression ratio	10.7:1 → 13.5 :1
At engine speed, min^{-1}	6000	Number of valves per cylinder	4
Rated engine torque, Nm	156	Intake valves open	336° bTDC
At engine speed, min^{-1}	4000	Intake valves close	108° bTDC
Fuel system	Multipoint injection	Exhaust valves open	156° aTDC
Bore, mm	78.0	Exhaust valves close	308° bTDC

In AVL BOOST software for biogas was used mixture of 35 % carbon dioxide and 65 % methane. Also for increase of SIE, working with biogas, performance hydrogen was added and produced three fuel mixtures with 5 %, 10 % and 15 % of H_2. All compounds were mixed by mass. All mixtures have different *LHV* witch could be calculated according to the formula:

$$LHV_{fuelmix} = \frac{LHV_1 \times \%_{fuelcomp.1}}{100} + \frac{LHV_2 \times \%_{fuelcomp.2}}{100} \quad (1)$$

where:
LHV_1, LHV_2 – lower heating values of the fuel components, MJ/kg;
$\%_{fuelcomp.1}$, $\%_{fuelcomp.2}$ – percentage of a certain fuel component in a fuel mixture, % (by mass).

Lower heating values of methane, hydrogen and calculated biogas/hydrogen mixtures showed in Fig. 5. Highest *LHV* of 120 MJ/kg has hydrogen which increase all biogas mixtures lower heating values.

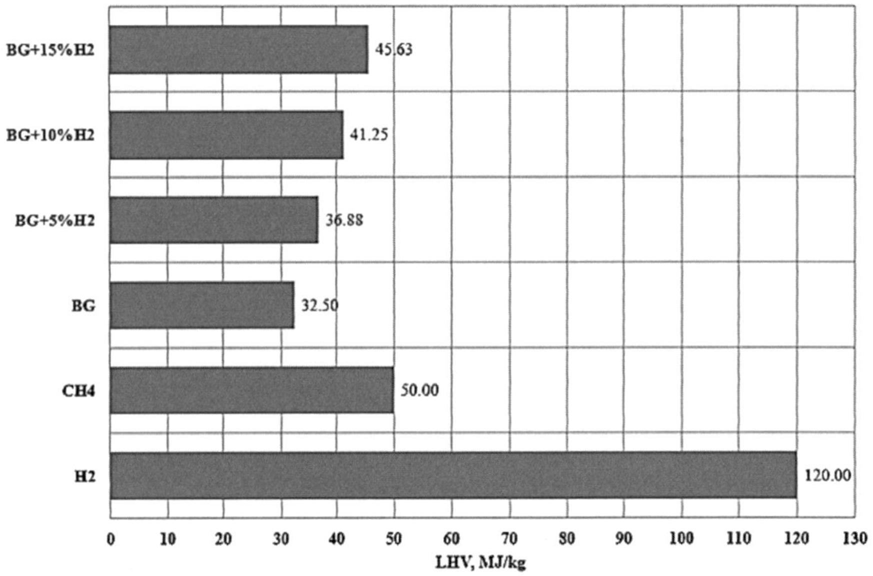

Fig. 5. LHV of methane, hydrogen, biogas and biogas/hydrogen mixtures

The simulation of engine work cycles was performed at fully opened throttle and changing engine crankshaft rotation speeds: $n_{e1} = 1500$, $n_{e2} = 3000$, $n_{e3} = 4500$, $n_{e4} = 6000$ rpm, engine working on biogas and biogas/hydrogen mixture where hydrogen mass fraction was 5 %, 10 % and 15 %. For engine crankshaft rotation speed 1500 rpm start of combustion was set to be 17.8 °CA before top dead center (BTDC) and air excess ratio (AER) – 0.950. Increased engine crankshaft rotation speed to 3000 rpm start of combustion was set to be 18.8 °CA BTDC and AER – 0,950. For engine crankshaft rotation speed of 4500 rpm start of combustion was set to be 20.0 °CA BTDC and AER – 0.900. Engine working at 6000 rpm start of combustion was set to be 20.0 °CA BTDC and AER – 0.866. For all mixtures ignition timing and AER was set respectively.

3 Results and discussions

Modelling engine work cycles for biogas all four crankshaft rotation speeds have set angle of combustion start. While adding hydrogen same angle is maintained. As showed in Fig.6 indicated pressure in cylinder reach maximum value of 9.857 MPa when engine working on biogas, while adding hydrogen indicated pressure drops in all measured point. The higher concentration of hydrogen, the lower indicated pressure. Comparing maximum pressure of BG and mixture with 15 % H_2 its decrease by 1,88 %. Pressure drop could be caused by constant angle of combustion start, because mixtures with hydrogen have faster diffusion velocity, a lower ignition and a faster flame speed [16].

Adding hydrogen with LHV of 120 MJ/kg increase all mixtures lower heating value. Biogas and 5 % hydrogen mixture LHV increased by 4.38 MJ/kg, with 10 % H_2 – 8.75 MJ/kg and with 15 % H_2 by 13.13 MJ/kg comparing with BG. Because of higher mixture LHV maximum burning temperature increased (Fig.7). While engine working with biogas reach maximum temperature of 2611.21 K adding 5 %, 10 % and 15 % of hydrogen increased maximum temperature by 44.03 K, 81.04 K and 110,75 K compering with biogas.

Modelling engine work cycles, while engine working with biogas, best brake specific fuel consumption (BSFC) achieved when engine crankshaft rotation speed was 3000 rpm, then BSFC was 334.7 g/kWh (Fig. 8). Adding hydrogen to BG increase mixtures LHV and in all measured points BSFC decrease. Highest difference noticeable with BG and 15 % of hydrogen. Then BSFC decreased by 90.21 g/kWh (26.9 %) at engine crankshaft rotation speed 3000 rpm, at the same conditions 5 % and 10 % of hydrogen lowered BSFC respectively 10.9 % and 19.7 %.

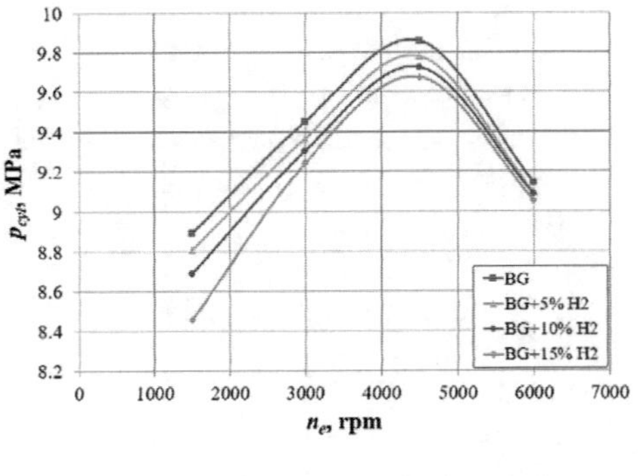

Fig. 6. Indicated pressure in cylinder

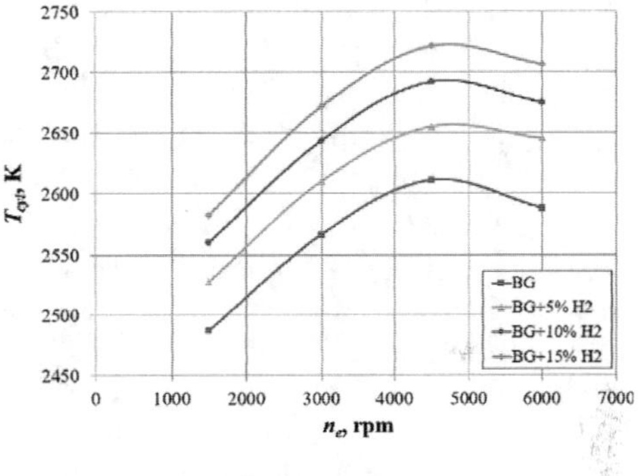

Fig. 7. Maximum temperature in cylinder

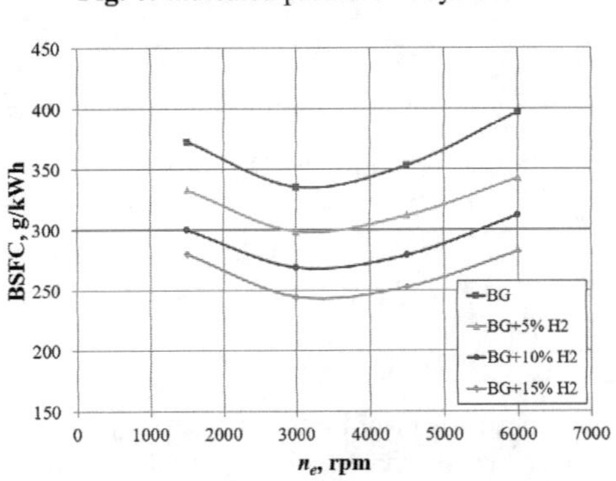

Fig. 8. Brake specific fuel consumption

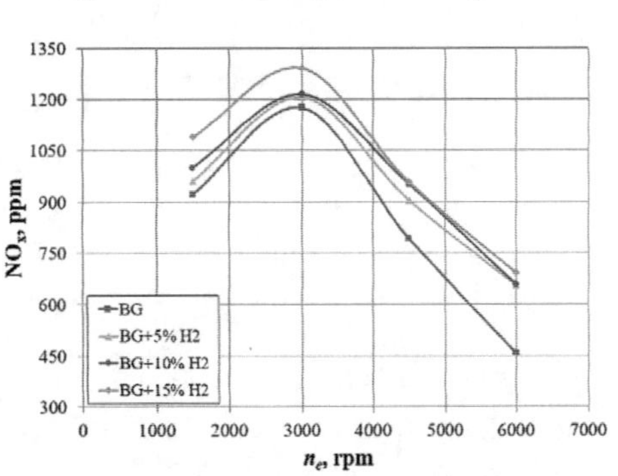

Fig. 9. Nitrogen oxides emission

Fig. 9 presents NO_x concentration in the exhaust gas. Minimum concentration established when engine working with biogas. Adding hydrogen increases NO_x concentration, which depends on maximum in cylinder temperature [17]. As shown in Fig. 6 H_2 raised temperature in cylinder at all measured points. Highest NO_x concentration when engine working with BG (1175.42 ppm) achieved at 3000 rpm. Increasing H_2 concentration leads to increase of NO_x emission, adding 5 %, 10 % and 15 % of hydrogen respectively increased nitrogen oxides by 2.65 % 3.34 % and 9.89 % comparing to BG.

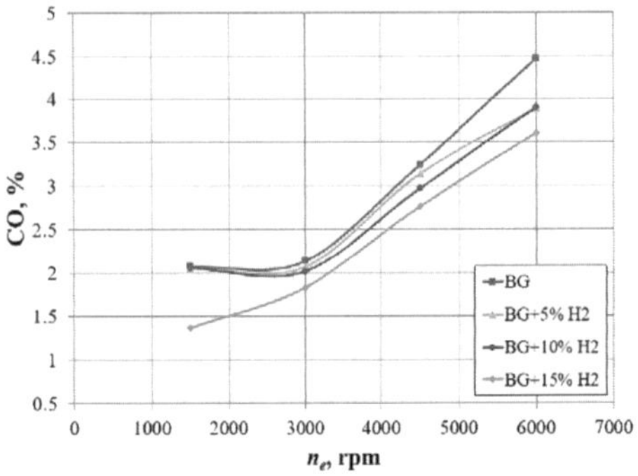

Fig. 10. Carbon monoxide emission

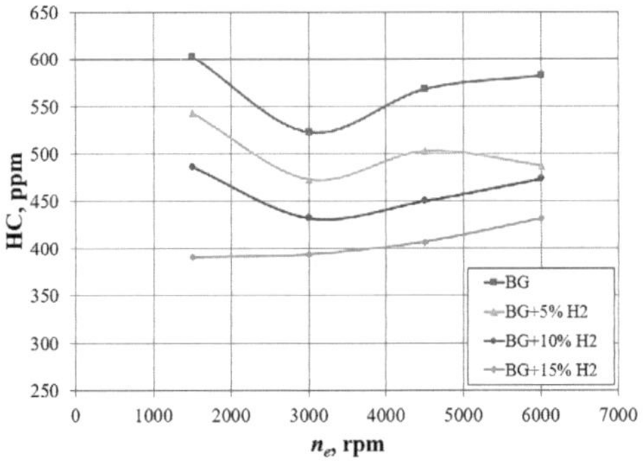

Fig. 11. Hydrocarbons emission

Fig. 10 shown CO emission dependence on engine speed with and without addition of H_2. It can be noted that CO emissions have decreased at all engine speeds adding 5 %, 10 % and 15 % of hydrogen. Biggest influence in CO emission could be seen when engine working with BG and 15 % H_2 mixture. At engine speed of 1500 rpm CO concentration decreased by 0.707 %, 3000 rpm by 0.301 %, 4500 rpm by 0.483 %, and at 6000 rpm by 0.852 % comparing when engine works with biogas. Emission reductions are due to the fact that hydrogen is carbon-free fuel and reduces fuel mixture C/H ratio, furthermore H_2 have high flame rate, increases in-cylinder temperature and combustion efficiency [18], [19].

The assessment of biogas and hydrogen mixtures revealed what increased amount of hydrogen decrease HC emission as can be seen in Fig. 11. While adding H_2, concentration of hydrocarbons decreased in all measured point. Engine working with biogas at 3000 rpm HC concentration in exhaust gasses was 522.48 ppm, added 5 % of hydrogen reduced hydrocarbons concentration by 9.51 %, 10 % of H_2 reduced by 17.28 % and 15 % of H_2 reduced by 24.78 % comparing at same engine speed with emission when engine working with biogas. HC reduction causes by increased in-cylinder temperature and more complete combustion because of added hydrogen and reduced carbon content of the fuel [20].

4 Conclusions

Numerical simulations of *Nissan Qashqai HR16DE* engine with increased compression ratio and working with biogas/hydrogen mixtures analysis lead to conclusions:

1. Engine working with biogas (65 % CH_4/ 35 % CO_2) and additionally supplied hydrogen has lower indicated pressure, because hydrogen have faster diffusion velocity, a lower ignition and a faster flame speed. To maintain indicated pressure angle of combustion start and air excess ratio should be adjusted.
2. Adding hydrogen with LHV of 120 MJ/kg increase in-cylinder temperature in all measuring points. Bigger part of biogas replaced with hydrogen increases LHV of the mixture and in-cylinder temperature. Maximum recorder temperature of 2721.96 K fixated with 15 % of hydrogen.
3. In all measuring points BSFC decreased while adding hydrogen, because of increased mixtures LHV and more efficient combustion process. Engine working with BG and 15 % H_2 BSFC was 279,52 g/kWh, 244,53 g/kWh, 252,48 g/kWh, and 281,71 g/kWh while increasing engine speed from 1500 rpm to 6000 rpm.
4. Adding hydrogen to biogas improved CO and HC emissions in all measured points, because of increased in-cylinder temperature, more complete combustion and reduced

carbon content of the fuel. However, adding hydrogen increased NO_x concentration because of increased in-cylinder temperature.

Acknowledgement. The results of the research, described in the article, were obtained by using a virtual internal engine simulation tool AVL BOOST, acquired by signing the Cooperation Agreement between AVL Advanced Simulation Technologies and Faculty of Transport Engineering of Vilnius Gediminas Technical University.

References

1. R. Hannah, M. Roser, *CO_2 and other Greenhouse Gas Emissions.* [interactive]. [accessed Apr 13 2018]; Available from: https://ourworldindata.org/co2-and-other-greenhouse-gas-emissions
2. EEA, *The European environment — state and outlook 2015: synthesis report.* European Environment Agency, Copenhagen, (2015)
3. I. Konstantinavičiūtė, T. Juraitė, et al., *Lithuania's national inventory report 2018.* Greenhouse gas emissions, 1990-2016. Vilnius, (2018)
4. *Car CO_2 and fuel economy figures.* [interactive]. FleetNews [accessed Mar 02 2018]; Available from: https://www.fleetnews.co.uk/cars/Car-CO2-and-fuel-economy-figures ?CarType=&Manufacturer=&Model=&CO2To=&EquaMpgFrom=&SortBy=CO2& SortDesc=False&FuelType=petrol (2017)
5. Lithuanian Department of Statistics. 2017. *Fuel and energy balance.* Lietuva, Vilnius (2016)
6. K. Mustafa, S. Suat, D. Emrah, C. Burak, *The effect of the CO_2 ratio in biogas on the vibration and performance of a spark ignited engine.* Fuel **214**, 634–639 (2018)
7. M. Herout, J. Malaťák, L. Kučera, T. Dlabaja, *Biogas composition depending on the type of plant biomass used.* In Research in Agricultural Engineering **57**, 137–143 (2011)
8. G. Belgiornoa, G. Di Blasioa, C. Beatricea, *Parametric study and optimization of the main engine calibration parameters and compression ratio of a methane-diesel dual fuel engine.* Fuel **222**, 821–840 (2018)
9. R. B. Gupta, *Hydrogen fuel: production, transport, and storage.* Taylor & Francis Group: CRC Press (2009)
10. E. Porpatham, A. Ramesh, B. Nagalingam, *Experimental studies on the effects of enhancing the concentration of oxygen in the inducted charge of a biogas fuelled spark ignition engine.* Energy **142**, 303–312 (2018)
11. E. C. Kwon, K. Song, M. Kim, Y. Shin, S. Choi, *Performance of small spark ignition engine fuelled with biogas at different compression ratio and various carbon dioxide dilution.* Fuel **196,** 217–224 (2017)
12. C. L. Tang, Y. J. Zhang, Z. H. Huang, *Progress in combustion investigations of hydrogen enriched hydrocarbons.* Renew Sustain Energy Rev., 195–216 (2014)
13. S. Singh, S. Jain, A. K. Tiwari, M. R. Nouni, J. K. Pandey, et al., *Hydrogen: a sustainable fuel for future of the transport sector.* Renew Sustain Energy Rev., 623–633 (2015)
14. T. Leung, I. Wierzba, *The effect of hydrogen addition on biogas non-premixed jet flame stability in a co-flowing air stream.* Int J Hydrogen Energy, 3856-3862 (2008)
15. E. Porpatham, A. Ramesh, B. Nagalingam, *Effect of compression ratio on the performance and combustion of a biogas fuelled spark ignition engine.* Fuel **95**, 247–256 (2012)

16. T. Su, S. Ji, S. Wang, X. Cong, L. Shi, *Improving the combustion performance of a gasoline rotary engine by hydrogen enrichment at various conditions*. International journal of hydrogen energy **43**, 1902–1908 (2018)
17. R. Egnell, *Combustion diagnostics by means of multizone heat release analysis and NO calculation*. Michigan: SAE International Spring Fuels and Lubricants Meeting and Exposition Dearborn; (1998)
18. J. Heywood, *Internal combustion engine fundamentals*. New York: McGraw-Hill, Inc.; (1988)
19. Y. Karag, O. Balcı, E. Orak, M. S. Habib, *Effect of hydrogen addition using on-board alkaline electrolyser on SI engine emissions and combustion*. International journal of hydrogen energy **43**, 11275–11285 (2018)
20. J. Zheng, E. Hu, Z. Huang, D. Ning, J. Wang, *Combustion and emission characteristics of a spray guided direct-injection spark-ignition engine fuelled with natural gas-hydrogen blends*. Int J Hydrog Energy, 11155–11163 (2011)

Application of the methodology related to the emission standard to specific railway line in comparison with parallel road transport: a case study

Iwona Rybicka[1,*], *Ondrej* Stopka[2], *Vladimír* Ľupták[2], *Mária* Chovancová[2], *Paweł* Droździel[1]

[1]Lublin University of Technology, Faculty of Mechanical Engineering, Nadbystrzycka Street 36, 20-618 Lublin, Poland
[2]Department of Transport and Logistics, Faculty of Technology, Institute of Technology and Business in České Budějovice, Okružní 517/10, 370 01 České Budějovice, Czech Republic

Abstract. At present, the impact of transport on the environment constitutes a serious problem. This mainly concerns energy consumption and production of greenhouse gases (GHG) that via their participation in greenhouse effect intensification contribute to global warming. In this paper, the calculations provided by the EN 16258: 2012 methodology will be applied to two modes of transport: railway and road transport. Subsequently, the methodology will be applied to a case study of a selected non-electrified railway line in comparison with parallel road transport. In particular, energy consumption and production of greenhouse gases will be monitored depending on traveling the distance between selected cities in the Czech Republic.

Keywords: emission, transport, railway line

1 Introduction

Transport represents an essential part of the economic stability and development of every state, but also a means for easier cultural exchange or preservation of genetic diversity. Transport has become so much a part of us that at present few people can imagine life without a car or the opportunity to use public transport at any time. The growing importance of transport and the demands we make on it lead to the world being flooded with traffic. Moreover, the amount of traffic tends to permanently increase and the severity of its negative impacts on human health and the environment is growing. In the last thirty years, the European policy has sought to limit the production of emissions from transport and industry to a minimum, and has also managed to ensure a significant reduction in the production in certain aspects. The need for a political fight against emissions from transport is also evidenced by the increasing urgency of the global warming trend, caused mainly by greenhouse gases generated predominantly by transport. It is stated that transport emits around 23% of all anthropogenic CO_2 production on Earth [1].

---

* Corresponding author: i.rybicka@pollub.pl

Reviewers: *Kateryna* Kravchenko, *Pavol* Kukuča

The paper focuses on this negative aspect of transport and offers a comparison of two modes of transport on a chosen transport route in order to comprehend all the side effects and load levels that these possibilities bring. The individual calculations of GHG emissions will be based on the methodology set out by the EN 16258: 2012 standard, which is a full-fledged part of the European policy for combating emissions from transport [2, 3]. In addition, the paper presents the main producers of transport emissions and describes their significant impact on the environment. It also includes graphically illustrated results of the comparison of the railway and bus passenger transport on the given transport route with respect to their production of emissions and energy consumption depending on the number of transported people [1].

2 European standard EN 16258:2012

This European standard specifies and describes the methodology for calculating and determining energy consumption and the number of GHG emissions concerning any mode of transport. It can be applied to both different modes of transport (road, sea, etc.) and various transport items (passengers, cargo, combinations) [6]. The methodology details the general steps, necessary definitions, relationships, calculation methods, divisions, and recommended information as support for an accurate and standardized calculation that presents energy consumption and greenhouse gases production in all modes of transport. Every calculation of energy consumption and greenhouse gases production requires the necessary amount of information about the situation [5].

For the calculations of energy consumption and GHG emissions in transport services, the methodology introduced in the EN 16258: 2012 standard was used. The calculations based on this methodology naturally take into account the following aspects: all vehicles used for the performance of transport services (including subcontracting), the total fuel consumption of each energy carrier used by single vehicles, and all journeys performed by both full and empty vehicles [4, 7, 8].

Thus, every calculation must produce results for four unknown variables that are:

- well-to-wheels energy consumption (Ew);
- well-to-wheels GHG emissions (greenhouse gas emissions) (Gw);
- tank-to-wheels energy consumption (Et);
- tank-to-wheels GHG emissions (greenhouse gas emissions) (Gt).

The final calculations of these unknown variables give us information on energy consumption and greenhouse gas emissions, both in the sum of primary and secondary emissions and in the value of primary emissions only [9].

Steps of calculating energy consumption and greenhouse gas emissions for a specified transport service.

The calculation for a specified transport service must be performed according to the following steps:

- step 1 - Identification of the various journeys (legs) that make up the specified transport service;
- step 2 - Calculation of energy consumption and greenhouse gas emissions for each leg;
- step 3 - Sum of the results for each leg [19, 20].

Secondary steps of calculating energy consumption and greenhouse gas emissions for a leg of a specified transport service

The calculation for a leg of a specified transport service must be performed using the following steps:

- secondary step 1 – Determination of the vehicle operation system (VOS);

- secondary step 2 – Calculation of total fuel consumption for the VOS;
- secondary step 3 – Calculation of total energy consumption and GHG emissions for the VOS;
- secondary step 4 - Allocating the leg a share from each of the four results of the partial step [19, 20].

3 Categories of values used for the calculation

The calculation method set out by the EN 16258: 2012 standard uses two kinds of value categories. The first category includes general values that include general data. The second category refers to preselected values, the values of which are clearly stated in the EN 16258: 2012 standard and the use of their declared values is essential for achieving results [2].

The category of general values includes fuel consumption, distance, fuel consumption with respect to distance, cargo, load factor, and vehicle capacity [10].

The category of preselected values includes such values that are stated in the EN 16258: 2012 documentation. It is advisable to check whether the latest version of the standard is available. The values should correspond to the mode of transport and to the available general values to make the results as accurate as possible [11].

3.1 Calculation principles at the level of vehicle operation system (VOS)

Secondary step 1 Determination of VOS related to a specified leg – In order to calculate each leg of the transport route, it is necessary to start by selecting the VOS for a particular leg. As a minimum requirement, the VOS must be a consistent group of vehicles operation relevant to the particular leg. In determining the VOS, the factors that affect the range and composition of the VOS (the number of vehicles and the time of operation of these vehicles) should be taken into account. The choice of the VOS is an arbitrarily chosen quantity relating to the selected criteria, but it must also consider journeys performed by empty vehicles [12, 13, 14].
Secondary step 2 Calculation of total fuel consumption for the VOS - The above mentioned value category is used for the calculation. If a vehicle uses more than one type of energy, each energy carrier must be calculated separately.

Secondary step 3 Calculation of total energy consumption and GHG emissions for the VOS - the recalculation from the total fuel consumption for the VOS to the amount of consumed energy and the greenhouse gas emissions is to be done using the following formulas:

- for well-to-wheels energy consumption VOS: Ew (VOS) = F (VOS) x ew;
- for well-to-wheels GHG emissions VOS: Gw (VOS) = F (VOS) x gw;
- for tank-to-wheels energy consumption VOS: Et (VOS) = F (VOS) x et;
- for tank-to-wheels GHG emissions VOS: Gt (VOS) = F (VOS) x gt.

where:

- F (VOS) is overall fuel consumption used for the VOS;
- ew well-to-wheels energy factor for fuel used;
- gw well-to-wheels GHG emissions for fuel used;
- et tank-to-wheels energy factor for fuel used;
- gt tank-to-wheels GHG emissions for fuel used.

Secondary step 4 Calculation for one leg - After the implementation of secondary steps 1, 2 and 3, we must allocate the transport service legs the share of Ew (VOS), Gw (VOS), Et (VOS), Gt (VOS) corresponding to its relative share of transport activity performed within VOS [12, 13, 14].

Therefore, the relevant formulas are as follows:

- S (leg) = T (leg) / T (VOS);

- Ew (leg) = Ew (VOS) x S (leg);
- Gw (leg) = Gw (VOS) x S (leg);
- Et (leg) = Et (VOS) x S (leg);
- Gt (leg) = Gt (VOS) x S (leg).

where:

- S (leg) is the factor used for calculating the share of energy consumption and emissions of the VOS that is allocated to a transport service for a particular leg. This share is based on the ratio of transport activity for the leg and for the associated VOS;
- T (leg) it is the transport service of transport activity for a specified leg;
- T (VOS) refers to transport performance of the VOS that concerns the leg;
- T (leg) a T (VOS) must have the same allocation of parameters and units [15].

4 Application of the european standard EN 16258: 2012: a case study

The main objective of the case study is to determine energy consumption and greenhouse gas emissions for two modes of transport (railway passenger transport and road passenger transport). After obtaining the results, two modes of transport will be compared and the more ecological alternative will be chosen. The case study will be applied to Jihlava-Brno transport section [16].

The length of Jihlava-Brno route by rail is 104 km. The capacity of a railway car is of 98 passengers. The railway car uses engine diesel oil as a fuel and a fuel consumption for developing enough power and travelling the distance is calculated at the level of 59.43 l/100 km.

Due to the fact that several types (814, 841, 842 and 854) of motor railway cars (integrated railway units) are deployed on this particular transport section, thus for objectification within this research study, calculation of the average values of fuel consumption of individual deployed motor units and the average capacity value of individual units are taken into consideration. Consumption of individual units is determined based on the technical parameters and specifications declared by the manufacturer.

Even though the bus service transports passengers between the same two cities - Jihlava and Brno - it uses road communications. The final distance is 92 km. The vehicle is fully occupied and therefore carries 50 passengers. The average power consumption for the route is 32 l/100 km. The line bus also uses diesel oil as a fuel.

Table 1. Final values of energy consumption

	Final values of rail transport 1st leg + 2nd leg (50% of passengers)	**Final values of road transport 1st leg + 2nd leg (50% of passengers)**	**Final values of rail transport 1st leg + 2nd leg (no passengers)**	**Final values of road transport 1st leg + 2nd leg (no passengers)**
E_w(TS)	80.76 MJ	75.42 MJ	2666.09 MJ	1282.22 MJ
E_t(TS)	73.57 MJ	68.71 MJ	2428.82 MJ	1168.12 MJ

The values from Table 1 indicate that railway transport has higher energy requirements for transport. In all conditions, rail transport requires more energy for its movement. When considered, this result has also a logical justification; although it is true that the friction generated by the movement along the rails is lower than that of the tire movement on the asphalt road, the weight of the motor railway car is more than doubled. The energy consumption to move more than 50 tons is significant.

Table 2. Final values of GHG emissions

	Final values of rail transport 1st leg + 2nd leg (50% of passengers)	Final values of road transport 1st leg + 2nd leg (50% of passengers)	Final values of rail transport 1st leg + 2nd leg (no passengers)	Final values of road transport 1st leg + 2nd leg (no passengers)
$G_w(TS)$	6.13 $kgCO_{2e}$	5.72 $kgCO_{2e}$	202.30 $kgCO_{2e}$	97.20 $kgCO_{2e}$
$G_t(TS)$	5.05 $kgCO_{2e}$	4.71 $kgCO_{2e}$	166.71 $kgCO_{2e}$	80.17 $kgCO_{2e}$

In this case, railway transport produces higher levels of greenhouse gas emissions in all examined values. Therefore, in terms of the amount of greenhouse gas emissions, it is a less environment-friendly mode of transport on Jihlava-Brno transport route. The reason is that in the case study we took into account the values of the motor railway car. If we were to use a unit of driving car and propelled car as a railway vehicle, it would probably reduce the amount of emissions calculated according to the analytical method because the increased number of passengers would reduce the ratio of the amount of pollutants to passenger-kilometers [17, 24, 25].

The case study deals with the significance of the number of transported passengers as an aspect affecting the final results of energy consumption and greenhouse gas emissions. Before the implementation of the application part, it was not exactly clear how significant the influence of the number of transported people was on the calculation process and on the final results. In the calculations two variants of the return route were chosen, which varied only in the number of passengers transported by the vehicle. Using this simulation, it was possible to present the maximum possible impact of passengers, as the maximum and minimum occupancy rates of the vehicle were included in the calculation. The table below shows the differences in values caused only by the change of the passengers who were part of the transport [17, 21].

Table 3. Final results and the influence of the number of passengers

	Final values of rail transport 1st leg + 2nd leg (50% of passengers)	Final values of road transport 1st leg + 2nd leg (50% of passengers)	Final values of rail transport 1st leg + 2nd leg (no passengers)	Final values of road transport 1st leg + 2nd leg (no passengers)
$E_w(TS)$	80.76 MJ	75.42 MJ	2666.09 MJ	1282.22 MJ
$E_t(TS)$	73.57 MJ	68.71 MJ	2428.82 MJ	1168.12 MJ
$G_w(TS)$	6.13 $kgCO_{2e}$	5.72 $kgCO_{2e}$	202.30 $kgCO_{2e}$	97.20 $kgCO_{2e}$
$G_t(TS)$	5.05 $kgCO_{2e}$	4.71 $kgCO_{2e}$	166.71 $kgCO_{2e}$	80.17 $kgCO_{2e}$

As shown by Table 3, which contains all the final values analyzed in this study, the number of passengers has a significant impact on understanding the final results. We can say with exaggeration that the significant differences in values are only caused by the reasonability with which the movement of the means of transport was carried out. The transport was meaningful if it was able to attract some passengers, which led to the values on the left of the table that are considerably smaller. The values on the right are several times larger due to the fact that there was no passenger on board of the vehicle for half of the transport route. In principle, the production of emissions and energy consumption of the vehicle do not change too much. This methodology demonstrates that the production of GHG emissions can be understood from a completely different point of view [18, 22, 23].

5 Conclusions

Last but not least, it is necessary to focus on specification of the proposals concerning the case study and the data analyzed in this paper, where it is necessary to point out the need for gradual electrification of railways in the Czech Republic. At present, roughly a third of the railway network is electrified, which is below average compared to Western Europe. Unfortunately, the fact is that railway transport in the Czech Republic does not reach the required level. Consequently, the share of transportation by rail in this country is rather below average.

Important tools for developing high-quality public transport, which is an important means for reduction of world emissions from transport, are also integrated transport systems within territorial units. Simplification, connection, but also uniformity of carriers, where the use of one personal document enables the use of different modes of transport, leads to a future priority of mass transport and to reduction of congestion and other negative aspects of glutted transport.

A possible way to decrease production of emission gases is to use alternative power sources such as magnetic trains or hybrid power traction. The possibilities for road transport are even wider, ranging from electric vehicles to hydrogen-powered cars. However, many alternative technologies have not achieved the necessary development in order to, for acceptable financial investment, offer an adequate replacement of current means of transport.

References

1. E. Demir, W. Burgholzer, M. Hrušovský, E. Arıkan, W. Jammernegg, T.V. Woensel, *A green intermodal service network design problem with travel time uncertainty.* Transportation Research Part B: Methodological, vol. **93**, 789-807 (2016)
2. T. Skrucany, S. Semanova, T. Figlus, B. Sarkan, J. Gnap, *Energy intensity and GHG production of chosen propulsions used in road transport.* Communications - Scientific Letters of the University of Zilina, vol. **19**, no. 2, 3-9 (2017)
3. W. Zhu, M.P. Nowak, S.O. Erikstad, *Emission allocation issues in repositioning transporttation.* International Journal of Sustainable Transportation, vol. **10**, no. 4, 365-375 (2016)
4. F. Kellner, *Allocating greenhouse gas emissions to shipments in road freight transportation: Suggestions for a global carbon accounting standard.* Energy Policy, vol. **98**, 565-575 (2016)
5. F. Kellner, *Exploring the impact of traffic congestion on CO_2 emissions in freight distribution networks*. Logistics Research, vol. **9**, no. 1. (2016)
6. T. Skrucany, M. Kendra, M. Skorupa, J. Grencik, T. Figlus, *Comparison of chosen environmental aspects in individual road transport and railway passenger transport.* Procedia Engineering, 806 (2017)
7. F.Cavallaro, F.Maino,V.Morelli, *A new method for forecasting CO_2 operation emissions along an infrastructure corridor*. European Transport - Trasporti Europei, no. **55**, (2013)
8. Y. Cheng, Z. Liu, J. Yan, *Comparisons of PM10, PM2.5, particle number, and CO_2 levels inside metro trains traveling in underground tunnels and on elevated tracks.* Aerosol and Air Quality Research, vol. **12**, no. 5, 879-891 (2012)

9. G. Dalkic, O. Balaban, H. Tuydes-Yaman, T. Celikkol-Kocak, *An assessment of the CO2 emissions reduction in high speed rail lines: Two case studies from Turkey*. Journal of Cleaner Production, vol. **165**, 746-761 (2017)
10. M. Tomášiková, M. Lukáč, J. Caban, F. Brumerčík, Controllability and stability of a vehicle. Logi - Scientific Journal on Transport and Logistics, vol. **7**, no. 1, 136-142 ISSN 1804-3216 (2016)
11. G.J. Hull, C. Roberts, S. Hillmansen, *Simulation of energy efficiency improvements on commuter Railways*. IET Seminar Digest, (2016)
12. T. Kirschstein, F. Meisel, *GHG-emission models for assessing the eco-friendliness of road and rail freight transports*. Transportation Research Part B: Methodological, vol. **73**, 13-33 (2015)
13. M. Mohsen, M.B. Ahmed, J.L. Zhou, *Particulate matter concentrations and heavy metal contamination levels in the railway transport system of Sydney, Australia*. Transportation Research Part D: Transport and Environment, vol. **62**, 112-124 (2018)
14. X. Tao, Q. Wu, L. Zhu, *Mitigation potential of CO2 emissions from modal shift induced by subsidy in hinterland container transport*. Energy Policy, vol. **101**, 265-273 (2017)
15. Y. Ge, J. Long, F. Xiao, Q. Shi, *Traffic modeling for low-emission transport*. Transportation Research Part D: Transport and Environment, vol. **60**, 1-6 (2018)
16. F. Ma, W. Wang, Q. Sun, F. Liu, X. Li, *Ecological pressure of carbon footprint in passenger transport: Spatio-temporal changes and regional disparities*. Sustainability (Switzerland), vol. **10**, no. 2. (2018)
17. H. Novak, V. Lešić, M. Vašak, *Hierarchical coordination of trains and traction substation storages for energy cost optimization*. IEEE Conference on Intelligent Transportation Systems, Proceedings (2018)
18. W. Yang, S. Wang, X. Zhao, *Measuring the direct and indirect effects of neighborhood-built environments on travel-related CO2 emissions: A structural equation modeling approach*. Sustainability (Switzerland), vol. **10**, no. 5. (2018)
19. V. Hromádka, E. Vítková, H. Kovářová, P. Vařbuchta, *Environmental impacts in the evaluation of transport infrastructure projects*. International Multidisciplinary Scientific GeoConference Surveying Geology and Mining Ecology Management, SGEM, 201 (2016)
20. M. Kendra, *Integration of individual car transport and public passenger transport in cities*. OPT-i 2014 - 1st International Conference on Engineering and Applied Sciences Optimization, Proceedings. 1582 (2014)
21. M. Vojtek, M. Kendra, V. Zitrický, J. Daniš, *Principles of logistics applied to railway passenger transport*. MATEC Web of Conferences. (2017)
22. F. Brumerčík, Simulation of an unconventional vehicle transmission. Logi - Scientific Journal on Transport and Logistics, vol. **3,** no. **1**, 36 - 44 (2012)
23. R. Mielnik, M. Sułowicz, W. Jakubas, K. Ludwinek, M. Jaśkiewicz, *The reliability of critical systems in railway transport based on the track rail circuit. Analysis and simulation of electrical and computer systems,* The 13th Conference on Selected Problems of Electrical Engineering and Electronics WZEE, 219-228 (2016)
24. J. Caban, A. Marczuk, B. Sarkan, J. Vrabel, *Studies on operational wear of glycol-based brake fluid*. Przemysl chemiczny, vol. **94**, Issue **10**, 1802-1806 (2015)
25. M. Jurkovic, T. Kalina, A.F. Teixeira, *Possibilities of Using Alternative Fuels for Transport Solution in Brazil*. The 21tst International Scientific Conference, 724-728 (2017)

The peculiarities of monitoring road vehicle performance and environmental impact

Ivan Kuric[1], *Vasyl* Mateichyk[2,*], *Miroslaw* Smieszek[2], *Mykola* Tsiuman[3], *Nataliia* Goridko[3] and *Igor* Gritsuk[4]

[1]University of Zilina, Faculty of Mechanical Engineering, Department of Automation and Production Systems, 01026 Zilina, Univerzitna 8215/1, Slovakia
[2]Rzeszow University of Technology, Department of Quantitative Methods, 35-959 Rzeszow, al. Powstancow Warszawy, 8, Poland
[3]National Transport University, Faculty of Automotive and Mechanical Engineering, 01010 Kyiv, 1 Omelianovycha-Pavlenka str., Ukraine
[4]Kherson State Maritime Academy, Department of Operation of Ship Power Plants, 73000 Kherson, 20 Ushakova avenue, Ukraine

Abstract. The article describes the peculiarities of mathematical models for determining the performance indicators of operation and environmental impact of road vehicles. The models are the basis of intelligent monitoring systems for separate road transport objects: vehicle propulsion systems, separate vehicles in motion on the real route and traffic flows. The structures of these monitoring systems with the characteristics of input data, the features of simulating the main processes and output results of monitoring the performance indicators of operation and environmental impact of road vehicles are presented. The examples of current parameters monitoring of the car with a gasoline engine in three typical motion phases in the traffic flow are given.

Keywords: road vehicle, mathematical model, monitoring system, performance indicators of operation, environmental impact

1 Introduction

Road vehicles are currently the most efficient means of passenger and freight transportation, individual trips, etc. At the same time, road vehicles are the largest consumers of fuel and other resources. They are the sources of noise, vibrations, electromagnetic radiation and environmental pollution by different solid and gaseous substances.

The issues of road vehicle performance have been important throughout the history of automobiles. With increasing number of automobiles, the issues of environmental impact of road transport have become more urgent.

Many scientists [1-10] devoted their research to environmental impact of road network objects – separate vehicles and traffic flows – and the ways to improve their environmental safety. The investigations also focused on the development of methods to study and predict environmental safety of separate vehicles and roadside pollution by traffic flows. The

* Corresponding author: vmate@prz.edu.pl

Reviewers: *Kateryna* Kravchenko, *Dalibor* Barta

development of transport environmental monitoring systems was of great scientific interest as well.

The paper [1] is devoted to the study of environmental pollution by motor transport and methods for preventing such pollution. It results in the development of a "driver-vehicle-road-environment" system which enables to optimize the operation factors for reducing environmental pollution by motor transport.

The research [2] deserves particular attention. The research results include the systematized quantitative estimation of harmful emissions by different categories of vehicles, the estimation of urban environmental pollution by motor transport. Measures aimed at reducing air pollution by vehicles have also been analysed.

The issues of improving vehicle environmental safety are described in [7]. The research includes systems analysis methods for developing a mathematical model of "road vehicle" system. It enables to evaluate vehicle environmental safety and to find ways to improve it. The peculiarities of developing a mathematical model of "traffic flow-road" system and creating specific elements for intelligent monitoring system (IMS) of vehicle environmental safety in operation are described in [8, 9].

[10] describes the peculiarities of "Vehicle to Infrastructure" (V2I) information system and the process of monitoring and evaluating the vehicle technical condition in operation with the possibility of forecasting. The system is based on the general approach to investigate the "Vehicle - Driver - Operating Conditions - Vehicle Operation Infrastructure" system.

The issue of integrating IMS separate elements of vehicle environmental safety to monitor and control roadside pollution by traffic flows requires additional research.

The composition and the level of hardware and software for intelligent monitoring systems significantly depend on monitoring objects and monitoring tasks (Fig. 1). Vehicle propulsion systems, separate vehicles and traffic flows are considered to be monitoring objects in road transport. Depended on monitoring tasks intelligent monitoring systems can include monitoring systems of monitoring objects technical performance (such as the speed of motion, the mode of motion, operation process parameters, technical condition, traffic flow composition, etc) [11]. On the other hand, the monitoring systems of environmental performance are more science intensive because it significantly depends on both operational technical performance and design parameters of monitoring objects [12].

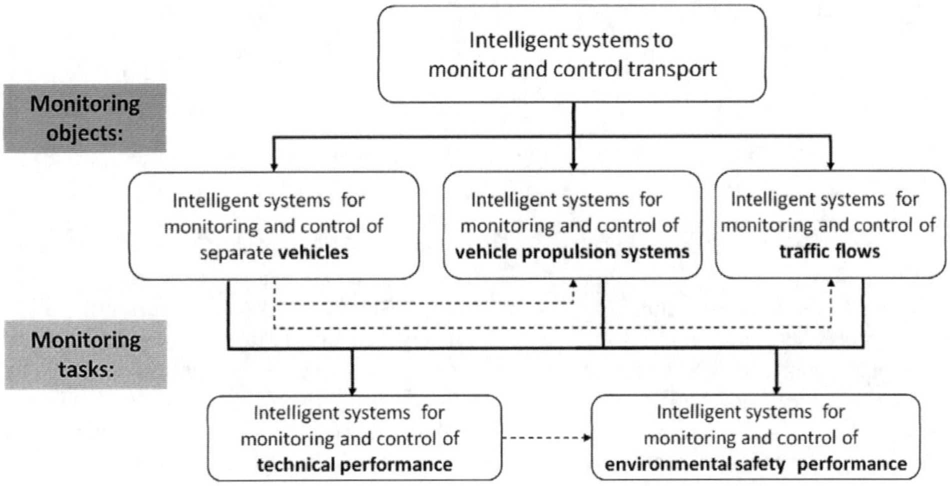

Fig. 1. The hierarchy of using intelligent monitoring systems in transport

2 The stages of vehicle monitoring process

2.1 Monitoring of vehicle propulsion system parameters

The first stage of vehicle monitoring is vehicle performance evaluation for standard driving cycles. This enables to evaluate the efficiency of certain means for improving fuel economy and environmental performance of vehicle propulsion systems according to specific vehicle operation conditions. The mathematical model of the "engine-catalytic converter" system [13] is used for this purpose. The mathematical model is based on volume balance method [14]. The model enables to determine fuel economy, engine power and environmental performance according to its design parameters, operation mode, fuel type and environmental parameters. The input data for modeling are the actual engine operation data obtained from in-cylinder pressure sensor; engine parameters in steady operation modes obtained in laboratory experiments; on-board diagnostics (OBD) system data for appropriate modes of vehicle motion (Fig. 2). The data set of in-cylinder working body parameters, fuel economy, engine power and environmental performance in operation modes according to modes of vehicle motion in a driving cycle is determined during simulation. The vehicle fuel economy and environmental performance in the driving cycle are the simulation results.

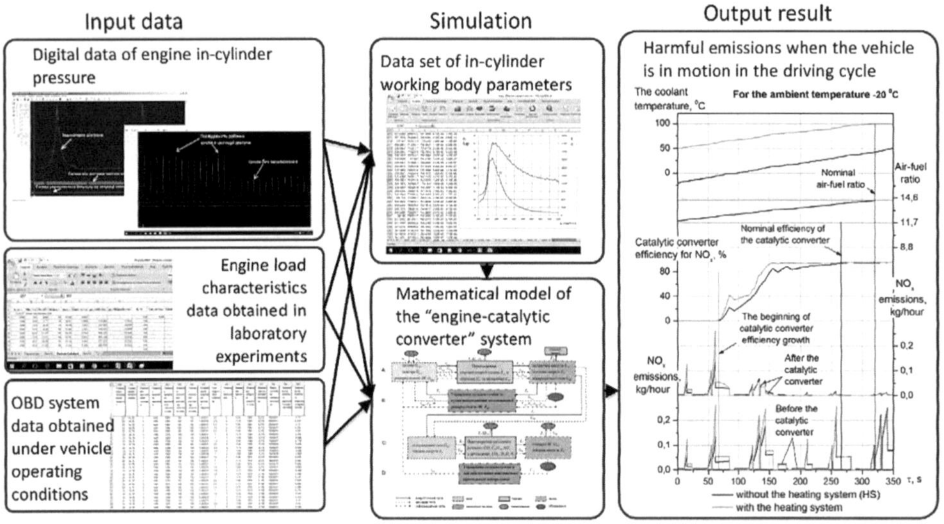

Fig. 2. Monitoring of vehicle propulsion system parameters

2.2 Monitoring of vehicle parameters

The second stage of vehicle monitoring is environmental safety evaluation when the vehicle is in motion on the real route. It is possible using OBD data and vehicle current location parameters based on Global Positioning System (GPS) data. Figure 3 shows the enlarged functional scheme of vehicle performance monitoring under operating conditions. The data set of real characteristics of the vehicle route is determined according to the vehicle current location data. This is the basis for determining real resistance forces of the vehicle motion on the route. Using the mathematical model of the "road vehicle" system [7], the environmental safety is evaluated when the vehicle is in motion on a certain route. The obtained data of vehicle motion parameters and environmental safety parameters are transferred to remote databases for further analysis using data transmission systems.

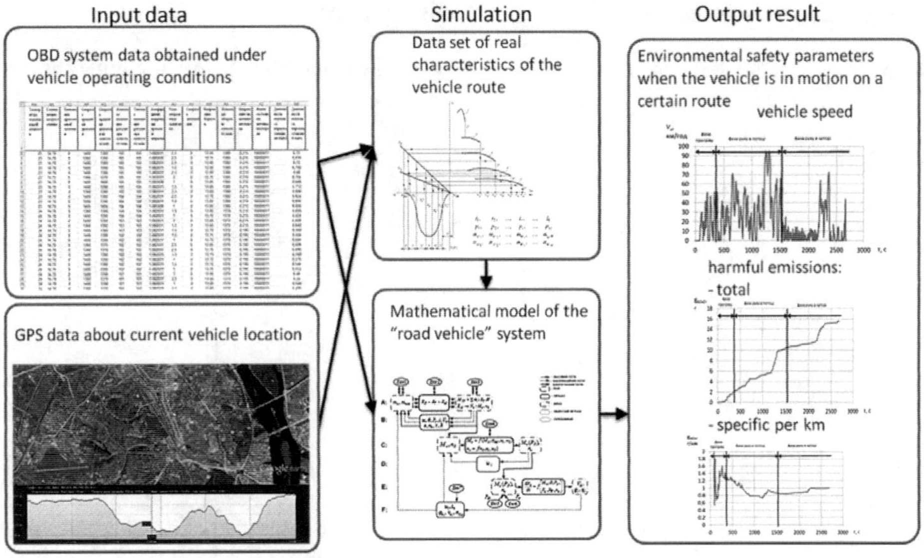

Fig. 3. Monitoring of vehicle parameters

2.3 Monitoring roadside pollution by traffic flow

Traffic flow parameters can be monitored and controlled according to the information from vehicle parameters database (Fig. 4). Thus, the traffic flow structure is created according to the vehicle categories, fuel types, vehicle environmental classes and traffic flow intensity. Route parameters and weather conditions are determined by the monitoring system according to the vehicle current location data. The simulation of vehicle environmental safety in the traffic flow and the evaluation of roadside pollution by the traffic flow are based on the determined traffic flow parameters. The use of geographical information systems (GIS) enables to visualize and analyze roadside pollution by traffic flows.

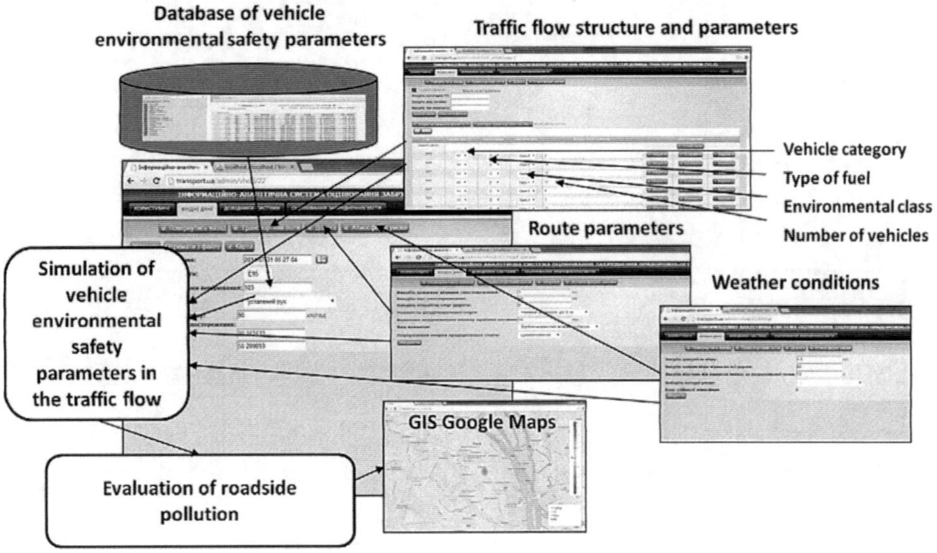

Fig. 4. Monitoring roadside pollution by traffic flow

3 The examples of vehicle monitoring results

Fig. 5, 6 show the examples of current parameters of the car with a gasoline engine moving in the city traffic flow and a separate vehicle environmental safety.

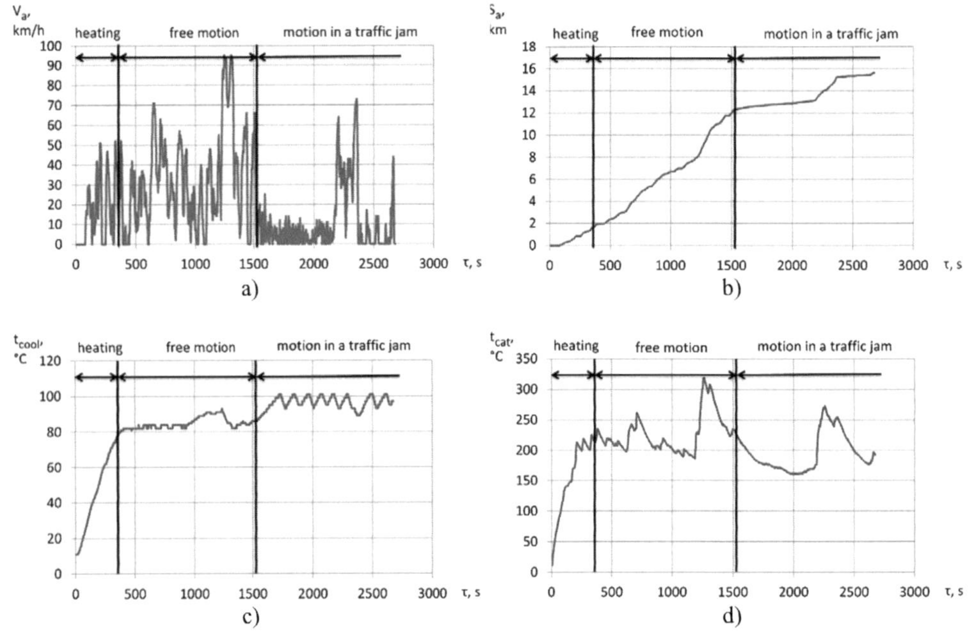

Fig. 5. Vehicle motion parameters in the traffic flow: a) speed; b) the distance travelled; c) coolant temperature; d) catalytic converter temperature

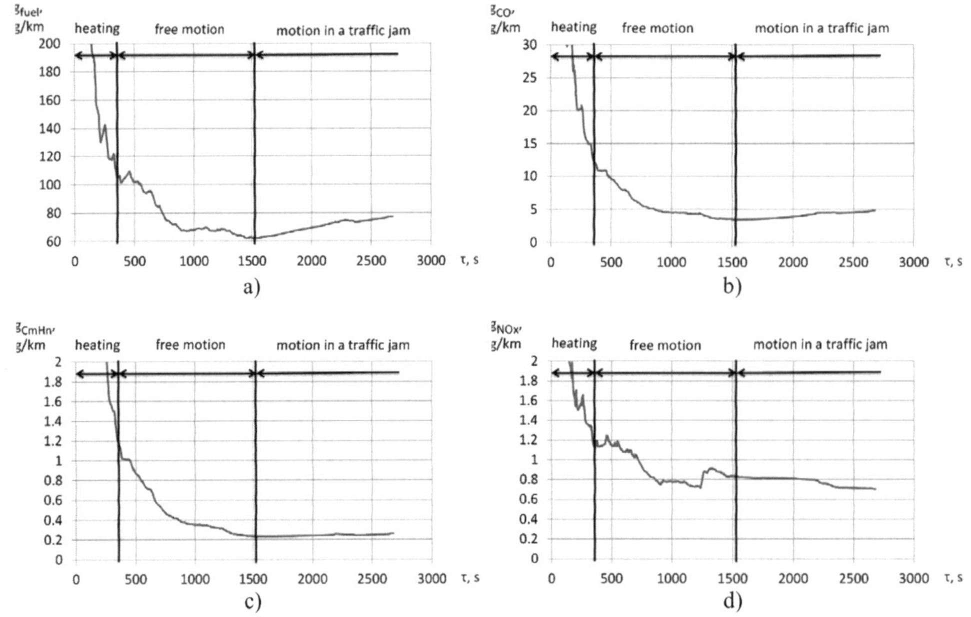

Fig. 6. A separate vehicle environmental safety performance:
a) fuel consumption; b) CO emission; c) C_mH_n emission; d) NO_x emission

The analysis of the obtained vehicle motion parameters (Fig. 5) enables to determine three typical motion phases, namely: heating, free motion and motion in a traffic jam. The heating phase is characterized by low speed of motion V_a, increasing the coolant temperature t_{cool} and catalytic converter temperature t_{cat} to operating values, the increased values of specific fuel consumption g_{fuel}, carbon monoxide emission g_{CO}, hydrocarbons emission g_{CmHn}, nitrogen oxides emission g_{NOx} (Fig. 6).

The phase of free motion is characterized by increasing the speed mode (Fig. 5 a), increasing the distance travelled S_a (Fig. 5 b), normal values of the coolant temperature (Fig. 5 c) and catalytic converter temperature (Fig. 5 d). This contributes to the decreased values of specific fuel consumption and harmful emission (Fig. 6).

The temperature mode of the cooling system in the phase of motion in a traffic jam is characterized by maximum values of the coolant temperature and periodic turning on the radiator cooling fan (Fig. 5 c). In contrast, the temperature mode of the catalytic converter is characterized by the decreased temperature (Fig. 5 d). This is related to low engine load. The values of specific fuel consumption and harmful emission increase in this phase (Fig. 6).

4 Conclusion

Thus, the approach to evaluate the efficiency of road vehicle operation and environmental impact is shown. The vehicle performance in standard driving cycles, based on fuel economy, power and environmental performance of the vehicle propulsion systems, is taken into account in this approach. It also implies the vehicle current location and motion parameters under real route conditions, total roadside pollution by the traffic flow according to the determined emissions of separate vehicles in traffic. The examples of current parameters monitoring of the car with a gasoline engine in three typical motion phases in the traffic flow are given. It is shown that the phases of heating and motion in a traffic jam are characterized by increased fuel consumption and, consequently, increased harmful emissions.

References

1. Yu.F. Gutarevych, *Reducing vehicle emissions under operating conditions: Monograph*. Vyshcha shkola, Kyiv, (1991)
2. V.N. Lukanin, Yu.V. Trofimenko, *Environmental burden decrease in motor transport operation*. The outcomes of science and technology. Ser. Automobile and city transport, VINITI, Moscow, (1996)
3. V.A. Zvonov, A.V. Kozlov, V.F. Kutenev, *Environmental safety of a car in the complete life cycle*. NAMI, Moscow, (2001)
4. P.M. Kanilo, *Motor transport. Fuel and environmental problems and prospects: Monograph*. KhNAHU, Kharkiv, (2013)
5. P. Eastwood, *Particulate Emissions from Vehicles*. John Wiley & Sons Ltd, Chichester, (2008)
6. A. Faiz, C.S. Weaver, M.P. Walsh, *Air Pollution from Motor Vehicles: Standards and Technologies for Controlling Emissions*. The World Bank, Washington, D.C., (1996)
7. V.P. Mateichyk, *Evaluation methods and methods for improving environmental safety of road vehicles: Monograph*. NTU, Kyiv, (2006)
8. V.P. Mateichyk, M.P. Tsiuman, H.O. Waigang, *Simulation of the "traffic flow – road" system*. Inter-university collected articles "Scientific notes", **46**, 371-381, (2014)

9. V.P. Mateichyk, M.P. Tsiuman, *Creating the structure of intelligent monitoring system for environmental safety indicators of vehicles at various stages of life cycle*. The Herald of National Transport University, **32**, 193-200, (2015)
10. I.V. Gritsuk, V. Volkov, V. Mateichyk, Yu. Grytsuk et al., *Information Model of V2I System of the Vehicle Technical Condition Remote Monitoring and Control under Operating Conditions*. SAE Technical Paper **2018-01-0024** (2018)
11. V.P. Volkov, P.B. Komov, O.B. Komov, I.V. Gritsuk, *The peculiarities of transport and information monitoring system "KhNADU-TESA"*. Automobile and Electronics. Modern Technologies, **4**, 106-110, (2013)
12. V.P. Mateichyk, H.O. Waigang, M. Smieszek, *Information monitoring system of roadside pollution by traffic flows*. Automobile and Electronics. Modern Technologies, **4**, 74-77, (2013)
13. I. Gritsuk, V. Volkov, V. Mateichyk, Yu. Gutarevych, M. Tsiuman, N. Goridko, *The Evaluation of Vehicle Fuel Consumption and Harmful Emission Using the Heating System in a Driving Cycle*. SAE International Journal of Fuels and Lubricants, **10(1)**, doi:10.4271/2017-26-0364, (2017)
14. V.G. Diachenko, *Internal combustion engines: Theory: Manual*. NTU "KhPI", Kharkiv, (2008)

Engine with unconventional crank mechanism FIK_1

Pavol Kukuča[1], *Dalibor* Barta[1,*] , *Robert* Labuda[1], *Tsvetomir* Gechev[2]

[1]University of Zilina, Department of Transport and Handling Machines, Faculty of Mechanical Engineering, Univerzitná 1, 010 26 Zilina, Slovakia
[2]Technical University – Sofia, Faculty of Transport, 8 Kliment Ohridski Blvd., Block 9, 1000 Sofia, Bulgaria

Abstract: The geometry and kinematics of unconventional engine mechanism FIK_1 are described in the paper. Transformation of the piston linear movement to rotational movement is done by a swinging board, a baseboard and a crankshaft. The paper describes the trajectory and acceleration of an observed point on the swinging board in all three planes as well as the kinematics of the piston movement. The principle of the unconventional mechanism FIK_1 is protected by patents No. 283742 and No. 283743. A mathematical model of the mechanism and some significant results are presented.

Keywords: Combustion engine, unconventional crank mechanism, kinematics, swinging board

1 Introduction

The effort to reduce the production costs of individual vehicle parts and to ensure their dimensional compactness while increasing reliability leads to the development of new structures and mechanisms. With the development and improvement of material and fuel technologies new and uncommon engines can also find application [1-4]. Some of these unconventional engines use an alternative construction of their mechanical parts and mechanisms to acquire certain advantages in comparison to the conventional solutions. A typical example is the axial engine for Stirling [5, 6] or IC application, which has a low frontal area, very good balance and great compactness [7-10]. Another unconventional type of engine mechanism which kinematics was solved by Shih [11] is the cycloidal internal combustion engine mechanism.

One of the mechanisms, applicable in such engines is the FIK_1, whose geometry and kinematics are described in this paper.

The base of the engine with unconventional mechanism is the swinging board D_K, supported on the crank pin and rolling by its perimeter around the immovable base board D_P (see Fig. 1, left) [12]. Both boards have an angle gearing at their perimeter (resp. at the optional suitable diameter) for exact and safe mutual rolling. The position of the gearing on

[*] Corresponding author: dalibor.barta@fstroj.uniza.sk

Reviewers: *Pawel* Drozdziel, *Saugirdas* Pukalskas

the base and swinging board may not be in praxis situated on their perimeters [13]. Because the swinging discs′ movement presents a relative consecutive rolling of two cones (see Fig. 1 right, there is shown the trajectory of a point on the swinging board perimeter, acquired from the virtual model), the gear can be positioned practically on the whole cone length from the bottom to the top.

The relative angle between both boards is φ (see Fig. 1, left). From the kinematics point of view this is the angle between the bottoms of the rolling boards. The top angle of the meshing cones does not need to be generally the same. In case of the identical cones their top angle value is ($180° - \varphi$).

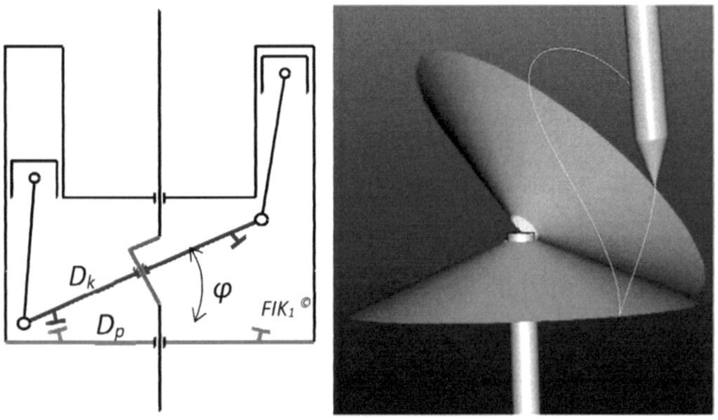

Fig. 1. Scheme of the unconventional engine mechanism

2 Description of the mechanism geometry

Based on the previous considerations, the center of the swinging board makes planar motion along a circle which is parallel to the plane of the immovable base board. Because each point on the swinging board makes a space movement, it is necessary to define and solve this system spatially (in three axes). We choose the coordinate system so that its vertical axis **Z** goes through the axis of the crank shaft and its center lies on the horizontal plane **XY** (see Fig. 2). The initial crank position for the rotation angle $\alpha = 0°$ lies on the axis **Y**. The observed point on the swinging board lies at the connecting line of the board center and its perimeter, it means opposite to the crank. The circle radius r_{kr}, which describes the center of the swinging board (hereinafter referred as "crank circle"), is given by the following expression:

$$r_{kr} = R \cdot (1 - \cos \phi), \qquad (1)$$

where R – radius of the swinging (generating), resp. immovable base board,
φ - angle between boards which is equal to the crank angle (see Fig. 2).

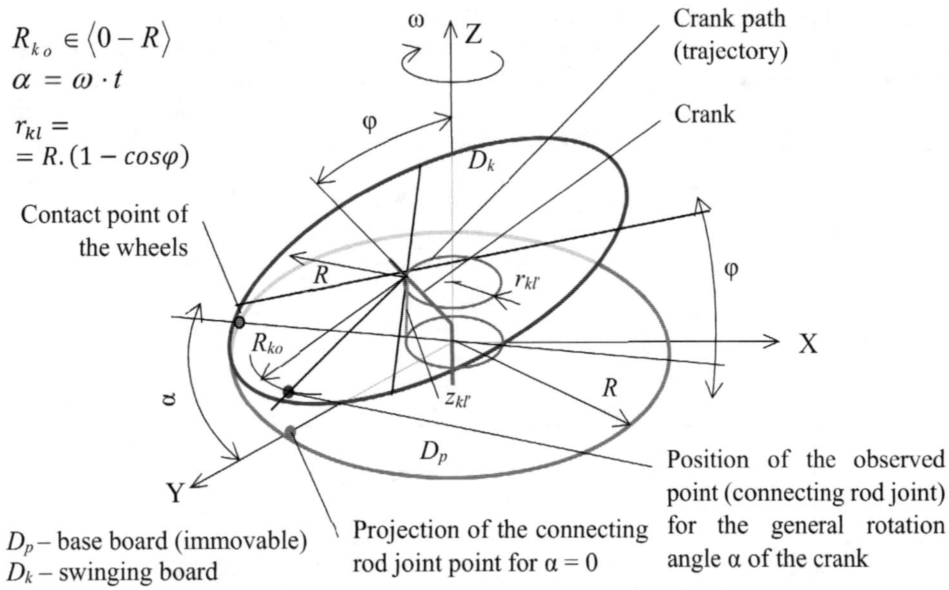

Fig. 2. Geometrical scheme of the unconventional mechanism FIK_1

Distance between the planes of the crank circle and the immovable base board is given by expression

$$z_{kl'} = R \cdot \sin \phi \,. \tag{2}$$

By those assumptions each point of the swinging board, which lies outside of its center makes a general movement in three axes. In order to describe the kinematics of the piston movement, it is first necessary to describe the kinematics of a random point on the swinging board which lies on the radius R_{ko} (see Fig. 2), in the interval $R_{ko} \in \langle 0, R \rangle$. Further, it will be considered the possibility of changing the angle φ between both boards, which can theoretically be in the range $\varphi \in \langle 0, 90° \rangle$. Given the requirement, it is clear, that when the angle $\varphi = 0$ the swinging board makes no movement and when the angle φ is close to the 90°, then the engine design overall is unrealistic. Practically it is impossible to use higher angle values than 50°.

In relation to the piston movement it is most important to define mainly the vertical path of the random point on the swinging board (in direction of **Z** axis). This path, depending on the path of the point in the plane **XY** and on the connecting rod length, determines the value of the piston stroke which is in the direction of **Z** axis. According to the presented above, the path of the observed point on the swinging board in each axis depends on:

- radius R of the swinging, resp. base board,
- angle φ, between the planes of both boards,
- radius R_{ko}, where the observed point is located,
- rotation angle α of the crank as an independent variable.

3 Trajectory estimation of observed point on the swinging board

The path of the observed point is based upon a specific case when the angle value between both boards $\varphi = 90°$, the point lies on the perimeter of the swinging board i.e. $R_{ko} = R$, and the crank shaft has the same crank angle $\varphi = 90°$. In such conditions the observed point performs a movement, whose projections to the individual axes (resp. planes) could be determined more simply, because the center of the swinging board is moving around the circle with a radius $r_{kl'} = R$. Its coordinates in four basic crank positions can be defined directly from the geometry of the physical model (Table 1).

Table 1. Position of the observed point

φ (°)	α (°)	x	y	z
90	0	0	R	0
	90	$-R$	R	R
	180	0	$-R$	$2R$
	270	R	R	R
	360	0	R	0

With so defined parameters the swinging board is situated permanently in the vertical plane and its perimeter is rolling around the perimeter of the immovable base board (see Fig. 3). At the same time the center of the swinging board describes a circle with radius $r_{kl'} = R$ in the horizontal plane (**XY**) at the height $z_{kl'} = R$. The crank is rotating around its axis with an angle α by an angular velocity ω.

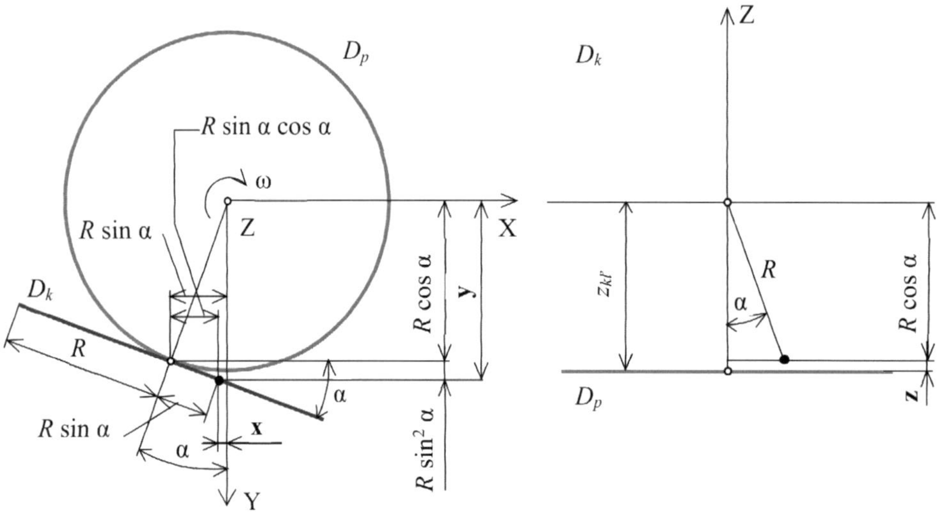

Fig. 3. System scheme for the angle φ = 90°

The position of the observed point at each random crank angle α can be determined from the geometric relationships with the following expressions (see Fig. 3):

$$x = R \cdot \sin\alpha \cdot (\cos\alpha - 1), \tag{3}$$

$$y = R \cdot (\cos\alpha + \sin^2\alpha), \tag{4}$$

$$z = R \cdot (1 - \cos\alpha). \tag{5}$$

If we want to observe movement of the point at an optional radius of the swinging board in the range of $R_{ko} \in \langle 0, R \rangle$, based on the similar consideration, by transforming (3), (4) and (5) we get expressions (6), (7) and (8):

$$x = \sin\alpha \cdot (R_{ko} \cdot \cos\alpha - R), \tag{6}$$

$$y = R \cdot \cos\alpha + R_{ko} \cdot \sin^2\alpha, \tag{7}$$

$$z = R - R_{ko} \cdot \cos\alpha \cdot \tag{8}$$

In the next step, it is necessary to define the path of the observed point at the swinging board's perimeter ($R_{ko} = R$), provided that the angle φ between swinging and base board is random, but not 90°. Similar as in the previous case it is possible to simply define its location at four base locations of the crank from the physical model geometry. Values are presented in the Table 2.

Table 2. Position of the observed point

φ (°)	α (°)	x	y	z
≠90	0	0	R	0
	90	$- r_{kl'}$	R	$R \sin \varphi$
	180	0	$R - 2\, r_{kl'}$	$2R \sin \varphi$
	270	$r_{kl'}$	R	$R \sin \varphi$
	360	0	R	0

Even in this case, the swinging board, inclined under an angle φ, is rolling with its perimeter around the perimeter of the base board. The path of the observed point in the relation to the angle of the crank rotation α in each axis is affected by the inclination of the swinging board. Based on this, the path of the observed point in each axis is described by the following equations:

$$x = R \cdot \sin\alpha \cdot (\cos\alpha - 1) \cdot (1 - \cos\phi), \tag{9}$$

$$y = R \cdot (\cos\alpha + \sin^2\alpha) \cdot (1 - \cos\phi) + R \cdot \cos\phi, \tag{10}$$

$$z = R \cdot (1 - \cos\alpha) \cdot \sin\phi . \tag{11}$$

In the most general case, when the angle between the two boards φ is in the range $\varphi \in \langle 0, 90° \rangle$ and the general point position on the swinging board D_K is in the range $R_{ko} \in \langle 0, R \rangle$ the expressions (9), (10) and (11) are presented in their universal form:

$$x = \sin\alpha \cdot (R_{ko} \cdot \cos\alpha - R) \cdot (1 - \cos\phi), \tag{12}$$

$$y = (R \cdot \cos\alpha + R_{ko} \cdot \sin^2\alpha) \cdot (1 - \cos\phi) + R_{ko} \cdot \cos\phi, \tag{13}$$

$$z = (R - R_{ko} \cdot \cos\alpha) \cdot \sin\phi . \tag{14}$$

The following images show the visual interpretation of the solution of equations (12), (13) and (14) for one crank revolution and for the parameters of the physical model $R = 0{,}1$m,

R_{ko} = 0, 072m and φ = 30°. Graphical interpretation is done for each axis and in relation to the rotation angle of the crank α as well as in each of the three planes.

Fig. 4 shows the path of the observed point in the horizontal plane (**XY**). For better presentation the graph displays the half of the base board. Fig. 5 presents the path of the point in the vertical plane (**XZ**) and Fig. 6 – the point path in the vertical plane (**YZ**), whereas the axis of the engine cylinder is visible in the figure. The path of the observed point in all of the three axes in relation to the angle of the crank rotation α for one revolution is shown on the Fig. 7. The single curves indicate that the movement in the axes **X** and **Y** is inharmonic and movement in the axis **Z** is harmonic.

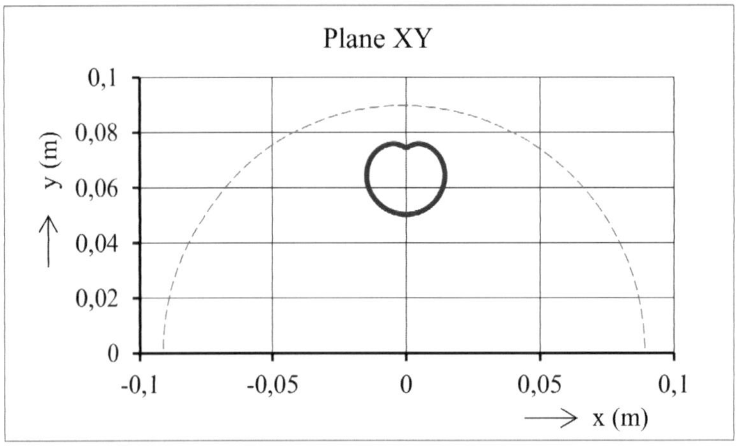

Fig. 4. Path of the observed point in the plane XY (shape of the immovable board is dashed)

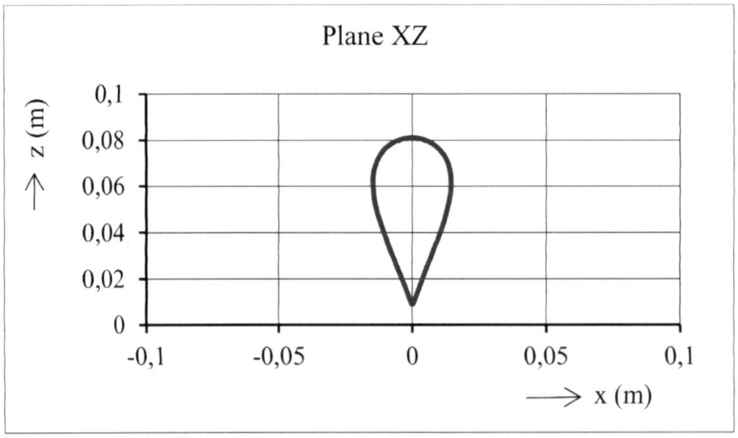

Fig. 5. Path of the observed point in the plane XZ

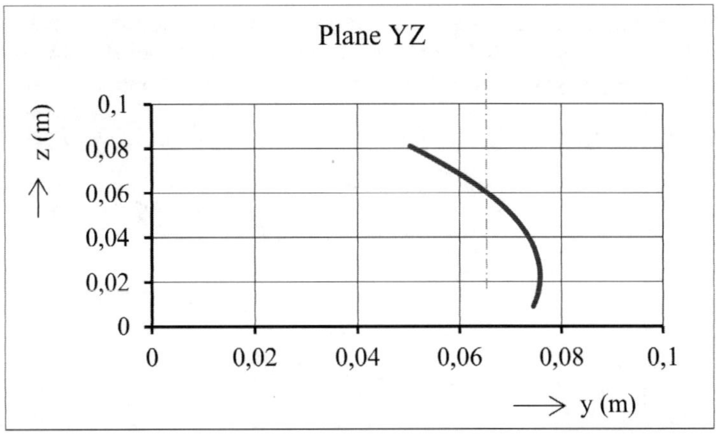

Fig. 6. Path of the observed point in the plane YZ (cylinder axis is dot-dashed)

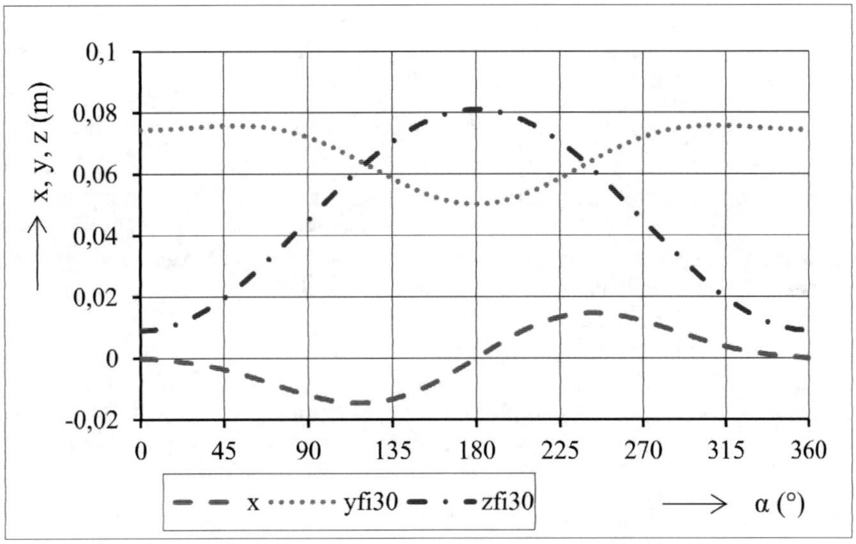

Fig. 7. Path of the observed point in relation to the angle of the crank rotation α

4 Velocity estimation of observed point

The velocity graphs of the observed point in each axis are obtained as a time derivative from the equations (12), (13) and (14). The following expressions represent them:

$$c_x = \frac{dx}{dt} = \frac{dx}{d\alpha} \cdot \frac{d\alpha}{dt} = \omega \cdot (1 - \cos\phi) \cdot \left[R_{ko} \cdot \left(1 - 2 \cdot \sin^2 \alpha\right) - R \cdot \cos\phi\right], \quad (15)$$

$$c_y = \frac{dy}{dt} = \frac{dy}{d\alpha} \cdot \frac{d\alpha}{dt} = \omega \cdot (1 - \cos\phi) \cdot \left(2 \cdot R_{ko} \cdot \sin\alpha \cdot \cos\alpha - R \cdot \sin\alpha\right), \quad (16)$$

$$c_z = \frac{dz}{dt} = \frac{dz}{d\alpha} \cdot \frac{d\alpha}{dt} = \omega \cdot R_{ko} \cdot \sin\alpha \cdot \sin\phi \,. \quad (17)$$

Calculated velocities of the observed point on the swinging board in each of the three axes with crank revolutions n = 4000 min^{-1} are shown in the graph (Fig. 8). From the periodic and harmonic point of view, velocities have similar properties to paths.

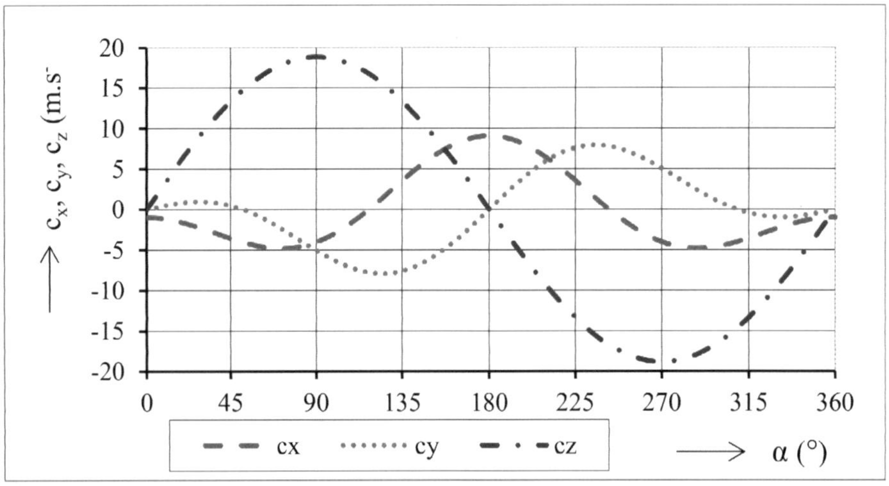

Fig. 8. Velocity of the observed point in relation to the angle of the crank rotation α

From the engine piston kinematics point of view, the dominant velocity is in the **Z** direction. Its maximal value is in the direction of cylinder axis and is about 4 m.s^{-1} higher, in comparison to the maximal velocity of the crank joint of a classical engine with comparable piston stroke. When calculating this velocity to the velocity of the piston, it must be taken into account, that the kinematics of the piston movement is affected by the wave movement of the connecting rod. Having a classic crank mechanism, the connecting rod makes the planar wave movement in the range equal to the piston stroke. In the case of engine with the swinging board, the connecting rod performs a spatial movement in significantly smaller range, in this case approximately only 40% of the piston stroke (see Fig. 4 and 6).

5 Acceleration of observed point

The acceleration graph of the observed point in each axis is obtained as the second derivative of equations (12), (13) and (14) with respect to time. After derivation and appropriate modification, the acceleration in axes X, Y, Z is:

$$a_x = \frac{d^2x}{dt^2} = \frac{d^2x}{d\alpha^2} \cdot \left(\frac{d\alpha}{dt}\right)^2 = \omega^2 \cdot (1 - \cos\phi) \cdot (R \cdot \sin\alpha - 4 \cdot R_{ko} \cdot \sin\alpha \cdot \cos\alpha)\,, \quad (18)$$

$$a_y = \frac{d^2y}{dt^2} = \frac{d^2y}{d\alpha^2} \cdot \left(\frac{d\alpha}{dt}\right)^2 = \omega^2 \cdot (1 - \cos\phi) \cdot \left[2 \cdot R_{ko} \cdot (1 - 2 \cdot \sin^2\alpha) - R \cdot \cos\alpha\right], \quad (19)$$

$$a_z = \frac{d^2z}{dt^2} = \frac{d^2z}{d\alpha^2} \cdot \left(\frac{d\alpha}{dt}\right)^2 = \omega^2 \cdot R_{ko} \cdot \cos\alpha \cdot \sin\phi\,, \quad (20)$$

Calculated accelerations of the observed point on the swinging board in each of the three axes with crank revolutions n = 4000 min^{-1} are shown in the graph (Fig. 9). From the periodic and harmonic point of view, accelerations have similar properties to paths or velocities.

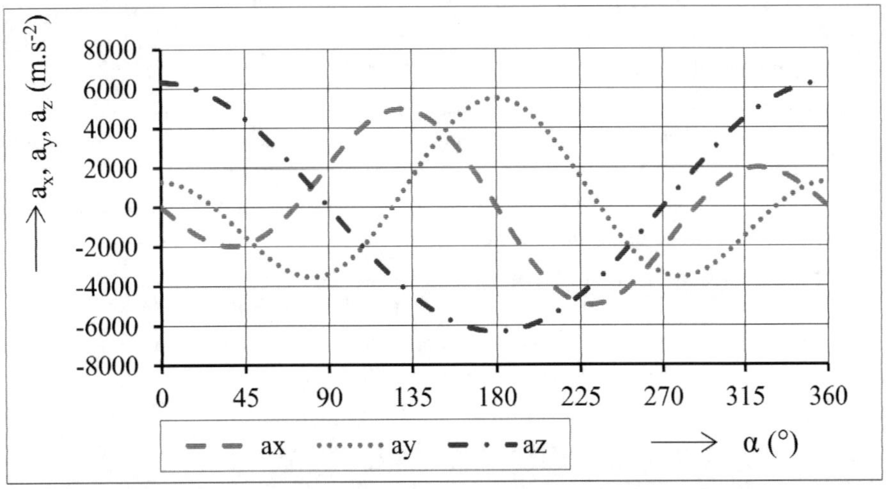

Fig. 9. Acceleration of the observed point in relation to the angle of the crank rotation α

In this case the dominant acceleration is in the Z direction as well. Its maximal value is in the direction of cylinder axis and is about 90 m.s^{-2} higher, in comparison to the maximal acceleration of the crank joint of a classical engine with comparable piston stroke. Percentagewise it is about 1,5% higher.

6 Conclusion

Based on the analytical result, that kinematics properties of the unconventional crank mechanism of the FIK_1 engine are comparable to the kinematics properties of the classical piston combustion engine with the same value of the piston stroke. It is possible to influence piston kinematics parameters by adjusting the distance of the cylinder axis from the axis of the crankshaft. A specific disadvantage of this mechanism is that the connecting rod performs a space movement, in contrast to the classical crank mechanism. That demands better design solution of its connection with the swinging board and piston.

This paper was realized as a part of the project KEGA 022ŽU-4/2017 – "Implementation of on-line education in the area of precise technologies with an impact on educational process to increase skills and flexibility of students of engineering fields of study" and as a result of the project implementation: "Modern methods of teaching of control and diagnostic systems of engine vehicles", ITMS code 26110230107, supported by the Operational Programme Education.

References

1. V. Morozov, A. Zhdanov, *Self-stopping actuating mechanism with high coefficient of efficiency of the forward stroke for linear motion drives*, MATEC Web of Conferences 75, 09001 (2016)
2. S. Erkaya, İ. Uzmay, *Experimental investigation of joint clearance effects on the dynamics of a slider-crank mechanism*, Multibody System Dynamics, vol. **24**, 1, p. 81–102, (2010)

3. S. J. Bo, T. C. Chen, Q. H. Ye, *Dynamics Research of rigid-flexible model of crank-rocker mechanism with Multi Clearance*, MATEC Web of Conferences **63**, 02027 (2016)
4. C. Toro, N. Lior, *Analysis and comparison of solar-heat driven Stirling, Brayton and Rankine cycles for space power generation*, Energy, vol. **120**, 549-564, (2017)
5. P. Baran, P. Kukuča, D. Barta, *Simulations of non-convetional mechanisms for stirling engine,* Proceeding of BulTrans – 2014, 167, (2014)
6. P. Baran, P. Stastniak, M. Brezani, *Dynamic analysis and comparison of balancing systems of nonconventional piston machine FIK_1*. Proceeding of conference: Dynamics of rigid and deformable bodies 2016,. 9-15, (2016)
7. *Axial Internal – Combustion Engines*, http://www.douglasself.com/MUSEUM/POWER/unusualICeng/axial-ICeng/axial-IC.htm, (2008)
8. E. García-Cansecoa, A. Alvarez-Aguirreb, J. M.A. Scherpenc, *Modeling for control of a kinematic wobble-yoke Stirling engine*, Renewable Energy, vol. **75**, 808-817, (2015)
9. Yu. Pogulyaev, O. Nikishin, A. Zheltov, *The kinematics of the swashplate engine with two rotating pairs*, Procedia Engineering, vol. **206**, 1722-1727, (2017)
10. Ł. Jedliński, J. Caban, L. Krzywonos, S. Wierzbicki, F. Brumerčik, *Application of vibration signal in the diagnosis of IC engine valve clearance,*. VP, vol. **3**, 14-19, (2014)

A. J. Shih, *Kinematics of the Cycloidal Internal Combustion Engine Mechanism,* Journal of mechanical design, vol. **115 (4)**, p. 953-959, (2008)

11. P. Fitz, R. Isteník, P. Kukuča, *Mechanizmus piestového stroja*, Vestník ÚPV SR, patent č. 283743, (12/2003)
12. R. Isteník, *Simulácia mechanických a termomechanických procesov v spaľovacích motoroch a vozidlách*, Habilitačná práca, KKVMZ SjF Žilinská univerzita, (2002)

Analysis of the market of electric tractors in agricultural production

Jacek Caban[1,*], *Jan* Vrabel[2], *Branislav* Šarkan[2], *Janusz* Zarajczyk[1], *Andrzej* Marczuk[1]

[1]University of Life Sciences in Lublin, Faculty of Production Engineering, Głęboka 28, 20-612 Lublin, Poland
[2]University of Žilina, Faculty of Operation and Economics of Transport and Communications, Univerzitna 1, 010 26, Žilina, Slovakia

Abstract. In Poland, the market of electric field tractors is practically non-existent. There are individual models in offers dedicated to the agriculture made by foreign producers. However, these offers are presented mainly at agricultural fairs. The article presents the research on the needs of farms for electric tractors and presents the possibilities of developing electro mobility in this sector of the economy. Questionnaire was presented, data were collected from those working in the agricultural sector. The data will be used to gauge attitudes and opinions towards alternative power systems implemented in agriculture.

Keywords: agricultural vehicles, electric drive, electric motors, energy intensity

1 Introduction

The reduction of oil sources, the increase of environment pollution and the intensification of climate changes have led the whole world looking for alternative fossil energy sources for more efficient usage [1]. Energy intensity of transport is a current topic around the world [2]. At present, the drives of vehicles use energy that is largely bounded in fossil fuels [3, 4]. Based on the ever-increasing demands that are placed on each of the transport sector, these resources are gradually depleting and are not unlimited [5-7]. Renewable energy sources in agriculture more and more are being talked about. Incorporation of electric drives in tractor and agricultural machinery presents advantages in terms of increased energy efficiency and expanded functionalities [8]. Possibilities to improve work efficiency and reduce fuel consumption, and thus reduce exhaust emissions from agricultural tractors, were presented in research [9-11]. Magalhaes et al. in the paper [12] described the main researches with proposals to overcome the problems of implementing electric tractors, as supply and electricity storage. They present technical and economic comparisons between conventional tractors with ICE, and moreover electric tractors are discussed. Tomasikova et al. [13] in research work describes modelling and simulation of the vehicle physical model and analysis of transport mechatronic system properties. The model includes also other blocks, which are used to show measured outputs and control blocks, that make the vehicle mechatronic

* Corresponding author: jacek.caban@up.lublin.pl
Reviewers: *Oleksandr* Kravchenko, *Dalibor* Barta

systems controllable according the driver model inputs [13]. Agricultural technology based on the achievements of the automotive industry and implemented in cars technologies is often used nowadays in agricultural vehicles. Two development trends of agricultural tractors can be seen. One of them is a hybrid drive combining the advantages of an internal combustion engine with electric motors. The second development trend is the production of tractors with purely electric drive. That is why complex technologies used in them justify the cost of the vehicle – in most cases the cost of the hybrid vehicle is two times bigger than that of a conventional vehicle and matches the cost of the electric vehicle. Prototypes of electric machines have been created for several years. Some manufacturers believe that in the future electric tractors will become cheaper than traditional tractors due to environmental requirements, which are becoming stricter. Mcfadzean and Butters [14] reported on the basis of calculations that, the cost of charging an all-electric system capable of producing the required output of the practical tests would be higher than the cost of diesel for the same output. In pursuit to lower the levels of toxic exhausts, Fendt, German manufacturer of agricultural equipment has shown the public on recent Agritechnica fair in Hanover their first electric tractor, model e100 Vario [15]. First examples of e100 Varios are believed to be on the roads – or farms – in the year 2018 [15]. John Deer has been testing prototypes for several years and also has an electric tractor. More details about diesel-electric and electric agricultural tractors we can found in this references [15-17]. We can expect a real war in the market of electric tractors in the coming years. Because investments in electric drive are another natural step among manufacturers.

One of the advantages of the electric drive is, for example, the lack of emissions at the place of use. An electric motor compared to a traditional internal combustion engine generates less vibration and noise. This is a special advantage mainly at night and in the residential area. The electric drive allows, above all, precision in the control of the vehicle. However, the ideal electric tractor, despite many attempts to make and test it, still lies in the sphere of challenges. The main problem that is hindered by is the electrical capacity of the battery, because a few hours of one-time operation is too short for most of the hosts to carry out their field tasks. The necessity of recharging the battery causes a stoppage at work, during which the farmer exposes himself to a waste of time, which is also connected with agro-technical periods.

In Poland, the market of electric agricultural tractors is practically non-existent. There are individual models in offers dedicated to the agriculture of foreign producers. However, these offers are presented mainly at agricultural fairs. On foreign markets one can meet agricultural machines produced by large concerns equipped with electric drives, such as the Massey Ferguson 9280 Delta combine harvester, Fendt X harvester, CASE IH ProHybrid EECVT tractor, Caterpilar D7WE crawler tractor. In contrast, in the case of passenger cars, every major automotive company currently has such a vehicle produced in series. It is a good prognosis for electric drives to have a significant market share in the future in the agricultural machinery sector as well.

Currently, in Poland, the Energy and Development Departments is working together on the act on electro mobility, which is to regulate the entire market. Pilot projects are being carried out to develop electro mobility in individual voivodships. In the opinion of many experts, electric vehicles can become popular in Poland, if macroeconomic conditions are created for this purpose. This applies not only to the individual vehicle market but also to the public transport buses. Electric vehicles will also be a vital element in future smart urban transport systems [18]. In the case of electric vehicles, energy intensity of traffic constitutes smaller part of the consumed energy than in the case of vehicles with ICE [19]. Fuel savings in hybrid vehicles oscillate between approx. 30% in comparison with the vehicles with internal combustion engines [20].

Another issue is the need to promote and awaken awareness among farmers about the benefits of using electric agricultural tractors. Tax incentives would also encourage farmers to invest in such machines. Undoubtedly, investments in charging infrastructure and tax incentives for zero-emission car buyers are necessary for the development of electro-mobility in Poland. Most authors agree that innovative technologies can effectively facilitate the reduction of emissions and lead to sustainable development [18]. At the same time, technological progress needs to be supplemented by behavioural and cultural changes concerning mobility patterns and preferences [18]. This last issue seems to be the most difficult task.

2 Agricultural tractors market

Overall farm tractors market is highly fragmented in nature characterized by more than 300 companies with numerous brands and joint ventures across the globe [21]. Major farm tractor manufacturers have been striving extensively in order to maintain their market position. Entering into joint ventures with local tractor manufacturers has been one of the key strategies adopted by tractor manufacturers from North America and Europe regions [21]. The competition is expected to become more intense with the revival of farm tractors market during the forecast period from 2018 to 2026 according to the research study conducted by Credence Research [21].

This study (Fig. 1) presents the main division due to the tractor's power output without distinguishing the type of drive source. However, the presented graph shows another important part of the market (autonomous tractors). In this segment, the drive sources can be varied, but the electric drive is particularly important, ensuring the perfect precision of vehicle control.

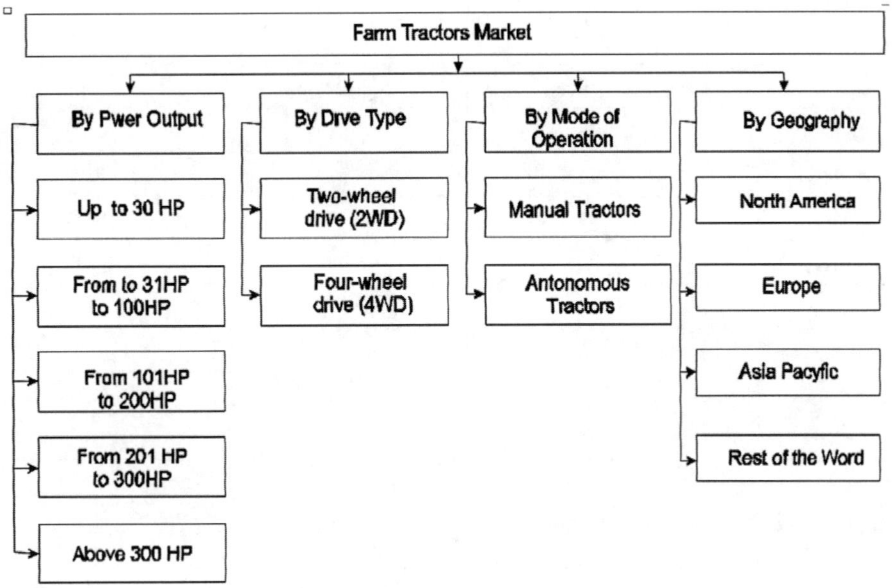

Fig. 1. World agricultural tractors market [21]

In Table 1., structure of the agricultural machinery industry in the European Union in 2014 was presents.

Table 1. Structure of the agricultural machinery industry in the European Union in 2014 [22]

Country	Turnover (€m)	Number of companies	Number of employees
Germany	12,348.7	637	40,166
Italy*	9,224.4	1,908	29,120
France	5,124.9	589	16,806
United Kingdom**	2,978.8	478	6,362
Nethelads	2,384.2	337	7,400
Belgium	2,205.5	142	4,980
Austria	2,058.5	116	6,271
Poland	1,748.0	478	16,420
Sweden	1,551.9	200	4,527
Spain	1,370.5	701	7,163
Finland	1,096.5	144	4,346
Czech Republic	693.2	291	7,405
Denmark	564.2	160	3,561
Hungary	483.4	145	5,196
Slovakia	177.7	52	1,381
Ireland***	147.2	36	1,051
Slovenia	146.5	59	1,201
Portugal	124.7	192	1,238
Greece	103.3	358	1,661
Romania	98.5	69	2,205
Croatia	91.0	58	1,098
Estonia	79.2	27	648
Bulgaria	28.7	54	1,103
Latvia	27.0	21	645
Lithuania	23.3	11	486
Cyprus	5.9	11	55
Luxenburg	0.0	0	0
Malta	0.0	0	0
Total EU (28)	**44,886**	**7,274**	**172,495**
Number of employees (* Italy: 2013, ** United Kingdom 2010, *** Ireland 2013) – Source Eurostat			

The agricultural machinery sector reveals a highly specialised Europe with 10 countries totalising 80% of the number of companies and jobs in the manufacturing sector and generating 91% of turnover in the European Union [22]. The first place in the agricultural machinery industry in the European Union production is taken by Germany with a turnover of €12 bn, and more than 40,000 employees. Second one is Italy with a total of €9 bn, and 29,000 employed. France is the third producing country with a turnover of €5 bn, and 16,000 of employees. Poland with the number of 16,000 employees is the fourth country in terms of highest number of employees, but in eight place regarding turnover - €1.74 bn, in the agricultural machinery industry in the European Union in 2014 (Table 1.).

Figure 2 presents the European production of agricultural tractors divided by the tractors power between 2007÷2014. European production of agricultural tractors amounted €7.9 bn in 2014 decreased by 7.9% compared to 2013 [22]. This reduction affects all power segments. This category represents 25% of the total value of agricultural machinery produced in Europe [22]. As we can see, the largest production was achieved in 2008, just before the global banking crisis.

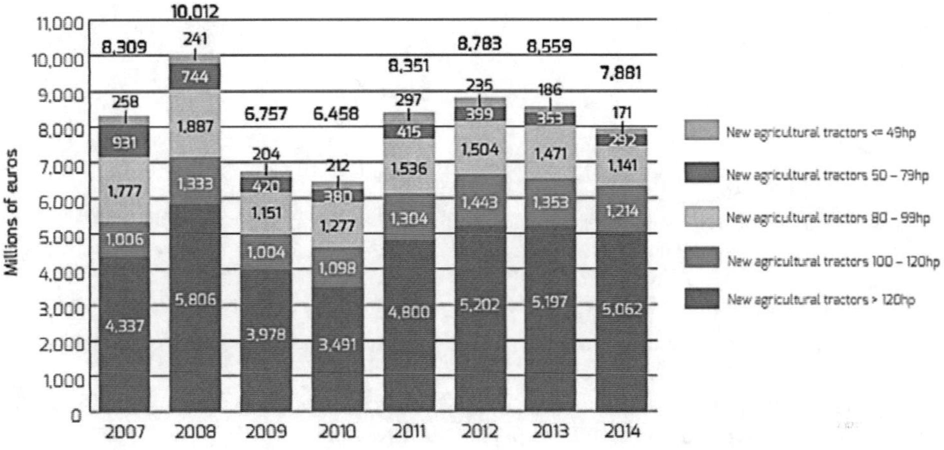

Fig. 2. European production of agricultural tractors [22]

The global agricultural tractors market is trying to recover from its current state of regression with a global market value of US$ 8.15 bn in 2017 and anticipated to exhibit a growth with a CAGR of 5.7% from 2018 to 2026. The global demand for food grains is expected to rise from 3 billion tonnes in 2010 to a staggering 4.5 billion tonnes in 2050. In 2017, Deere and Co., world's second largest farm equipment and tractor manufacturer (by sales) registered growth in sales in past four years. The aforementioned incidents are clearly highlighting the revival of the tractor market in the following years.

3 Methodology

The purpose of the work was to analyse the awareness of farmers and the needs of farms for electric tractors. In addition, a view on the development of electro mobility in Poland was presented. The following research methods were used in the work:

- a questionnaire survey using a paper questionnaire was carried out on 40 farm owners of both sexes, in different age groups,
- analysis of collected data and their presentation by graphical method.

The research part was elaborated on the basis of a survey among farmers in the Lubelskie and Świętokrzyskie voivodships. The survey was conducted on a group of 40 respondents and contained 12 questions with the possibility of single and multiple choice. The respondents participating in the study were of both sexes and belonged to different age categories. Respondents had the opportunity to choose the answers from among the listed and one open question, in which the respondents could express their opinion on electric tractors. In addition, the research part includes a presentation of the results of the survey, their analysis, and a summary.

4 Analysis of the survey results

The following figures present a graphic interpretation of the results of the survey. Figure 3a) presents the results regarding the age of the respondents, and in Figure 3b) the division by gender. The questionnaire was attended by seven people aged 20÷30, five people between 31÷40, seven people between 41÷50, seventeen people between 51÷60, and four above 61 years. 40 people took part in the study, of which five were women and thirty-five were men.

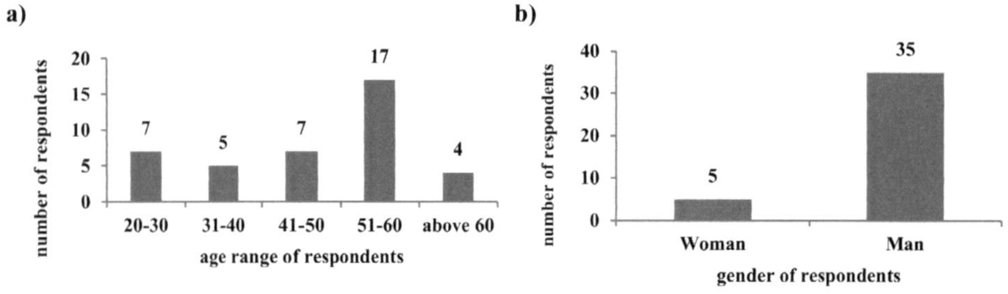

Fig. 3. Share of survey participants relative to: a) Age of respondents, b) Gender of respondents

Figure 4a) shows the type of education held by the respondents. Figure 4b) shows the period of running a farm by respondents expressed in years.

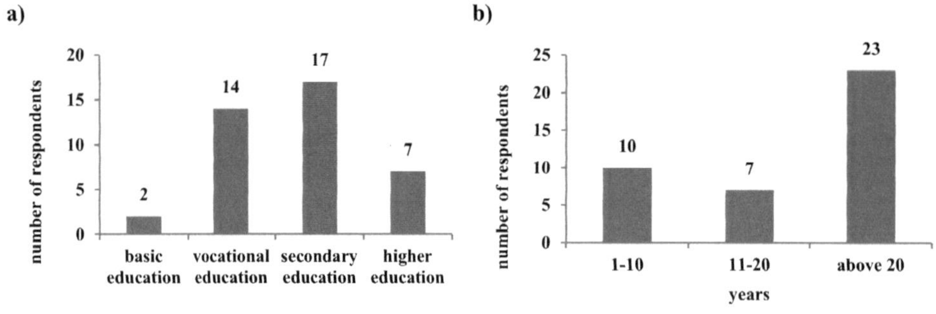

Fig. 4. Share of survey participants relative to: a) Type of education held by the study participants, b) Period of running a farm

As it can be seen in Figure 4a), two of the respondents have basic education, fourteen vocational, seventeen secondary and seven higher. Data analysis regarding the period of activity (Fig. 4b)) is presented as follows: in the 1 ÷ 10 year range, the farm is run by ten people, seven in the range of 11 ÷ 20 years, and twenty-three people over 20 years.

Figure 5a) presents data on the size of the farm. In turn, Figure 5b) contains data on the production profile of farms run by respondents.

Analysing the data presented in Figure 5a), it can be seen that the largest number of people, thirty two respondents work on the farm of size from 1 ÷ 30 ha, five in the range of 31 ÷ 60 ha, one 61 ÷ 90 ha and two above 90 ha. In most of the farm production profile (Fig.

5b)) is directed at field crops, as many as thirty-two farms. Five farms specialize in horticultural crops, two in fruit and shrubs and eight specializing in animal husbandry.

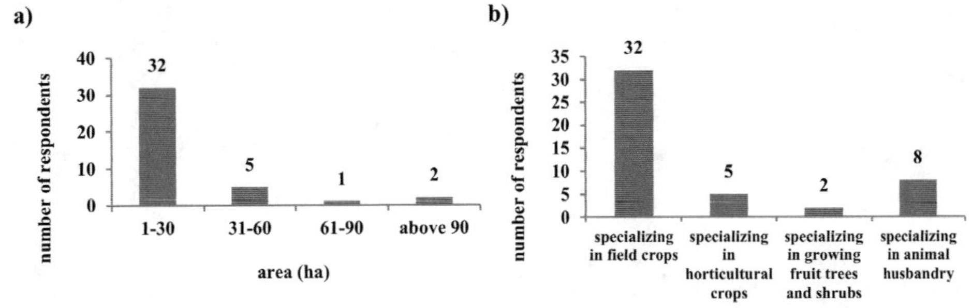

Fig. 5. Share of survey participants relative to: a) The size of the farm given in [ha], b) The production profile of farms

Figure 6a) presents the number of tractors used on the farm. In the Figure 6b) presents data on the planned purchase of a tractor this year by respondents.

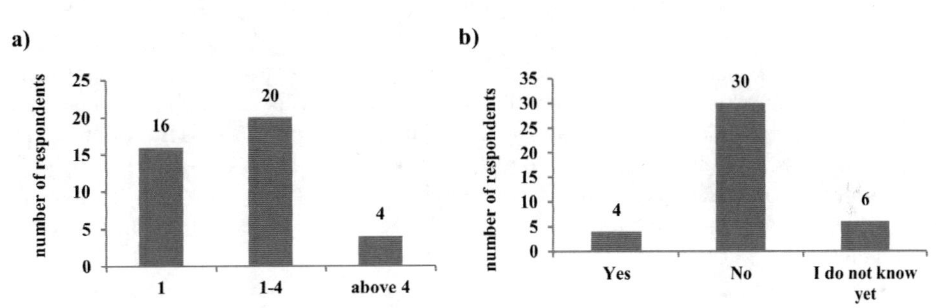

Fig. 6. Share of survey participants relative to: a) Number of tractors used on the farm, b) Planned purchase of a tractor this year

Depending on the profile of the farm and its area, farmers use machines of various power and sometimes need several agricultural tractors. As shown in Figure 6a), up to 4 machines are mostly used in the analysed farms. Among the respondents, sixteen people use one tractor, in the range of 1 to 4 tractors use twenty respondents, and above four tractors, four people.

The analysis of the plans for the purchase of a new tractor in the current year is presented in Figure 6b). The purchase of a new agricultural tractor was confirmed by four people, thirty respondents said that they are not planning to buy, and six respondents have not yet made a decision.

The following figures (Fig. 7 and Fig. 8), the results of the questionnaire regarding purchase preferences of agricultural tractors by respondents and their knowledge of electrically driven tractors are presented.

When buying a new tractor (Fig. 7a)), sixteen people by the engine power, eleven looks at innovative solutions, and twenty-two people looks at the price. According to twenty-seven respondents, the most important criterion that guides the purchase of a new tractor is the comfort of work. It is very good that the majority of respondents are comfortable working. This shows their high awareness of health protection in the work environment.

Answers to the question on knowledge of electric tractors among farmers is shown in Figure 7b). Fourteen people met with the concept of electric tractors, but only three knew what this type of drive consists of. Most respondents, that is twenty-six people met with this issue for the first time.

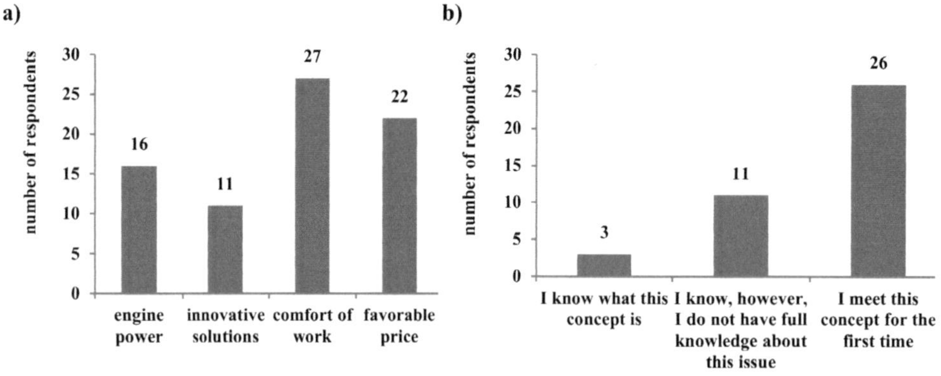

Fig. 7. Share of survey participants relative to: a) Categories that the respondents follow when buying a new tractor, b) Knowledge about electric tractors

Response to the question about the feelings of respondents in connection with investments concerning the introduction of agricultural electric tractors for use are presented in Figure 8a). For sixteen respondents, this investment is extremely useful and necessary for farms, for eight the investment is unnecessary due to the purchase of traditional tractors. Eleven people have no opinion on the subject, while five people had different opinions. They complained about the lack of specific information about electric tractors, the poor range of their batteries and too high price for tractor that can work only a few hours.

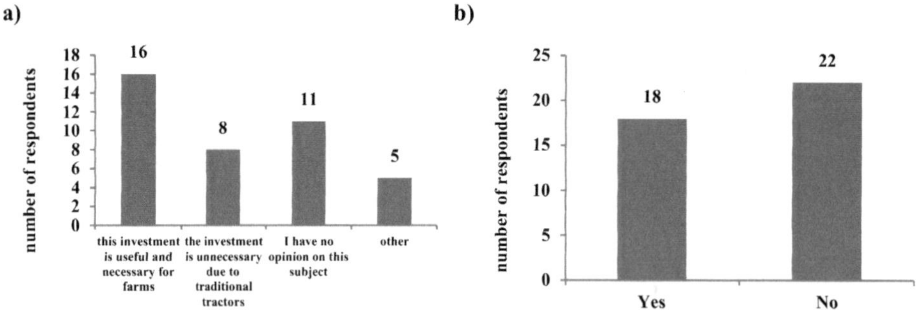

Fig. 8. Share of survey participants relative to: a) Respondents' feelings caused by the investment involving the introduction of electric tractors, b) Decision to buy an electric tractor

Asking the respondents the question "Would you decide to buy an electric tractor?", the authors of the survey expected a negative answer, due to the earlier answers about buying a new tractor, the majority of respondents did not plan to buy a new machine (see the results in Figure 6b)). The result turns out to be different. 18 people have expressed a desire to buy an electric tractor. This is a surprising result and can be an example of the fact that Polish farms are ready and open to implement the concept of electro mobility in agriculture.

5 Conclusions

In the case of electric tractors, in order to be able to talk about a real competition for diesel tractors, they would have to provide a full working day without interruption to recharging. Current solutions that allow only a few hours of work are sufficient only for orchards or green-houses. Moreover, analysing the answers of the respondents when asked whether they are planning to buy a new tractor in the near future, the majority, as many as 75% of respondents replied firmly – no. This is due to the fact that they have already invested in machines with a traditional drive. However, This does not mean, that electric-powered tractors are a distant future on farms. They are an optimistic herald of changes in the market moving towards the environmental performance and energy efficiency of farms, and the elimination of current restrictions is probably a matter of a few years. John Deere expects that the electric motors will require much less maintenance than a diesel engine [16]. They will also provide and last longer [16].

The decision on whether to invest in an electric tractor is a decision that everyone should take on their own. If someone needs a zero-emission ecological tractor for horticultural or greenhouse works, the choice may seem right, and guaranteeing a significant increase in demand for electric tractors should prompt domestic and foreign investors to build an industry that generates electric vehicles and their components in Poland. As was shown at the beginning of the work (Table 1), Poland is one of the leading countries in terms of employment in this sector.

References

1. S. Pukalskas, A. Rimkus, M. Melaika, R. Peceliunas, *Comparison of conventional and hybrid cars exploitation costs*. Advances in Science and Technology Research Journal, **12**, 1, 221-227 (2018)
2. T. Skrucany, V. Harantova, M. Kendra, D. Barta, *Reducing energy consumption by passenger car with using of non-electrical hybrid drive technology*. Advances in Science and Technology Research Journal, **11**, 1, 166-172 (2017)
3. D. Barta, M. Mruzek, R. Labuda, T. Skrucany, L. Gardyński, *Possibility of increasing vehicle energy balance using coasting*. Advances in Science and Technology Research Journal, **12**, 1, 228-235 (2018)
4. J. Caban, P. Ignaciuk. *Technical-economic aspects of CNG gas usage in buses of urban communication*. Engineering For Rural Development, Jelgava, Latvia, 23.-25.05.2018, 2129-2133 (2018)
5. D. Barta, M. Mruzek, *Factors influencing the hybrid drive of urban public transport buses*. Management Systems in Production Engineering, **4**(20), 213-218 (2015)
6. J. Dizo, M. Blatnicky, *Use of multibody system dynamics as a tool for rail vehicle behaviour diagnostics*. Diagnostyka, **17**(2), 9-16 (2016)
7. T. Skrucany, J. Gnap, *The effect of the crosswinds on the stability of the moving vehicles*. Applied Mechanics and Materials, **617**, 296-301, (2014)
8. G.P. Moreda, M.A. Munoz-Garcia, P. Barreiro, *High voltage electrification of tractor and agricultural machinery - A review*. Energy Conversion And Management, **115**, 117-131 (2016)
9. F. Bauer, P. Portes, D. Slimarik, J. Cupera, M. Fajman, *Observation of load transfer from fully mounted plough to tractor wheels by analysis of three point hitch forces during ploughing*. Soil & Tillage Research, **172**, 69-78 (2017)

10. J. Engstrom, O. Lagnelof, *Battery electric autonomous agricultural machine - Simulation of all operations on a Swedish farm*. Land, Technik Ageng 2017: The Forum For Agricultural Engineering Innovations (VDI-MEG 2017), VDI Berichte, 75th Conference of LAND, TECHNIK - AgEng 2017, Hannover, Germany, Nov. 10-11, 2017, **2300**, 15-22 (2017)
11. T. Tunka, J. Cupera, *The engine combustion analysis of newly developing diesel tractor engine Zetor Z1727 with Common-Rail system in a first firing week*. 22nd International PhD Students Conference, Mendel Univ, Fac Agron, Brno, Czech Republic Nov. 11-12, 2015, Mendelnet 2015, Edited by: Polak O., Cerkal R., Belcredi N.B., 562-567 (2015)
12. R.O. Magalhaes, M.V. da Assuncao, J.P.M. Santos, E.V. da Silva, L.D.G. Ferreira, R.R. Magalhaes, D.D. Ferreira, *Review on applications of electric vehicles in the countryside*. Ciencia Rural, **47**, 7, 1-9 (2017)
13. M. Tomasikova, M. Tropp, T. Gajdosik, L. Krzywonos, F. Brumercik, *Analysis of transport mechatronic system properties*. 12th International Scientific Conference of Young Scientists on Sustainable, Modern and Safe Transport, High Tatras, Slovakia, May 31-Jun 02, 2017, Edited by: Bujnak J., Guagliano M., In: Procedia Engineering, **192**, 881-886 (2017)
14. B. Mcfadzean, L. Butters, *An investigation into the feasibility of hybrid and all-electric agricultural machines*. Scientific Papers-Series A-Agronomy, **60**, 500-511 (2017)
15. J. Šujica, *Fend pioneering with electric tractor*. https://plugin-magazine.com/driving/fendt-pioneering-with-electric-tractor/
16. F. Lambert, *John Deere unveils latest all-electric tractor prototype for zero-emission agriculture*. https://electrek.co/2016/12/05/john-deere-electric-tractor-prototype/
17. J. Scott, *Improving tractors, implements with electrification*. https://www.agriculture.com/machinery/tractors/improving-tractors-implements-with-electrification
18. J. Zawieska, J. Pieriegud, *Smart city as a tool for sustainable mobility and transport decarbonisation*. Transport Policy, **63**, 39-50 (2018)
19. E. Siemionek, M. Dziubiński, *Testing energy consumption in the trolleybus and the bus on a chosen public transport line in Lublin*. Advances in Science and Technology Research Journal, **9**(26), 152-160 (2015)
20. M. Dziubiński, A. Drozd, M. Adamiec, E. Siemionek, *Energy intensity of the electric vehicle*. Advances in Science and Technology Research Journal, **11**(4), 81-84 (2017)
21. http://www.credenceresearch.com/report/farm-tractors-market
22. http://www.axema.fr/EspacePresse/Lists/Publications/AXM_RE-EN_2015_BD.PDF Economic Report 2015

Piston kinematics of a combustion engine with unconventional crank mechanism

Pavol Kukuča[1], *Dalibor* Barta[1,1,] *Ján* Dižo[1], *Jacek* Caban[2]

[1]University of Žilina, Department of Transport and Handling Machines, Faculty of Mechanical Engineering, Univerzitná 1, 010 26 Žilina, Slovakia
[2]University of Life Science in Lublin, Faculty of Production Engineering, 28 Głęboka Street 20-612 Lublin, Poland

Abstract: The article deals with a theoretical solution of kinematic variables describing the movement of a combustion engine piston with unconventional mechanism FIK protected by patents No. 283742 and No. 283743 [1]. The course of the path, the velocity, and the acceleration were calculated to assess the dynamic properties of the unconventional mechanism, which transform the piston linear movement to rotational movement by using a swinging board, a baseboard and a crankshaft.

Keywords: Combustion engine, unconventional crank mechanism, piston, kinematics, swinging board

1 Introduction

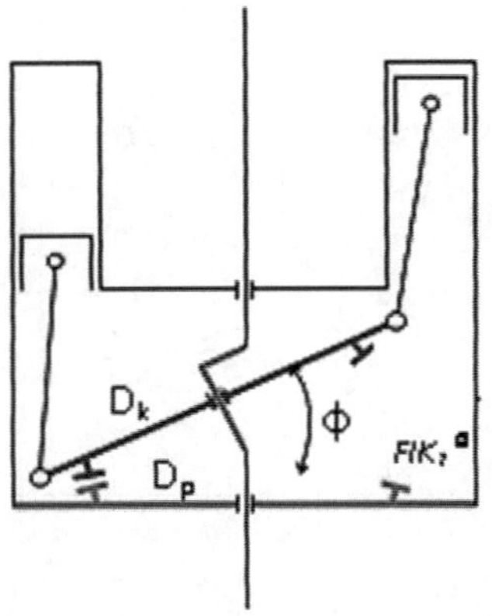

Fig. 1. Scheme of the unconventional engine mechanism

[1] Corresponding author: dalibor.barta@fstroj.uniza.sk
Reviewers: *Pawel* Drozdziel, *Plamen* Punov

Current trends lead to the use of various innovative technologies in vehicle drives. They aim to reduce the weight and dimensions of engines, as well as their vibration and noise.

In combination with the development and improvement of material and fuel technologies, new and uncommon solutions find application [2-4]. The applications of unconventional solutions are often utilized in stationary machines like generators, cogeneration units etc. before being implemented in automobiles, where the requirements for reliability and comfort is much higher. Stirling engine [5, 6], cycloidal internal combustion engine mechanism of Shih [7], which have a low frontal area, very good balance and great compactness [8-10] are examples of using of unconventional mechanisms. Kinematic parameters of unconventional mechanism FIK, also applicable in such engines, are described in this paper.

2 Kinematics of the piston movement

Piston path is mostly specified by the path of the observed point on the swinging board, where the articulated connecting rod is mounted. Its size, similar to the classical crank mechanism, is affected by the connecting rod length and position of the cylinder axis in relation to the crankshaft axis. For the next considerations the cylinder axis is situated parallel to the axis of the crank shaft and simultaneously in the plane of the crank (see Fig. 1). The cylinder axis intersects the plane **XY** approximately in the center of the observed point trajectory in the same plane. In such a position ($\alpha = 0$) the piston is in the bottom dead centre. The cylinder axis can practically be situated in a random position and it may not be parallel to the axis of the crankshaft.

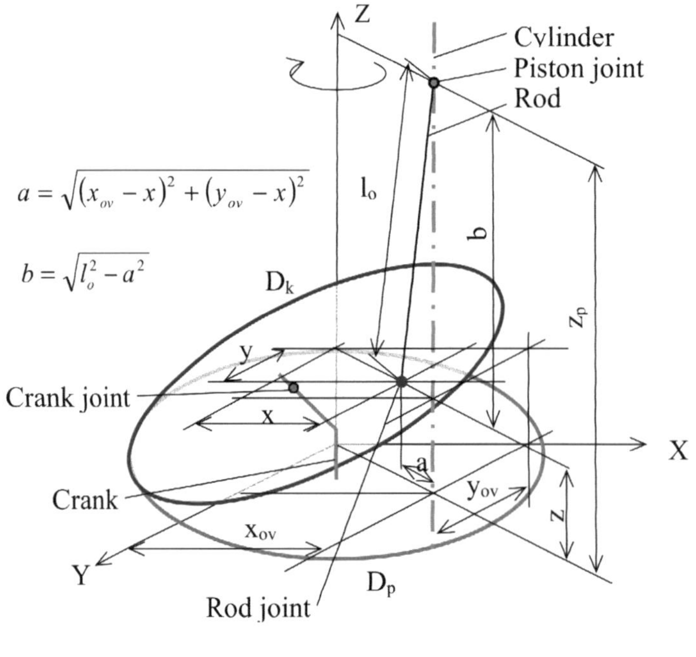

Fig. 2. Scheme of the unconventional engine mechanism FIK$_1$

The graphical scheme of the unconventional mechanism on Fig. 2 is the basis for the following deduction of the piston path expression.

The scheme on the Fig. 2 shows, that the piston position (path of the piston joint) in the **Z** axis direction is given by the sum of the rod joint position z and the distance b according to the equation:

$$z_p = z + b = z + \sqrt{l_o^2 - (x_{ov} - x)^2 - (y_{ov} - y)^2}\,, \tag{1}$$

where
x_{ov} – position of the cylinder axis in the X direction,
y_{ov} – position of the cylinder axis in the Y direction,
l_o - rod length.

After derivation with respect to time of expression and appropriate modification, the velocity of the piston joint is

$$c_p = \frac{dz_p}{dt} = c_z + \frac{c_x \cdot (x_{ov} - x) + c_y \cdot (y_{ov} - y)}{\sqrt{l_o^2 - (x_{ov} - x)^2 - (y_{ov} - y)^2}}\,, \tag{2}$$

where

c_x, c_y, c_z - the velocities of observed point in axes X, Y, and Z defined as:

$$c_x = \frac{dx}{dt} = \frac{dx}{d\alpha} \cdot \frac{d\alpha}{dt} = \omega \cdot (1 - \cos\phi) \cdot \left[R_{ko} \cdot (1 - 2 \cdot \sin^2\alpha) - R \cdot \cos\phi\right], \tag{3}$$

$$c_y = \frac{dy}{dt} = \frac{dy}{d\alpha} \cdot \frac{d\alpha}{dt} = \omega \cdot (1 - \cos\phi) \cdot (2 \cdot R_{ko} \cdot \sin\alpha \cdot \cos\alpha - R \cdot \sin\alpha), \tag{4}$$

$$c_z = \frac{dz}{dt} = \frac{dz}{d\alpha} \cdot \frac{d\alpha}{dt} = \omega \cdot R_{ko} \cdot \sin\alpha \cdot \sin\phi\,\cdot \tag{5}$$

The piston joint acceleration is obtained as a time derivative of the velocity. After appropriate modification the relation is:

$$a_p = \frac{dc_p}{dt} = a_z + \frac{(x_{ov} - x) \cdot a_x + (y_{ov} - y) \cdot a_y - (c_x^2 + c_y^2)}{\sqrt{l_o^2 - (x_{ov} - x)^2 - (y_{ov} - y)^2}} -$$

$$- \frac{\left[(x_{ov} - x) \cdot c_x + (y_{ov} - y) \cdot c_y\right]^2}{\left[\sqrt{l_o^2 - (x_{ov} - x)^2 - (y_{ov} - v)^2}\right]^3}\,, \tag{6}$$

where

a_x, a_y, a_z - the accelerations of observed point in axes X, Y, and Z defined as:

$$a_x = \frac{d^2x}{dt^2} = \frac{d^2x}{d\alpha^2} \cdot \left(\frac{d\alpha}{dt}\right)^2 = \omega^2 \cdot (1 - \cos\phi) \cdot (R \cdot \sin\alpha - 4 \cdot R_{ko} \cdot \sin\alpha \cdot \cos\alpha) \tag{7}$$

$$a_y = \frac{d^2y}{dt^2} = \frac{d^2y}{d\alpha^2} \cdot \left(\frac{d\alpha}{dt}\right)^2 = \omega^2 \cdot (1 - \cos\phi) \cdot \left[2 \cdot R_{ko} \cdot (1 - 2 \cdot \sin^2\alpha) - R \cdot \cos\alpha\right] \tag{8}$$

$$a_z = \frac{d^2 z}{dt^2} = \frac{d^2 z}{d\alpha^2} \cdot \left(\frac{d\alpha}{dt}\right)^2 = \omega^2 \cdot R_{ko} \cdot \cos\alpha \cdot \sin\phi \tag{9}$$

The results of calculations for the same basic parameters as previously stated are on the following pictures. Fig. 3 shows the course of the piston joint path z_p in relation to the angle of the crank rotation α and also in comparison with the connecting rod joint path z (observed point on the swinging board). By comparing both curves it is clear, that the swinging movement of the connecting rod causes small change in the course of the piston stroke in comparison to the stroke of the connecting rod joint. The velocity of the piston joint c_p together with the velocity of the connecting rod joint c_z is shown in Fig. 4, in order to review the influence of the swinging movement of the connecting rod and cylinder axis position on the piston velocity. Comparing the velocities, it is possible to see, that the design parameters influence also in this case the piston joint velocity to slightly higher values. It is possible to partially eliminate this influence by the appropriate rod length selection, appropriate cylinder axis location, as well as correct rod joint position on the swinging board.

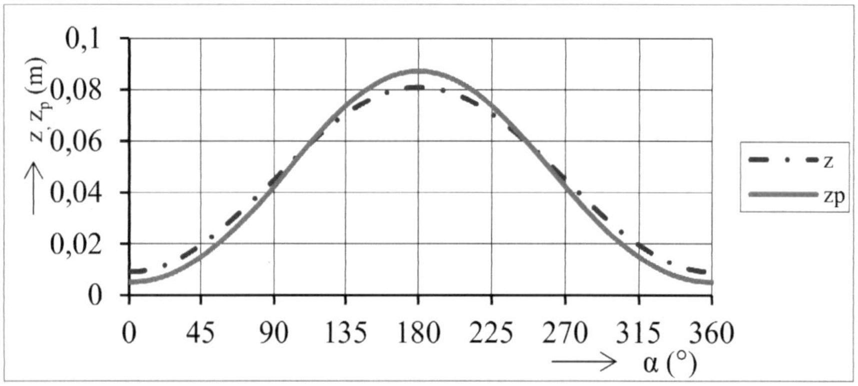

Fig. 3. Path of the piston joint z_p point in relation to the angle of the crank rotation α

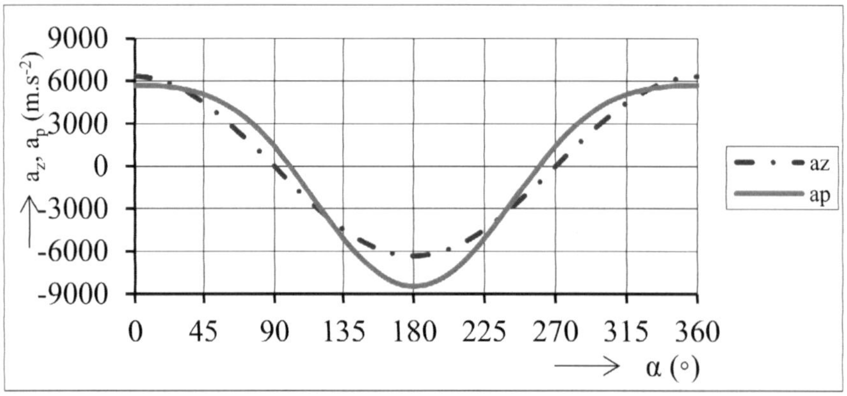

Fig. 4. Velocity of the piston joint c_p in relation to the angle of the crank rotation α

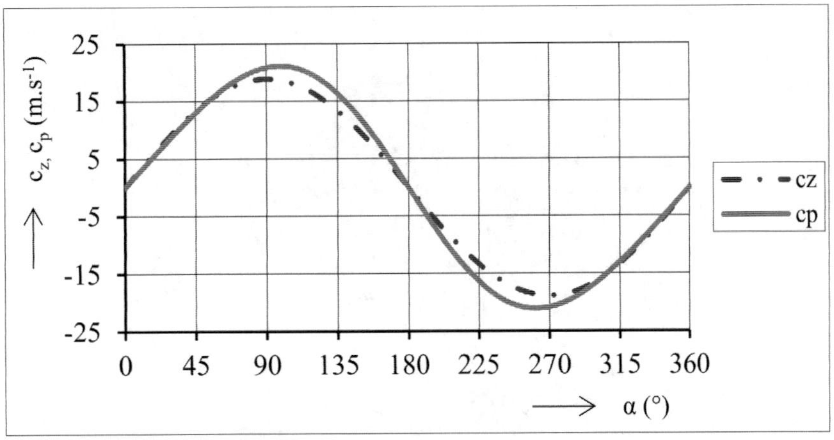

Fig. 5. Acceleration of the piston joint a_p in relation to the angle of the crank rotation α

Fig. 5 shows acceleration a_p of the piston joint in dependence on the angle of the crank shaft rotation. Simultaneously, for better presentation, the acceleration a_z of the connecting rod joint on the swinging board is shown in the diagram. By comparing both accelerations it can be concluded that, in this case the design parameters also have some influence on the piston joint acceleration, which has a slight value increase in comparison to rod joint acceleration. Similar to the velocity case, it is possible to partially eliminate this influence by adopting appropriate mechanism design.

Particular examples of the visual solution of such an engine are shown in Fig. 6, Fig. 7 and Fig. 8.

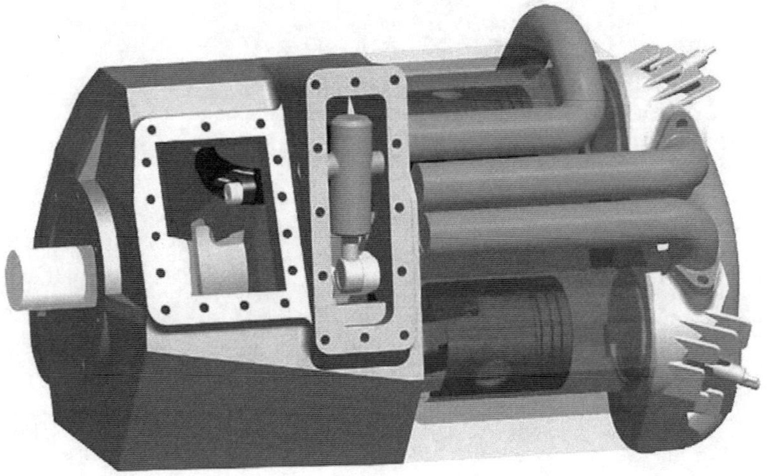

Fig. 6. Virtual model of the engine with unconventional mechanism FIK

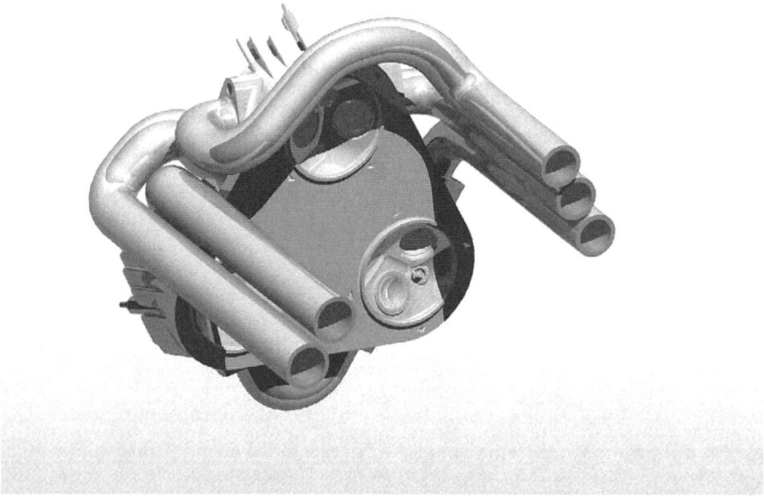

Fig. 7. Internal arrangement of the engine with FIK mechanism

The view on the internal arrangement of the FIK mechanism, where the swinging plate is replaced by a special swinging plate with three arms for the movable attachment of connecting rods, is shown in Fig. 7.

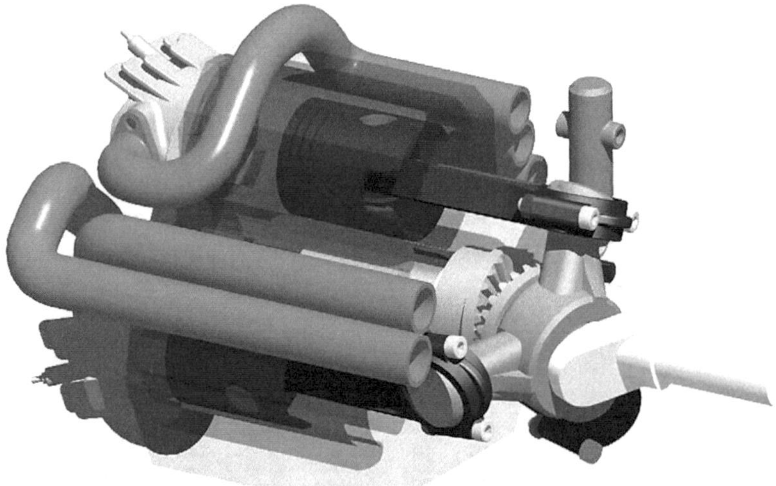

Fig. 8. The cylinder head with intake and exhaust pipes of the engine with FIK mechanism

3. Conclusion

As shown the calculations, the kinematic properties of the engine with FIK unconventional crank mechanism are comparable to the kinematic properties of the classical piston combustion engine with the same value of the piston stroke. This solution has the potential to reduce the wear of the cylinder and the noise caused by the tilting of the piston due to the small deflection of the connecting rod from the vertical axis of the piston. The piston kinematics parameters can be influenced by adjusting the distance of the between the crankshaft axis and the axis of the cylinder. When comparing the kinematic values of a classical engine and a FIK engine with the same stroke and at the same revolutions, the values for the FIK engine are about 10% higher.

This paper was realized as a part of the project KEGA 022ŽU-4/2017 - Implementation of on-line education in the area of precise technologies with an impact on educational process to increase skills and flexibility of students of engineering fields of study and as a result of the project implementation: "Modern methods of teaching of control and diagnostic systems of engine vehicles", ITMS code 26110230107, supported by the Operational Programme Education.

References

1. P. Fitz, R. Istenik, P. Kukuca, *Mechanizmus piestového stroja*, Vestník ÚPV SR, **patent č. 283742**, (12/2003)
2. S. Erkaya, İ. Uzmay, *Experimental investigation of joint clearance effects on the dynamics of a slider-crank mechanism*, Multibody System Dynamics, **vol. 24**, Issue 1, p. 81–102, (2010)
3. Z. Bo, S. Fei, *A numerical coupling model for a multibody system with multiple lubricated clearance joints*, MATEC Web of Conferences 108, 15006, (2017)
4. M. A.A.Fenelon, T. Furukawa, *Design of an active flapping wing mechanism and a micro aerial vehicle using a rotary actuator*, Mechanism and Machine Theory, **vol. 45**, p. 137-146, (2010)
5. P. Baran, P. Stastniak, M. Brezani, *Dynamic analysis and comparison of balancing systems of nonconventional piston machine FIK*. Proceeding of conference: Dynamics of rigid and deformable bodies 2016. Usti nad Labem, p. 9-15, (2016)
6. D. Barta, J. Saniga, M. Mruzek, M. Kendra, *Design of selected parts of non-conventional Stirling engine with FIK mechanism,* TOJSAT, **3**, Issue 2, (2013)
7. J. Shih, *Kinematics of the Cycloidal Internal Combustion Engine Mechanism,* Journal of mechanical design, **vol. 115 (4)**, p. 953-959, (2008)
8. *Axial Internal – Combustion Engines*, http://www.douglas-self.com/MUSEUM/POWER /unusualICeng/axial-ICeng/axial-IC.htm, (2008)
9. R. Isteník, *Simulácia mechanických a termomechanických procesov v spaľovacích* motoroch a vozidlách, Habilitation thesis, KKVMZ SjF Žilinská univerzita, (2002)
10. P. Punov, T. Evtimov, *Research influence of various geometrical parameters on the gasoline engine performance*, Proceedings of 15th International conference of Quality and reliability on technical systems, Nitra, Slovakia, p. 42–51, (2010).

Non-destructive testing of automotive heat exchangers

Katarzyna Peta[1,*], *Jan* Żurek[1], *Adam* Patalas[1]

[1]Poznan University of Technology, Faculty of Mechanical Engineering and Management, 60-965 Poznan, Poland

Abstract. The paper presents the results of non-destructive testing to the final control of automotive heat exchangers, which must meet a number of technological and exploitative requirements resulting from their working conditions. For the observation of images of heat exchangers, verification of geometrical dimensions and identification of surface and volume defects, the used methods were: computed tomography (high-resolution microtomograph Phoenix v|tome|x), three-dimensional optical scanning (3D GOM ATOS III optical scanner), coordinate measuring technique (coordinate measuring machine Hexagon Global Performance 122210). The effectiveness of non-destructive testing in industrial conditions was assessed and the directions of further research in this area were indicated.

Keywords: heat exchangers, non-destructive testing, computed tomography, 3D optical scanning, quality control

1 Introduction

Non-destructive testing allows the assessment of technical objects without causing changes in their microstructure and exploitative properties, also allow to determine their geometrical dimensions, detect material defects and appoint the properties of the used materials. Their aim is to improve the effectiveness of detecting irregularities in products and to minimize production time and the costs of control tests [1-3].

Non-destructive testing methods are of particular importance in the control of automotive industry products and contribute to ensuring the reliability and safety of motor vehicles. The size of production in the automotive industry induces to look for new solutions in the field of widely understood product control, including automotive heat exchangers, which final assessment is of key importance in the proper functioning of the vehicle. Detection of their irregularities prevents the leakage of cooling liquid and the failure of the engine cooling system or car air conditioning. Therefore, the identification of possible defects at the production stage affects their reliable usage [4].

The main technology used during the manufacturing of automotive heat exchangers is brazing in a furnace in controlled nitrogen atmosphere. Consequently, special attention during the control is paid to their assembly quality, which essentially decide about final

* Corresponding author: katarzyna.peta@put.poznan.pl

Reviewers: *Oleksandr* Kravchenko, *Vladimír* Stuchlý

durability. The development of brazing technology poses new challenges for a discipline of non-destructive testing of automotive heat exchangers. Research methods which are adequate to industrial conditions has been distinguished [5-7]:

- computed tomography,
- three-dimensional optical scanning,
- coordinate measuring technique.

In the paper [8] a comparison of the results of research cube measurements made for the analysis using the methods of coordinate measuring technique, optical scanning and computed tomography were presented. Each of these methods enables effective measurement of the research cube and it was considered deliberate to carry out further research in order to search for areas of non-destructive testing methods. The authors [9] pointed out to the legitimacy of using optical measurement systems to control products in automated production and the rightness of development of this field of metrology. In the paper [10] a thermovision and computed tomography for the evaluation of products were discussed. Insufficient number of studies on effective and successful methods of their control, in particular those produced in special processes has been indicated. The authors [11] presented a comparison of the results of geometry measurements of the object obtained by destructive micrographic method and non-destructive three-dimensional optical scanning method. The first one provides more information about the material properties of the product but from the point of view of industrial conditions the latter is characterized by greater utility. The works [5-11] indicate the purposefulness of using non-destructive testing, which application in industrial practice is insufficiently discussed.

2 Research methodology

Non-destructive tests were carried out on car radiators, which core was made of cladded tapes: aluminium-manganese alloy AA3003 (core material) and aluminum-silicon alloy AA4343 (clad material). Additional parts of the radiator are tanks and gaskets clinched on its core made from polyamide PA6 and terpolymer EPDM. Radiators were made in a controlled atmosphere of nitrogen in a radiation-convection furnace according to the CAB technology (Controlled Atmosphere Brazing) and these steps followed sequentially: thermal degreasing, fluxing, heating, brazing, cooling in nitrogen atmosphere, cooling in the air. The basic mechanical properties and chemical composition of materials used in radiators production is given in table 1.

Table 1. Basic mechanical properties and chemical composition of aluminium alloys AA3003/AA4343

Alloy	Fe	Si	Cu	Mg	Mn	Zn	Al	Tensile strength N/mm^2	Yield point N/mm^2	Elongation %
AA3003 (core)	0.7	0.6	0.05-0.2	0.05	1.0-1.5	0.1	Re.	115.00	58.00	37.00
AA4343 4% (clad)	0.8	6.8-8.2	0.25	0.05	0.1	0.2	Re.			

The major methods of non-destructive testing including computed tomography, three-dimensional optical scanning, coordinate measuring technique used for verification of geometrical dimensions and identification of possible defects in the automotive car radiators are compared in table 2. For each method, the five measurements of the analysed object were made.

Table 2. The methods of non-destructive testing of car radiators

Method	Measuring device	Software	Measurement accuracy
Coordinate measuring technique	Hexagon Global Performance 122210	PC-DMIS	from 3.4 μm to 3.7 μm
Three-dimensional optical scanning	3D GOM ATOS III	Atos Professional V8	from 12.4 μm to 14.9 μm
Computed tomography	Phoenix v|tome|x	VGStudio MAX 2.2	from 10.2 μm to 11.45 μm

The non-destructive assessment of car radiators included the verification of its geometrical dimensions using a Hexagon Global Performance 122210 coordinate measuring machine with a measurement range of 1200 x 2200x 1000 mm and accuracy of measurements 3.4-3.7μm. The coordinate measuring machine receives the set of points of the measured object in coordinates X, Y, Z, which location enables the contact measuring head, and their position - measuring straightedges in each axis. PC-DMIS computer software allows to obtain the characteristic dimensions of the measured object.

The results obtained coordinate measuring technique was compared with the results received three-dimensional optical scanning. The optical scanner 3D GOM ATOS III, as opposed to measurements of single points in coordinate method, captures the geometry of whole objects as clouds of points or a triangle grid. The 3D optical scanner operates on the basis of Triple Scan technology, in which the measurement takes place using two cameras. On the surface of the measured object, patterns consisting of stripes are displayed, and the whole is recorded by two cameras on a stereoscopic basis. The points of the three-dimensional surface are designated from intersections of the beams left and right visual cameras. This solution enables to measure light reflective surfaces and objects with complicated shapes, including indentations. The measurements were made in Blue Light Technology, in which the optical scanner uses blue narrowband light to filter out disturbing light coming from the surroundings. Depending on the measuring field, the measurement accuracy is from 12.4 μm to 14.9 μm. The Atos Professional V8 software was used for dimensional analysis of the car radiators.

The verification of the car radiator dimensions were complemented by computed tomography method using a high-resolution microtomograph Phoenix v|tome|x. The tomographic image occurs as a result of measurements of the radiation absorption penetrating of the examined object, the solid of which is divided into cells called voxels with identical linear coefficients of radiation absorption. The cross-sectional image after reconstruction is a quantitative map of linear radiation absorption in the voxels located in the scanned layer. On the basis of the obtained point cloud, the dimensional characteristic of the analyzed object, including its internal structure, are obtained. The Phoenix v|tome|x microtomograph is equipped with a 300kV single-pole microfocus light with the power of 500W of an open type and a liquid cooling system. The minimum voxel size is 2 μm, while the detection of details is less than 1μm. The device has a GE DXR digital detector with a resolution of 2000 x 2000 pixels, where the size of a single pixel is 200 μm. The VGStudio MAX 2.2 software was used for dimensions evaluation of the car radiators.

3 Research results

In order to verify the legitimacy and the possibility of using non-destructive tests in the production process of automotive heat exchangers, the geometrical dimensions of car radiators specified in control documentation and checked in the current production of these products were measured. The obtained measurements results of car radiator diameters are presented in figures (Fig. 1 - the results obtained using the coordinate measuring technique, Fig. 2 - three-dimensional optical scanning, Fig. 3 - computed tomography).

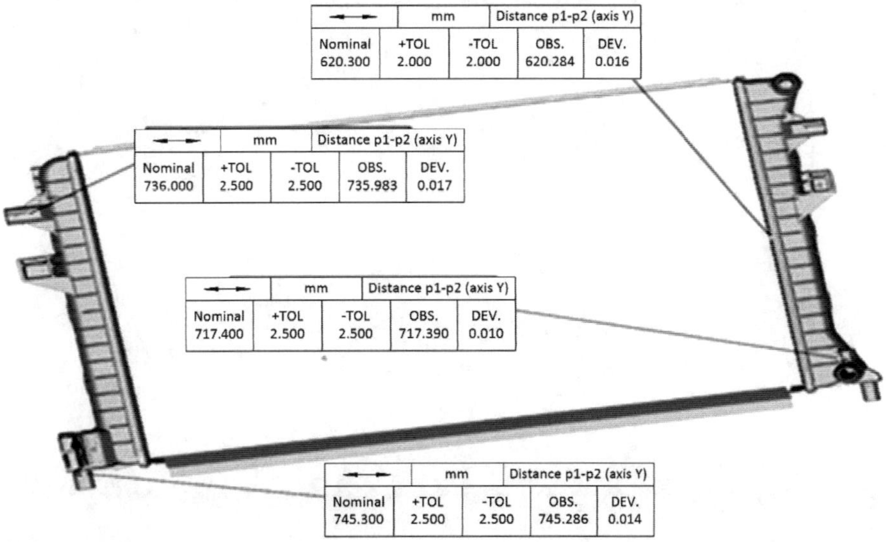

Fig. 1. The results of measurements of car radiator diameters obtained by coordinate measuring technique

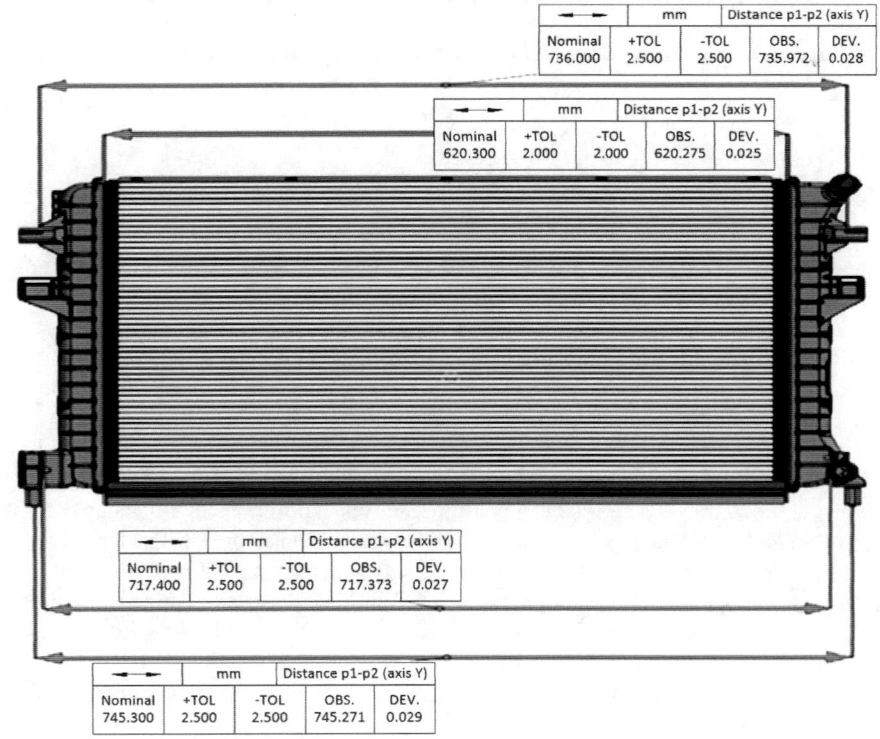

Fig. 2. The results of measurements of car radiator diameters obtained by three-dimensional optical scanning

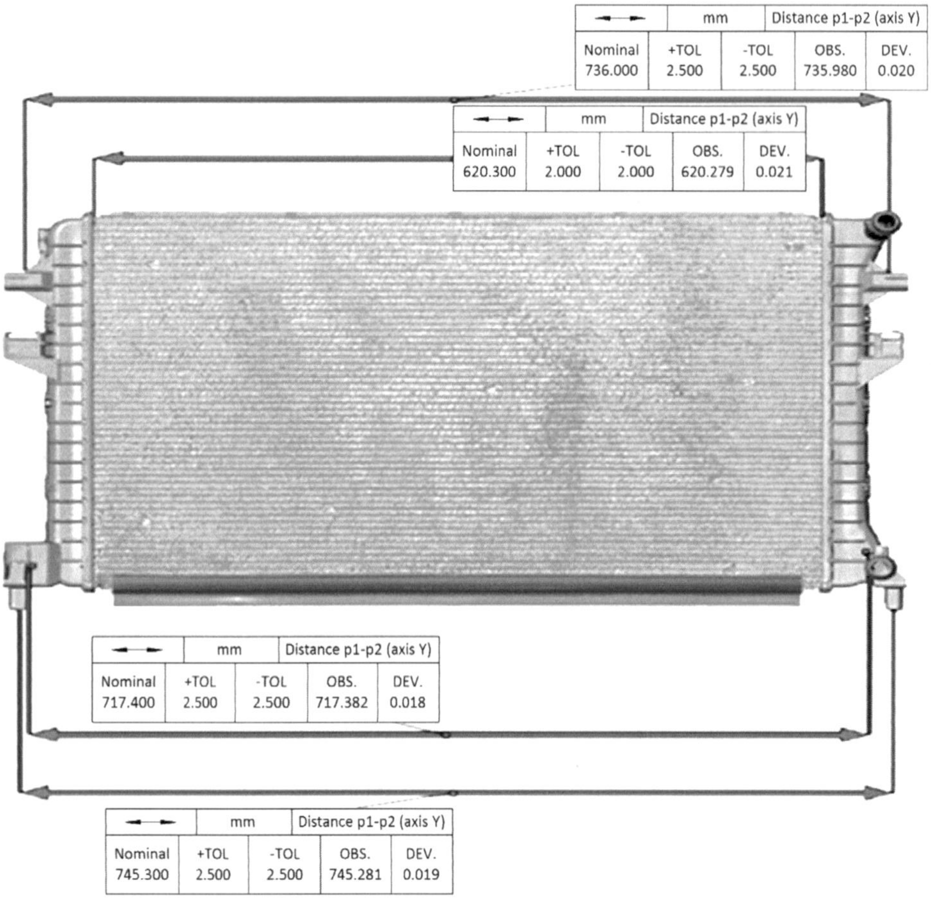

Fig. 3. The results of measurements of automotive car radiator diameters obtained by computed tomography

The comparison of obtained measurements results automotive car radiators geometrical dimensions with nominal sizes together with an indication of the location on the analysed object are presented in Table 3. The average measurements results are compiled for three methods: coordinate measuring technique, three-dimensional optical scanning and computed tomography. The car radiators measurements were repeated 5 times, specifying the mean values and standard deviations.

The measurements results of geometrical dimensions of automotive car radiators, depending on the measuring device which was used, they differ from nominal values from 0.010 mm to 0.029 mm. These results are within the fields of tolerance dimensions specified in the technical documentation (T = ± 2.000 mm for spacing of header plates and T = ± 2.500 mm for spacing of the plastic tank key fragments), providing the stability of the manufacturing process of these products. Taking into account the obtained measurement results and the accuracy of devices specified in their specifications (coordinate measuring technique - from 3.4 μm to 3.7 μm, three-dimensional optical scanning - from 12.4 μm to 14.9 μm, computed tomography - from 10.2 μm to 11.45 μm) it allows to use of the non-destructive testing in industrial practice.

Table 3. The results of measurements of car radiator diameters obtained by non-destructive methods

Measured diameter	Nominal value mm	Measurement method	Average measurement result mm	Standard deviation mm
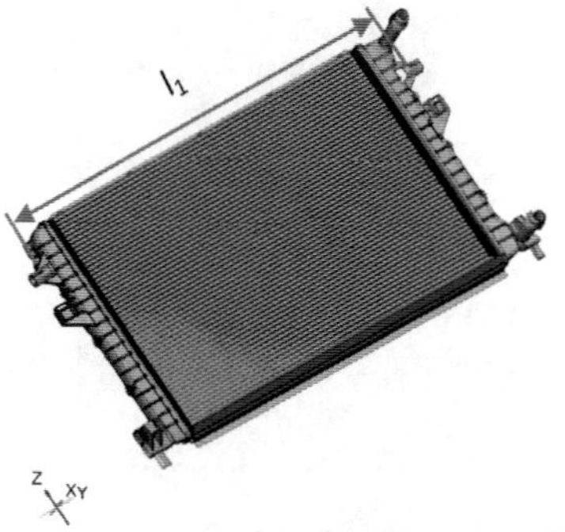	$l_1 = 736.000$ $T = \pm 2.500$	Coordinate measuring technique	$l_1 = 735.983$	$s_1 = 0.017$
		Three-dimensional optical scanning	$l_1 = 735.972$	$s_1 = 0.028$
		Computed tomography	$l_1 = 735.980$	$s_1 = 0.020$
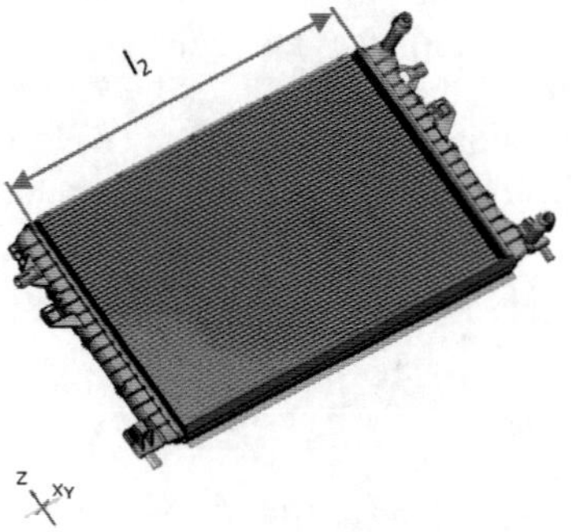	$l_2 = 620.300$ $T = \pm 2.000$	Coordinate measuring technique	$l_2 = 620.284$	$s_2 = 0.016$
		Three-dimensional optical scanning	$l_2 = 620.275$	$s_2 = 0.025$
		Computed tomography	$l_2 = 620.279$	$s_2 = 0.021$
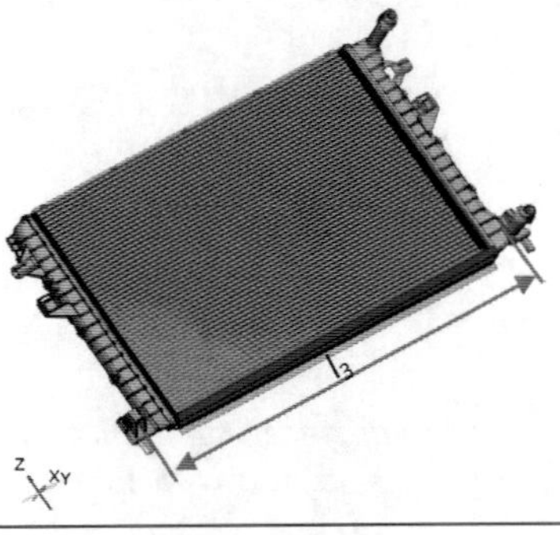	$l_3 = 717.400$ $T = \pm 2.500$	Coordinate measuring technique	$l_3 = 717.390$	$s_3 = 0.010$
		Three-dimensional optical scanning	$l_3 = 717.373$	$s_3 = 0.027$
		Computed tomography	$l_3 = 717.382$	$s_3 = 0.018$
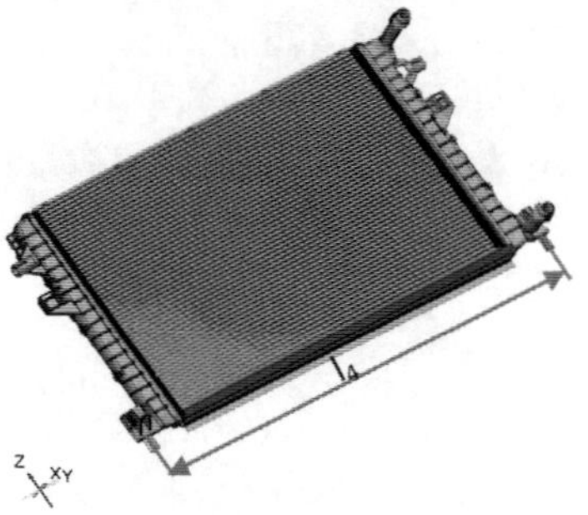	$l_4 = 745.300$ $T = \pm 2.500$	Coordinate measuring technique	$l_4 = 745.286$	$s_4 = 0.014$
		Three-dimensional optical scanning	$l_4 = 745.271$	$s_4 = 0.029$
		Computed tomography	$l_4 = 745.281$	$s_4 = 0.019$

4 Conclusions

Non-destructive testing allows to effectively detect the irregularities in automotive heat exchangers, while their time-consumption is comparable to destructive micrographic control. Thus, they allow to evaluate products without changing their microstructure and exploitative properties.

The results of measurements of geometric dimensions of automotive radiators were in the tolerance fields included in the design documentation, which confirms the stability of their production processes.

All of the used methods are also applicable in the production process, especially for random control of geometrical dimensions and potential defects of analysed products. The most noteworthy device for detecting irregularities in products is the computer microtomograph that allows not only obtaining a measurement accuracy comparable to a three-dimensional optical scanner and a coordinate measuring machine, but also surface and volume evaluation of automotive heat exchangers with the possibility of imaging its interiors. Computed tomography is cost-consuming (purchase and exploitation of the device), but it allows to carry out a wide range of automotive heat exchanger tests.

The literature study and the obtained research results justify the continuation of work related to the practical application of non-destructive testing. This will include the comparison of the destructive micrograph method with the non-destructive computed tomography tests, the three-dimensional optical scanning and coordinate measuring technique. Therefore, a comprehensive assessment of heat exchangers is planned, including the characteristics of geometric dimensions and identification of defects (according to the technical documentation), using only the non-destructive testing.

References

1. J.P. Kruth, M. Bartscher, S. Carmignato, R. Schmitt, L. De Chiffre, A. Weckenmann, *Computed tomography for dimensional metrology*. CIRP Ann-Manuf. Techn., **60**, 821-842 (2011)
2. S.K. Dwivedi, M. Vishwakarma, A. Soni, *Advances and Researches on Non Destructive Testing: A Review*. Mater. Today-proc., **5**, 3690-3698 (2018)
3. R. Hanke, T. Fuchs, N. Uhlmann, *X-ray based methods for non-destructive testing and material characterization*. Nucl. Instrum. Meth. A, **591**, 14-18 (2008)
4. A. Dubey, *Investigation on Suitability of Aluminium to Copper in a Radiator*. Adv. Mater. Res-switz., **3**, 16-23 (2015)
5. S. Kumar, D. Mahto, *Recent trends in industrial and other engineering applications of non-destructive testing: a review*. International Journal of Scientific & Engineering Research, **4**, 183-195 (2013)
6. J. Kastner, B. Harrer, H.P. Degischer, *High resolution cone beam x-ray computed tomography of 3D microstructures of cast Al-alloys*. Mater. Charact., **62**, 99-107 (2011)
7. P. Chaudhuri, P. Santra, S. Yoele, A. Prakash, D. Reddy, L. Lachhvani, G. Jagannathan, Y. Saxena, *Non-destructive evaluation of brazed joints between cooling tube and heat sink by IR thermography and its verification using FE analysis*. Non-destructive Testing and Evaluation International, **39**, 88-95 (2006)
8. B. Gapiński, M. Wieczorowski, L. Marciniak-Podsadna, B. Dybala, G. Ziolkowski, *Comparison of Different Method of Measurement Geometry using CMM, Optical Scanner and Computed Tomography 3D*. Procedia Engineer., **69**, 255 – 262 (2014)

9. J. Jaworski, R. Kluz, T. Trzepieciński, *Research on Accuracy of Automatic System for Casting Measuring*. Archives of Foundry Engineering, **16**, 49-54 (2016)
10. K. Żaba, S. Nowak, M. Kwiatkowski, M. Nowosielski, P. Kita, A. Sioma, *Application of non-destructive methods to quality assessment of pattern assembly and ceramic mould in the investment casting elements of aircraft engines*. Arch. Metall. Mater., **59**, 1517-1524 (2014)
11. H. Lend, K. Ratas, P. Peetsalu, *Measurement of Geometry of Small Axisymmetric Sheet Metal Component after Forming*. Mater. Sci-Medzg+, **21**, 473-478 (2015)

AUTHOR INDEX

AUTHOR INDEX